THERMODYNAMICS AND CONTROL OF OPEN QUANTUM SYSTEMS

The control of open quantum systems and their associated quantum thermodynamic properties is a topic of growing importance in modern quantum physics and quantum chemistry research. This unique and self-contained book presents a unifying perspective of such open quantum systems, first describing the fundamental theory behind these formidably complex systems, before introducing the models and techniques that are employed to control their quantum thermodynamics processes. A detailed discussion of real quantum devices is also covered, including quantum heat engines and quantum refrigerators. The theory of open quantum systems is developed pedagogically, from first principles, and the book is accessible to graduate students and researchers working in atomic physics, quantum information, condensed matter physics, and quantum chemistry.

GERSHON KURIZKI has held the G.W. Dunne Professorial Chair in Theoretical Quantum Optics at the Weizmann Institute of Science in Israel since 1998. He was the recipient of the W.E. Lamb Medal in Laser Science and Quantum Optics (USA) in 2008 and the Humboldt-Meitner Award (Germany) in 2009 for pioneering contributions to the theory of quantum measurements and decoherence control in open quantum systems. As Fellow of the Optical Society of America, the American Physical Society, and the UK Institute of Physics, he has coauthored more than 300 scientific publications.

ABRAHAM G. KOFMAN is Research Consultant at the Weizmann Institute of Science. He was the recipient of the Maxine Singer Prize for Outstanding Research at the Weizmann Institute of Science in 2005 and received the 'Highlights of 2013' citation from the *New Journal of Physics*. He has coauthored more than 100 scientific publications related to various fields of theoretical physics and chemistry, including quantum optics, quantum measurements, quantum information processing, atomic and molecular physics, condensed matter, and chemical reactions.

THERMODYNAMICS AND CONTROL OF OPEN QUANTUM SYSTEMS

GERSHON KURIZKI

Weizmann Institute of Science, Israel

ABRAHAM G. KOFMAN

Weizmann Institute of Science, Israel

CAMBRIDGE
UNIVERSITY PRESS

CAMBRIDGE
UNIVERSITY PRESS

University Printing House, Cambridge CB2 8BS, United Kingdom

One Liberty Plaza, 20th Floor, New York, NY 10006, USA

477 Williamstown Road, Port Melbourne, VIC 3207, Australia

314–321, 3rd Floor, Plot 3, Splendor Forum, Jasola District Centre, New Delhi – 110025, India

103 Penang Road, #05–06/07, Visioncrest Commercial, Singapore 238467

Cambridge University Press is part of the University of Cambridge.

It furthers the University's mission by disseminating knowledge in the pursuit of education, learning, and research at the highest international levels of excellence.

www.cambridge.org
Information on this title: www.cambridge.org/9781107175419
DOI: 10.1017/9781316798454

© Gershon Kurizki and Abraham G. Kofman 2022

First published 2022

Printed in the United Kingdom by TJ Books Limited, Padstow Cornwall

A catalogue record for this publication is available from the British Library.

ISBN 978-1-107-17541-9 Hardback

Contents

Preface

Scope and Motivation

This book serves two purposes:

1. It provides a common framework for two hitherto disparate, rapidly emerging disciplines: control of open quantum systems and their nonequilibrium thermodynamics.
2. It applies the unifying principles of these two disciplines to a wide scope of topics at the forefront of current research, with a focus on systems of atoms and spins interacting with classical or quantized electromagnetic and mechanical-strain fields.

Its intended readership is graduate students and scientists in a broad range of areas, primarily in physics (atomic, molecular and optical physics, condensed matter physics, quantum information and thermodynamics, and chemical physics).

Synopsis

Quantum System–Bath Interactions and Their Control (Part I)

In Part I of the text we introduce and discuss in depth the fundamental concepts and applications pertaining to open quantum systems and their dynamical control. Our starting point is the problem of equilibration in large but closed quantum systems (Ch. 1). In Chapter 2 we discuss the thermalization conditions for a small subsystem of an equilibrated large system that is coupled to a much larger subsystem (environment). Such coupled subsystems are often referred to as an open system and a bath, respectively. The focus is on a unified approach to processes involving quantized matter (atomic, atom-like, and spin systems) that are subject to control by electromagnetic fields while interacting with quantized fields that constitute their environment ("baths") consisting of phonons, plasmons, photons, polaritons, and spins, to name a few (Ch. 3).

The importance of studying such processes from a common perspective stems from the ubiquitous contact of quantum systems with environments described as thermal or nonthermal baths (Ch. 4): with very few exceptions, quantum systems are inherently open, and their dynamics reflects their strong or weak coupling to the bath (Ch. 5). The weak-coupling limit of spin-1/2 systems to a bosonic bath gives rise to irreversible decoherence or relaxation of the system state (Ch. 6). In multi-spin systems, decoherence and relaxation acquire cooperative features, and so do their resonant energies (Ch. 7).

An alternative motivation for such studies may be colloquially summarized as follows: "if you can't fight the bath, join it." By this we mean that control may take advantage of bath effects, particularly of virtual quanta exchange via the bath (self-energy), in the form of cooperative Lamb shifts and dipole–dipole interactions. Such effects may preserve or even reinforce the "quantumness" of the system and thereby turn the bath into a potentially useful resource for quantum technologies (Ch. 8).

Measurements of quantum systems are commonly effected by detectors that act as baths for most purposes (Ch. 9). Therefore, the ability of dynamical control to suppress detrimental effects of the bath, namely, decoherence or dissipation, on the quantum system of interest is a prerequisite for the successful implementation of quantum measurements and emerging technologies that rely on quantum coherence or entanglement: quantum information processing, quantum sensing, and metrology. Yet dynamical control of open systems has its limitations: it must be faster than the correlation time of the bath (i.e., it should act on a non-Markovian timescale). We study the dynamical control of open quantum systems within a unified framework that allows for any type of action on the system, be it pulsed, continuous, or projective (Chs. 10–14). This universal approach optimizes the control for the bath and task at hand (Chs. 11 and 12). Among possible applications of dynamical control are its use as a means of reliably processing, storing, and transferring quantum information (Chs. 13 and 14). The underlying paradigms are either the dynamical suppression of the system–bath coupling, dubbed the quantum Zeno effect, or, conversely, the enhancement of such coupling, alias the anti-Zeno effect (Chs. 10–13).

Quantum Thermodynamic Processes and Their Control (Part II)

We then apply the insights and tools acquired in Part I to the elucidation of a problem that lies at the heart of quantum thermodynamics, namely: To what extent can dynamical control enhance the thermodynamic performance of devices that display quantum features, particularly heat machines with quantum ingredients? The ability to obtain new or improved functionalities of such machines by harnessing dynamical control to our advantage is studied based on methods developed in Part I. The underlying fundamental issue is the rapport of thermodynamics and quantum mechanics. In order to shed light on this formidable problem, we discuss the applicability of the principles of thermodynamics in the quantum domain and revisit its tenets for open quantum systems under dynamical control that acts on either non-Markovian or Markovian timescales (Chs. 15 and 16). The ability to harness information acquired by measurements on open quantum systems as a thermodynamic resource is discussed (Ch. 17).

Certain models of quantum heat machines reproduce the standard thermodynamic bounds, such as the Carnot bound on efficiency. In others, those bounds appear to be violated at the quantum level. Hence the need for a clarification of the general principles of quantum heat machines, starting with appropriate definitions of their work and power output that safeguard their analysis against inconsistencies with the laws of thermodynamics (Chs. 18–22). These studies primarily address the following key questions: (a) Is the Carnot bound on efficiency valid in the quantum domain? (b) Are work or power bounds of quantum heat machines different from those of their classical counterparts? We show that the nonpassivity of a quantized piston in a heat machine is an indispensable thermodynamic resource (Ch. 22). A nonthermal quantum bath can be another nonpassive resource that transforms the heat engine into a thermomechanical machine to which the Carnot bound does not apply (Ch. 18). The answer to the second question is that entanglement of two-level systems (Ch. 21) may cause quantum heat machines to produce much higher power than their nonentangled counterparts.

In Chapter 23, we show that bath dispersion affects the scaling of cooling speed with temperature attainable by quantum coolers or refrigerators. The compatibility of this scaling with Nernst's Third Law is an open issue.

Chapter 24 discusses other types of heat machine, dubbed quantum heat managers. These include heat diodes or heat transistors that rectify or amplify, respectively, the heat flow between different heat baths, provided the baths are spectrally filtered. The Conclusions summarize the main results and the open issues.

To sum up, the book shows that the unification of quantum dynamical control theory with that of quantum thermodynamics provides us with a powerful and versatile toolbox for resolving both conceptual and practical issues related to the controllability of open quantum systems and their possible applications in quantum technologies.

Part I
Quantum System–Bath Interactions and Their Control

1

Equilibration of Large Quantum Systems

The tendency of large systems to evolve to an equilibrium, namely, a stationary state that does not depend on their initial state, is called thermalization. The stationary state of a closed system is described by the *microcanonical ensemble* corresponding to a narrow energy distribution.

Large-system thermalization is commonplace. Yet it appears to contradict the unitarity of quantum mechanics, which requires symmetry with respect to time inversion. This contradiction has not yet been fully reconciled with the fundamental reversibility of quantum mechanics, despite a continuing endeavor that may be traced back to von Neumann's work in 1929. The central paradigm is the eigenstate thermalization hypothesis (ETH) put forward by Srednicki. It relies on the conjecture that a typical (randomly picked) eigenstate of a realistic many-body Hamiltonian yields the same mean value for any reasonable observable as predicted by a microcanonical ensemble with the same total energy. The ETH predicts that almost any superposition of such eigenstates relaxes at long times to a state that is practically indistinguishable from a thermal equilibrium state. Studies of thermalization in closed quantum many-body systems are aimed at bridging unitarity with irreversibility, ergodicity, and the onset of thermodynamic behavior in complex or open systems. Yet the quantum thermalization mechanism and the route to the bridging of quantum and classical descriptions of the world by this mechanism remain enigmatic and are still being debated. However, under generic conditions, one can show that the observables of a large system are governed at long times by a canonical density operator, as detailed in this chapter.

1.1 From Quantum Dynamics to Thermodynamics

In an isolated system with a large but countable number of degrees of freedom (DOF), $1 \ll f < \infty$, governed by an autonomous Hamiltonian H, the spectrum is discrete (quantized). Its (typically infinitely many) eigenstates $|n\rangle$ $(n = 0, 1, \dots)$ possess eigenvalues E_n, ordered as

$$E_0 \leq E_1 \leq E_2 \leq \dots, \tag{1.1}$$

with a bounded ground state energy, $E_0 > -\infty$. The Hamiltonian can then be written as

$$H := \sum_n E_n |n\rangle \langle n|. \tag{1.2}$$

In the presence of energy degeneracy, we may use the projectors onto subspaces of degenerate energies $E_m = E_n$,

$$P_{E_n} := \sum_{E_m = E_n} |m\rangle \langle m|, \tag{1.3}$$

to rewrite the Hamiltonian (1.2) as

$$H = \sum_{E_n} E_n P_{E_n}, \tag{1.4}$$

where $\sum_{E_n}$ is a summation over all *mutually different* E_n values.

1.1.1 Thermodynamic Variables

The number of energy levels below any given E is

$$N(E) := \sum_n \theta(E - E_n), \tag{1.5}$$

the Heaviside (step) function $\theta(x)$ being equal to 1 for $x > 0$ and 0 for $x \leq 0$. The *entropy* associated with this number of levels is defined as

$$S(E) := k_B \ln N(E), \tag{1.6}$$

where k_B is Boltzmann's constant. Commonly, this entropy is an extensive quantity, since it scales for a system with f DOF as

$$S(E)/k_B = O(f). \tag{1.7}$$

Equation (1.5) implies that for macroscopic $f = O(10^{23})$, the level density is staggering even on extremely small energy scales. Hence, the step function $\theta(x)$ in (1.5) may be assumed to be washed out. The level number $N(E)$ then becomes a smooth function of E, whose well-defined derivative represents the density of states

$$\Omega(E) = \sum_n \delta(E - E_n), \tag{1.8}$$

the delta-function $\delta(x) = \theta'(x)$ being also assumed to be washed out over many energy levels.

The *coarse-grained entropy* defined by (1.6) leads to the definition of *temperature*, which applies *whether the system is at equilibrium or not*:

$$T(E) := 1/S'(E). \tag{1.9}$$

In accordance with Nernst's *third law of thermodynamics*, the entropy and temperature converge to zero as the energy approaches the ground-state value, $E \to E_0$. For macroscopic values of $E - E_0$, the dependence of S on E is logarithmic. It then follows from (1.7) and (1.9) that

$$k_{\mathrm{B}} T(E) = O\left(\frac{E - E_0}{f}\right), \tag{1.10}$$

so that, for any macroscopic energy change ΔE,

$$T(E + \Delta E) = T(E)\left[1 + O\left(\frac{\Delta E}{E - E_0}\right)\right]. \tag{1.11}$$

All these relations may fail at extremely low temperatures, which are beyond our consideration here (but cf. references in this chapter).

1.1.2 States and Dynamics

A quantum mechanical state of the system is given by a density operator $\rho(t)$, whose evolution,

$$\rho(t) = U_t \, \rho(0) \, U_t^\dagger, \tag{1.12}$$

is governed by the unitary propagator

$$U_t := \exp(-i H t/\hbar) = \sum_n \exp(-i E_n t/\hbar)|n\rangle\langle n|. \tag{1.13}$$

Equations (1.12) and (1.13) yield, for an arbitrary initial state $\rho(0)$,

$$\rho(t) = \sum_{m,n} \rho_{mn}(0) e^{-i(E_m - E_n)t/\hbar} |m\rangle\langle n|, \tag{1.14}$$

where $\sum_{m,n}$ is a summation over all $m, n = 0, 1, 2, ...$, $\rho_{mn}(t) := \langle m|\rho(t)|n\rangle$ being the matrix elements of $\rho(t)$.

The ensemble-averaged occupation probability p_{E_n} of an eigenvalue E_n is given by the expectation value of the projector (1.3) onto the corresponding eigenspace,

$$p_{E_n} := \mathrm{Tr}\,[P_{E_n}\rho(t)] = \sum_{E_m=E_n} \rho_{mm}(t) = \sum_{E_m=E_n} p_m, \tag{1.15}$$

where the level population p_n is the time-independent expectation value of the observable $|n\rangle\langle n|$,

$$p_n := \mathrm{Tr}\,[|n\rangle\langle n|\rho(t)] = \rho_{nn}(t) = \rho_{nn}(0), \tag{1.16}$$

normalized by

$$1 = \mathrm{Tr}\,\rho(t) = \sum_n \rho_{nn}(t) = \sum_n p_n = \sum_{E_n} p_{E_n}. \tag{1.17}$$

In what follows, we shall employ the energy basis in which all the non-diagonal elements of $P_{E_n}\rho(0)P_{E_n}$ vanish,

$$\rho_{mn}(0) = 0 \quad \text{if} \quad m \neq n \quad \text{and} \quad E_m = E_n. \tag{1.18}$$

1.2 The Problem of Equilibration for Physical Observables

The statistical ensemble $\rho(t)$ is not stationary at short t if $\rho(0)$ is out of equilibrium. Yet, if the right-hand side of (1.14) initially depends on t, it cannot approach at large t any time-independent "equilibrium ensemble." Furthermore, any mixed state $\rho(t)$ returns arbitrarily "near" its initial state $\rho(0)$ at certain times t (analogously, but not identically, to pure-state Poincaré recurrences). In what follows, we examine the apparent contradiction of such recurrences with equilibration.

According to (1.14), there exists at least one $\rho_{mn}(0) \neq 0$ with

$$\omega := (E_n - E_m)/\hbar \neq 0. \tag{1.19}$$

We consider observables represented by Hermitian operators

$$X = \sum_{m,n} X_{mn}|m\rangle\langle n|, \quad X_{mn} := \langle m|X|n\rangle, \tag{1.20}$$

with expectation values

$$\langle X\rangle(t) := \mathrm{Tr}[\rho(t)X]. \tag{1.21}$$

For the observable that represents an interlevel transition,

$$X = \hat{X} + \hat{X}^\dagger, \quad \hat{X} := |m\rangle\langle n|/\rho_{mn}(0), \tag{1.22}$$

we find from (1.14) that

$$\mathrm{Tr}\,[\rho(t)X] = 2\cos(\omega t). \tag{1.23}$$

Thus, the mean value of X in the ensemble $\rho(t)$ exhibits *permanent oscillations*, allowing us to conclude that *quantum mechanics and equilibration are, in general, incompatible.* Nevertheless, as shown below, equilibration can approximately hold true for a restricted class of observables X and initial conditions $\rho(0)$.

A measurement of an observable X may be assumed to yield a finite range of possible outcomes,

$$\Delta_X := \max_{\mathcal{H}} \langle \psi | X | \psi \rangle - \min_{\mathcal{H}} \langle \psi | X | \psi \rangle = x_{\max} - x_{\min}, \qquad (1.24)$$

where the maximization and minimization are over all normalized vectors in the pertinent Hilbert space $\mathcal{H}$, $|\psi\rangle \in \mathcal{H}$, so that $x_{\max}$ and $x_{\min}$ are the largest and smallest eigenvalues of X, respectively.

1.2.1 Equilibration Conditions

The key requirement on the initial condition $\rho(0)$ is that the ensemble-averaged level populations p_n can be split into a locally averaged level *population density* $h(E)$ and "unbiased fluctuations" δp_n, whose average within the interval around E is vanishingly small compared to $h(E)$,

$$p_n = h(E_n) + \delta p_n. \qquad (1.25)$$

This requirement should hold within any energy interval, which contains many levels E_n, but is still exceedingly small on any experimentally resolvable scale.

This initial condition is the result of a preparation process, during which the system was still entangled with the outside world. The reduced initial state (at $t = 0$) of the system (obtained by tracing out the outside world) must therefore be a mixed state. Any time-dependent system Hamiltonian will cause the spreading of occupation probabilities over neighboring energy levels. Since the levels are so dense, the spreading *randomizes the p_n's* in accordance with (1.25). This preparation process stands in contrast to a "sudden" (discontinuous) parametric change of the Hamiltonian, dubbed *quantum quench*.

1.2.2 Energy Density

Let us define the ensemble-averaged energy density

$$\rho(E) := \langle \delta(E - H) \rangle, \qquad (1.26)$$

$\rho(E) dE$ being the probability to find an energy value between E and $E + dE$. From (1.16) it follows that

$$\rho(E) = \sum_n p_n \delta(E - E_n). \qquad (1.27)$$

The delta-functions in (1.27) are assumed to be "washed out" over many energy levels so that they give rise to a well-defined, smooth energy density. Consistently with (1.25) one then finds that

$$\rho(E) = h(E)\Omega(E). \tag{1.28}$$

Namely, the probability $\rho(E)dE$ of finding an energy between E and $E + dE$ is given by the locally averaged energy-level population $h(E)$ multiplied by the local level density $\Omega(E)$ times the interval dE. It is important that the locally averaged population-density $h(E)$ be independent of the specific choice of the energy interval around E.

1.2.3 Maximal Level Population

Even if the energy levels are populated nonuniformly, we expect from (1.25)–(1.28) that

$$\max_n p_n \simeq 10^{-O(f)}, \tag{1.29}$$

so that $\max_n p_n$ is extremely small.

According to (1.22), the spectrum of X consists of the eigenvalues $x_\pm = \pm|\rho_{nm}(0)|^{-1}$ and, for $\dim\mathcal{H} > 2$, of the eigenvalue $x_0 = 0$. From (1.24) we then have

$$\Delta_X = 2|x_\pm| > \frac{2}{\max_n p_n}. \tag{1.30}$$

We can deduce from (1.29) that $\Delta_X \geq O(10^f)$. Although the observable (1.22) exhibits persistent oscillations (1.23), such oscillations are beyond the conceivable resolution limit for macroscopic systems ($f \gg 1$). Hence, any realistic measurement will yield one of the three outcomes, $x_\pm$ or x_0.

1.2.4 Equilibrium Ensemble

For an arbitrary $\rho(0)$ evolving according to (1.14), we can show that the corresponding equilibrium ensemble is described by the density operator of the time-averaged ensemble $\rho(t)$,

$$\rho_{\text{eq}} := \overline{\rho(t)}, \tag{1.31}$$

where ρ_{eq} is a nonnegative, Hermitian operator of unit trace.

In the energy basis employed in (1.18), one finds from (1.16) and (1.31) that

$$\rho_{\text{eq}} = \sum_n \rho_{nn}(0)|n\rangle\langle n| = \sum_n p_n|n\rangle\langle n|, \tag{1.32}$$

so that ρ_{eq} is the (time-independent) diagonal part of $\rho(t)$ from (1.14).

From (1.21) and (1.31) we obtain that the average of $X(t)$ over all times $t > 0$ equals

$$\overline{\langle X \rangle(t)} = \text{Tr}\,(\rho_{\text{eq}}X). \tag{1.33}$$

Since this equality holds for $X = I$, the unity operator, it follows that the average of $\rho(t)$ over all times $t > 0$ is indistinguishable from ρ_{eq}.

We now estimate the derivation of $\langle X \rangle(t)$ from its time average (1.33). To this end, we define $\tilde{\Delta}_X$, the minimal range of the eigenvalues of the difference between X and *any energy-diagonal operator* $Y := \sum_n y_n |n\rangle\langle n|$ with arbitrary real coefficients $y_0, y_1, \ldots$. Explicitly,

$$\tilde{\Delta}_X := \min_Y [\max_{\mathcal{H}}\langle\psi|X - Y|\psi\rangle - \min_{\mathcal{H}}\langle\psi|X - Y|\psi\rangle]. \tag{1.34}$$

By definition,

$$\tilde{\Delta}_X = 0 \quad \text{if } X = Y. \tag{1.35}$$

From (1.24) and (1.34) we then find that

$$\tilde{\Delta}_X \leq \Delta_X. \tag{1.36}$$

The mean-square deviation of $\langle X \rangle(t)$,

$$\sigma_X^2 := \overline{\left[\langle X \rangle(t) - \overline{\langle X \rangle(t)}\right]^2}, \tag{1.37a}$$

can then be shown to be bounded by

$$\sigma_X^2 \leq \tilde{\Delta}_X^2\,\text{Tr}\,\rho_{\text{eq}}^2. \tag{1.37b}$$

The factor $\text{Tr}\,\rho_{\text{eq}}^2$ in (1.37b) is the purity of the *time-independent* part of $\rho(t)$: $\text{Tr}\,\rho_{\text{eq}}^2 = \sum_n \rho_{nn}^2(0)$ according to (1.32). While $\rho(t)$ may be a pure state, the purity of ρ_{eq} may be as small as $10^{-O(f)}$ according to (1.29). We then finally obtain

$$\sigma_X^2 \leq \tilde{\Delta}_X^2 \max_n p_{E_n}, \tag{1.38}$$

where p_{E_n} [see (1.15)] is the occupation probability of E_n.

Let us consider $\text{Tr}\,[\rho(t)X]$ as a *random variable*, generated by sampling the observable at random times t. The probability for $\langle X \rangle(t)$ to deviate from its time average $\overline{\langle X \rangle(t)}$ by more than an infinitesimal quantity δX can then be proven to be bounded by

$$\text{Prob}\{|\text{Tr}\,[\rho(t)X] - \text{Tr}\,(\rho_{\text{eq}}X)| \geq \delta X\} \leq \left(\frac{\sigma_X}{\delta X}\right)^2. \tag{1.39}$$

Upon replacing (1.38) by its estimate according to (1.29) and considering (1.39), one arrives at the main result of the present chapter:

$$\text{Prob}\{|\text{Tr}\,[\rho(t)X] - \text{Tr}\,(\rho_{\text{eq}}X)| \geq \delta X\} \leq \left(\frac{\tilde{\Delta}_X}{\delta X}\right)^2 10^{-O(f)}. \tag{1.40}$$

This general expression defines the resolution limit δX of $\langle X\rangle(t)$ deviations from equilibrium.

1.3 From Equilibration to Thermalization

According to the present discussion, all observable expectation values (1.21) become practically indistinguishable from

$$\text{Tr}\,(\rho_{\text{eq}}X) = \sum p_n X_{nn} \tag{1.41}$$

after initial transients have died out. Hence, the arguments raised above imply that for realistic typical observables of macrosystems the problem of equilibration can be considered as settled.

Upon adopting the arguments that support equilibration of experimentally realistic (typical) observables, as per (1.41), we next turn to the key question: To what extent is the equilibrium expectation value of X from (1.41) in agreement with that predicted by the microcanonical ensemble, namely

$$\text{Tr}\,(\rho^{\text{mic}}X) = \sum p_n^{\text{mic}} X_{nn}? \tag{1.42}$$

Accordingly, are the level populations p_n^{mic} simply equal to a normalization constant if E_n is contained within a small energy interval $[E - \Delta E, E]$? Alternatively, we may ask: Under what conditions does the microcanonical formalism of equilibrium statistical mechanics break down?

The main condition for the microcanonical formalism to be valid is that only E_n within a small energy interval have a nonvanishing occupation probability, namely, the system energy is known to a high precision. We specifically assume that the system energy is known within an uncertainty ΔE that is small, but experimentally realistic.

Another validity condition of the microcanonical formalism is that the expectation values (1.42) must be (practically) independent of the exact choice of the energy interval (i.e., of its upper limit E and its width ΔE). The same conclusion follows from the equivalence of the microcanonical and canonical ensembles (for all energies E), considered as a self-consistency condition for equilibrium statistical mechanics. Clearly, if p_n values are irrelevant, the expectation values (1.41) and (1.42) are practically indistinguishable.

To ensure this equivalence of the microcanonical and canonical ensembles, we assume that the expectation values $X_{nn} = \langle n|X|n\rangle$ are the same within any small energy interval ΔE. This assumption amounts to *coarse graining*. The same assumption underlies Srednicki's "eigenstate thermalization hypothesis" (ETH), whereby each individual energy eigenstate $|n\rangle$ behaves like the equilibrium ensemble.

1.3.1 Integrals of Motion and Recurrence

Consider a many-body quantum system characterized by a set $\{X_1, X_2, \ldots, X_f\}$ of nontrivial integrals of motion. These operators commute with each other and with the Hamiltonian H_0 of the system, which may be integrable (solvable). These integrals of motion result in high degeneracy of the eigenstates of H_0; namely, many states $|\alpha, i\rangle$ (where $i = \{i_1, \ldots, i_f\}$), corresponding to different eigenvalues x_{ji_j} of the respective operators X_j, have the same energy E_α. Such a system does not thermalize, but approaches a state described by a generalized Gibbs ensemble (GGE).

This picture is idealized. In reality, the Hamiltonian $H = H_0 + V$ usually contains a perturbation V that does not commute with X_j. This perturbation lifts the degeneracy, thereby splitting the energy levels, $E_\alpha \rightarrow E_{\alpha i}$.

Consider a system prepared in a nonstationary state,

$$|\Psi(0)\rangle = \sum_{\alpha,i} C_{\alpha i}|\alpha, i\rangle, \tag{1.43}$$

with the average energy E and the energy uncertainty δE. The expectation value $\bar{X}(t) = \langle\Psi(t)|X|\Psi(t)\rangle$ of an observable X evolves in time as

$$\bar{X}(t) = \sum_{\alpha,i} |C_{\alpha i}|^2 \langle\alpha, i|X|\alpha, i\rangle$$

$$+ \sum_{\alpha,i\neq l} C_{\alpha l}^* C_{\alpha i} \langle\alpha, l|X|\alpha, i\rangle \exp[i(E_{\alpha l} - E_{\alpha i})t/\hbar]$$

$$+ \sum_{\alpha\neq\beta,i,l} C_{\beta l}^* C_{\alpha i} \langle\beta, l|X|\alpha, i\rangle \exp[i(E_{\beta l} - E_{\alpha i})t/\hbar]. \tag{1.44}$$

The hypothesis of quantum typicality states that the first term in Eq. (1.44) gives the thermal average of X at the temperature related to the total energy E via the equation of state. The two other terms describe the process of approaching the equilibrium. The last term, containing the frequencies $\omega_{\alpha\beta} = (E_\alpha - E_\beta)$, $\beta \neq \alpha$, vanishes on a timescale $\tau_E \sim 1/|\omega_{\alpha\beta}|$. If the perturbation V is weak enough, then the timescale τ_V of the decay of the second term is much longer, $\tau_V \gg \tau_E$. In other words, the system *prethermalizes* on the timescale $\tau_E \ll t \ll \tau_V$ and approaches the slowly decaying state characterized by the expectation value given by the GGE:

$$\bar{X}_{\mathrm{GGE}} = \sum_{\alpha,i,l} C_{\alpha l}^* C_{\alpha i} \langle \alpha, l | X | \alpha, i \rangle = \langle \Psi(0) | \tilde{X} | \Psi(0) \rangle. \qquad (1.45)$$

Here $\tilde{X} = \sum_\alpha P_\alpha X P_\alpha$, where $P_\alpha = \sum_i |\alpha, i\rangle\langle\alpha, i|$ is the projection operator on the subspace of the nearly degenerate states corresponding to E_α.

Quantum typicality implies that the time dependence of (1.44) does not essentially depend on the initial state $|\Psi(0)\rangle$. Therefore, repeating an experiment many times, averaging over experimental realizations, and subsequently Fourier transforming the measured time dependence of the observable and its moments allows for the estimation of the structure of the Hilbert space of the many-body system (how much are the states degenerate to the zeroth approximation split via the perturbation).

In a finite-size system all the unperturbed frequencies $\omega_{\alpha\beta}$ may be commensurate to the lowest of them, ω_0, that is not small (i.e., lies in a range accessible to measurements). Therefore, at times equal to an integer multiple of ω_0 a partial recurrence of the initial state will be observed. The recurrence is partial, since the exact eigenstates are not degenerate, and the perturbation-induced interaction blurs the recurrence, as we can see from substituting $\omega_{\alpha\beta}t$ in (1.44) by an integer multiple of 2π. Nevertheless, its observation would demonstrate, for quantum systems with commensurate-frequency spectra, that thermalization merely masks reversible, unitary features that can be revived.

1.4 Discussion

We have seen that for the overwhelming majority of sampling times $t > 0$ and any realistic observable X in a large system, the difference between $\mathrm{Tr}\,[\rho(t)X]$ and $\mathrm{Tr}\,(\rho_{\mathrm{eq}}X)$ is far below any conceivable resolution limit. It then follows that the steady-state ensemble ρ_{eq} *appears to be* adequate at almost any time $t \geq 0$, although the actual density operator $\rho(t)$ is rather different, allowing for $\langle X \rangle(t)$ oscillation. This difference between $\rho(t)$ and ρ_{eq} explains the apparent discrepancy between unitarity and equilibration.

These conclusions do not require a macroscopic number f of degrees of freedom: Systems with $f \sim 3$ exhibit the trend to equilibrate and thermalize.

The resolution limit of (1.40) is based on the exact quantum mechanical time evolution (1.12)–(1.14) and thus obeys the time-inversion symmetry required by quantum mechanics. In particular, (1.40) allows for recurrences of $\mathrm{Tr}\,[\rho(t)X]$, but such large deviations from the equilibrium state ρ_{eq} are extremely improbable. The same behavior is obtained if one propagates $\rho(0)$ *backward* in time. An initial condition $\rho(0)$ that is far from ρ_{eq} is therefore a rare deviation for any choice of $t = 0$.

Deviations of $\mathrm{Tr}\,[\rho(t)X]$ from the *apparent equilibrium* value $\mathrm{Tr}\,(\rho_{\mathrm{eq}}X)$ are not expected to exhibit time-inversion symmetry. Yet, the *probabilities* of such excursions are expected to satisfy a detailed-balance symmetry with respect to time inversion.

Thus, *quantum mechanical time-inversion symmetry is preserved. However, under out-of-equilibrium initial conditions, a "time arrow" emerges with extremely high probability.*

2

Thermalization of Quantum Systems Weakly Coupled to Baths

In general, it is impossible to establish thermalization without resorting to coarse graining, as discussed in Chapter 1. Here we focus on the case of an isolated composite system, decomposed into a small (sub)system that is *weakly coupled* to a much bigger (sub)system that is dubbed an "environment" or a "heat bath." Then the steady state of the small system generally thermalizes (i.e., is described by a canonical ensemble characterized by the bath temperature). However, this kind of decomposition may fail in many cases, thus precluding thermalization.

2.1 Division into System and Bath

We assume a system, labeled by 'S', ruled by Hamiltonian H_S, in an f_S-dimensional Hilbert space $\mathcal{H}_S$ that interacts with a bath (labeled by 'B') governed by a Hamiltonian H_B in an f_B-dimensional Hilbert space $\mathcal{H}_B$. The bath is taken to be much "bigger" than the system,

$$f_B \gg f_S. \tag{2.1}$$

The system and the bath are coupled by an interaction Hamiltonian $\tilde{H}_{SB}$ scaled by a coupling parameter ϵ. The system-plus-bath complex (alias the total system or supersystem) with

$$f = f_S + f_B \tag{2.2}$$

degrees of freedom (DOF) occupies the product space $\mathcal{H} := \mathcal{H}_S \otimes \mathcal{H}_B$ and is governed by the total Hamiltonian

$$H(\epsilon) = H_S \otimes \hat{I}_{\mathcal{H}_B} + \hat{I}_{\mathcal{H}_S} \otimes H_B + \epsilon \tilde{H}_{SB}, \tag{2.3}$$

where $\hat{I}_{\mathcal{H}_\mathrm{S}}$ indicates the identity on $\mathcal{H}_\mathrm{S}$, and similarly for $\hat{I}_{\mathcal{H}_\mathrm{B}}$. We are mainly interested in observables of the *system only*, of the form

$$X = X^\mathrm{S} \otimes \hat{I}_{\mathcal{H}_\mathrm{B}}. \tag{2.4}$$

We assume that all conditions for equilibration of the isolated quantum system-plus-bath complex (Sec. 1.2.4) are fulfilled by the observables (2.4). The present quantum-mechanical treatment of equilibration can be linked with the *thermodynamic limit*, where, instead of $f_\mathrm{B} \gg 1$ DOF of an N-particle bath confined in a volume V, we characterize the infinite-volume bath by the density N/V, and $f_\mathrm{B} \to \infty$.

2.1.1 Weak Coupling Condition and Thermal Equilibrium

The weak-coupling assumption is a precondition for the formation of the canonical ensemble (Sec. 1.3). We may require ϵ in (2.3) to be so small that the eigenvectors $|n\rangle$ and eigenvalues E_n of $H(\epsilon)$ deviate very little from those of the noninteracting $H(0)$, in accordance with ordinary perturbation theory, whereby these deviations are of the order $\epsilon \langle m|\tilde{H}_\mathrm{SB}|n\rangle/(E_m - E_n)$. However, $E_m - E_n$ may be of the order of $10^{-O(f)}$ joules (see Ch. 1), so that the admissible ϵ-values would be too small for any realistic model to satisfy the condition. We therefore assume an alternative weak-coupling condition: *After equilibration of the system-plus-bath compound, a reversible (adiabatically slow) decoupling of the system from the bath does not result in any experimentally resolvable changes.*

This weak-coupling condition refers to *time-averaged* expectation values of the observables (2.4) and the energy density (1.26) of the system-plus-bath complex. It is thus consistent with the ability of time-dependent expectation values $\langle X \rangle(t)$, even after equilibration, to exhibit large deviations from equilibrium, albeit at exceedingly rare times t (Sec. 1.2.4).

Typically, systems in contact with a heat bath satisfy the weak coupling condition, but its proof for any given model and initial conditions $\rho(0)$ of the system-plus-bath complex may be difficult. The difficulty is seen for observables of the form (2.4), which are expected, following time averaging, to be described by the (Gibbs) canonical ensemble ρ_T for small ϵ,

$$\mathrm{Tr}(\rho_T X) = \mathrm{Tr}\left[\frac{e^{-H(\epsilon)/k_\mathrm{B}T}}{Z(\epsilon)} X\right] = \mathrm{Tr}_\mathrm{S}\left(\frac{e^{-H_\mathrm{S}/k_\mathrm{B}T}}{Z_\mathrm{S}} X_\mathrm{S}\right), \tag{2.5}$$

with Tr and Tr_S indicating the traces over the Hilbert spaces $\mathcal{H}$ and $\mathcal{H}_\mathrm{S}$, respectively, and $Z(\epsilon)$ and Z_S being the standard partition sums, normalizing the respective (canonical) density operators. Although relation (2.5) appears to be self-evident in the "weak coupling" limit, it is *unproven* for small coupling strengths ϵ. More detailed analysis of the spin-boson system–bath equilibration will be presented in Chapter 6.

The same is true for thermal equilibrium in the weak-coupling limit of the grand canonical ensemble, which is expected to be described by a Gibbs state with the modified Hamiltonian

$$H(\epsilon, \mu) = H(\epsilon) - \mu \hat{N}, \tag{2.6}$$

with μ being the chemical potential and $\hat{N}$ the bath particle-number operator.

2.1.2 Weak Coupling in the Adiabatic Limit

If ϵ in (2.3) is time dependent, then the density operator obeys the Liouville–von Neumann equation

$$i\hbar\dot{\rho}(t) = [H(\epsilon(t)), \rho(t)]. \tag{2.7}$$

In the case of slow (quasi-static) parameter changes, the adiabatic theorem can be invoked, yielding

$$\rho(t) = \sum_{m,n} \rho_{mn}(0) e^{-i \int_0^t \omega_{mn}(\epsilon(s))ds} |m(\epsilon(t))\rangle \langle n(\epsilon(t))|, \tag{2.8}$$

where $\omega_{mn}(\epsilon) = [E_m(\epsilon) - E_n(\epsilon)]/\hbar$, $E_n(\epsilon)$ and $|n(\epsilon)\rangle$ being the eigenvalues and eigenvectors of $H(\epsilon)$. Here we have tacitly assumed the case of *nondegenerate* energy levels.

Specifically, let us assume that $\epsilon(t)$ in (2.8) is adiabatically slowly switched off to zero and remains zero for all later times t until infinity. Then, the time average of the density operator in (2.8) is governed by the infinitely long time period with $\epsilon(t) = 0$, that is,

$$\overline{\rho(t)} = \sum_n \rho_{nn}(0) |n(0)\rangle \langle n(0)|. \tag{2.9}$$

The weak-coupling condition then implies that the time-averaged expectation values are practically indistinguishable from those obtained from (2.9), that is,

$$\overline{\langle X\rangle(t)} = \sum_n p_n \langle n(0)|X|n(0)\rangle, \tag{2.10}$$

where $p_n := \rho_{nn}(0) = \langle n(\epsilon(0))|\rho(0)|n(\epsilon(0))\rangle$ are the level populations of the system at $t = 0$ [cf. (1.16)]. Namely, *although the true system–bath coupling strength ϵ is finite, we can assume the zero-coupling limit for the time-averaged expectation values.*

The adiabatic evolution (2.8) further implies that the level populations, $\langle n(\epsilon(t))|\rho(t)|n(\epsilon(t))\rangle$, are identical with $p_n = \rho_{nn}(0)$ at all times t and are independent of ϵ. In the relation $p_n = h(E_n(\epsilon)) + \delta p_n$ [cf. (1.25)], the right-hand side must then be ϵ independent. By averaging over many n-values, we find that $h(E_n(\epsilon))$ must be ϵ-independent as well for all n, so that relation (1.25) becomes

$$p_n = h_0(E_n(0)) + \delta p_n, \tag{2.11}$$

where h_0 and $E_n(0)$ are labeled by $\epsilon = 0$. Likewise, the density of states (1.8) and the energy density (1.27) then satisfy the relation

$$\rho_0(E) = h_0(E)\Omega_0(E). \tag{2.12}$$

Thus, the weak coupling condition implies the *indistinguishability* of the energy density from that obtained after decoupling the system and the bath, that is,

$$\rho_\epsilon(E) = \rho_0(E). \tag{2.13}$$

2.2 System–Bath Separability and Non-separability

In the limit of vanishing coupling strength $\epsilon \to 0$, $|n\rangle_S$ and E_n^S that denote the eigenvectors and eigenvalues of H_S, respectively, and $|m\rangle_B$ and E_m^B that denote those of H_B [cf. (2.3)] become *separable*,

$$|nm\rangle := |n\rangle_S |m\rangle_B, \tag{2.14}$$
$$E_{nm} := E_n^S + E_m^B. \tag{2.15}$$

Accordingly, the ϵ-independent level populations of the system–bath complex are denoted by p_{nm}. For observables of the form (2.4), the time-averaged expectation value (2.10) can then be rewritten as

$$\overline{\langle X \rangle(t)} = \sum_{mn} p_{nm}\, {}_S\langle n|X^S|n\rangle_S = \sum_n p_n^S\, {}_S\langle n|X^S|n\rangle_S, \tag{2.16}$$

$$p_n^S := \sum_m p_{nm}. \tag{2.17}$$

We may rewrite the p_{nm} according to (2.11) as

$$p_{nm} = h_\epsilon(E_{nm}(\epsilon)) + \delta p_{nm}. \tag{2.18}$$

One can then conclude (cf. Sec. 1.2.1) that for any n, the fluctuation δp_{nm} with the bath index $m = 0, 1, 2, \ldots$ is unbiased, that is,

$$\sum_m p_{nm} = \sum_m h_\epsilon(E_{nm}(\epsilon)). \tag{2.19}$$

Namely, the energies E_{nm} given by (2.15), with n arbitrary but fixed and m variable, are still extremely dense.

From (2.11), (2.17), and (2.19) we obtain that $p_n^{\mathrm{S}} = \sum_m h_0(E_{nm})$, and from (2.12) and (2.13) that

$$p_n^{\mathrm{S}} = \sum_m \frac{\rho(E_{nm})}{\Omega(E_{nm})}, \tag{2.20}$$

where $\rho(E)$ denotes the energy density $\rho_\epsilon(E)$ of the system-plus-bath complex evaluated at finite coupling ϵ.

2.2.1 Additivity of Entropy

Here, we revisit the entropy [cf. (1.6)] and temperature [cf. (1.9)] defined in Section 1.1 for the case of Hamiltonian (2.3). For $\epsilon = 0$, the difference with the definitions in Section 1.1 is that the single index n is now replaced by the double indices nm [see (2.14) and (2.15)].

For $\epsilon = 0$ in (2.3), the indices 'S' and 'B' label the separate, isolated system and bath. The following relations hold between the quantities pertaining to the separate system and bath and those of the system-plus-bath compound (2.3) for $\epsilon = 0$:

$$E_{\mathrm{S}}(E) := E - E_{\mathrm{B}}(E), \tag{2.21}$$

$$\mathcal{S}(E) = \mathcal{S}_{\mathrm{B}}(E_{\mathrm{B}}(E)) + \mathcal{S}_{\mathrm{S}}(E_{\mathrm{S}}(E)), \tag{2.22}$$

$$T(E) = T_{\mathrm{B}}(E_{\mathrm{B}}(E)) = T_{\mathrm{S}}(E_{\mathrm{S}}(E)). \tag{2.23}$$

These relations are asymptotically exact in the thermodynamic limit $f_{\mathrm{B}} \to \infty$.

In other words, in the limit of zero coupling the entropies of the system-plus-bath complex are additive, provided the sum $\mathcal{S}_{\mathrm{B}}(E') + \mathcal{S}_{\mathrm{S}}(E - E')$ is maximized. The consequence is that all three temperatures in Eq. (2.23) are identical, that is, the *equilibrium condition (or zeroth law of thermodynamics)*.

For low-dimensional systems it is appropriate to resort to the von Neumann (VN) entropy

$$\mathcal{S}_{\mathrm{VN}} = -k_{\mathrm{B}} \mathrm{Tr}(\rho \ln \rho), \tag{2.24}$$

which is an invariant of the density operator ρ. The VN entropy is nonnegative and bounded from below by the Shannon entropy

$$\mathcal{S}_{\mathrm{SH}} = -k_{\mathrm{B}} \sum_n p_n \ln p_n. \tag{2.25}$$

The VN entropy is maximized for a fixed mean energy $E = \mathrm{Tr}(\rho H)$ in the Gibbs state.

2.3 Thermal Equilibrium and Correlation Functions

The description of quantum baths and their interactions with smaller quantum systems cannot be accomplished by detailed solutions of their Schrödinger or Liouville equations on account of their huge dimensionalities, up to infinity in the thermodynamic limit. The alternative in this limit is to resort to Kubo's multi-time correlation (or Green) functions of the pertinent observables in the equilibrium state. In this book we shall only consider two-time correlations (auto-correlation functions) of the same observable, $\epsilon \tilde{H}_{SB} \sim \hat{S} \cdot \hat{B}$, where $\hat{S}$ and $\hat{B}$ are operators pertaining to the system and the bath, respectively. These functions are

$$\Phi(t, 0) = \epsilon^2 \text{Tr}[\rho \tilde{H}_{SB}(t) \tilde{H}_{SB}(0)], \tag{2.26}$$

where $\tilde{H}_{SB}(t)$ is the interaction-picture form of $\tilde{H}_{SB}$ and ρ is the density matrix of the equilibrium state of system S.

Hereafter, we shall simplify the notation to $\Phi(t)$. Assuming that the Hamiltonian spectrum is discrete, $\Phi(t)$ is a *quasi-periodic function* that after sufficient time returns arbitrarily close to its initial value, thus satisfying the quantum analog of classical Poincaré recurrences.

However, if the description of ρ as a Gibbs state at temperature T is valid, the functions $\Phi(t)$ can be extended by analytic continuation to the complex domain ($t \to z$), where they satisfy the following Kubo–Martin–Schwinger (KMS) condition at any time,

$$\Phi(-t) = \Phi(t - i\hbar\beta), \tag{2.27}$$

with $\beta = 1/(k_B T)$ being the inverse temperature. As shown in Chapter 4, for typical bosonic or fermionic quantum baths, the KMS condition implies the decay of $\Phi(t)$ to zero as $t \to \infty$ (i.e., to the loss of recurrences and irreversibility). Explicitly, for a bosonic or fermionic bath,

$$\Phi(t) \propto \int_{-\infty}^{+\infty} d\omega \, e^{-i\omega t} \{ [1 - (\mp)n(\omega)]\theta(\omega) + n(|\omega|)\theta(-\omega) \}. \tag{2.28}$$

Here $n(\omega)$ is the average number of quanta with the energy $\hbar\omega$, which in the case of a thermal bath without chemical potential is given by

$$n(\omega) = \frac{1}{e^{\hbar\beta\omega} \mp 1}, \tag{2.29}$$

with the $-$ sign corresponding to bosons and the $+$ sign to fermions. The function (2.28) has the structure of a Fourier transform, consistently with the fact that in the thermodynamic limit correlations decay to zero at long times, without exhibiting

Poincare recurrences. For generic infinite baths, the inverse Fourier transform $\tilde{\Phi}(\omega)$ exists, so that the KMS condition (2.27) implies the relation

$$\tilde{\Phi}(-\omega) = e^{-\hbar\beta\omega}\tilde{\Phi}(\omega), \tag{2.30}$$

which will be extensively discussed in the context of open-system control (Chs. 7, 11, and 12).

The KMS condition defines thermal equilibrium states for infinite baths. These states are stable under local perturbations. Finite baths at a given temperature are characterized by a unique Gibbs state that satisfies the KMS condition. On the other hand, in an infinite bath at a given temperature, many states obeying the KMS condition may coexist below a certain critical temperature. This multiplicity of states complying with the KMS condition arises at a *phase transition*, a notion that can only be defined in the thermodynamic limit.

2.4 Discussion

The thermalization of a finite subsystem embedded in a much larger, typically infinite-dimensional system dubbed a "bath" (i.e., its adherence to a canonical ensemble) appears to be universal. Hence, one may conclude that such thermalization is ubiquitous at asymptotically long times. Yet this conclusion is false for a variety of system classes in quantum optics, some of which are discussed in Chapter 5. Is there a common pattern behind their *non-thermalization*? In subsequent chapters we shall examine three possible mechanisms for non-thermalization:

(a) System–bath strong coupling and the bath-continuum approximation failure: One of the conditions for thermalization is weak coupling of the system and the bath. Hence, we can expect non-thermalization when the weak-coupling condition fails,

$$\langle \tilde{H}_{SB}^2(E)\rangle^{1/2}\rho(E) \gg 1, \tag{2.31}$$

where the first factor is the mean system–bath coupling and the second is the bath mode density. Counterintuitively, the opposite limit,

$$\langle \tilde{H}_{SB}^2(E)\rangle^{1/2}\rho(E) \ll 1, \tag{2.32}$$

may also lead to thermalization failure, because the bath-continuum approximation breaks down (Ch. 5).

(b) Nonanalytic bath spectra: Does thermalization hold if the spectrum has band-edges or cutoffs at which the system–bath coupling vanishes abruptly? We shall consider this issue in Section 5.2, where such scenarios are shown to lead to *partial decay* and hence *non-thermalization*.

(c) Cooperative effects: Can thermalization be blocked by cooperative (multipartite) features of the system? Since their cooperative (Dicke) states are degenerate only in the absence of system–bath coupling, the answer is not straightforward (Sec. 7.2).

3

Generic Quantum Baths

In the strict quantum-mechanical approach, expounded in Chapters 1 and 2, the system and its environment taken together form an isolated entity that is governed by a Hamiltonian and may evolve unitarily, depending on its initial state. The division of this entity into a system with a limited number of degrees of freedom (DOF) and its environment with a much larger number of DOF (infinite number, in the thermodynamic limit) is a matter of expediency: we choose a system that we are able to observe and control, as opposed to the unobserved and uncontrolled environment, either for lack of interest or due to its complexity or inaccessibility (or both).

It is convenient and plausible to view the environment as a collection of "reservoirs" ("baths") that hardly change as a result of their interaction with the system, because such change is distributed over many, possibly infinitely many, DOF of the baths. The standard bath is a thermostat (i.e., a thermal ensemble with a constant temperature), as assumed in Chapter 2, but contemporary theory and experiment allow for a much more detailed characterization of bath quantum states and bath spectra, as detailed below. These properties are key to the dynamical control of quantum systems in contact with a bath.

In general, the quantum state of the bath is mixed because of the lack of knowledge concerning this state, although a bath may also be found in a pure state, particularly its ground state. All the baths discussed in this book are composed of *identical* quantum objects (particles or quasiparticles) that obey either bosonic or fermionic statistics and may collectively undergo elementary excitations with nontrivial spectra.

3.1 Bosonic Bath Models

The Hamiltonian of a bosonic bath is a quadratic functional of a bosonic bath (field) operator $B(x)$, that is, it ignores higher-order nonlinearities in $B(x)$. Such higher-order nonlinearities may lead to diverse effects, which are not discussed here (e.g., phonon anharmonicity, corrections to magnon dispersion, or nonlinear-optics effects for photons).

We here assume that each normal mode in a bosonic bath is associated with a quantized harmonic oscillator. The resulting decomposition of a bosonic bath operator $B(x)$ has the form

$$B(x) = \sum_{\Lambda} [a_{\Lambda}\phi_{\Lambda}(x) + a_{\Lambda}^{\dagger}\phi_{\Lambda}^{*}(x)], \tag{3.1}$$

where a_{Λ} $(a_{\Lambda}^{\dagger})$ denotes the annihilation (creation) operator of elementary excitations in a mode Λ described by wave function $\phi_{\Lambda}(x)$. The operators a_{Λ} and $a_{\Lambda}^{\dagger}$ obey the boson commutation relations,

$$[a_{\Lambda}, a_{\Lambda'}^{\dagger}] = \delta_{\Lambda\Lambda'}, \quad [a_{\Lambda}, a_{\Lambda'}] = [a_{\Lambda}^{\dagger}, a_{\Lambda'}^{\dagger}] = 0, \tag{3.2}$$

where $\delta_{\Lambda\Lambda'}$ is the Kronecker symbol.

The choice of $B(x)$ for a given bath is not unique. Thus, for an electromagnetic (photonic) bath, one may choose to work with the electric and magnetic field operators. However, it is often appropriate to adopt $B(x)$ obtainable by quantizing the vector potential (Sec. 3.1.1) for reasons explained in Chapter 4.

The same form of $B(x)$ and its normal-mode decomposition apply to various elementary excitations: photons, phonons, polaritons, and magnons. The geometry and composition of the bath medium may dictate different choices of normal modes, as opposed to plane waves $\phi_k(x) \sim e^{ik \cdot x}$ that are appropriate for free space, where k is the mode wave vector.

One such choice may be imposed by the *spherical symmetry* of the bath medium, which yields the expansion of $B(x)$ in terms of normal modes labeled by $\Lambda = (\omega, l, m, \lambda)$, where ω denotes the mode frequency, l and m are angular indices, and λ is the mode type. In particular, the electromagnetic (EM) field in spherical coordinates consists of transverse electric (TE) and transverse magnetic (TM) waves, which can be expressed, respectively, through the two Debye potentials and their derivatives. The two Debye potentials are solutions of the scalar wave equation. For a given Λ, they are different linear combinations of the spherical waves

$$\left\{ \begin{array}{c} \cos(m\phi) \\ \sin(m\phi) \end{array} \right\} P_l^m(\cos\theta)z_l(kr). \tag{3.3}$$

Here $k = n_r\omega/c$ is the wave number, n_r is the refractive index of the medium, r, θ, ϕ are the spherical coordinates, $P_l^m(\cos\theta)$ are the associated Legendre polynomials, $0 \leq m \leq l$, and $z_l(kr)$ is either of the two kinds of linearly independent spherical Bessel functions of order l.

If the bath modes possess *spatial periodicity*, as in a natural lattice structure, the expansion may be performed in terms of periodic (Bloch-wave) mode functions. The wave functions in (3.1) then have the form

$$\phi_k(x) = e^{ik \cdot x} u_k(x), \tag{3.4}$$

where the $u_k(x)$ expresses the parts of the Bloch functions that are periodic in the lattice, whereas $\hbar k$ is the quasimomentum that is restricted to the first Brillouin zone.

The lattice periodicity divides the bath spectrum into bands separated by forbidden band gaps. These band gaps will be shown to profoundly affect system–bath interactions (Chs. 4 and 5).

In what follows we briefly survey the generic bosonic bath models that recur in this book.

3.1.1 Photonic Baths

The electromagnetic (EM) bath operator that is derived from the canonical Lagrangian quantization in a big box of the volume $\mathcal{V}$ is the vector potential (in the Gaussian units)

$$\boldsymbol{B}(x) \equiv \boldsymbol{A}(x) = \sum_{k,\lambda} \left(\frac{2\pi \hbar c}{k \mathcal{V}} \right)^{1/2} \left(a_{k\lambda} \boldsymbol{\epsilon}_{k\lambda} e^{ik \cdot x} + a_{k\lambda}^{\dagger} \boldsymbol{\epsilon}_{k\lambda}^{*} e^{-ik \cdot x} \right), \tag{3.5}$$

where $\boldsymbol{A}$ is the vector potential, c is the light velocity in vacuum, and $\boldsymbol{\epsilon}_{k\lambda}$ is a unit polarization vector. The three-dimensional (3d) plane-wave decomposition of this bath operator is appropriate only for free-space EM baths.

Cavity Baths

In the case of a high-Q closed cavity, where Q is the quality factor of the cavity, it is natural to decompose the EM bath in terms of standing-wave discrete modes $\phi_{ij}(x)$ labeled by two indices. For an open cavity, by contrast, most of the modes have low Q and merge into a continuum. If the cavity is very large compared with the wavelength, as typically happens in optics, a few modes may have low losses, due to the directionality of the radiation.

In the standard *optical Fabry–Pérot cavity*, light bounces back and forth between two parallel flat mirrors that are made slightly transparent, to allow a coupling with external space. To obtain the mode frequencies of the high-Q modes of the open Fabry–Pérot cavity, the latter can be approximately considered as a closed cavity. For a cavity of length d (along the z-axis) and a square cross section $(2a)^2$ (in the xy-plane), the mode frequencies have the form

$$\omega = \frac{\pi c}{n_r} \left[\left(\frac{q}{d}\right)^2 + \left(\frac{r}{2a}\right)^2 + \left(\frac{s}{2a}\right)^2 \right]^{\frac{1}{2}}, \tag{3.6}$$

where q, r, and s are positive integers and n_r is the refractive index of the medium in the cavity. Each mode corresponds to a plane wave bouncing back and forth between the walls and is characterized by the direction cosines. For waves that propagate at a small angle with the cavity axis, q is approximately the total number of half wavelengths contained in d, a very large value, whereas r and s are small. Then, if the ratio $2a/d$ is not too small, (3.6) can be written as

$$\omega = \frac{\pi c}{n_r} \left(\frac{q}{d} + \frac{r^2 d}{8qa^2} + \frac{s^2 d}{8qa^2} \right). \tag{3.7}$$

The energy of a nearly axial wave will decay within a time

$$t_{\text{decay}} = \frac{n_r d}{(1 - \alpha_q)c}, \tag{3.8}$$

where α_q is the mirror reflectivity. The quality factor is then defined as $Q = \omega t_{\text{decay}} = (kc/n_r)t_{\text{decay}}$, where k is the wave number of the mode ω, or

$$Q = \frac{kd}{1 - \alpha_q}. \tag{3.9}$$

For $\alpha_q = 0.999$, $d = 1$ cm, and $k = 10^5$ cm^{-1}, $Q = 10^8$.

Two modes whose q values differ by 1 will, according to (3.7), have frequencies separated by

$$\Delta\omega_q \approx \frac{\pi c}{n_r d}. \tag{3.10}$$

If $\alpha_q \approx 1$, then (3.8) implies that $\Delta\omega_q \gg 1/t_{\text{decay}}$, so that successive axial modes are discrete. However, lateral modes with the same value of q are almost degenerate and merge within the mode width.

Without the sidewalls, the cavity is a true Fabry–Pérot resonator. For small values of r and s, the modes are then nearly the same as those of the closed cavity, but the Q factors of all modes are lowered, the effect increasing with the values of r and s: oblique waves, after being reflected back and forth, gradually walk off the end mirrors.

Essentially, lateral modes (with k perpendicular to the cavity axis) are free-space plane waves. Therefore, even stable modes (labeled by q) have finite decay time: they allow EM field leakage to the mode continuum outside the cavity, which may be described by a Lorentzian spectral profile of linewidth $\Gamma_q = 1/t_{\text{decay}}(\omega_q)$ that broadens the mode frequencies,

$$\delta(\omega - \omega_q) \rightarrow \frac{\Gamma_q/\pi}{(\omega - \omega_q)^2 + \Gamma_q^2}. \tag{3.11}$$

Waveguide Baths

A *photonic waveguide* supports a 1d continuum of modes along its axis z. The field modes are confined in the directions (x, y) transverse to the axis z. There are two different types of photonic waveguides in terms of their mode structure:

(a) A hollow waveguide supports electric field modes with normalized mode functions

$$u^\lambda_{mnk} = \frac{1}{\sqrt{L}} E^\lambda_{mnk}(x, y) e^{ikz}. \tag{3.12}$$

Here the quantization length along z is L, λ is the polarization index, and m and n are the transverse-mode indices. The transverse modes obey the dispersion relation

$$\omega_{mnk} = c\sqrt{k_{mn}^2 + k^2}, \tag{3.13}$$

where ck_{mn} is the lower-cutoff frequency of the mode. There are two possible polarizations for the transverse modes: transverse electric (TE) and transverse magnetic (TM).

As an example, we consider the rectangular hollow metal waveguide (the lengths of the sides being a and b) whose mode frequencies are much lower than the plasma frequency of the metal, so that it is nearly lossless. By normalizing their transverse profiles, that is, demanding

$$\int_0^a dx \int_0^b dy \, E^{\lambda*}_{mnk}(x, y) \cdot E^\lambda_{mnk}(x, y) = 1, \tag{3.14}$$

we obtain the mode functions; see Table 3.1, where $A = ab$ is the transverse area of the waveguide. Note that the transverse profile of the TE modes does not depend on the wave number k (or on frequency). The mode cutoff wave numbers are

$$k_{mn} = \pi \sqrt{(m/a)^2 + (n/b)^2}, \tag{3.15}$$

where m and n are nonnegative integers such that $m + n \geq 1$ or $m + n \geq 2$ for TE or TM modes, respectively. The lowest mode is TE_{01} or TE_{10}. The lowest TM mode is TM_{11}; it has a higher cutoff than the lowest TE mode.

Another example is a cladded fiber, a waveguide that is designed for optical and near-infrared photons. It has a fundamental TE mode with a lower cutoff frequency. As shown in Chapter 4, it is expedient to have a periodic dielectric index (Bragg grating) along the fiber in order to better control light dispersion and mode density (Fig. 3.1).

(b) *A coaxial cable or a coplanar waveguide*, designed for microwave radiation, is characterized by the fundamental TEM mode, which does not have a lower

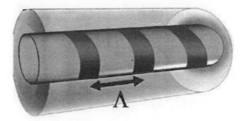

Figure 3.1 Optical-fiber waveguide with dielectric Bragg grating.

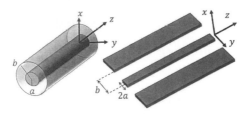

Figure 3.2 Left – coaxial waveguide. Right – coplanar waveguide with the same mode structure.

cutoff frequency and is dispersion-free, as a free-space plane wave (Fig. 3.2). Its normalized mode functions are

$$u_k^{\text{TEM}} = \frac{1}{\sqrt{L}} E^{\text{TEM}}(x, y) e^{ikz}, \tag{3.16}$$

where $k = \pm\omega/c$. Here $E^{\text{TEM}}(x, y)$ has the form of the static electric field of a capacitor with the same shape as the waveguide. In a coplanar waveguide (CPW) cavity for microwaves, the stable modes are surface modes with an exponentially small evanescent tail of the confined field away from the metallic surface.

Whispering-Gallery Mode Baths

A *spherical*, disk-shaped, or ring-shaped cavity possesses stable whispering-gallery modes (WGM). Here we consider a dielectric sphere of radius a and refractive index n_r, whose WGM are determined from the Mie scattering theory for an incident wave that travels along the z-axis with the electric vector polarized along the x-axis.

The general solution of the wave equations for the electric and the magnetic fields are superpositions of TE and TM modes, which can be expressed, respectively, through the Debye potentials ψ and φ and their first and second derivatives.

We here focus on the TE modes. We cast the incident ψ_{inc}, scattered ψ_{sca}, and internal ψ_{int} waves in terms of the spherical waves (3.3); see Table 3.1, where j_l

Table 3.1

$E_{mnk}^{\mathrm{TM}}(x,y)$	$\frac{2c}{\sqrt{A}}\left\{\frac{k_{mn}}{\omega_{mnk}}\sin\left(\frac{m\pi}{a}x\right)\sin\left(\frac{n\pi}{b}y\right)\boldsymbol{e}_z+\frac{ik}{k_{mn}\omega_{mnk}}\right.$
	$\left.\times\left[\frac{m\pi}{a}\cos\left(\frac{m\pi}{a}x\right)\sin\left(\frac{n\pi}{b}y\right)\boldsymbol{e}_x+\frac{n\pi}{b}\sin\left(\frac{m\pi}{a}x\right)\cos\left(\frac{n\pi}{b}y\right)\boldsymbol{e}_y\right]\right\}$
$E_{mn}^{\mathrm{TE}}(x,y)$	$\frac{2}{\sqrt{A}k_{mn}}\left[-\frac{n\pi}{b}\cos\left(\frac{m\pi}{a}x\right)\sin\left(\frac{n\pi}{b}y\right)\boldsymbol{e}_x+\frac{m\pi}{a}\sin\left(\frac{m\pi}{a}x\right)\cos\left(\frac{n\pi}{b}y\right)\boldsymbol{e}_y\right]$
ψ_{inc}	$\sum_{l=1}^{\infty}(-i)^l\frac{2l+1}{l(l+1)}j_l(kr)P_l^1(\cos\theta)\sin\phi$
ψ_{sca}	$\sum_{l=1}^{\infty}(-i)^l\frac{2l+1}{l(l+1)}b_l(k)h_l^{(1)}(kr)P_l^1(\cos\theta)\sin\phi$
ψ_{int}	$\sum_{l=1}^{\infty}(-i)^l\frac{2l+1}{l(l+1)}a_l(k)j_l(n_rkr)P_l^1(\cos\theta)\sin\phi$
$a_l(k)$	$\frac{h_l^{(1)}(ka)j_l'(ka)-j_l(ka)h_l^{(1)'}(ka)}{n_rh_l^{(1)}(ka)j_l'(n_rka)-j_l(n_rka)h_l^{(1)'}(ka)}$
$b_l(k)$	$\frac{j_l(n_rka)j_l'(ka)-n_rj_l(ka)j_l'(n_rka)}{n_rh_l^{(1)}(ka)j_l'(n_rka)-j_l(n_rka)h_l^{(1)'}(ka)}$

and $h_l^{(1)}$ are spherical Bessel and Hankel functions, respectively, $k=\omega/c$, and n_r is the refractive index of the sphere.

Explicitly, the radial dependence of the Debye potential for the wave with a given value of l takes the form

$$z_l=\begin{cases}a_l(k)j_l(n_rkr), & r<a,\\ j_l(kr)+b_l(k)h_l^{(1)}(kr), & r>a.\end{cases}\tag{3.17}$$

To find the coefficients $a_l(k)$ and $b_l(k)$, we make use of boundary conditions at the sphere surface $r=a$. These coefficients are shown in Table 3.1, where j_l' and $h_l^{(1)'}$ are the first derivatives of j_l and $h_l^{(1)}$ with respect to the argument of the function. The imaginary parts of the scattering coefficients $a_l(k)$ and $b_l(k)$ determine the line widths of the WGM Mie resonances. In dielectric spheres, Mie resonances associated with $l\gg1$ WGM correspond to high-Q "quasimodes" localized near the sphere surface with evanescent "tails" outside. These quasimodes can be viewed as nearly discrete, *broadened* $(2l+1)$-degenerate mode multiplets. The radial wave function outside the dielectric sphere is then approximated by

$$R_{kl}(r)=Ch_l^{(1)}(k_{ln}r),\tag{3.18}$$

where C is the normalization factor of the outgoing spherical Hankel function $h_l^{(1)}$, and k_{ln} belongs to a discrete set of Mie-resonance wave numbers.

Photonic-Crystal Baths

We next consider *structures with 3d periodicity* of the dielectric index $\epsilon_d(\boldsymbol{x})$, nicknamed photonic crystals. The dielectric index can be expanded as a Fourier series,

$$\epsilon_{\mathrm{d}}(\boldsymbol{x}) = \sum_{g} \epsilon_g \, e^{i\boldsymbol{g}\cdot\boldsymbol{x}}. \tag{3.19}$$

Here the sum is over the reciprocal-lattice vectors $\boldsymbol{g}$. The shortest vectors $\boldsymbol{g} \neq 0$ have the length $2\pi/d$, where d is the shortest structure period.

The mode functions are labeled by the quasimomentum $\hbar\boldsymbol{k}$, which is restricted to the first Brillouin zone $|\boldsymbol{k}| \leq \pi/d$:

$$\boldsymbol{\phi}_{\Lambda}(\boldsymbol{x}) = e^{i\boldsymbol{k}\cdot\boldsymbol{x}} \sum_{g} \boldsymbol{C}_{g\Lambda} e^{i\boldsymbol{g}\cdot\boldsymbol{x}}, \tag{3.20}$$

where the mode labels $\Lambda = (\boldsymbol{k}, n, \lambda)$ consist of $\boldsymbol{k}$, the band index n, and the polarization λ. Band gaps are formed in such structures whenever the wave equation for $\boldsymbol{E}(\boldsymbol{x}, \omega_{\Lambda})$ near the edge of the Brillouin zone $|\boldsymbol{k}| \simeq \pi/d$ has only complex $\boldsymbol{k}$ solutions *for any direction and polarization*. The modes merge into a continuum in the limit of high density of modes (DOM), $\sum_{\Lambda} \to V \sum_{n,\lambda} \int d^3\boldsymbol{k}/(2\pi)^3$, where the integration extends over the first Brillouin zone.

In photonic crystals without absorption, the edge of a band gap is a singularity of the DOM, where $dk/d\omega$ changes discontinuously and the DOM vanishes abruptly. A 3d *photonic crystal (PC)* has two sets of photonic bands, one for TE and the other for TM solutions of the respective wave equations. The TE- and TM-wave band structures have, in general, spectrally distinct photonic bands.

Let us apply these general results to derive a model DOM distribution of a 3d-periodic PC. The Taylor expansion of the dispersion relation $\omega(\boldsymbol{k})$ near, say, an upper photonic-band edge ω_{U} to the second order yields the effective mass approximation,

$$\omega \approx \omega_{\mathrm{U}} + \sum_{i=x,y,z} \frac{(k - k_{\mathrm{U}})_i^2}{2m_i}, \tag{3.21}$$

where the reciprocal effective mass is $1/m_i = \hbar^{-1}(\partial^2\omega/\partial k_i^2)|_{k=k_{\mathrm{U}}}$ and k_{U} satisfies the Bragg condition $k_{\mathrm{U}} = \pi/d$. In this approximation, the DOM may be parametrized as

$$\rho(\omega) \propto (\omega - \omega_{\mathrm{U}})^{(1-D)/2}\theta(\omega - \omega_{\mathrm{U}}), \tag{3.22}$$

where D is the dimensionality of the Brillouin-zone surface spanned by band-edge modes with vanishing group velocity. In realistic photonic structures $D \leq 2$, $D = 2$ corresponding to completely isotropic dispersion (spherical Brillouin-zone surface), and $D = 0$ corresponding to anisotropic three-dimensional Brillouin zone. Both cases can be represented as the limits $\epsilon \to 0$ and $\epsilon \to \infty$ of the function

$$\rho(\omega) \propto \frac{\sqrt{\omega - \omega_{\mathrm{U}}}}{\omega - \omega_{\mathrm{U}} + \epsilon}\theta(\omega - \omega_{\mathrm{U}}), \tag{3.23}$$

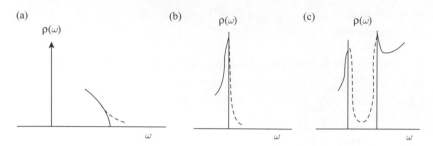

Figure 3.3 The density of modes near the edge of a tightly bound band (a), near a cutoff (b), or near a gap between two bands (c). The solid curves are replaced by dashed curves in the presence of disorder, which smooths out the singularities. (Adapted from Kurizki, 1990. © 1990 The American Physical Society.)

where ϵ is the "cutoff-smoothing" parameter. Figure 3.3 depicts different DOM shapes near a cutoff, a band edge, or a band gap.

Photonic-Mode Populations

If the photonic spectrum is thermal (see Ch. 6), then, regardless of the confining structure, the photon-mode occupancy (i.e., the average number of photons) depends on the inverse temperature $\beta = 1/k_\mathrm{B}T$, k_B being the Boltzmann constant, and the mode frequency ω, in accordance with the Bose–Einstein distribution,

$$n(\omega) = \frac{1}{e^{\hbar\beta\omega} - 1},\qquad(3.24)$$

wherein only photon modes with $\omega \lesssim k_\mathrm{B}T/\hbar$ are appreciably populated. Thus, near-infrared modes with frequency $\omega \gtrsim 10^{14}$ Hz are populated at a temperature $T > 1500$ K that is present in sunlight. By contrast, ambient temperatures $T \simeq 300$ K correspond to negligible populations of such modes, so that we may set the ambient temperature to be $T \to 0$ as far as optical or near-infrared frequency modes are concerned. Ambient temperature, however, suffices to populate microwave-frequency modes with $\omega \lesssim 1$ GHz.

3.1.2 Phonon Baths

Phonon baths have many similarities to photon baths, both representing bosonic-field multimode excitations. However, phonons obey a scalar wave equation, unlike photons, which possess the gauge freedom of four-vector potentials.

The upper cutoff of phonon spectra in crystals is the Debye frequency $\omega_D = v_\mathrm{s}K/2$, where v_s is the sound velocity and $K/2$ is the edge of the first Brillouin zone, which is a sphere in the Debye approximation. This cutoff may strongly affect system–bath interactions (Chs. 5 and 8). Confining structures of mechanical strain may act as cavities or waveguides that shape phonon spectra analogously to the photonic dielectric structures surveyed above.

Phonons in Isotropic Crystals

Here we consider long-wavelength acoustic phonons in isotropic crystals. In an isotropic crystal of density ρ, we assume cyclic boundary conditions over a cube of unit volume. For a mode with wave vector k, we choose one of the cube axes, x_l, parallel to k. Then the Hamiltonian is taken to be quadratic in the phonon-mode displacement Q_k and the corresponding momentum P_k:

$$H = \frac{1}{2\rho} \sum_{k,\lambda} P_k^\lambda P_{-k}^\lambda + \sum_k k^2 \left(\frac{A}{2} Q_k^l Q_{-k}^l + \frac{B}{2} \sum_\lambda Q_k^\lambda Q_{-k}^\lambda \right), \qquad (3.25)$$

where A and B are determined by the elastic moduli of the medium. The index λ labels the polarization of the particle displacement relative to the wave vector, l denotes the longitudinal phonon ($Q_k \parallel k$), and the other two choices of λ label transverse phonons.

The second-quantized boson operators for phonons then have the form

$$a_{k\lambda}^\dagger = -i(2\hbar\rho\omega_{k\lambda})^{-1/2} P_k^\lambda + (\hbar\omega_{k\lambda})^{-1/2}(A\delta_{\lambda l} + B)^{1/2} k Q_{-k}^\lambda,$$
$$a_{k\lambda} = i(2\hbar\rho\omega_{k\lambda})^{-1/2} P_{-k}^\lambda + (\hbar\omega_{k\lambda})^{-1/2}(A\delta_{\lambda l} + B)^{1/2} k Q_k^\lambda, \qquad (3.26)$$

where

$$\omega_{k\lambda} = [(A\delta_{\lambda l} + B)/\rho]^{1/2} k. \qquad (3.27)$$

In a crystal with s inequivalent ions per unit cell there are three branches of the vibrational spectrum whose frequencies approach zero as $k \to 0$, known as *acoustical modes*. The additional $3(s-1)$ branches that possess a nonzero frequency as $k \to 0$ are *optical modes*.

For example, in an ionic crystal with two ions per primitive cell, such as NaCl, the three optical-mode branches at long wavelengths (small k) consist of one longitudinal and two transverse branches. The limiting frequency ω_l of the longitudinal branch as $k \to 0$ is higher (because of electrostatic repulsion) than its counterpart ω_t of the transverse branches: they are related by $\omega_l^2 = (\epsilon_0/\epsilon_\infty)\omega_t^2$, where ϵ_0 is the static dielectric constant and ϵ_∞ is the square of the optical refractive index.

Appreciable electromagnetic coupling exists between photons and long-wavelength transverse optical phonons. Due to this coupling, a band gap arises between ω_t and ω_l, which corresponds to a strong optical reflection.

At $k = 0$ the ions in each cell move in unison (i.e., with the same amplitudes and phases). Three of the modes correspond to a uniform translation of the cell as a whole (i.e., have zero frequency). The other $3(s-1)$ modes are the $k = 0$ optical

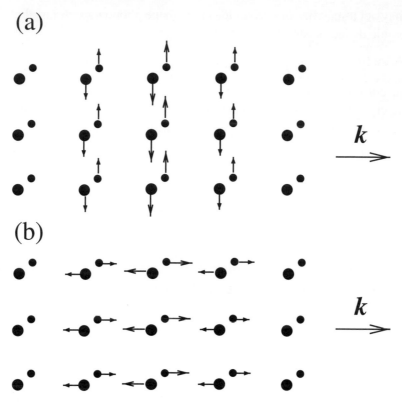

Figure 3.4 (a) Transverse and (b) longitudinal optical vibrational modes. The arrows denote the displacements of individual ions: circles – negative ions, and dots – positive ions.

modes, with nonzero frequencies in general, corresponding to ions moving relative to one another in a cell (Fig. 3.4).

Phonons in a Condensed Boson Gas

Phonons may be excited in a system of a large number of weakly interacting bosons described by the Hamiltonian

$$H = \sum_k E_k a_k^\dagger a_k + \frac{1}{2} \sum v(\mathbf{k}_1 - \mathbf{k}_1') a_{k_1}^\dagger a_{k_2}^\dagger a_{k_2'} a_{k_1'} \delta(\mathbf{k}_1 + \mathbf{k}_2 - \mathbf{k}_1' - \mathbf{k}_2'), \quad (3.28)$$

where $v(\mathbf{k}_1 - \mathbf{k}_1')$ is the matrix element of the interaction between two particles and the δ-function ensures wave vector (momentum) conservation. Without interaction, $v = 0$, the ground state has all particles condensed, in accordance with the Bose statistics in the lowest energy state, normally with $\mathbf{k} = 0$. A weak and short-range interaction renders most of the particles in the ground state, and a few ($n \ll N$) will be in excited states with $k > 0$.

3.2 Polaritons: Photon Interactions with Optical Phonons

The interaction of optical phonons with photons is appreciable when the frequencies and wave vectors of both coincide, which signifies the crossover of the dispersion relation for photons, $\omega = ck$, and the dispersion relation for an optical phonon branch.

This crossover is found by simultaneously solving the Maxwell equations for the electric field of the light and the equations for the medium polarization induced by the lattice displacement (phonon field), which is, in turn, driven by the light field. The resulting solutions express the *hybridization* of photon and phonon modes, characterized by the polarization dispersion. In particular, in an ionic crystal with two ions per primitive cell, the dispersion relation is

$$\omega^2 = \frac{\omega_t^2 \epsilon_0 + c^2 k^2}{2\epsilon_\infty} \pm \left[\frac{(\omega_t^2 \epsilon_0 + c^2 k^2)^2}{4\epsilon_\infty^2} - \frac{\omega_t^2 k^2 c^2}{\epsilon_\infty} \right]^{1/2}. \tag{3.29}$$

As $k \to 0$, the polaritonic frequencies reduce to

$$\omega^2 = \omega_t^2 (\epsilon_0 / \epsilon_\infty) = \omega_l^2 \tag{3.30}$$

and

$$\omega^2 = (c^2 / \epsilon_0) k^2. \tag{3.31}$$

Each of these solutions has double degeneracy, associated with two independent directions of the electric field E in the plane normal to k. For high k, the dispersion has two branches,

$$\omega^2 = c^2 k^2 / \epsilon_\infty; \qquad \omega^2 = \omega_t^2. \tag{3.32}$$

The lower branch is photon-like at low k and phonon-like at high k with frequency ω_t, whereas the upper branch exhibits the opposite trend: it is phonon-like at small k with frequency ω_l and photon-like at high k (Fig. 3.5).

There are no transverse solutions at frequencies between ω_t and ω_l; nor are there longitudinal-branch solutions in this range, since the photon field is transverse in isotropic media. Thus, a band gap exists between ω_t and ω_l (hampered by damping), which gives rise to high optical reflectivity.

The foregoing results were obtained by treating the electromagnetic field and the lattice vibrations classically. The same results follow from the corresponding quantum mechanical treatment. To this end, the normal-mode expansion of the photonic vector potential (3.5) and the electric polarization field in a dielectric medium is performed under periodic boundary conditions (over unit volume). The Hamiltonian is then diagonalized by the introduction of the annihilation polaritonic operator $\hat{\alpha}_k$ that is a linear combination of the photonic operators $a_k, a_{-k}^\dagger$, and their

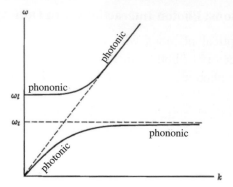

Figure 3.5 Polaritonic (hybridized) dispersion arising from the coupling of photons and transverse optical phonons in an ionic crystal. Broken lines – the dispersion without interaction.

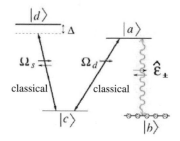

Figure 3.6 Atomic-level scheme aimed at trapping the field $\hat{\mathcal{E}}_\pm$ in a photonic band gap. (Adapted from Friedler et al., 2005. © 2005 The American Physical Society.)

electric polarization field counterparts b_k, $b^\dagger_{-k}$. The diagonalizing operators satisfy the same relation as the photonic annihilation operators:

$$[\hat{\alpha}_k, H] = \hbar\omega_k\hat{\alpha}_k. \tag{3.33}$$

The solution of the eigenvalue problem (3.33) is the dispersion relation, which coincides with (3.29).

3.2.1 Cold-Atom Optical-Polariton Baths

We consider a medium composed of cold alkali atoms with level configuration as shown in Figure 3.6.

The atoms, taken to be optically pumped to the ground states $|b\rangle$, resonantly interact with a running-wave classical field that drives the atomic transition $|c\rangle \rightarrow |a\rangle$ with the Rabi frequency Ω_{d}. Near the two-photon (Raman) resonance $|b\rangle \rightarrow |c\rangle$, the atomic medium then becomes transparent through an effect known as electromagnetically induced transparency (EIT) for a weak (quantum) signal field $\hat{\mathcal{E}}$ that acts on the transition $|b\rangle \to |a\rangle$.

A classical signal pulse of duration t_s in the atomic medium, under EIT conditions, is slowed down to group velocity v_s and spatially compressed, by a factor of $v_s/c \ll 1$, to the length $z_{loc} \approx v_s t_s$. In a medium of length L, such that $z_{loc} < L$, the signal pulse is converted into a standing-wave polaritonic excitation (Fig. 3.6), provided the driving field is adiabatically switched off and the signal pulse is stopped in the medium. The atoms then dispersively interact with a standing-wave classical field having the Rabi frequency $\Omega_s(z) = 2\Omega_s \cos(k_s z)$ and detuning $\delta \gg \Omega_s$ from the atomic transition $|c\rangle \to |d\rangle$. This field induces a spatially periodic ac Stark shift of level $|c\rangle$ and a corresponding modulation of the refractive index for the signal field,

$$\delta n_s(z) = \frac{c}{v_s} \frac{4\delta_s}{\omega_{ab}} \cos^2(k_s z),\qquad(3.34)$$

where $\delta_s = \Omega_s^2/\delta$ is the ac Stark-shift amplitude, $v_s \propto |\Omega_d|^2$, and ω_{ab} is the $|a\rangle \leftrightarrow |b\rangle$ transition frequency.

A forward-propagating quantized probe field $\hat{\mathcal{E}}_+$ with a carrier wave vector k near $k_s = \omega_s/c$ is Bragg-scattered into the backward-propagating field $\hat{\mathcal{E}}_-$ with a carrier wave vector $-k$. The two counterpropagating quantized fields form a standing wave that modifies the photonic density of states in the medium such that light propagation is forbidden near ω_s – a photonic band gap (PBG) arises.

In the slowly varying envelope approximation, the Maxwell–Bloch (operator) equations of motion for a weak (quantum) probe field read

$$\left(\frac{\partial}{\partial t} \pm c\frac{\partial}{\partial z}\right)\hat{\mathcal{E}}_\pm = igN\hat{\sigma}_{ba}^\pm.\qquad(3.35)$$

Here N is the number of atoms, and the $|b\rangle \to |a\rangle$ atomic polarization operators $\hat{\sigma}_{ba}^\pm$ evolve according to the Heisenberg–Langevin equations. These equations are solved perturbatively in the small parameter g/Ω_d and in the adiabatic approximation, upon neglecting noise and dissipation, yielding the following relations between the $|b\rangle \to |c\rangle$ and $|b\rangle \to |a\rangle$ polarization operators,

$$\hat{\sigma}_{ba}^\pm \simeq -\frac{i}{\Omega_d}\left(\frac{\partial}{\partial t} - 2i\delta_s\right)\hat{\sigma}_{bc}^\pm,$$

$$\hat{\sigma}_{bc}^\pm \simeq -\frac{g\hat{\mathcal{E}}_\pm}{\Omega_d}.\qquad(3.36)$$

This coupled set of Eqs. (3.35) and (3.36) is solved in terms of the new quantum fields (dark-state polaritons), defined by the canonical transformations

$$\hat{\Psi}_\pm = \cos\theta\hat{\mathcal{E}}_\pm - \sin\theta\sqrt{N}\hat{\sigma}_{bc}^\pm,\qquad(3.37)$$

where θ is the mixing angle, defined by $\tan\theta = g\sqrt{N}/\Omega_d$. The photonic component of the polariton field is proportional to $\cos\theta$ and the atomic component to $\sin\theta$. The Maxwell-Bloch equations of motion [(3.35) and (3.36)] are then transformed in the polaritonic basis into

$$\left(\frac{\partial}{\partial t} \pm v_s \frac{\partial}{\partial z}\right)\hat{\Psi}_\pm \simeq ir_c\hat{\Psi}_\mp, \tag{3.38}$$

where $r_c = \delta_s \sin^2\theta$ is the coupling rate of the forward and backward propagating components of the polariton. Here the linear phase modulation has been eliminated via the unitary transformation $\hat{\Psi}_\pm \rightarrow \hat{\Psi}_\pm e^{2ir_ct}$, whereas absorption, noise, and dissipation have been neglected.

It follows from (3.38) that a probe pulse containing modes with $|q| \ll r_c/v_s$ will be trapped inside the medium, periodically cycling between the forward- and backward-propagating polariton components, satisfying

$$\hat{\Psi}_+(z, t) = \hat{\Psi}_+(z, 0)\cos(r_ct)$$
$$\hat{\Psi}_-(z, t) = \hat{\Psi}_+(z, 0)\sin(r_ct). \tag{3.39}$$

The medium (bath) is then described by the effective polaritonic Hamiltonian

$$H_{B,eff} = -\frac{\hbar r_c}{L}\int dz(\hat{\Psi}_+^\dagger \hat{\Psi}_- + \hat{\Psi}_-^\dagger \hat{\Psi}_+). \tag{3.40}$$

The plane-wave expansion of the polariton operators,

$$\hat{\Psi}_\pm(z) = \sum_q \hat{\psi}_\pm^q e^{\pm iqz}, \tag{3.41}$$

in terms of the plane-wave mode operators that obey the bosonic commutation relations $[\hat{\psi}_i^q, \hat{\psi}_j^{q'\dagger}] = \delta_{ij}\delta_{qq'}$, yields the local commutation relations

$$[\hat{\Psi}_i(z), \hat{\Psi}_j^\dagger(z')] \simeq L\delta_{ij}\delta(z-z') \quad (i, j = \pm). \tag{3.42}$$

Thus, in the presence of a dispersively formed standing-wave field in the atomic setup of Figure 3.6, the medium excitations have a polaritonic character as per (3.37)–(3.42).

3.3 Magnon Baths

The low-lying energy states of spin systems coupled by exchange interactions give rise to quantized spin waves. The spin-wave quanta are known as *magnons*.

The simplest bath Hamiltonian that gives rise to magnons is that of N identical spin-S particles with nearest-neighbor interactions in a ferromagnetic spin lattice. It has the form,

$$H_{\mathrm{B}} = -J \sum_{j,j'} \hat{\boldsymbol{S}}_j \cdot \hat{\boldsymbol{S}}_{j'} - 2\mu_0 \mathcal{B}_0 \sum_j \hat{S}_{jz}, \qquad (3.43)$$

where $\hat{\boldsymbol{S}}_j$ is the jth spin operator, $\hat{\boldsymbol{S}}_{j'}$ are the spin operators of its nearest neighbors on a lattice, J is the positive exchange integral, $2\mu_0$ is the magnetic moment of a particle, and $\mathcal{B}_0 \geq 0$ is the static magnetic field that aligns the spins along the z axis.

To study this bath, it is convenient to resort to the Holstein–Primakoff transformation of the spin operators to bosonic creation and annihilation operators $a_j^\dagger$, a_j, which has the form

$$\hat{S}_j^+ = \hat{S}_{jx} + i\hat{S}_{jy} = (2S)^{1/2} \left(1 - \frac{a_j^\dagger a_j}{2S} \right)^{1/2} a_j,$$

$$\hat{S}_j^- = \hat{S}_{jx} - i\hat{S}_{jy} = (2S)^{1/2} a_j^\dagger \left(1 - \frac{a_j^\dagger a_j}{2S} \right)^{1/2},$$

$$\hat{S}_{jz} = S - a_j^\dagger a_j. \qquad (3.44)$$

We next introduce the collective spin-wave (magnon) operators $a(\boldsymbol{k})$, $a^\dagger(\boldsymbol{k})$, that satisfy

$$a_j = \frac{1}{\sqrt{N}} \sum_{\boldsymbol{k}} e^{-i\boldsymbol{k} \cdot \boldsymbol{x}_j} a(\boldsymbol{k}), \qquad (3.45)$$

where $\boldsymbol{x}_j$ is the position vector of the jth spin particle. We can rewrite the bath Hamiltonian in terms of these collective operators as

$$H_{\mathrm{B}} = H_{\mathrm{M}} + H_{\mathrm{M}}^{(1)}. \qquad (3.46)$$

Here

$$H_{\mathrm{M}} = \hbar \sum_{\boldsymbol{k}} \omega(\boldsymbol{k}) \left[a^\dagger(\boldsymbol{k}) a(\boldsymbol{k}) + \frac{1}{2} \right], \qquad (3.47)$$

and $H_{\mathrm{M}}^{(1)}$ represents (*higher-order*) terms in magnon variables, which may be viewed as *non-linear* interactions between magnons. At low temperatures when the excitation is low and $|\boldsymbol{k} \cdot \boldsymbol{d}| \ll 1$, where $\boldsymbol{d}$ is the shortest lattice vector, the nonlinear term $H_{\mathrm{M}}^{(1)}$ can be neglected. The magnons are then equivalent to bosonic excitations governed by the Hamiltonian H_M, whereby the dispersion law is quadratic,

$$\hbar\omega(\boldsymbol{k}) \cong \mathfrak{D}k^2 + 2\mu_0 \mathcal{B}_0, \qquad (3.48)$$

with $\mathfrak{D} = 2JSd^2$, d being the lattice constant. The term $2\mu_0 \mathcal{B}_0$ is the contribution of the external field to the magnon energy. It implies that the projection of the magnon magnetic moment on the field axis is $-2\mu_0$.

3.3.1 *XX Spin-Chain Bath*

A quantum bath consisting of a spin-1/2 chain with nearest-neighbor exchange interactions is defined as XX chain if it is governed by the Hamiltonian

$$H_{\mathrm{B}} = \hbar J \sum_{i=1}^{N-1} (\hat{S}_i^+ \hat{S}_{i+1}^- + \hat{S}_i^- \hat{S}_{i+1}^+). \tag{3.49}$$

Here $\hat{S}^\pm = \hat{S}^x \pm i\hat{S}^y$ are spin-1/2 raising and lowering (excitation and de-excitation, respectively) operators and J is the spin-spin coupling strength. This Hamiltonian conserves the total magnetization.

It can be transformed into the Hamiltonian of non-interacting spinless fermions that hop between neighboring sites of the form,

$$H_{\mathrm{B}} = \hbar J \sum_{i=1}^{N-1} (c_i^\dagger c_{i+1} + c_{i+1}^\dagger c_i), \tag{3.50}$$

by using the Jordan–Wigner transformation,

$$c_i = e^{i\pi \sum_{j=1}^{i-1} \hat{S}_j^+ \hat{S}_j^-} \hat{S}_i^-. \tag{3.51}$$

The Hamiltonian (3.50) conserves the number of fermions. It simplifies upon performing the discrete sine transform,

$$\hat{f}_q = \sqrt{\frac{2}{N+1}} \sum_{j=1}^{N} \sin\left(\frac{jq\pi}{N+1}\right) c_j \equiv \sum_{j=1}^{N} U_{qj} c_j \qquad (q = 1, 2, \ldots, N). \tag{3.52}$$

The transform matrix $U = \{U_{qj}\}$ is real, Hermitian, and unitary, $U = U^\dagger$, $U^2 = I$. Correspondingly, the inverse transform of (3.52) is

$$c_j = \sum_{q=1}^{N} U_{jq} \hat{f}_q. \tag{3.53}$$

Moreover, the unitarity of U ensures that $\hat{f}_q$ and $\hat{f}_q^\dagger$ satisfy the commutation relations for the fermionic annihilation and creation operators, respectively.

Inserting (3.53) into (3.50) yields a Hamiltonian for uncoupled fermionic modes,

$$H_{\mathrm{B}} = \hbar \sum_{q=1}^{N} \omega_q \hat{f}_q^\dagger \hat{f}_q, \tag{3.54}$$

where

$$\omega_q = 2J \cos \frac{q\pi}{N+1}. \tag{3.55}$$

For odd N, there is a single zero-energy fermionic mode in this quantum bath, corresponding to $q = (N+1)/2$.

3.4 Discussion

We have presented generic types of quantized (bosonic and fermionic) baths, focusing on their spectra, dispersion relations and mode density. The spectra and eigenfunctions of photon and phonon baths, solid-state magnons, polaritons, and their atomic medium analogs have been discussed. These features will serve us in studies of quantum system-bath interactions: system-bath quantum correlations, decay and oscillations of system excitations in the presence of these baths, system-state decoherence induced by the baths, and the ability to dynamically control these processes.

4

Quantized System–Bath Interactions

Having characterized in Chapter 3 the modes, mode-density spectra, and elementary excitations of common (photon, phonon, polariton, and magnon) baths, we here study the consequences of their interaction with generic two-level systems. The system–bath interaction Hamiltonians are derived by quantizing the appropriate Lagrangians. These derivations are not presented here.

4.1 Spin-Boson Models

4.1.1 Atom–Photon Interaction

The interaction Hamiltonian of atoms with the electromagnetic field bath is denoted here by

$$\epsilon \tilde{H}_{SB} \equiv H_{SB} \equiv H_I = \sum_i \left[-\frac{e_i}{2m_i c} (\boldsymbol{p}_i \cdot \boldsymbol{A}_i + \boldsymbol{A}_i \cdot \boldsymbol{p}_i) + \frac{e_i^2}{2m_i c^2} \boldsymbol{A}_i^2 \right]$$

$$= \sum_i \left(-\frac{e_i}{m_i c} \boldsymbol{A}_i \cdot \boldsymbol{p}_i + \frac{e_i^2}{2m_i c^2} \boldsymbol{A}_i^2 \right). \tag{4.1}$$

Here $\boldsymbol{A}_i = \boldsymbol{A}(\boldsymbol{r}_i)$ is the vector potential at the position $\boldsymbol{r}_i$ of a particle i with the charge e_i and the mass m_i, and $\boldsymbol{p}_i$ is the momentum canonically conjugate to the coordinate $\boldsymbol{r}_i$. The replacement of $\boldsymbol{p}_i \cdot \boldsymbol{A}_i$ by $\boldsymbol{A}_i \cdot \boldsymbol{p}_i$ in the second line of Eq. (4.1) follows from the gauge condition $\boldsymbol{\nabla}_i \cdot \boldsymbol{A}_i = 0$. Equation (4.1) describes the interaction of moving charges with the electromagnetic field, but does not account for the interaction of their spin moments with magnetic fields.

The quantization of the free-atom Hamiltonian combined with the free-field Hamiltonian and the interaction Hamiltonian (4.1) is performed by subjecting the particles' $\boldsymbol{r}_i$ and $\boldsymbol{p}_i$ to the standard commutation relations and quantizing the radiation field, as in Eq. (3.5). The longitudinal electric field E_L does not provide any additional freedom in this quantization, being completely determined through the Maxwell equation $\nabla \cdot E_L = \rho_e$ by the charge density $\rho_e(x, t)$.

The interaction H_I in Eq. (4.1) is commonly treated as a perturbation that causes transitions between the states of the free Hamiltonian. The interaction (4.1) contains a term quadratic in the vector potential that gives rise to two-photon processes within first-order perturbation theory (i.e., emission, absorption or scattering of two photons). However, as the quadratic term is typically small, it is neglected below. The first term in (4.1) is treated below in the common electric dipole (or long-wavelength) approximation, which neglects the spatial variation of $A(x)$. The dependence of A on x is responsible for magnetic interactions and higher-order effects that are not treated here.

4.1.2 Radiative Transitions in Atoms

Here we consider transitions between two electronic states of an atom, $|e\rangle$ and $|g\rangle$, resulting in the emission or absorption of one photon, via the interaction (4.1). The matrix element for single-photon emission that corresponds to the term linear in A in (4.1), expanded as per (3.5), is given by

$$\langle g, n_\lambda(k) + 1 | H_I | e, n_\lambda(k) \rangle = -\frac{e}{m} \left(\frac{2\pi\hbar}{\mathcal{V}\omega_k} \right)^{1/2} [n_\lambda(k) + 1]^{1/2}$$
$$\times \langle g | \epsilon_{k\lambda}^* \cdot \sum_i e^{-ik \cdot r_i} p_i | e \rangle, \qquad (4.2)$$

where $n_\lambda(k)$ is the initial photon number in the mode (k, λ), k being the wave vector and λ the polarization, ω_k is the photon frequency at a given k, $\mathcal{V}$ is the quantization volume, $\epsilon_{k\lambda}$ is a unit polarization vector of the photon, e and m are the electron charge and mass, whereas r_i and p_i are the position and momentum of the atomic electron i. The corresponding transition probability per unit time is then

$$w_\lambda d\Omega_s = \frac{e^2 \omega_a d\Omega_s}{2\pi m^2 \hbar c^3} [n_\lambda(k) + 1] \left| \epsilon_{k\lambda}^* \cdot \langle g | \sum_i e^{-ik \cdot r_i} p_i | e \rangle \right|^2, \qquad (4.3)$$

where $\omega_a = (E_e - E_g)/\hbar$ and E_e and E_g are the energies of the excited and ground atomic (electronic) states, Ω_s being the solid angle of the emission. Equation (4.3) can be adapted to the absorption of a photon upon replacing the factor $[n_\lambda(k) + 1]$ by $n_\lambda(k)$.

The electric dipole approximation is valid provided we can approximate the exponential factors in Eqs. (4.2) and (4.3) by unity:

$$e^{-i\mathbf{k}\cdot\mathbf{r}_i} \approx 1. \tag{4.4}$$

This holds if the wavelength $2\pi/k$ of the photon is very large compared to the size R of the atom, as in the case of optical atomic transitions. Then the wave functions of $|e\rangle$ and $|g\rangle$ restrict the values of r_i to $|\mathbf{k}\cdot\mathbf{r}_i| \lesssim kR \ll 1$. The equation of motion

$$i\hbar\dot{\mathbf{r}}_i = [\mathbf{r}_i, H_{\mathrm{a}}], \tag{4.5}$$

where H_{a} is the atomic Hamiltonian, yields

$$\langle g|\mathbf{p}_i|e\rangle = m\langle g|\dot{\mathbf{r}}_i|e\rangle = -im\omega_{\mathrm{a}}\langle g|\mathbf{r}_i|e\rangle. \tag{4.6}$$

The electric dipole operator $\mathbf{d} = e\sum_i \mathbf{r}_i$ in this two-state basis may be represented by

$$\mathbf{d} = \sum_{j,l=e,g} |j\rangle\langle j|\mathbf{d}|l\rangle\langle l| = \sum_{j,l=e,g} \wp_{jl}\sigma_{jl}, \tag{4.7}$$

where $\wp_{jl} = \langle j|\mathbf{d}|l\rangle$ is the electric-dipole transition matrix element. The transition operators $\sigma_{jl} = |j\rangle\langle l|$ form the set

$$\sigma_z = |e\rangle\langle e| - |g\rangle\langle g|,$$
$$\sigma_+ = |e\rangle\langle g|,$$
$$\sigma_- = |g\rangle\langle e|, \tag{4.8}$$

where σ_+, σ_-, and σ_z satisfy the spin-1/2 algebra of the Pauli matrices, that is,

$$[\sigma_-, \sigma_+] = -\sigma_z,$$
$$[\sigma_-, \sigma_z] = 2\sigma_-. \tag{4.9}$$

Here the σ_- operator takes an atom from the upper state $|e\rangle$ to the lower state $|g\rangle$, and σ_+ does the opposite.

Under the common assumptions that $\wp_{jj} = 0$, whereas $i\wp_{ge} \equiv \wp$ and $\epsilon_{k\lambda}$ are real, the quantized interaction Hamiltonian (4.1) assumes the form

$$H_{\mathrm{I}} = \hbar \sum_{k,\lambda} \eta_{k\lambda}(\sigma_+ + \sigma_-)(a_{k\lambda} + a_{k\lambda}^\dagger), \tag{4.10}$$

where the field-atom coupling strength,

$$\eta_{k\lambda} = \left(\frac{2\pi}{V\hbar\omega_k}\right)^{1/2} \omega_{\mathrm{a}}\epsilon_{k\lambda}\cdot\wp, \tag{4.11}$$

is determined by the dipole matrix element $\wp$ and the field quantization volume [cf. (4.2)].

The interaction energy in (4.10) consists of four terms: (i) The term $a_{k\lambda}^\dagger\sigma_-$ describes a transition of the atom from the upper state to the lower state accompanied by a creation of a photon of mode $(\mathbf{k}, \lambda)$. (ii) The term $a_{k\lambda}\sigma_+$ describes the opposite process. Both terms are energy conserving. (iii) By contrast, the term $a_{k\lambda}\sigma_-$ describes an atomic transition from the upper to the lower level along with photon annihilation, resulting in the loss of approximately $2\hbar\omega_a = 2(E_e - E_g)$. (iv) Similarly, the term $a_{k\lambda}^\dagger\sigma_+$ describes a process that results in the gain of $2\hbar\omega_a$. The latter two energy-nonconserving terms correspond to processes that do not conform to the *rotating-wave approximation* (RWA) and are only allowed by time-energy uncertainty on timescales $t \lesssim 1/(2\omega_a)$.

If energy-nonconserving (non-RWA or antiresonant) fast-scale processes are ignored, we may rewrite (4.10) in the RWA form,

$$H_I = \hbar \sum_{k,\lambda} \eta_{k\lambda}(\sigma_+ a_{k\lambda} + \sigma_- a_{k\lambda}^\dagger). \tag{4.12}$$

This RWA Hamiltonian describes the interaction of a two-level atom with a multimode free-space field. More generally, the RWA Hamiltonian of the TLS interaction with a bosonic bath is

$$H_I = \hbar \sum_\Lambda (\eta_\Lambda \sigma_+ a_\Lambda + \eta_\Lambda^* \sigma_- a_\Lambda^\dagger), \tag{4.13}$$

where η_Λ is the generally complex strength of the coupling of the system to the Λth mode of the bath. The Hamiltonian (4.12) or (4.13) is the starting point of many calculations in this book. It is, however, inconsistent with multiple-atom, non-dipolar, or non-RWA interactions (Chs. 5, 7, and 8).

If selection rules forbid the transition $|e\rangle$ to $|g\rangle$ via the electric dipole interaction, it may still occur via higher terms in the multipolar expansion

$$e^{-i\boldsymbol{k}\cdot\boldsymbol{r}_i} = 1 - i\boldsymbol{k}\cdot\boldsymbol{r}_i + \dots. \tag{4.14}$$

If the second term is accounted for, the expression within the modulus sign in Eq. (4.3) becomes

$$\boldsymbol{\epsilon}_{k\lambda} \cdot \langle g| \sum_i (-i\boldsymbol{k}\cdot\boldsymbol{r}_i)\boldsymbol{p}_i|e\rangle = -i\sum_{\alpha=1}^{3}\sum_{\beta=1}^{3} E_{k\lambda\alpha}k_\beta\langle g|\sum_i r_{i\beta}p_{i\alpha}|e\rangle, \tag{4.15}$$

where α, β (= 1, 2, 3) label the Cartesian components of the vectors $\boldsymbol{\epsilon}_{k\lambda}$, $\boldsymbol{k}$, $\boldsymbol{r}_i$, and $\boldsymbol{p}_i$. The matrix element in (4.15) can be written as

$$\langle g|\sum_i r_{i\beta}p_{i\alpha}|e\rangle = \frac{1}{2}\left[\langle g|\sum_i (r_{i\beta}p_{i\alpha} - r_{i\alpha}p_{i\beta})|e\rangle \right.$$
$$\left. + \langle g|\sum_i (r_{i\beta}p_{i\alpha} + r_{i\alpha}p_{i\beta})|e\rangle \right]. \tag{4.16}$$

The first, antisymmetric, term involves the angular momentum operator that corresponds to the magnetic dipole interaction and generally must be augmented by the spin part. The second, symmetric, term corresponds to the electric quadrupole interaction.

4.2 Polaronic System–Bath Interactions

4.2.1 Opto-Mechanical Polaron Interaction with a Bath

We consider the basic opto-mechanical Hamiltonian that governs an optical cavity mode (denoted by O) that is coupled to a photonic bath and to a mechanical oscillator (denoted by M). The total Hamiltonian then has the form

$$
\begin{aligned}
H_{\text{Tot}} &= H_{\text{O+M}} + (O^\dagger + O) \otimes B; \\
H_{\text{O+M}} &= \omega_\text{O} O^\dagger O + \Omega_\text{M} M^\dagger M + g O^\dagger O (M + M^\dagger).
\end{aligned}
\tag{4.17}
$$

Here $O^\dagger$, O and $M^\dagger$, M are the creation and annihilation operators of the cavity mode and the oscillator, respectively; ω_O, Ω_M and g are their respective frequencies and coupling rate; and B is the photonic-bath operator (Fig. 4.1).

We transform these operators to the basis of hybridized optical-mechanical modes that diagonalize $H_{\text{O+M}}$ without changing their frequency. Namely,

$$
H_{\text{O+M}} = \tilde{H}_\text{O} + \tilde{H}_\text{M}, \quad \tilde{H}_\text{O} = \omega_\text{O} \tilde{O}^\dagger \tilde{O} - (g \tilde{O}^\dagger \tilde{O})^2 \frac{1}{\Omega_\text{M}}, \quad \tilde{H}_\text{M} = \Omega_\text{M} \tilde{M}^\dagger \tilde{M},
$$

$$
\tilde{M} = M + \frac{g}{\Omega_\text{M}} O^\dagger O, \quad \tilde{O} = O e^{\frac{g}{\Omega_\text{M}} (M^\dagger - M)}.
\tag{4.18}
$$

The new variables can be expressed in terms of the unitary ("polaron") transformation

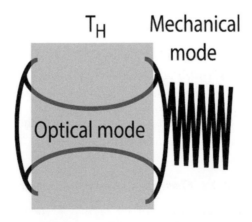

Figure 4.1 Schematic opto-mechanical setup: an optical cavity mode is coupled to a photonic bath and to the mechanical oscillator. (Adapted from Gelbwaser-Klimovsky and Kurizki, 2015.)

$$U = e^{\frac{g}{\Omega_M}(M^+ - M)O^\dagger O}. \tag{4.19}$$

as $\tilde{O} = U^\dagger O U$ and $\tilde{M} = U^\dagger M U$. Then, the interaction between the optical mode and the photonic bath is found to *indirectly* affect the mechanical oscillator.

We shall restrict ourselves to low excitations of the transformed number operators $\hat{n}_{\tilde{O}} = \tilde{O}^\dagger \tilde{O}$ and $\hat{n}_{\tilde{M}} = \tilde{M}^\dagger \tilde{M}$ and to the weak optomechanical-coupling regime. Namely, we shall assume

$$\left(\frac{g}{\Omega_M}\right)^2 \langle n_{\tilde{M}} \rangle \ll 1, \quad \frac{g^2}{\Omega_M} \langle n_{\tilde{O}} \rangle^2 t \ll 1, \tag{4.20}$$

where $\langle n_{\tilde{M}} \rangle$ and $\langle n_{\tilde{O}} \rangle$ are the mean numbers of quanta in the $\tilde{M}$ and $\tilde{O}$ degrees of freedom, respectively.

The Heisenberg-picture Fourier decomposition of $O^\dagger + O$ to lowest-order approximation with respect to a small parameter g/Ω_M is obtained from

$$
\begin{aligned}
O^\dagger(t) &= e^{iHt} O^\dagger e^{-iHt} \\
&= e^{-i(g\tilde{O}^\dagger \tilde{O})^2 \frac{1}{\Omega_M} t} e^{i\omega_O t} \tilde{O}^+ e^{\frac{g}{\Omega_M}(\tilde{M}^+ e^{i\Omega_M t} - \tilde{M} e^{-i\Omega_M t})} e^{i(g\tilde{O}^\dagger \tilde{O})^2 \frac{1}{\Omega_M} t} \\
&\approx e^{-i(g\tilde{O}^\dagger \tilde{O})^2 \frac{t}{\Omega_M}} \left[\tilde{O}^+ e^{i\omega_O t} + \frac{g}{\Omega_M} \left(\tilde{O}^\dagger \tilde{M}^\dagger e^{i(\omega_O + \Omega_M)t} - O^\dagger \tilde{M}^\dagger e^{i(\omega_O - \Omega_M)t} \right) \right] \\
&\quad \times e^{i(g\tilde{O}^\dagger \tilde{O})^2 \frac{t}{\Omega_M}}.
\end{aligned}
\tag{4.21a}
$$

The approximation made in (4.21a) is valid under (4.20). We then have in the interaction picture,

$$O^\dagger(t) \approx \tilde{O}^\dagger \left[e^{i\omega_O t} + \frac{g}{\Omega_M} \left(\tilde{M}^\dagger e^{i(\omega_O + \Omega_M)t} - \tilde{M} e^{i(\omega_O - \Omega_M)t} \right) \right]. \tag{4.21b}$$

Thus, the polaron transformation detailed in (4.19)–diagonalizes the optomechanical Hamiltonian by hybridizing the optical- and mechanical-mode creation and annihilation operators. The transformation entails *nonlinear quantum effects* when eigenstates of the noninteracting Hamiltonian ($g = 0$) are subjected to the action of the Hamiltonian (4.18) with non-negligible g/Ω_M.

4.2.2 Electron–Phonon Interaction

The electron–phonon interaction can be shown to bear analogy to the optomechanical interaction in Section 4.2.1. In media where the electron–phonon interaction is weak, the deformation-potential method may be applied to long-wavelength phonons. Whereas in an unstrained (cubic) covalent crystal the electron energy band may be assumed spherical,

$$E_0(\boldsymbol{k}) = \frac{\hbar^2 k^2}{2m^*}, \tag{4.22}$$

m^* being the effective mass of the conduction electron, a small (uniform) static deformation yields for low k,

$$E(\boldsymbol{k}) \simeq E_0(\boldsymbol{k}) + C_{\mathrm{d}} \delta \mathcal{V}, \tag{4.23}$$

where $\delta\mathcal{V}$ is the dilation (relative volume change) given by the trace of the strain tensor, and

$$C_{\mathrm{d}} = \frac{\partial E(0)}{\partial (\delta\mathcal{V})} = -\frac{2}{3} E_{\mathrm{F}} \tag{4.24}$$

for a free electron gas, E_{F} being the Fermi energy.

For long-wavelength acoustic phonons, we have, instead of (4.23),

$$E(\boldsymbol{k}, \boldsymbol{x}) = E_0(\boldsymbol{k}) + C_{\mathrm{d}} \delta\mathcal{V}(\boldsymbol{x}). \tag{4.25}$$

We then expand the Hamiltonian of the dilation perturbation in phonon operators (3.26),

$$\tilde{H}_{\mathrm{d}} = \int d^3x \, \hat{\Psi}^\dagger(\boldsymbol{x}) C_{\mathrm{d}} \delta\mathcal{V}(\boldsymbol{x}) \hat{\Psi}(\boldsymbol{x}) = \sum_{k',k} c_{k'}^\dagger c_k \langle \boldsymbol{k}' | C_{\mathrm{d}} \delta\mathcal{V} | \boldsymbol{k} \rangle$$

$$= i C_{\mathrm{d}} \sum_{k',k} c_{k'}^\dagger c_k \sum_q |\boldsymbol{q}| \sqrt{\frac{\hbar}{2\rho\omega_q}} \left[a_q \int d^3x \, u_{k'}^* u_k e^{i(k-k'+q)\cdot x} \right.$$

$$\left. - a_q^\dagger \int d^3x \, u_{k'}^* u_k e^{i(k-k'-q)\cdot x} \right]. \tag{4.26}$$

Here ρ is the density, the unperturbed one-electron Bloch states $|\boldsymbol{k}\rangle$ and $|\boldsymbol{k}'\rangle$ have the form $|\boldsymbol{k}\rangle = e^{i\boldsymbol{k}\cdot\boldsymbol{x}} u_k(\boldsymbol{x})$, $u_k(\boldsymbol{x})$ being lattice-periodic, and the creation and annihilation operators $a_q^\dagger$, a_q pertain to longitudinal phonons of wave vector $\boldsymbol{q}$. The electron-field operator $\hat{\Psi}(\boldsymbol{x})$ is written here in terms of Bloch states and electron annihilation operators c_k,

$$\hat{\Psi}(\boldsymbol{x}) = \sum_k c_k \, e^{i\boldsymbol{k}\cdot\boldsymbol{x}} u_k(\boldsymbol{x}), \tag{4.27}$$

Since the product $u_{k'}^*(\boldsymbol{x}) u_k(\boldsymbol{x})$ in (4.26) is periodic in the lattice, the integrals in (4.26) vanish unless

$$\boldsymbol{k} - \boldsymbol{k}' \pm \boldsymbol{q} = \boldsymbol{K}, \tag{4.28}$$

$\boldsymbol{K}$ being a reciprocal-lattice vector.

If we restrict ourselves to $K = 0$ processes, in the first Brillouin zone, and approximate $\int d^3x\, u_{k'}^* u_k$ by unity, then the dilation (deformation potential) perturbation (4.26) is quantized as follows,

$$\tilde{H}_d = iC_d \sum_{k,q} |q| \sqrt{\frac{\hbar}{2\rho\omega_q}} (a_q\, c_{k+q}^\dagger\, c_k - a_q^\dagger\, c_{k-q}^\dagger\, c_k). \qquad (4.29)$$

Alternatively, it may be rewritten as

$$\tilde{H}_d = iC_d \sum_{k,q} |q| \sqrt{\frac{\hbar}{2\rho\omega_q}} (a_q - a_{-q}^\dagger)c_{k+q}^\dagger\, c_k. \qquad (4.30)$$

The existence of the electron–phonon coupling $\tilde{H}_d$ means that an electron in a state k with no phonons excited is not an eigenstate of the system: a cloud of virtual phonons surrounds the electron. The composite particle comprised of an electron and a lattice deformation is called a polaron (Fig. 4.2).

Whereas the state $|k0\rangle$ involves the zero-phonon state and an electronic Bloch state, the state perturbed to first order in $\tilde{H}_d$ is given by

$$|k0\rangle^{(1)} = |k0\rangle + \sum_q |k-q; 1_q\rangle \frac{\langle k-q; 1_q|\tilde{H}_d|k0\rangle}{E_k - E_{k-q} - \omega_q}. \qquad (4.31)$$

Using (4.29), we find

$$|\langle k-q; 1_q|\tilde{H}_d|k0\rangle|^2 = \frac{C_d^2|q|}{2\rho c_s}, \qquad (4.32)$$

c_s being the longitudinal velocity of sound.

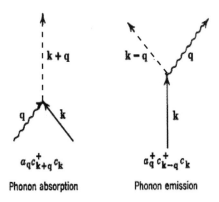

Figure 4.2 Electron–phonon first-order scattering processes.

The denominator of (4.31) evaluates to

$$E_{\boldsymbol{k}} - E_{\boldsymbol{k}-\boldsymbol{q}} - \hbar\omega_{\boldsymbol{q}} = \frac{\hbar^2(2\boldsymbol{k}\cdot\boldsymbol{q} - q^2)}{2m^*} - \hbar c_s q. \tag{4.33}$$

For slow electrons we may neglect $\boldsymbol{k}$ compared with $\boldsymbol{q}$.

4.2.3 Polaronic Interaction of a Two-Level System with a Phonon Bath

This model consists of a driven two-level system (TLS) whose σ_z operator is coupled to a (dephasing) bath, while its σ_x operator is coupled to another (energy-exchange) bath. The Hamiltonian is then

$$H = H_{\text{S}} + H_{\text{SB}} + H_{\text{B}}, \tag{4.34}$$

where

$$H_{\text{S}} = \frac{\hbar\omega_0}{2}\sigma_z + \frac{\hbar\Omega}{2}(\sigma_+ e^{-i\omega_l t} + \sigma_- e^{i\omega_l t}),$$

$$H_{\text{SB}} = \sigma_z \otimes \hbar \sum_k (g_k a_k^\dagger + g_k^* a_k) + \sigma_x \otimes \hbar \sum_k (\eta_k b_k^\dagger + \eta_k^* b_k),$$

$$H_{\text{B}} = \hbar \sum_k \omega_k a_k^\dagger a_k + \hbar \sum_k \tilde{\omega}_k b_k^\dagger b_k. \tag{4.35}$$

Here Ω is the Rabi frequency of the driving field, g_k and η_k are coupling strengths of the TLS to the mode k of the corresponding bath, and $a_k^\dagger$, a_k ($b_k^\dagger$, b_k) are the creation and annihilation operators of the mode k of the corresponding bath. Due to the driving, the system Hamiltonian is not diagonal in the σ_z basis, thereby allowing energy exchange with the dephasing bath.

The system dynamics can be studied upon applying to (4.35) the polaron transformation $e^{\mathcal{T}}$, where

$$\mathcal{T} = \sigma_Z \otimes \sum_k (\alpha_k a_k^\dagger - \alpha_k^* a_k), \quad \alpha_k = \frac{g_k}{\omega_k}. \tag{4.36}$$

This transformation shifts the equilibrium position of the dephasing bath oscillators by a factor proportional to the TLS energy. The transformed Hamiltonian has the form

$$\tilde{H} = e^{\mathcal{T}} H e^{-\mathcal{T}} = \tilde{H}_{\text{S}} + \tilde{H}_{\text{SB}} + \tilde{H}_{\text{B}}, \tag{4.37}$$

where

$$\tilde{H}_{\text{S}} = \frac{\hbar\omega_0}{2}\sigma_z + \frac{\hbar\Omega_{\text{r}}}{2}(\sigma_+ e^{-i\omega_l t} + \sigma_- e^{i\omega_l t}),$$

$$\tilde{H}_{\text{SB}} = \frac{\hbar\Omega}{2}\left[e^{-i\omega_l t}\sigma_+ \otimes (A_+ - A) + e^{i\omega_l t}\sigma_- \otimes (A_- - A)\right]$$

$$+ (\sigma_+ \otimes A_+ + \sigma_- \otimes A_-) \otimes \hbar \sum_k (\eta_k b_k^\dagger + \eta_k^* b_k),$$

$$\widetilde{H}_{\mathrm{B}} = \hbar \sum_k \omega_k a_k^\dagger a_k + \hbar \sum_k \tilde{\omega}_k b_k^\dagger b_k. \tag{4.38}$$

Here $A_\pm = \Pi_k D(\pm 2\alpha_k)$, $D(\alpha_k) = e^{\alpha_k a_k^\dagger - \alpha_k^* a_k}$ is the k-mode displacement operator,

$$A = \langle A_\pm \rangle = e^{-2 \sum_k \left| \frac{g_k}{\omega_k} \right|^2 \coth\left(\frac{\beta\omega_k}{2} \right)} \tag{4.39}$$

is the mean (multimode) displacement, $\Omega_r = \Omega A$, and β is the inverse equilibrium temperature of the transformed dephasing bath. The terms on the Hamiltonian proportional to the identity have been neglected.

In the transformed system–bath coupling Hamiltonian $\widetilde{H}_{\mathrm{SB}}$, the (multimode) bath operators are $A_+ - A$ and $A_- - A$, instead of $a_k^\dagger$, a_k. The transformed baths then interact with the TLS through two distinct operators,

$$\widetilde{F}_1(t) = \frac{\Omega}{2} \left[A_-(t) - A \right], \quad \widetilde{F}_2 = A_-(t) \otimes \sum_k \left[\eta_k b_k^\dagger(t) + \eta_k^* b_k(t) \right]. \tag{4.40}$$

The autocorrelation function of the first operator is

$$\Phi_1(t) = \langle \widetilde{F}_1(t)^\dagger \widetilde{F}_1(0) \rangle = \left(\frac{\Omega A}{2} \right)^2 \left(e^{4 \sum_k \frac{\Lambda_k(t)}{\omega_k^2}} - 1 \right), \tag{4.41}$$

where

$$\sum_k \Lambda_k(t) = \langle F_1^\dagger(t) F_1(0) \rangle = \sum_k |g_k|^2 \left[\cos(\omega_k t) \coth\left(\frac{\beta \hbar \omega_k}{2} \right) - i \sin(\omega_k t) \right] \tag{4.42}$$

is the autocorrelation function of the original system–bath coupling operator,

$$F_1(t) = \sum_k \left[g_k a_k^\dagger(t) + g_k^* a_k(t) \right]. \tag{4.43}$$

The system–bath coupling spectra that govern the evolution are derived from the correlations of the transformed operators,

$$\widetilde{G}_i(\omega) = \int_{-\infty}^{\infty} e^{it\omega} \langle \widetilde{F}_i(t)^\dagger \widetilde{F}_i(0) \rangle dt, \quad i \in 1, 2. \tag{4.44}$$

In Figure 4.3, the dependence on the coupling strength, g, of the coupling spectrum in the original basis, $G_1(\omega) = \int_{-\infty}^{\infty} e^{it\omega} \langle F_1(t)^\dagger F_1(0) \rangle dt$ (dotted line), and in the transformed basis, $\widetilde{G}_1(\omega)$ (continuous line), is shown. This coupling strength is proportional to the square of the coupling only as long as it is weak. Similarly, the operator $A_\pm(t)$ restricts $\widetilde{G}_2(\omega)$ to the weak coupling regime as long as $g \sim \eta$.

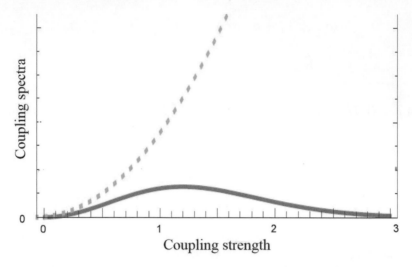

Figure 4.3 System–bath coupling spectra in the original basis (dotted line) and in the transformed basis (solid line) as a function of the coupling strength g/ω_0.

The Kubo–Martin–Schwinger (KMS) equilibrium condition

$$\widetilde{G}_1(-\omega) = e^{-\beta\omega}\widetilde{G}_1(\omega) \qquad (4.45)$$

(Sec. 2.3) is satisfied, which implies that the rate between absorption and emission processes is equal to the bath-temperature Boltzmann factor, independent of the frequency.

In the transformed basis, the second spectrum involves exchange of excitation with *both baths* (via the operators $b_k^{(\dagger)}$ and $A_{\pm}$, respectively). Energy exchange involving both baths prevents their separability and breaks the KMS condition, as well as the validity of other thermodynamic principles. The ratio between absorption and emission rates must then be expressed in terms of frequency-dependent ("local") inverse temperatures $\beta(\omega)$.

4.2.4 Optically Induced Polarons

Atomic Bose–Einstein condensates (BECs) are suitable for the exploration of elementary excitations, describable as quasiparticles. Here we consider a composite quasiparticle that consists of a moving "impurity" atom (an atom in a ground-state sublevel that differs from the rest of the BEC) "dressed" by a cloud of virtual phonons due to deformation of its vicinity. This quasiparticle is the BEC analog of the standard solid-state polaron. It is an example of the response of a many-body quantum system to the presence of a probe particle.

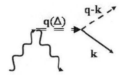

Figure 4.4 The Feynman diagram for the formation and decay of the polaron induced by Raman beams (curved lines). The unstable polaron (double dashed line) decays into a bare impurity (dashed line) and BEC Bogoliubov excitation (solid line). (Adapted from Mazets et al. 2005. © 2005 The American Physical Society.)

We shall consider polaron formation by Raman scattering off an atom in the BEC, as shown in Figure 4.4. Two laser beams are arranged to transfer momentum $\hbar q$ and energy $\hbar \delta$ to the atom. The Rabi frequency of the Raman two-photon transition is Ω_R. The laser-beam frequencies and polarizations are chosen to ensure that the atom is transferred to another sublevel of the ground state, thereby leaving the condensate and forming an impurity atom. We assume blue two-photon laser detuning, $\delta > \hbar q^2/(2m)$, $\delta \gtrsim \Omega_R$.

The Hamiltonian

$$H = H_M + H_I \tag{4.46}$$

consists of H_M that pertains to the matter system, including the BEC–impurity interaction, and H_I that describes the matter–laser interaction,

$$H_I = \hbar \Omega_R \sqrt{n} (e^{-i\delta t} \beta_k^\dagger + e^{i\delta t} \beta_k). \tag{4.47}$$

Here the creation and annihilation operators of the impurity atoms are denoted by $\beta_k^\dagger$, β_k.

We define the dressed (polaronic) states as eigenstates of the matter–system Hamiltonian H_M and express the field-matter interaction Hamiltonian (4.47) in the basis of these states, obtaining

$$H_I = \frac{\hbar}{\sqrt{\mathcal{V}}} \sum_k [\mathcal{M}_{qk}(t) c_{q-k,k}^\dagger c_0 + \text{H.c.}], \tag{4.48}$$

where c_0 is the annihilation operator of a bosonic atom in the BEC state, and $c_{q-k,k}^\dagger$ is the creation operator of a correlated pair consisting of a dressed impurity atom with momentum $\hbar(q - k)$ and an elementary BEC (Bogoliubov) excitation with momentum $\hbar k$. The latter operator obeys the bosonic commutation rules as long as the population of the state $|q - k, k\rangle_d \equiv c_{q-k,k}^\dagger |0\rangle$ is much less than one.

The matrix element of the off-resonant two-photon transition that couples the initial vacuum state $|0\rangle$ to the state $|q - k, k\rangle_d$, denoted by $\mathcal{M}_{qk}(t)$, is proportional to $\Omega_R e^{-i\delta t}$. The operator Heisenberg equations that we obtain from (4.48) are

$$i\frac{\partial}{\partial t}c_0 = \sum_k \mathcal{M}^*_{qk} c_{q-k,k}, \tag{4.49}$$

$$i\frac{\partial}{\partial t}c_{q-k,k} = \hbar^{-1}\epsilon_{|q-k|,k}c_{q-k,k} + \mathcal{M}_{qk} c_0. \tag{4.50}$$

Equations (4.49) and (4.50) describe a discrete state coupled to a continuum. The rate of creation of correlated pairs, consisting of an impurity atom and a Bogoliubov excitation, is obtained upon treating these coupled equations as a weak perturbation of the discrete state by the continuum. This rate is given by

$$\Gamma(t) = \int \frac{d^3k}{(2\pi)^3}|\mathcal{M}_{qk}|^2\frac{2\sin(vt)}{v}, \tag{4.51}$$

where

$$v = \delta - \omega_k - \frac{\hbar(q-k)^2}{2m} \tag{4.52}$$

and

$$\omega_k = k\sqrt{[\hbar k/(2m)]^2 + c_{\rm s}^2} \tag{4.53}$$

is the frequency of the BEC elementary excitation with the momentum $\hbar k$, $c_{\rm s} = \sqrt{\mu/m}$ being the speed of sound in the BEC.

4.3 Two-Level System Coupling to Magnon or Spin Bath

We consider a TLS immersed in a periodic medium that serves as a bath. Following the quantization procedure used in Chapter 3, the local displacement of this medium is quantized in terms of normal-mode creation and annihilation operators (bath excitations/de-excitations), as per (3.1),

$$B(x) = \frac{1}{\sqrt{\mathcal{V}}}\sum_k \frac{1}{\sqrt{\omega(k)}}[\phi_k(x)a^\dagger(k) + {\rm H.c.}]. \tag{4.54}$$

The couplings of a charged or dipolar TLS to this bath are, to leading order, determined by the gradient of the displacement operator

$$\nabla B(x) = \frac{-i}{\sqrt{\mathcal{V}}}\sum_k \frac{1}{\sqrt{\omega(k)}}[\nabla\phi_k(x)a^\dagger(k) - {\rm H.c.}]. \tag{4.55}$$

In the case of plane-wave normal modes, $\phi_k(x) = e^{-ik\cdot x}$, the corresponding coupling constant scales as

$$|\eta_k| \sim k\big/\sqrt{\omega(k)}. \tag{4.56}$$

In the case of spatially periodic $\phi_k(x)$ (Bloch waves) the coupling-constant scaling is easily found from (3.20).

Two typical cases can then be discerned:

- The dispersion relation of *acoustic phonons* reads $\omega(k) \simeq v|k|$, v being the sound velocity. Their coupling to the system qubit scales linearly with the frequency,

$$|\eta_k|^2 \sim k^2/\omega(k) \sim \omega. \qquad (4.57)$$

- Spin-wave low-temperature excitations of a ferromagnetic spin lattice with nearest-neighbor interactions (*magnons*) are bosons by virtue of the Holstein–Primakoff transformation (Sec. 3.3).

The coupling strength of a spin-1/2 particle (TLS) to a magnon bath satisfies, from Section 3.3,

$$|\eta_k|^2 = \frac{Ck^2}{\mathfrak{D}k^2 + 2\mu_0\mathcal{B}_0}, \qquad (4.58)$$

where C is a constant characterizing the coupling strength. The small k behavior in (4.58) is well approximated by $|\eta_k|^2 \approx Ck^2/(2\mu_0\mathcal{B}_0)$, as long as $k \ll \kappa_{\mathrm{m}}$, where κ_{m} is the magnetization parameter, defined as

$$\kappa_{\mathrm{m}} = \sqrt{\frac{\mu_0\mathcal{B}_0}{JSd^2}}. \qquad (4.59)$$

Except for $\kappa_{\mathrm{m}} = 0$ (spontaneous magnetization), the dispersive coupling coefficient scales quadratically with small k, $|\eta_k|^2 \propto k^2$. Thus, the TLS coupling strength with magnons satisfying $|\eta_k|^2 \approx C/\mathfrak{D}$ corresponds to a classical phase transition from induced to spontaneous magnetization.

A simple, exactly solvable model for the coupling of a TLS to a bath of spin-1/2 particles is described by the (spin-spin) Hamiltonian

$$H = \frac{\hbar}{2}\omega_0\sigma_z + \hbar\sum_k \omega_k\sigma_z^{(k)} + \hbar\sigma_z\sum_k \eta_k\sigma_x^{(k)} + \frac{\hbar}{2}\Omega_{\mathrm{d}}(t)\sigma_x. \qquad (4.60)$$

The first term is the Hamiltonian for the TLS (represented by the Pauli matrices σ) with the energy difference $\hbar\omega_0$; $\pm\hbar\omega_k$ determine the (Zeeman) level-energies of the kth bath spin (represented by the vector of the Pauli matrices $\sigma^{(k)}$); the (real) η_k is the corresponding coupling strength of the kth spin to the TLS, and the last term in the Hamiltonian describes the driving of the TLS by a classical field with the Rabi frequency $\Omega_{\mathrm{d}}(t)$. Importantly, the drive and the bath act on perpendicular Bloch-sphere axes of the TLS, x and z, respectively. In the σ_z basis of the TLS ($|+\rangle, |-\rangle$), one can rewrite (4.60), on setting $\Omega_{\mathrm{d}}(t) = 0$, as

$$H = H_+|+\rangle\langle+| + H_-|-\rangle\langle-|, \qquad (4.61)$$

where $H_\pm$ are the bath operators,

$$H_\pm = \pm \frac{\hbar \omega_0}{2} + \hbar \sum_k \left(\omega_k \sigma_z^{(k)} \pm \eta_k \sigma_x^{(k)} \right). \tag{4.62}$$

The simple form of $H_\pm$ makes them easy to diagonalize, so that we obtain a closed-form equation for the time-evolution operator (in the absence of driving)

$$U(t) = e^{-iHt} = U_+(t)e^{-i\omega_0 t/2}|+\rangle\langle+| + U_-(t)e^{i\omega_0 t/2}|-\rangle\langle-|, \tag{4.63}$$

where

$$U_\pm(t) = \prod_k \left[\cos(\delta_k t)\hat{I} - i \frac{\sin(\delta_k t)}{\delta_k} \left(\omega_k \sigma_z^{(k)} \pm \eta_k \sigma_x^{(k)} \right) \right]. \tag{4.64}$$

Here $\hat{I}$ is the unity operator, and the renormalized eigenfrequencies of the bath spins are $\delta_k = (\omega_k^2 + \eta_k^2)^{1/2}$. Such renormalization speeds up the system–bath exchange dynamics.

4.4 Discussion

This chapter has discussed the coupling of a two-level system (TLS) to bosonic baths and their fermionic (spin) counterparts. The TLS coupling dispersion and excitation spectra have been analyzed for diverse bosonic (photon, phonon, polariton, polaron, and magnon) baths, and their possible realizations have been discussed for cold atoms in optical lattices, laser-driven atomic Bose–Einstein condensates, and spin impurities in solids.

5

System–Bath Reversible and Irreversible Quantum Dynamics

The unitary evolution of a quantized system–bath complex is expected to consist in reversible energy exchange between the two. Yet, an excited atom coupled to a free-space photonic bath undergoes irreversible decay. Full or partial reversibility in photonic bath–atom interactions may only be exhibited by an initially excited atom in a field-confining structure – a cavity or a photonic crystal, as shown in this chapter.

5.1 Wigner–Weisskopf Dynamics

We analyze here the evolution of an initially excited two-level atom coupled to an arbitrary electromagnetic (photonic) bath. The bath is characterized by the density-of-modes (DOM) spectrum $\rho(\omega)$ of the electromagnetic field, assumed to be in the vacuum state. The atom–field interaction in the RWA is given by (4.13). The spectral response of the bath, representing, according to Fermi's Golden Rule, the rate (divided by 2π) of spontaneous emission of the atom at frequency ω, is given by

$$G(\omega) = \sum_{\Lambda} |\eta_{\Lambda}|^2 \delta(\omega - \omega_{\Lambda}), \qquad (5.1)$$

where ω_{Λ} is the frequency of the Λth mode of the bath.

The function $G(\omega)$ is the spectrum of the autocorrelation function of the atom–bath interaction (cf. Ch. 2) at zero temperature. Namely, the time-domain Fourier transform of $G(\omega)$ (from 0 to ∞, since $\omega_{\Lambda} \geq 0$) is

$$\Phi(t) = \int_0^{\infty} d\omega\, G(\omega) e^{-i(\omega - \omega_a)t} = \sum_{\Lambda} |\eta_{\Lambda}|^2 e^{i(\omega_a - \omega_{\Lambda})t}, \qquad (5.2)$$

which can be recast as a correlation function,

$$\Phi(t) = \hbar^{-2} \langle e, \text{vac}|H_{\text{I}}(t) H_{\text{I}}|e, \text{vac}\rangle. \qquad (5.3)$$

Here $|vac\rangle$ stands for the bath vacuum state and $H_I(t) = e^{iH_0t}H_Ie^{-iH_0t}$ is the coupling Hamiltonian in the interaction representation, where

$$H_0 = \hbar\omega_a|e\rangle\langle e| + \sum_\Lambda \hbar\omega_\Lambda a_\Lambda^\dagger a_\Lambda \qquad (5.4)$$

is the sum of the free Hamiltonians of the system and the bath. The autocorrelation function $\Phi(t)$ is sometimes referred to as the memory kernel of the bath response.

From physical considerations, $G(\omega)$ can be divided into two parts,

$$G(\omega) = G_s(\omega) + G_b(\omega). \qquad (5.5)$$

Here $G_s(\omega)$ represents the sharply varying (nearly singular) part of the spectral distribution. This part is typical of field-confining structures: it may be associated with narrow cavity mode lines, the frequency cutoff in waveguides, or photonic band edges. The remaining part of the spectral response, $G_b(\omega)$, describes the broad portion of the DOM distribution (the "background" modes), which coincides with the free-space DOM $\rho(\omega) \sim \omega^2$ at frequencies well above the sharp spectral features. In an open field-confining structure (see below), $G_b(\omega)$ represents the coupling of an atom to the unconfined free-space modes.

Let us cast the excited-state amplitude in the form $\alpha(t)e^{-i\omega_a t}$, where ω_a is the atomic resonance frequency. Then, for sharply varying DOM spectra and coupling strengths, the Schrödinger equation for spontaneous emission (radiative decay) is reducible to the following integrodifferential equation in the interval from $t = 0$ to $t = \tau$,

$$\dot\alpha = -\int_0^\tau dt\,\Phi_s(\tau - t)\alpha(t). \qquad (5.6)$$

Here the autocorrelation function

$$\Phi_s(t) = \int_0^\infty d\omega\,G_s(\omega)e^{-i(\omega-\omega_a)t} \qquad (5.7)$$

is associated with the sharp spectral feature. Its finite extent in time is a measure of its non-Markovianity.

Let us apply this analysis to the case of a two-level atom coupled to a near-resonant cavity mode at the frequency ω_s. The mode is characterized by a Lorentzian line shape (Sec. 3.1.1),

$$G_s(\omega) = \frac{g_s^2\Gamma_s}{\pi[\Gamma_s^2 + (\omega - \omega_s)^2]}, \qquad (5.8)$$

g_s being the resonant coupling strength and Γ_s the linewidth (Fig. 5.1). Then, the evolution of $\alpha(t)$ in (5.6) is soluble within the rotating-wave approximation (RWA), yielding

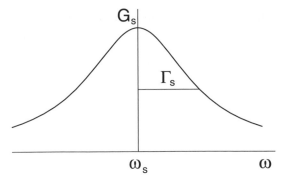

Figure 5.1 Lorentzian coupling spectrum (5.8) for a two-level atom coupled to a cavity mode. (Adapted from Kofman and Kurizki, 1996. © 1996 The American Physical Society.)

$$\alpha(t) = \frac{1}{2} e^{(i\delta - \Gamma_s)t/2} (\alpha_+ e^{\mathcal{D}t} + \alpha_- e^{-\mathcal{D}t}), \qquad (5.9)$$

where $\delta = \omega_a - \omega_s$ is the detuning of the atomic resonance and

$$\alpha_\pm = 1 \pm \frac{\Gamma_s - i\delta}{2\mathcal{D}}, \qquad \mathcal{D} = \sqrt{\frac{(\Gamma_s - i\delta)^2}{4} - g_s^2}. \qquad (5.10)$$

The *strong-coupling atom–bath regime*, characterized by underdamped Rabi oscillations, then corresponds to $2g_s \gg \Gamma_s + |\delta|$. The weak-coupling regime, characterized by overdamped Rabi oscillations (i.e., *irreversible nearly exponential decay*), corresponds to $2g_s \ll \Gamma_s + |\delta|$.

5.2 Photon–Atom Binding and Partial Decay

5.2.1 Equations of Motion for the Field and Atom Amplitudes

The evolution of a two-level atom in any zero-temperature (vacuum-state) photonic bath is describable by the joint field-atom wave function, which has the general RWA form,

$$|\Psi(t)\rangle = \alpha(t)|e, \{0_{k\lambda}\}\rangle + \sum_{k,\lambda} \beta_{k\lambda}(t)|g, 1_{k\lambda}\rangle. \qquad (5.11)$$

Here $|\{0_{k\lambda}\}\rangle$ stands for the totality of field-bath modes in the vacuum state and $|1_{k\lambda}\rangle$ for the field-bath state with single-photon occupation of the (k, λ)-mode. The corresponding Schrödinger equation can be solved under the initial condition of an initially excited atom,

$$|\Psi(0)\rangle = |e, \{0_{k\lambda}\}\rangle, \qquad (5.12a)$$

in the Laplace-domain form,

$$\hat{\alpha}(s) = [s + i\omega_a + \mathcal{G}(s)]^{-1}, \qquad \hat{\beta}_{k\lambda}(s) = -\frac{i\eta_{k\lambda}^*\hat{\alpha}(s)}{s + i\omega_{k\lambda}}. \qquad (5.12b)$$

Here the Laplace transform is denoted by

$$\hat{\alpha}(s) = \int_0^\infty dt\, \alpha(t)e^{-st} \quad (\mathrm{Re}s > 0), \qquad (5.13)$$

where the bath-response (or self-energy) term,

$$\mathcal{G}(s) = \int_0^\infty d\omega \frac{G(\omega)}{s + i\omega}, \qquad (5.14)$$

is derived from the bath-response (coupling) spectrum

$$G(\omega) = \sum_{k,\lambda} |\eta_{k\lambda}|^2 \delta(\omega - \omega_{k\lambda}) \rightarrow |\eta(\omega)|^2 \rho(\omega). \qquad (5.15)$$

This spectrum is the continuum limit of the (k, λ)-summation over modes, for a directionally independent (*isotropic*) DOM $\rho(\omega)$, namely,

$$\sum_{k,\lambda} \rightarrow \int_0^\infty d\omega\, \rho(\omega). \qquad (5.16)$$

We stress that (5.12b) is an *exact solution* of the atom–bath evolution problem under condition (5.12a) for *all times*, provided we can invert this solution from the Laplace to the time domain. Such inversion is, however, possible in a closed analytical form only for certain spectral bath-response forms $G(\omega)$, as illustrated below.

5.2.2 The Photon-Bound State and Incomplete Decay

In the following analysis of (5.12b), we aim at revealing the peculiar behavior of the atomic excitation decay $\alpha(t)$ and the corresponding emission and Lamb-shift spectra on account of $G(\omega)$ singular features. To this end, we consider a field-confining structure [e.g., a waveguide, a cavity or a photonic crystal (PC), as per Ch. 3] wherein $G(\omega)$ has one or several photonic band gaps (PBGs), separated by allowed photonic bands. Let us label each PBG by index n and its lower and upper cutoff frequencies by $\omega_{\mathrm{L}n}$ and $\omega_{\mathrm{U}n}$, respectively:

$$G(\omega) = 0 \quad \text{for } \omega_{\mathrm{L}n} < \omega < \omega_{\mathrm{U}n}. \qquad (5.17)$$

From Fermi's Golden Rule, we may expect an excited atom not to decay, if ω_a is within a PBG, or decay completely at $t \rightarrow \infty$ if ω_a is anywhere in an allowed band. Yet, neither conclusion is necessarily true, since the Golden Rule may break down in such scenarios.

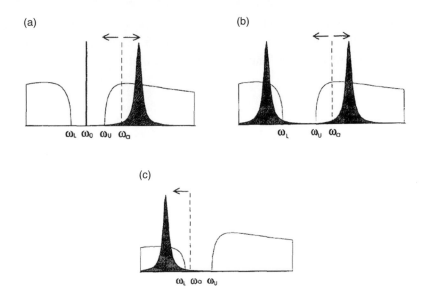

Figure 5.2 An atomic spectral line shifted and split on account of its proximity to a photonic band gap (PBG). (Reprinted from Kofman et al., 1994. © 1994 Taylor & Francis Ltd.)

In fact, *incomplete decay* up to long times occurs if $\hat{\alpha}(s)$ in (5.12b) has a purely imaginary pole, $s = -i\omega_n$, $\hbar\omega_n$ signifying a *stable* (real) energy level of the total (atom+bath) Hamiltonian. Such a pole requires that $G(\omega_n) = 0$ (i.e., ω_n is within a PBG). A real level $\hbar\omega_n$ corresponding to such a pole has to satisfy

$$\omega_n = \omega_a + \Delta(\omega_n). \tag{5.18}$$

Here

$$\Delta(\omega_n) = \int_0^{\omega_{Ln}} d\omega \, \frac{G(\omega)}{\omega_n - \omega} + \int_{\omega_{Un}}^{\infty} d\omega \, \frac{G(\omega)}{\omega_n - \omega}, \tag{5.19}$$

$\hbar\Delta(\omega_n)$ being the bath-induced Lamb (energy) shift of the atomic resonance $\hbar\omega_a$ (often it has been renormalized to account for the open-space Lamb shift). The two bath-induced Lamb-shift terms in (5.19) arise from the parts of $G(\omega)$ extending, respectively, below and above the nth PBG (Fig. 5.2). Whenever (5.18) holds, the corresponding term in $\alpha(t)$ [see (5.11)] oscillates as $\exp(-i\omega_n t)$, *without decay*. This feature signifies a *stable*, dressed, field-atom state formed by the *binding of the photon to the atom* or its confinement to the vicinity of the atom. Such confinement is caused by the nearly complete Bragg reflection of the emitted photon at frequencies within the PBG.

If ω_n is located in the nth PBG, then the first Lamb-shift term on the right-hand side of (5.19) is positive, whereas the second term therein is negative. This means that each allowed spectral band of the bath *repels the atomic level* from the corresponding PBG edge. A real ω_n, according to (5.18), occurs if and only if ω_a falls between the minimal and maximal frequencies admitted by this equation in the nth PBG, namely,

$$\omega_{Ln} - \Delta(\omega_{Ln}) < \omega_a < \omega_{Un} - \Delta(\omega_{Un}). \tag{5.20}$$

There may be at most one stable dressed field-atom state in a single PBG (Fig. 5.2). Yet, there can be *several* such states *simultaneously*, provided conditions (5.20) hold for several PBGs. It is remarkable that conditions (5.20) for incomplete decay can be fulfilled even if ω_a is in an *allowed band*, for example, when $\Delta(\omega_{Un})$ is negative or when $\Delta(\omega_{Ln})$ is positive [Figs. 5.2(a), 5.2(b)].

The converse possibility, of *complete decay for ω_a within a PBG* is equally counterintuitive. Such complete decay will occur, if one of the inequalities in (5.20) is violated, namely [Fig. 5.2(c)],

$$\omega_{Ln} < \omega_a < \omega_{Ln} - \Delta(\omega_{Ln}) \quad \text{if } \Delta(\omega_{Ln}) < 0, \tag{5.21a}$$

or, alternatively,

$$\omega_{Un} - \Delta(\omega_{Un}) < \omega_a < \omega_{Un} \quad \text{if } \Delta(\omega_{Un}) > 0. \tag{5.21b}$$

The existence of one or several discrete, stable, states yields the inverse Laplace transform of (5.12b) and the corresponding wave function (5.11) in the form

$$|\Psi(t)\rangle = \sum_n \sqrt{c_n}\,|\Psi_n\rangle e^{-\omega_n t} + |\Psi_c(t)\rangle. \tag{5.22}$$

Here we sum over all discrete dressed states with energies $\hbar\omega_n$; $|\Psi_c(t)\rangle$ is the decaying part of the wave function (wherein the atom eventually is in the ground state and a photon has been emitted); and $|\Psi_n\rangle$ is a discrete-state eigenfunction of the Hamiltonian, normalized to unity and weighted by the amplitude $\sqrt{c_n}$, where

$$c_n = \left[1 + \int_0^\infty d\omega \frac{G(\omega)}{(\omega - \omega_n)^2}\right]^{-1}. \tag{5.23}$$

Each normalized eigenfunction has the form,

$$|\Psi_n\rangle = \sqrt{c_n}\,|e, \{0_{k\lambda}\}\rangle + \sqrt{1 - c_n}\,|g, \psi_n\rangle, \tag{5.24}$$

where ψ_n is a (normalized to unity) one-photon wave packet,

$$|\psi_n\rangle = \sqrt{\frac{c_n}{1 - c_n}} \sum_{k,\lambda} \frac{\eta_{k\lambda}^*}{\omega_n - \omega_{k\lambda}}|1_{k\lambda}\rangle. \tag{5.25}$$

The discrete state in Eq. (5.24) consists of an excited-atom (zero photon) component and a bound-photon component associated with a ground-state atom. As seen from (5.25), the bound-photon state is a wave packet composed of single-photon contributions from all photonic modes.

The decayed part of the wave function can be expressed by means of (5.12b) in the form of a wave packet consisting of all photonic modes,

$$|\Psi_c(t \to \infty)\rangle = -i \sum_{k,\lambda} \eta_{k\lambda}^* \hat{\alpha}(-i\omega_{k\lambda}) e^{-i\omega_{k\lambda}t} |g, 1_{k\lambda}\rangle. \tag{5.26}$$

Equation (5.26) corresponds to the following spectral probability density of photon emission,

$$P(\omega) = \sum_{k,\lambda} |\eta_{k\lambda}\hat{\alpha}(-i\omega_{k\lambda})|^2 \Delta(\omega - \omega_{k\lambda}) = G(\omega)|\hat{\alpha}(-i\omega)|^2, \tag{5.27}$$

with

$$\hat{\alpha}(-i\omega) = \lim_{\eta \to 0^+} \hat{\alpha}(-i\omega + \eta) = \{i[\omega_a - \omega + \Delta(\omega)] + \gamma(\omega)\}^{-1}. \tag{5.28}$$

Here the first equality is an analytic continuation of $\hat{\alpha}(s)$ in (5.12b) to the imaginary axis, expressed in terms of the real (dissipative) and imaginary (dispersive) parts of the bath response, respectively,

$$\gamma(\omega) = \pi G(\omega), \quad \Delta(\omega) = P \int_0^\infty d\omega' \frac{G(\omega')}{\omega - \omega'}. \tag{5.29}$$

Completely generally, (5.27) can be expressed by means of (5.28) as a skewed Lorentzian (i.e., a Lorentzian whose height and width are ω-dependent),

$$P(\omega) = \frac{1}{\pi} \frac{\gamma(\omega)}{\gamma^2(\omega) + [\omega - \omega_a - \Delta(\omega)]^2}. \tag{5.30}$$

The area under the skewed-Lorentzian spectral emission probability is evaluated to be

$$\int_0^\infty d\omega\, P(\omega) = \langle \Psi_c(t)|\Psi_c(t)\rangle = 1 - \sum_n c_n, \tag{5.31}$$

where the right-hand side is the probability to emit a free (unbound) photon. The second equality in (5.31) follows from (5.22), allowing for the fact that $|\Psi(t)\rangle$ is normalized to unity and that $|\Psi_n\rangle$ are orthonormal and orthogonal to $|\Psi_c(t)\rangle$. The quantity in (5.31) must be nonnegative, hence we obtain the inequality

$$P_b = \sum_n c_n \leq 1, \tag{5.32}$$

where P_b is the probability for the spontaneous decay to result in a bound state.

The long-time population of the excited atomic state satisfies

$$|\alpha(t)|^2 = \sum_{n,n'} c_n c_{n'} \cos(\omega_n - \omega_{n'})t \quad \text{for } t \to \infty. \tag{5.33}$$

It reduces to a time-constant c_n^2 only if there is one discrete, stable, state. Yet (5.33) can also yield a *coherence* between two or more stable states. The excited-atom population then exhibits beats at the frequencies corresponding to the energy differences of the superposed stable states, due to energy exchange between the atom and the bath. The upper bound for the beats (5.33) is given by the inequality,

$$|\alpha(t)|^2 \leq \left(\sum_n c_n \right)^2, \tag{5.34}$$

whereas the time average of the beats is

$$\overline{|\alpha(t)|^2} = \sum_n c_n^2. \tag{5.35}$$

An example is that of two dressed stable states, analogous to the case of an excited atom coupled to a cavity mode, where the spontaneously induced coherence causes vacuum Rabi oscillations. Upon subtracting (5.35) from P_b in (5.32), one obtains the average probability for a bound photon to form,

$$\bar{P}_{bp} = \sum_n c_n(1 - c_n). \tag{5.36}$$

To sum up, we have seen that the excited state $|e\rangle$ can split into coherently superposed stable states, whose energies $\hbar\omega_n$ and amplitudes $\sqrt{c_n}$ *are controllable* by the atomic transition detuning from cutoff.

5.2.3 Complete Decay: Irreversible (Weisskopf–Wigner) Solutions

As follows from (5.20), incomplete excited-state decay cannot occur for ω_a well within an allowed band, far from a PBG. We then anticipate that long-time decay should obey, to a good approximation, the (open-space) Weisskopf–Wigner exponential decay law. In this section we discuss higher-order corrections to the standard Weisskopf–Wigner solution and obtain the conditions for its applicability.

The inverse Laplace transform of $\hat{\alpha}(s)$ is known to involve contributions from the poles of $\hat{\alpha}(s)$ in (5.12b). The Weisskopf–Wigner approximation applies when the contribution of one of the poles (call it s_p) is dominant in $\alpha(t)$. Assuming that the self-energy term $\mathcal{G}(s)$ is sufficiently small in the equation $s + i\omega_a + \mathcal{G}(s) = 0$ for the poles, s_p can be found perturbatively. Namely, we successively iterate the pole, with $s_p^{(0)} = -i\omega_a$ as the lowest approximation. The first-order iteration yields

$$s_p \approx s_p^{(1)} = -i(\omega_a + \Delta_a) - \gamma_a, \tag{5.37}$$

where we have abbreviated $\gamma_a = \gamma(\omega_a)$, and $\Delta_a = \Delta(\omega_a)$ [see Eqs. (5.29)].

If $\alpha(t)$ can be approximated by the contribution of a *single pole s_p* (known as the pole approximation), then the inverse Laplace transform of the first Eq. (5.12b) to first-order approximation for the pole yields an exponentially decaying amplitude

$$\alpha(t) \approx \exp[-(\gamma_a + i\tilde{\omega}_a)t], \tag{5.38}$$

where $\tilde{\omega}_a = \omega_a + \Delta_a$.

Further iterations yield higher-order corrections to the pole. An analysis shows that these corrections can be neglected and hence this regime holds if the function $G(\omega)$ is *smooth near* ω_a or, quantitatively, under the conditions

$$\text{(a)} \ |\gamma_a'|, |\Delta_a'| \ll 1; \quad \text{(b)} \ |\Delta_a \gamma_a'| \ll \gamma_a; \quad \text{(c)} \ \gamma_a, |\Delta_a| \ll |\omega_a - \omega_g|. \tag{5.39}$$

Here the primes denote differentiation with respect to ω_a,

$$\gamma_a' = \pi G'(\omega_a), \quad \Delta_a' = P \int_0^\infty d\omega \, \frac{G'(\omega)}{\omega_a - \omega}, \tag{5.40}$$

$G'(\omega)$ being the derivative of $G(\omega)$ and ω_g is the band-edge frequency nearest to ω_a.

Conditions (5.39) correspond to a locally smooth $G(\omega)$ and require the decay rate γ_a and the Lamb shift Δ_a to be sufficiently small. Equivalently, the detuning of the atomic resonance from the nearest band edge should be sufficiently large, and the atom–bath coupling sufficiently weak.

Under the conditions (5.39), the skewed-Lorentzian line shape (5.30) becomes Lorentzian,

$$P(\omega) \approx \frac{1}{\pi} \frac{\gamma_a}{\gamma_a^2 + (\omega - \tilde{\omega}_a)^2}. \tag{5.41}$$

In addition to the Weisskopf–Wigner solution (5.38), the present pole approximation describes also the case of ω_a within a PBG ($\gamma_a = 0$). We then obtain, taking into account the second-order correction,

$$\alpha(t) \approx (1 + \Delta_a') \exp(-i\tilde{\omega}_a t), \tag{5.42}$$

where now Δ_a' in Eq. (5.40) is evaluated to be

$$\Delta_a' = -\int_0^\infty d\omega \frac{G(\omega)}{(\omega - \omega_a)^2} < 0. \tag{5.43}$$

This quantity is negative since $G(\omega) \geq 0$. This result holds for $|\Delta_a'| \ll 1$, $|\Delta_a| \ll |\omega_a - \omega_g|$.

The pole-approximation solutions (5.38) and (5.42) do not generally hold at very short times. Moreover, (5.38) is violated at very long times (see Sec. 5.2.4). In particular, (5.42) implies *fast nonvanishing decay* within a gap, giving rise to the term Δ_a'.

5.2.4 Partial and Complete Decay near Cutoff

As per (3.23), the bath spectral response near the upper PBG cutoff ω_U can be modeled by

$$G(\omega) = \frac{\gamma_c^{3/2}}{\pi} \frac{\sqrt{\omega - \omega_U}}{\omega - \omega_U + \epsilon_c} \theta(\omega - \omega_U), \qquad (5.44)$$

where ϵ_c is the "cutoff-smoothing" parameter and γ_c characterizes the atom–bath coupling strength at cutoff. Depending on whether $\epsilon_c \ll \gamma_c$ or $\epsilon_c \gg \gamma_c$, the bath is close to the isotropic case $c_c = 0$ ($D = 2$) or the anisotropic case $\epsilon_c \to \infty$ ($D = 0$), respectively.

Upon inserting (5.44) into (5.29), we obtain the bath-induced skewed-Lorentzian linewidth

$$\gamma(\omega) = \frac{\gamma_c^{3/2} \sqrt{\omega - \omega_U}}{\omega - \omega_U + \epsilon_c} \theta(\omega - \omega_U), \qquad (5.45)$$

$\theta()$ being the Heaviside step function and the corresponding spectral shift

$$\Delta(\omega) = -\gamma_c^{3/2} \begin{cases} \sqrt{\epsilon_c}/(\omega - \omega_U + \epsilon_c), & \omega > \omega_U, \\ (\sqrt{\omega_U - \omega} + \sqrt{\epsilon_c})^{-1}, & \omega < \omega_U. \end{cases} \qquad (5.46)$$

It thus follows that the spectrum (5.30) can be strongly non-Lorentzian. The model (5.44) yields an exact analytical expression for the time dependence of spontaneous decay, from which we can infer the criteria for the two regimes of Sections 5.2.2 and 5.2.3:

(i) *Incomplete decay* [cf. (5.20)] is now obtained for (Fig. 5.3)

$$\omega_a < \omega_U + \frac{\gamma_c^{3/2}}{\sqrt{\epsilon_c}}. \qquad (5.47)$$

Equation (5.47) implies that an abrupt, singular cutoff of the DOM with $D = 2$ ($\epsilon_c \to 0$) allows for a discrete state at *any* ω_a, either inside or outside the PBG. The energy of the discrete state, $\hbar\omega_0$, which must lie within the PBG, is found, upon substituting the second of Eqs. (5.46) into (5.18), to be a real and positive root of the equation

$$\omega_0 = \omega_a - \frac{\gamma_c^{3/2}}{\sqrt{\omega_U - \omega_0} + \sqrt{\epsilon_c}}. \qquad (5.48)$$

(ii) The conditions (5.39) for *nearly exponential decay* (Fig. 5.4) can now be shown to amount to the requirement that ω_a be in the allowed zone sufficiently far from cutoff,

$$\omega_a - \omega_U \gg \begin{cases} \gamma_c, & \text{if } \epsilon_c \le \gamma_c, \\ \gamma_c^{3/2}/\sqrt{\epsilon_c}, & \text{if } \epsilon_c > \gamma_c. \end{cases} \qquad (5.49)$$

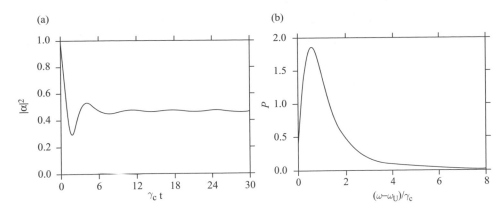

Figure 5.3 (a) Analytically evaluated incomplete decay of the excited state population as a function of $\gamma_c t$ at cutoff, $\omega_a = \omega_U$, for $\epsilon_c = 10^{-2}\gamma_c$. The beat period is $2\pi/(\omega_U - \omega_0)$. The nondecaying probability is $c_0^2 = 4/9$. (b) The corresponding emission spectrum. (Adapted from Kofman et al., 1994. © 1994 Taylor & Francis Ltd.)

Under this large-detuning condition, the intermediate-time exponential decay of the excited state is given to first approximation by (5.38), where $\gamma_a = \gamma(\omega_a)$ and $\Delta_a = \Delta(\omega_a)$ are obtained from (5.45) and (5.46). The atomic Lamb shift Δ_a is negative in the present case, vanishing for $\epsilon_c \to 0$ $(D = 2)$. The long-time behavior of $\alpha(t)$ can be shown to exhibit a tail diminishing as $t^{-3/2}$ and oscillating at the cutoff frequency ω_U (Fig. 5.4).

By increasing the smoothing parameter ϵ_c, we can further inhibit the decay for ω_a at or near the cutoff ω_U, since $G(\omega_a \simeq \omega_U)$ then becomes weaker and the stable-state probability c_0 [cf. (5.23)] correspondingly larger. On the other hand, at a large detuning of ω_a from cutoff, large-ϵ_c smoothing allows complete decay [cf. (5.47)] and an entirely Lorentzian spectrum.

5.3 Atomic Coupling to a High-Q Defect Mode in the PBG

A defect in the periodic structure can produce a narrow-linewidth local mode in the PBG akin to a high-Q cavity mode. In fact, this is the optimal way to create such a cavity mode. The spectral response is then describable by a Lorentzian,

$$G_d(\omega) = \frac{\gamma_d}{\pi} \frac{\Gamma_d^2}{\Gamma_d^2 + (\omega - \omega_d)^2}, \tag{5.50}$$

where γ_d expresses the coupling strength of the atomic dipole with the defect field, whereas ω_d and Γ_d represent the line center and width, respectively. Such a defect line has an additive effect on the parameters of the skewed Lorentzian line shape (5.30), as compared to (5.45) and (5.46),

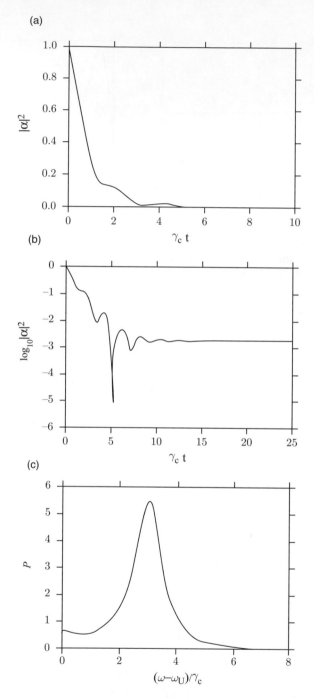

Figure 5.4 Similar to Figure 5.3, for a large detuning from cutoff within the allowed zone ($\epsilon_c = 10^{-2}\gamma_c$, $\omega_a - \omega_U = 3\gamma_c$). (a) Nearly complete decay ($c_0^2 \approx 10^{-3}$) modulated by beats with frequency $\omega_a - \omega_U$. (b) Logarithmic-scale decay exhibits beats at the non-exponential tail. (c) The corresponding spectrum is nearly Lorentzian but with a small peak near ω_U. (Adapted from Kofman et al., 1994. © 1994 Taylor & Francis Ltd.)

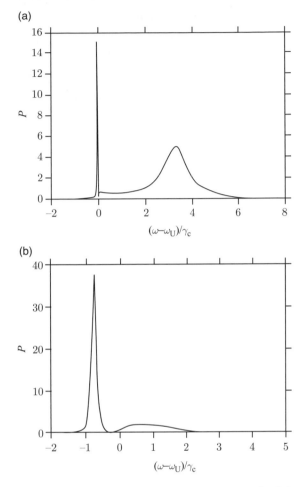

Figure 5.5 Spontaneous-emission spectra in the presence of a defect in the PBG ($\omega_d - \omega_U = -10^2\gamma_c$, $\gamma_d = \gamma_c$, $\Gamma_d = 30\gamma_c$, and $\epsilon_c = 10^{-2}\gamma_c$). (a) At large detuning, $\omega_a - \omega_U = 3\gamma_c$, the oscillator strength is mostly in the Lorentzian peak about $\tilde{\omega}_a$. (b) At cutoff, $\omega_a = \omega_U$, the oscillator strength is equally shared between the distribution above cutoff [as in Fig. 5.3(b)] and the defect-induced peak at ω_0. (Adapted from Kofman et al., 1994. © 1994 Taylor & Francis Ltd.)

$$\gamma(\omega) \to \gamma(\omega) + \frac{\gamma_d \Gamma_d^2}{\Gamma_d^2 + (\omega - \omega_d)^2},$$

$$\Delta(\omega) \to \Delta(\omega) + \frac{\gamma_d \Gamma_d (\omega - \omega_d)}{\Gamma_d^2 + (\omega - \omega_d)^2}. \tag{5.51}$$

The nonvanishing DOM in the PBG due to a defect causes spontaneous emission within the PBG and thus broadens the discrete state ω_0, rendering it metastable (see Fig. 5.5).

5.4 Discussion

The Wigner–Weisskopf solution for the dynamics of an excited two-level system (TLS) coupled to a bath allows for either irreversible decay or reversible, oscillatory evolution, depending on the coupling spectrum and strength, which are in turn determined by the field-confining geometry. The most extreme case of reversible evolution is that of an atom whose resonant transition is in a photonic band gap. Such evolution can give rise to a bound photon state. Remarkably, by varying the parameters of the confining structure (a high-Q cavity or a photonic crystal) we can control reversibility.

Atomic resonance detuning from photonic band edges has been shown to allow for *spontaneous coherence control*. This effect opens interesting perspectives for three-level atoms with a resonant transition near a band edge of a PC. Processes that rely on coherences between atomic levels, such as lasing without inversion (LWI), nonlinear photon switching, or electromagnetically induced transparency (EIT), may be spontaneously and controllably induced in such systems.

6

System–Bath Equilibration via Spin-Boson Interaction

As shown in Chapter 2, the eigenstates of the total Hamiltonian $H \equiv H_{\text{tot}}$ (2.3) exhibit entanglement between the system and the bath, due to their interaction H_{SB}. The same is true of thermal states of the combined system–bath complex $(Z^{-1}e^{-\beta H_{\text{tot}}})$, at low temperatures and for weak coupling.

Here we consider a scenario wherein the system and the bath undergo a quench, namely, they are initially in a factorizable eigenstate of $H_0 = H_S + H_B$ and then are abruptly exposed to H_{tot}, the Hamiltonian that includes the system–bath interaction term. They will consequently evolve into non-separable (or entangled) system–bath states, at any temperature. If, by contrast, the interaction is turned on adiabatically, an H_0 eigenstate will asymptotically evolve into an eigenstate of H_{tot}.

6.1 System–Bath Non-separability or Entanglement near Thermal Equilibrium

Here we further study the spin-boson model (Chs. 4 and 5), now considering a TLS coupled to an oscillator bath *without the RWA* at finite temperature. Our goal is to allow for the bath temperature when evaluating the system–bath correlations near their thermal equilibrium.

The TLS with energy-level distance $\hbar\omega_{\text{a}}$, governed by the Hamiltonian

$$H_S = \hbar\omega_{\text{a}} |e\rangle \langle e| , \tag{6.1}$$

is assumed to be at thermal equilibrium at temperature T. The equilibrium TLS state is diagonal in the σ_z (energy) basis and can be written as

$$\rho_S = \frac{1}{2}(\hat{I} + P_{\text{Eq}}\sigma_z). \tag{6.2}$$

Here $\hat{I}$ is the identity operator and $P_{\text{Eq}} = \langle \sigma_z \rangle = \rho_{ee} - \rho_{gg}$ is the spin polarization, where $\rho_{ii} = \langle i|\rho_S|i \rangle$. The parameter P_{Eq} characterizes the TLS purity at equilibrium, so that $|P_{\text{Eq}}| = 1$ ($|P_{\text{Eq}}| < 1$) for pure (non-pure) states. Assuming *vanishing coupling* with the bath, the equilibrium-value purity of the TLS is

$$P_{\text{Eq}}^0 = \rho_{ee}^0 - \rho_{gg}^0 = -\tanh(\beta\hbar\omega_a/2), \qquad (6.3)$$

where $1/\beta = k_B T$, k_B being the Boltzmann constant, and the ground and excited populations, respectively, are given by

$$\rho_{ee}^0 = (1 + P_{\text{Eq}}^0)/2 = \frac{1}{1 + \exp(\beta\hbar\omega_a)}, \qquad (6.4)$$

$$\rho_{gg}^0 = (1 - P_{\text{Eq}}^0)/2 = \frac{1}{1 + \exp(-\beta\hbar\omega_a)}. \qquad (6.5)$$

These expressions are consistent with the factorized system–bath state

$$\rho_0 = \frac{1}{Z_0} e^{-\beta H_0} = \rho_S^{(0)} \otimes \rho_B^{(0)}, \qquad (6.6)$$

where

$$H_0 = H_S + H_B. \qquad (6.7)$$

H_B is the bath Hamiltonian, whereas $\rho_S^{(0)}$ and $\rho_B^{(0)}$ are the equilibrium density matrices for the system and the bath in the absence of coupling, with the partition function

$$Z_0 = \text{Tr}\, e^{-\beta H_0}. \qquad (6.8)$$

For nonzero system–bath coupling, the equilibrium state of the TLS and the bath is non-separable (which in many cases is equivalent to being entangled). In order to find the mean mixedness (impurity) of the TLS state at a temperature T, we have to characterize the equilibrium density matrix for the total system

$$\rho_{\text{Eq}} = \frac{1}{Z} \exp(-\beta H_{\text{tot}}), \qquad (6.9)$$

with the partition function

$$Z = \text{Tr} \exp(-\beta H_{\text{tot}}) \qquad (6.10)$$

corresponding to the total Hamiltonian of the interacting system and bath

$$H_{\text{tot}} = H_S + H_B + \epsilon \tilde{H}_{\text{SB}}. \qquad (6.11)$$

The explicit second-quantized form (Chs. 3 and 4) for an oscillator bath coupled to the TLS via σ_x is given by

$$H_B = \hbar \sum_k \omega_k a_k^\dagger a_k, \qquad (6.12)$$

ω_k and a_k being the frequency and the annihilation operator, respectively, of mode k, and

$$\epsilon \tilde{H}_{\text{SB}} = \hbar \sigma_x \sum_k (\eta_k a_k + \eta_k^* a_k^\dagger), \qquad (6.13)$$

η_k being the k-mode coupling amplitude.

The system–bath coupling spectrum at temperature T is given by (see below, Ch. 11)

$$G_T(\omega) = G_0(\omega)[n_T(\omega) + 1] + G_0(-\omega) n_T(-\omega). \qquad (6.14a)$$

It is expressed in terms of the average occupation number at inverse temperature β,

$$n_T(\omega) = \frac{1}{\exp(\beta \hbar \omega) - 1}, \qquad (6.14b)$$

and the zero-temperature bath-coupling spectrum

$$G_0(\omega) = \sum_k |\eta_k|^2 \delta(\omega - \omega_k). \qquad (6.14c)$$

In order to evaluate the effects of the weak system–bath coupling $\epsilon \tilde{H}_{\text{SB}}$ (Chs. 2 and 4) on the system purity, we resort to the finite-temperature Heims quantum-statistical perturbation theory by expanding ρ_{Eq} as

$$\rho_{\text{Eq}} = \frac{1}{Z} e^{-\beta(H_0 + \epsilon \tilde{H}_{\text{SB}})} = \frac{e^{-\beta H_0}(1 - \epsilon S_1 + \epsilon^2 S_2 + \dots)}{Z_0 - \epsilon \text{Tr}(e^{-\beta H_0} S_1) + \epsilon^2 \text{Tr}(e^{-\beta H_0} S_2) + \dots}, \qquad (6.15)$$

where the first- and second-order terms are

$$S_1 = \beta \int_0^1 dx \, e^{x \beta H_0} \tilde{H}_{\text{SB}} e^{-x \beta H_0}, \qquad (6.16a)$$

$$S_2 = \beta^2 \int_0^1 dx \int_0^x dy \, e^{x \beta H_0} \tilde{H}_{\text{SB}} e^{-(x-y)\beta H_0} \tilde{H}_{\text{SB}} e^{-y \beta H_0}. \qquad (6.16b)$$

Upon tracing out the bath degrees of freedom, the state of the system, $\rho_{\text{S}} = \text{Tr}_{\text{B}} \rho_{\text{Eq}}$, is obtained.

The Heims perturbation theory used above may become invalid at sufficiently low temperatures: as $\beta \to \infty$, the second-order terms tend to infinity in both the numerator and the denominator of the fraction in (6.15). That these terms diverge can be understood from the fact that the Heims perturbation theory is obtained from the quantum-mechanical time-dependent perturbation theory (TDPT) upon replacing time with the *imaginary time* $-i\hbar\beta$. Since the TDPT is generally valid only for sufficiently short times, its Heims counterpart is generally valid only for small β (high T).

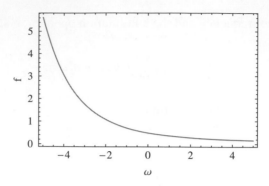

Figure 6.1 The dependence of the function $f(\omega)$, in units of $\hbar^2\beta^2$, on ω [in units of $(\hbar\beta)^{-1}$].

This obstacle can be overcome by expanding the fraction in (6.15) as a power series in the small system–bath coupling parameter ϵ. Then, evaluation of the TLS purity in (6.2) as a series up to second order $O(\epsilon^2)$ yields

$$P_{\text{Eq}} = P_{\text{Eq}}^0 + 2\epsilon^2 \rho_{ee}^0 \rho_{gg}^0 (M_- - M_+), \tag{6.17}$$

where the result is expressed in terms of the auxiliary functions

$$M_\pm = \int_{-\infty}^{\infty} d\omega \, G_T(\omega) f(\omega \pm \omega_{\text{a}}), \tag{6.18}$$

and (Fig. 6.1)

$$f(\omega) = \frac{e^{-\beta\hbar\omega} - 1 + \beta\hbar\omega}{\omega^2}. \tag{6.19}$$

In what follows, we omit the auxiliary parameter ϵ, which is equivalent to setting $\epsilon = 1$ and $\tilde{H}_{\text{SB}} = H_{\text{SB}}$.

Remarkably, the integrals in (6.18) are at the risk of diverging, since, in view of (6.14b), for $G_0(0) \neq 0$ the function in (6.14a) has a singularity,

$$G_T(\omega) \propto \frac{G_0(0)}{\beta|\omega|} \quad \text{for } \omega \to 0. \tag{6.20}$$

However, we may assume that for any *physical* bath-coupling spectrum

$$\lim_{\omega \to 0^+} G_0(\omega) = 0, \tag{6.21}$$

so that there is no singularity and the integrals in (6.18) converge, yielding finite $M_\pm$.

It can be checked that at $T = 0$ ($\beta \to \infty$) (6.17) yields for the excitation of the TLS,

$$\rho_{ee} = \frac{1 + P_{\text{Eq}}}{2} = \int_0^{\infty} d\omega \, \frac{G_0(\omega)}{(\omega + \omega_{\text{a}})^2}. \tag{6.22}$$

From (6.22) it can be seen that even at $T = 0$, TLS purity is incomplete, $|P_{\text{Eq}}| < 1$, due to the system–bath non-separability, which is often tantamount to entanglement.

In the opposite, high-temperature limit, $\beta \to 0$, we obtain

$$M_{\pm} \approx \beta \hbar J \left(1 \mp \frac{\beta \hbar \omega_a}{3} \right),\tag{6.23}$$

where

$$J = \int_0^{\infty} d\omega \, \frac{G_0(\omega)}{\omega}.\tag{6.24}$$

This integral too converges at the lower limit under the condition (6.21). As $\beta \to 0$, we have $\rho_{ee}^0 \approx \rho_{gg}^0 \approx 1/2$; (6.17), (6.3), and (6.23) then imply that the purity tends to zero with β,

$$P_{\text{Eq}} \approx P_{\text{Eq}}^0 \approx -\beta \hbar \omega_a / 2,\tag{6.25}$$

whereas the purity change due to the system–bath non-separability (or entanglement) behaves as

$$P_{\text{Eq}} - P_{\text{Eq}}^0 = 2(\rho_{ee} - \rho_{ee}^0) \approx \beta^2 \hbar^2 \omega_a J / 3.\tag{6.26}$$

More generally, we find that the system–bath coupling decreases the system (TLS) purity at any finite temperature. This can be shown from the inequalities

$$M_- > M_+ > 0 \quad \text{for } \beta > 0,\tag{6.27}$$

which follow from the fact that $f(\omega)$ is a positive, decreasing function (see Fig. 6.1). Taking into account that P_{Eq} and P_{Eq}^0 are negative (since there is no population inversion in equilibrium), (6.17) and (6.27) yield

$$P_{\text{Eq}}^0 < P_{\text{Eq}} < 0 \quad \text{for } \beta > 0\tag{6.28}$$

and

$$|P_{\text{Eq}}| < |P_{\text{Eq}}^0| \quad \text{for } \beta > 0,\tag{6.29}$$

that is, the system–bath coupling decreases the TLS purity at $T < \infty$.

As an illustration, we provide numerical calculations for the coupling spectrum of the form

$$G_0(\omega) = \frac{C\Gamma^2}{\Gamma^2 + (\omega - \omega_0)^2} \frac{\omega^2}{\epsilon^2 + \omega^2} \theta(\omega).\tag{6.30}$$

Here, as before, $\theta(\omega)$ is the Heaviside step function. The ϵ-dependent fractional factor ensures the validity of condition (6.21). For $0 < \epsilon \ll \Gamma, \omega_0$, the function (6.30) is close to a Lorentzian of width Γ, with cutoff at $\omega = 0$. Then C is approximately the maximal value of $G_0(\omega)$ obtained at $\omega \approx \omega_0$. Figure 6.2 shows $G_0(\omega)$

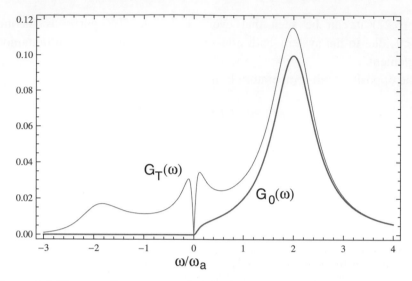

Figure 6.2 The coupling spectra $G_0(\omega)$ [Eq. (6.30)] and $G_T(\omega)$ [for $\beta = (\hbar\omega_a)^{-1}$], in units of the TLS resonance frequency ω_a. Here the memory time of the bath $t_c = 1/\Gamma = 2/\omega_a$, peak of the Lorentzian-like bath spectrum $\omega_0 = 2\omega_a$, maximal system–bath coupling $C = \omega_a/10$, $\epsilon = 10^{-2}\omega_a$.

and $G_T(\omega)$ for specific values of the parameters. In particular, the peaks of $G_T(\omega)$ near $\omega = 0$ result from the singular behavior of $n_T(\omega)$ in (6.14a). Nevertheless, $G_T(\omega)$ does not diverge, since $G_0(\omega)$ tends sufficiently fast to zero ($\propto \omega^2$) for $\omega \to 0$.

As illustrated by Figure 6.3, the difference (6.26) is positive, in accordance with (6.28), and tends to zero as β^2 for $\beta \to 0$. Yet, its dependence on β is non-monotonic, as shown in Figure 6.3 for the Lorentzian-like coupling spectrum in Figure 6.2. This example demonstrates that in order to achieve an appreciable system–bath non-separability, the system–bath coupling should be rather strong, as measured by the ratio of $G_0(\omega)$ to ω_a.

6.2 Mean Energies at Equilibrium

The mean interaction energy, to second order in the Heims perturbation theory, is given by

$$\langle H_{SB}\rangle_{Eq} = -\hbar(\rho_{ee}^0 \tilde{M}_- + \rho_{gg}^0 \tilde{M}_+), \qquad (6.31a)$$

where we have defined

$$\tilde{M}_\pm = \int_{-\infty}^{\infty} d\omega\, G_T(\omega)\, \frac{1 - \exp[-\beta\hbar(\omega \pm \omega_a)]}{\omega \pm \omega_a}. \qquad (6.31b)$$

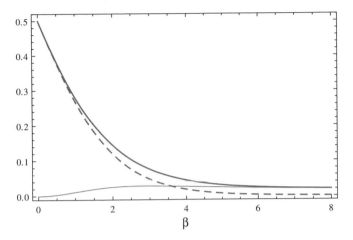

Figure 6.3 TLS excitation probability at equilibrium as a function of the inverse temperature β [in units of $(\hbar\omega_a)^{-1}$]: with (thick solid line) or without (dashed line) the effect of the system–bath interaction, as well as their difference $\rho_{ee} - \rho_{ee}^0$ (thin solid line). Same parameters are as in Figure 6.2.

None of the terms in (6.31a) *diverges* as $\beta \to \infty$, so that it is applicable to any temperature. Both factors in the integrand in (6.31b) are positive, so that $\tilde{M}_\pm > 0$. Together with (6.31a), this proves the *negativity* of the mean system–bath interaction (coupling) energy in equilibrium, $\langle H_{SB}\rangle_{Eq} < 0$. The system and the bath equilibrate to a stable, bound state with lower energy than their factorized state.

The average energies of the system and the bath are also affected by the coupling. For the system,

$$\langle H_S\rangle_{Eq} = \mathrm{Tr}\,(\rho_S H_S) = \hbar\omega_a\,\rho_{ee} = \langle H_S\rangle_{Eq}^{(0)} + \langle H_S\rangle_{Eq}^{(2)}, \tag{6.32}$$

where

$$\langle H_S\rangle_{Eq}^{(0)} = \hbar\omega_a\,\rho_{ee}^0, \tag{6.33}$$

$$\langle H_S\rangle_{Eq}^{(2)} = \hbar\omega_a\,\rho_{ee}^0\rho_{gg}^0(M_- - M_+), \tag{6.34}$$

upon allowing for (6.4) and (6.17). For the bath,

$$\langle H_B\rangle_{Eq} = \mathrm{Tr}\,(\rho_{eq}H_B) = \langle H_B\rangle_{Eq}^{(0)} + \langle H_B\rangle_{Eq}^{(2)}, \tag{6.35}$$

where

$$\langle H_B\rangle_{Eq}^{(0)} = \mathrm{Tr}\,(\rho_B^{(0)} H_B), \tag{6.36}$$

$$\langle H_B\rangle_{Eq}^{(2)} = \rho_{ee}^0\mathcal{M}_- + \rho_{gg}^0\mathcal{M}_+. \tag{6.37}$$

Here we have defined

$$\mathcal{M}_\pm = \frac{\hbar}{4}\int_{-\infty}^{\infty} d\omega\, G_0(|\omega|)\frac{|\omega|}{\sinh^2(\beta\hbar\omega/2)}f(\omega \pm \omega_a), \tag{6.38}$$

$f(\omega)$ being given by (6.19).

The above results yield also the corrections to the mean values of H_0 and H_{tot},

$$\langle H_0 \rangle_{Eq}^{(2)} = \langle H_S \rangle_{Eq}^{(2)} + \langle H_B \rangle_{Eq}^{(2)},$$

$$\langle H_{tot} \rangle_{Eq}^{(2)} = \langle H_0 \rangle_{Eq}^{(2)} + \langle H_{SB} \rangle_{Eq}. \tag{6.39}$$

Equations (6.31a), (6.34), (6.37), and (6.39), upon allowing for (6.27) and (6.38), prove that (Fig. 6.4)

$$\langle H_S \rangle_{Eq}^{(2)} > 0, \quad \langle H_B \rangle_{Eq}^{(2)} > 0, \quad \langle H_0 \rangle_{Eq}^{(2)} > 0, \quad \langle H_{SB} \rangle_{Eq} < 0. \tag{6.40}$$

These inequalities are strict, except that $\langle H_S \rangle_{Eq}^{(2)}$ vanishes at the high-temperature limit $\beta = 0$.

At $T = 0$, the mean interaction energy is found from (6.31) to be *twice the bath-induced Lamb shift* of the TLS ground level,

$$\langle H_{SB} \rangle_{Eq} = -2\hbar \int_0^\infty d\omega \, \frac{G_0(\omega)}{\omega + \omega_a}, \tag{6.41}$$

whereas from (6.34) and (6.37) we find at $T = 0$

$$\langle H_S \rangle_{Eq}^{(2)} = \hbar \omega_a \int_0^\infty d\omega \, \frac{G_0(\omega)}{(\omega + \omega_a)^2}, \quad \langle H_B \rangle_{Eq}^{(2)} = \hbar \int_0^\infty d\omega \, \frac{G_0(\omega)\omega}{(\omega + \omega_a)^2}. \tag{6.42}$$

It then follows that at $T = 0$ $\langle H_{tot} \rangle_{Eq}^{(2)}$ equals the *dispersive bath-induced Lamb shift*, to be discussed in Chapter 8.

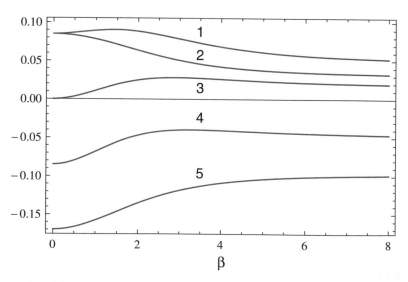

Figure 6.4 Mean energies at equilibrium (in units of $\hbar\omega_a$) as a function of β [in units of $(\hbar\omega_a)^{-1}$]: $\langle H_0 \rangle_{Eq}^{(2)}$—curve 1, $\langle H_B \rangle_{Eq}^{(2)}$—curve 2, $\langle H_S \rangle_{Eq}^{(2)}$—curve 3, $\langle H_{tot} \rangle_{Eq}^{(2)}$—curve 4, $\langle H_{SB} \rangle_{Eq}$—curve 5. Same parameters as in Figure 6.2.

Figure 6.4 shows that $\langle H_{\text{tot}}\rangle_{\text{Eq}}^{(2)}$ is negative at any temperature and has a maximum at a finite temperature.

In the high-temperature limit $\beta = 0$, we obtain

$$\langle H_0\rangle_{\text{Eq}}^{(2)} = \langle H_{\text{B}}\rangle_{\text{Eq}}^{(2)} = -\langle H_{\text{SB}}\rangle_{\text{Eq}}^{(2)}/2 = -\langle H_{\text{tot}}\rangle_{\text{Eq}}^{(2)} = \hbar J, \tag{6.43}$$

where J is given by (6.24). Remarkably, the relation

$$\langle H_{\text{tot}}\rangle_{\text{Eq}}^{(2)} = \langle H_{\text{SB}}\rangle_{\text{Eq}}^{(2)}/2 = -\langle H_0\rangle_{\text{Eq}}^{(2)} \tag{6.44}$$

holds at both infinite and zero temperatures, but not at temperatures between zero and infinity.

The inequalities in (6.40) can be generalized by showing that the average coupling energy is nonpositive to the first nonvanishing order,

$$\langle H_{\text{SB}}\rangle_{\text{Eq}} \leq 0, \tag{6.45}$$

provided $\text{Tr}(\rho_0 H_{\text{SB}}) = 0$. The inequality (6.45) holds for *arbitrary* systems, baths, their interactions, and temperatures. This follows from (6.15) and (6.16a), whereby to the lowest nonvanishing order

$$\langle H_{\text{SB}}\rangle_{\text{Eq}} = -\text{Tr}(\rho_0 S_1 H_{\text{SB}}) = -\beta \int_0^1 dx\ \text{Tr}[\rho(x)A(x)]. \tag{6.46}$$

Here $\rho(x) = e^{-(1-x)\beta H_0}/Z_0$ is an (unnormalized) density matrix and $A(x) = H_{\text{SB}}e^{-x\beta H_0}H_{\text{SB}}$ is a positive operator, since $e^{-x\beta H_0}$ is positive and H_{SB} is Hermitian. Hence, the trace under the integral is nonnegative, which proves the inequality (6.45).

6.3 System Evolution toward Equilibrium

We here investigate the evolution of the system toward equilibrium following the quench caused by turning on H_{SB} at $t = t_0$. In the notation of Section 6.1, the system–bath coupling Hamiltonian in the interaction picture has the form

$$\begin{aligned}
H_{\text{I}}(t) &\equiv e^{iH_0(t-t_0)}H_{\text{SB}}e^{-iH_0(t-t_0)} \\
&= \hbar[\sigma_+ e^{-i\omega_{\text{a}}(t-t_0)} + \sigma_- e^{i\omega_{\text{a}}(t-t_0)}] \\
&\quad \times \sum_k [\eta_k a_k e^{-i\omega_0(t-t_0)} + \eta_k^* a_k^\dagger e^{i\omega_0(t-t_0)}].
\end{aligned} \tag{6.47}$$

This Hamiltonian causes the combined system–bath state to evolve as

$$|\Psi_I(t)\rangle \equiv e^{-iH_0(t-t_0)}|\Psi(t)\rangle = U(t,t_0)|\Psi(t_0)\rangle. \tag{6.48}$$

The *exact* evolution operator $U(t, t_0)$ can be expanded in a series,

$$U(t, t_0) = 1 + \sum_{m=1}^{\infty} (-i)^m \int_{t_0}^{t} dt_1 \int_{t_0}^{t_1} dt_2 \ldots \int_{t_0}^{t_{m-1}} dt_m \, H_I(t_1) H_I(t_2) \ldots H_I(t_m)$$

$$\equiv \sum_{m=0}^{\infty} O_m(t). \tag{6.49}$$

Consider first the initial state $|\Psi(t_0)\rangle = |\mathbf{n}\rangle \otimes |g\rangle \equiv |\mathbf{n}, g\rangle$, where we denote the joint eigenstates of H_B and the total number operator $\hat{N}$ of the multimode bath quanta by

$$|\mathbf{n}\rangle \equiv \mathcal{N}_{\mathbf{n}} \prod_j (a_{k_j}^{\dagger})^{n_j} |0\rangle, \tag{6.50}$$

Here j labels the modes populated by n_j quanta, $|\mathbf{n}\rangle \equiv |\{n_j\}_j\rangle$ is the $\hat{N}$ eigenstate with eigenvalue n_{tot}, and $\mathcal{N}_{\mathbf{n}}$ is the normalization constant.

Each factor $H_I(t_i)$ in the perturbation expansion of $|\Psi_I(t)\rangle$, given by (6.48) and (6.49), has the effect of flipping the TLS state and changing the bath state into a sum of states with one more or one less excitation. Therefore, the even terms $O_{2m}(t)|\Psi(t_0)\rangle$ (counting the 1 as the zeroth term!) are superpositions of states with excitation numbers n_{tot}+even, all associated with the TLS state $|g\rangle$, whereas the odd ones correspond to the state $|e\rangle$. The resulting state initiated as $|\mathbf{n}, g\rangle$ is then

$$|\Psi_I(t)\rangle = U(t, t_0)|\mathbf{n}, g\rangle = |B_{\mathbf{n},g}^{\text{even}}(t)\rangle \otimes |g\rangle + |B_{\mathbf{n},g}^{\text{odd}}(t)\rangle \otimes |e\rangle \equiv |\Psi_{\mathbf{n},g}(t)\rangle. \tag{6.51}$$

Here, $B_{\mathbf{n},g}^{\text{even}}$ (respectively, $B_{\mathbf{n},g}^{\text{odd}}$) is a sum of $\hat{N}$-eigenstates $|\mathbf{n}'\rangle$, where $\mathbf{n}' = \{n_j'\}_j$, with n_j' differing from n_j by even (respectively, odd) numbers.

Similarly, if the initial state is $|\Psi(t_0)\rangle = |\mathbf{n}, e\rangle$, it evolves to

$$|\Psi_I(t)\rangle = U(t, t_0)|\mathbf{n}, e\rangle = |B_{\mathbf{n},e}^{\text{even}}(t)\rangle \otimes |e\rangle + |B_{\mathbf{n},e}^{\text{odd}}(t)\rangle \otimes |g\rangle \equiv |\Psi_{\mathbf{n},e}(t)\rangle. \tag{6.52}$$

Now let us consider that the system and bath are initially in a factorized Gibbs state of H_0, at inverse temperature β:

$$\rho_{\text{tot}}(t_0) = Z_{\text{tot}}^{-1} e^{-\beta H_0} = Z_S^{-1} e^{-\beta H_S} Z_B^{-1} e^{-\beta H_B} \tag{6.53}$$

$$= Z_{\text{tot}}^{-1} \sum_{\mathbf{n}} \left[e^{-\beta(\omega_g + \omega_{\mathbf{n}})} |\mathbf{n}, g\rangle\langle \mathbf{n}, g| \right.$$

$$\left. + e^{-\beta(\omega_e + \omega_{\mathbf{n}})} |\mathbf{n}, e\rangle\langle \mathbf{n}, e| \right]. \tag{6.54}$$

This initially factorized state evolves at time $t \gg t_0$ into:

$$\rho_{\text{tot}}(t) = Z_{\text{tot}}^{-1} \sum_{\mathbf{n}} \left[e^{-\beta(\omega_g + \omega_{\mathbf{n}})} |\Psi_{\mathbf{n},g}(t)\rangle\langle \Psi_{\mathbf{n},g}(t)| + e^{-\beta(\omega_e + \omega_{\mathbf{n}})} |\Psi_{\mathbf{n},e}(t)\rangle\langle \Psi_{\mathbf{n},e}(t)| \right].$$

$$\tag{6.55}$$

The resulting ρ_{tot} has acquired *off-diagonal* $|e\rangle\langle g|$ and $|g\rangle\langle e|$ elements (coherences) by virtue of (6.51) and (6.52).

Let us now assume that (despite the *initial* degeneracy of the eigenstates), adiabatic switching-on of H_{SB} allows the adiabatic theorem to be applied. We then find

$$\rho_{\text{tot}}(t \to \infty) = \sum_{\mathbf{n}} \sum_{m=g,e} e^{-\beta E_{\mathbf{n},m}} |\Psi_{\mathbf{n},m}(t)\rangle\langle\Psi_{\mathbf{n},m}(t)|, \qquad (6.56)$$

$E_{\mathbf{n},m}$ being the eigenvalues of H_0. This state has a Gibbs-like form, albeit with the original Boltzmann weights. Yet, these weights are irrelevant for the state parity.

For $\rho_{\text{tot}}(t) = |\Psi_{\mathbf{n},g}(t)\rangle\langle\Psi_{\mathbf{n},g}(t)|$, [Eq. (6.51)], we then obtain

$$(\rho_S)_{eg}(t) = \langle e|\rho_S(t)|g\rangle = \text{Tr}_B\langle e|\rho_{\text{tot}}(t)|g\rangle = \langle B_{\mathbf{n},e}^{\text{even}}(t)|B_{\mathbf{n},e}^{\text{odd}}(t)\rangle = 0. \qquad (6.57)$$

Thus, ρ_S is diagonal at any time t under adiabatic turn-on of the interaction.

The combined (system- and bath-) *equilibrium state satisfies*:

$$\rho_{\text{tot}} = Z^{-1} e^{-\beta H_{\text{tot}}} = Z^{-1} e^{-\beta[H_0 + O(H_{\text{SB}}^2)]}. \qquad (6.58)$$

Thus, for sufficiently weak coupling, the equilibrium state is separable. However, the inseparable (entangling) contribution to the system–bath combined state grows with their coupling.

6.4 Discussion

The above analysis shows that a finite (albeit small) coupling strength of the system to the bath, which is *necessary for their equilibration*, is incompatible with the notion of a thermal (Gibbs) state for the system and the bath separately: they become non-separable (entangled) and *nonadditive in the thermodynamic sense*. The quench resulting from turning on the system–bath coupling strength results in finite mixedness of an initially pure state of the system, even at zero temperature. Its equilibration following the quench is *non-monotonic* provided it is non-Markovian.

7

Bath-Induced Collective Dynamics

The collective coupling of a multipartite system to a single field mode is shown to yield cooperative oscillations. If the collective coupling is to a multimode bath, the result is cooperative decay of system oscillations. In either case, angular-momentum symmetry determines the features of these processes, which arise from quantum interferences among many-atom interactions with the field.

7.1 Collective TLS Coupling to a Single Field Mode

We here consider the simplest variant of a collective multiatom (spin) model: an ensemble of individual TLS coupled to a single-mode cavity, known as the Tavis–Cummings model. In the dipolar approximation and the RWA, the model Hamiltonian can be written as

$$H = \hbar\omega_c b^\dagger b + \frac{\hbar}{2}\sum_{j=1}^{N}\omega_j\sigma_j^z + \hbar\sum_{j=1}^{N}\left(\eta_j^*\sigma_j^- b^\dagger + \eta_j\sigma_j^+ b\right), \qquad (7.1)$$

where the bosonic creation (annihilation) operators $b^\dagger$ (b) describe the field of the cavity mode with frequency ω_c. The Pauli spin operators $\sigma_j^{\pm,z}$ correspond to the jth atom (spin) with resonant frequency ω_j and excited (e_j) and ground (g_j) states, as in (4.8).

The collective spin operators of N identical spins are

$$\hat{J}+ = \frac{\sum_{j=1}^{N}\eta_j\sigma_j^+}{\sqrt{\sum_{i=1}^{N}|\eta_i|^2}}, \qquad \hat{J}- = \frac{\sum_{j=1}^{N}\eta_j^*\sigma_j^-}{\sqrt{\sum_{i=1}^{N}|\eta_i|^2}}. \qquad (7.2)$$

The $\hat{J}+$ operator generates the superradiant ("bright") state acting on the collective N-spin ground state $|G\rangle$,

$$|R\rangle = \hat{J}^+ |G\rangle . \tag{7.3}$$

All other $(N - 1)$ collective states $|S\rangle$ in the single-excitation subspace are subradiant ("dark").

If the cavity mode and the spin ensemble are resonantly coupled (i.e., *all* $\omega_j = \omega_c$), then two polaritonic states arise,

$$|\pm\rangle = (|1\rangle_c |G\rangle \pm |0\rangle_c |R\rangle)/\sqrt{2}, \tag{7.4}$$

$|1\rangle_c$ and $|0\rangle_c$ denoting the single- and zero-photon states of the cavity mode. These symmetric and antisymmetric polaritonic states $|\pm\rangle$ are maximally entangled states of (0 and 1) cavity photons and the collective spin-ensemble states $|G\rangle$ and $|R\rangle$.

The energies of these hybridized spin-photon states are separated by

$$\hbar\Omega_R = 2\hbar \sqrt{\sum_{j=1}^{N} |\eta_j|^2}. \tag{7.5}$$

Here the vacuum Rabi splitting $\hbar\Omega_R$ that corresponds to the superradiant collectively enhanced interaction scales approximately as $\sqrt{N}$.

7.1.1 Inhomogeneously Broadened Tavis–Cummings Dynamics

The $N - 1$ degenerate (dark) subradiant states $|S\rangle$ described in Section 7.1 remain uncoupled from the vacuum and external fields. An excitation that is stored in the subradiant space will therefore be protected from dissipation in the cavity mode. However, this protection may fail in the presence of inhomogeneous spin broadening: the polariton modes and the subradiant states are then no longer decoupled from each other.

In general, inhomogeneously broadened large spin ensembles are describable by a normalized spectral density that has a continuum limit,

$$g(\omega) = \sum_{j=1}^{N} |\eta_j|^2 \delta(\omega - \omega_j) \Bigg/ \sum_{j=1}^{N} |\eta_j|^2 . \tag{7.6}$$

Their coupling with a bath may be enhanced by a factor of $\sqrt{N}$ as compared to the single-spin coupling η_j, allowing for the strong-coupling regime in the case of sufficiently large N.

Let us consider the effects of inhomogeneous broadening for a spin ensemble (SE) of $N \gg 1$ spins in a superposition of its collective ground state

$$|G\rangle = |g_1, g_2, \ldots, g_N\rangle \tag{7.7}$$

and the fully symmetrized single-excitation (Dicke) state as in (7.3),

Bath-Induced Collective Dynamics

$$|\Psi_e(0)\rangle = |R\rangle = N^{-1/2} \sum_j |j\rangle, \tag{7.8}$$

$|j\rangle = |g_1, g_2, \ldots, e_j, \ldots, g_N\rangle$ denoting a state with only spin j excited. If all the spins had the same resonant frequency ω_a, the single-excitation state would evolve in time $\tau > 0$ as

$$|\Psi_e(\tau)\rangle = N^{-1/2} \sum_j e^{-i\omega_a\tau}|j\rangle, \tag{7.9}$$

while $|G\rangle$ would remain unchanged. Yet, under inhomogeneous broadening, the initially symmetric state $|\Psi_e\rangle$ would evolve into the *asymmetric* state

$$|\tilde{\Psi}_1(\tau)\rangle = N^{-1/2} \sum_j e^{-i\omega_j\tau}|j\rangle. \tag{7.10}$$

The squared overlap of the state $|\Psi_e(\tau)\rangle$ with its inhomogeneously broadened counterpart, $|\tilde{\Psi}_1(\tau)\rangle$, defines the storage fidelity at time $\tau \geq 0$, given by

$$\Phi_I(\tau) \equiv |\langle\Psi_e(\tau)|\tilde{\Psi}_1(\tau)\rangle|^2 = \left|\frac{1}{N}\int \rho(\omega)e^{-i(\omega-\omega_a)\tau}d\omega\right|^2. \tag{7.11}$$

Here $\rho(\omega) = \sum_j^N \delta(\omega-\omega_j)$ is the spectral density of the TLS ensemble normalized to the total number of spins, $\int \rho(\omega)d\omega = N$. The function $\Phi_I(\tau)$, which is akin to the memory kernel of an oscillator bath response [(5.7)], quantifies the loss of fidelity of a state *encoded* in the SE, starting with $\Phi_I(0) = 1$ for any $\rho(\omega)$. For an ensemble with Lorentzian spectrum of width Δ,

$$\rho(\omega) = \frac{N/(\pi\Delta)}{1 + (\omega - \omega_a)^2/\Delta^2}, \tag{7.12}$$

the fidelity loss is exponential in time, $\Phi_I(\tau) = e^{-2\Delta\tau}$.

Let us assume that initially the cavity field has a single excitation (photon) at frequency $\omega_c = \omega_a$, ω_a being the central resonance frequency of the SE, and the SE is in the ground state $|G\rangle$. In the RWA, their combined state at time τ is given by

$$|\Psi(\tau)\rangle = \alpha(\tau)|1, G\rangle + \sum_j \beta_j(\tau)e^{-i(\omega_j-\omega_a)\tau}|0, j\rangle. \tag{7.13}$$

Here $|1, G\rangle$ denotes the state with a single photon and all the spins in the ground state, and $|0, j\rangle$ is the confined state of the field vacuum and the jth spin excited. According to the Schrödinger equation, the probability amplitudes α and β_j evolve in time as

$$\dot{\alpha} = -i\sum_j \eta_j\beta_j e^{-i(\omega_j-\omega_a)\tau},$$

$$\dot{\beta}_j = -i\eta_j^*\alpha e^{i(\omega_j-\omega_a)\tau}. \tag{7.14}$$

Upon integrating the second equation and substituting the result for β_j in the first equation, the following integrodifferential equation is obtained for α,

$$\dot{\alpha} = -\sum_j |\eta_j|^2 \int_0^\tau d\tau' \alpha(\tau') e^{-i(\omega_j - \omega_a)(\tau - \tau')}, \tag{7.15}$$

where we have assumed that η_j and ω_j are uncorrelated with each other.

When the spin-field couplings η_j are not correlated with the transition frequencies ω_j, we have

$$g(\omega) = N^{-1}\rho(\omega). \tag{7.16}$$

We then finally obtain

$$\dot{\alpha} = -\bar{\eta}^2 N \int_0^\tau d\tau' \alpha(\tau') F(\tau - \tau'), \tag{7.17}$$

where $\bar{\eta}^2 \equiv N^{-1} \sum_{j=1}^N |\eta_j|^2$ is the average of $|\eta_j|^2$ and

$$F(\tau) = N^{-1} \sum_{j=1}^N e^{-i(\omega_j - \omega_a)\tau} = \langle \Psi_e(\tau)|\tilde{\Psi}_1(\tau)\rangle, \tag{7.18}$$

so that the fidelity $\Phi_I(\tau) = |F(\tau)|^2$ [cf. (7.11)].

With $\omega_j = \omega_a$, we have $\Phi_I(\tau) = F(\tau) = 1$ for all τ, and Eq. (7.17) predicts Rabi oscillations between the field and the SE in the form

$$\alpha(\tau) = \cos(\bar{\eta}\sqrt{N}\tau). \tag{7.19}$$

Then, at time $\tau_{tr} = \pi/(\bar{\eta}\sqrt{N})$ there is a full retrieval of the excitation into the field, $|\alpha(\tau_{tr})|^2 = 1$, with the combined state $|\Psi(\tau_{tr})\rangle = -|1, G\rangle$ acquiring a π-phase shift. By contrast, inhomogeneous broadening causes $\Phi_I(\tau)$ to undergo damped Rabi oscillations. The transfer fidelity Δ at the excitation retrieval time is the value of $|\alpha(\tau_{tr})|^2$.

7.2 Cooperative Decay of N Driven TLS

Here we consider an ensemble of N field-driven TLS governed by the Hamiltonian

$$H_S(t) = \frac{\hbar\omega_a}{2} \sum_{j=1}^N \sigma_j^z + \hbar\Omega \sum_{j=1}^N \left(\sigma_j^+ e^{-i\nu t} + \sigma_j^- e^{i\nu t}\right), \tag{7.20}$$

2Ω and ν being the Rabi and the carrier frequencies of the driving field, respectively. The atoms are coupled to a bosonic bath (e.g., a photonic bath), as in (4.10), via a Hamiltonian written here without the RWA,

$$H_{SB} = \hbar \sum_{j=1}^{N} \sigma_j^x \otimes B,$$ (7.21)

where

$$B = \sum_{k} \eta_k (b_k + b_k^\dagger).$$ (7.22)

The system-bath coupling in (7.21) via σ_x operators that are off-diagonal in the TLS energy basis allows for excitation and deexcitation exchange between the atoms and the bath. As in the Tavis–Cummings model, in the absence of inhomogeneous broadening (Sec. 7.1), here too all atoms are indistinguishable and are *identically* coupled to the bath, which leads to *cooperative exchange* between the atomic ensemble and the bath.

By transforming the Hamiltonian (7.20) to the frame rotating with the field-mode frequency ν, we remove its time dependence and obtain

$$\bar{H}_S = -\hbar(\nu - \omega_a) \sum_{j=1}^{N} \sigma_j^z + \hbar\Omega \sum_{j=1}^{N} \left(\sigma_j^+ + \sigma_j^- \right).$$ (7.23)

We then convert the Hamiltonian to the collective (multitatom) basis by introducing the (dimensionless) angular-momentum operators $\hat{\boldsymbol{J}} = (\hat{J}_x, \hat{J}_y, \hat{J}_z)$ with

$$\hat{J}_i := \sum_{j=1}^{N} \frac{1}{2} \sigma_j^i \qquad (i \in \{x, y, z\})$$ (7.24)

or, equivalently,

$$\hat{J}_- := \sum_{j=1}^{N} \sigma_j^-$$ (7.25a)

$$\hat{J}_+ := \sum_{j=1}^{N} \sigma_j^+.$$ (7.25b)

The Hamiltonian (7.23) is then expressed in terms of these collective operators as

$$\bar{H}_S = -\hbar(\nu - \omega_a) \hat{J}_z + 2\hbar\Omega \hat{J}_x,$$ (7.26)

where the factor of 2 follows from the relation

$$\hat{J}_x = \frac{1}{2}(\hat{J}_+ + \hat{J}_-).$$ (7.27)

The advantage of this basis change is that the Hamiltonian (7.26) becomes block-diagonal,

$$\bar{H}_S = \bigoplus_{k=1} \bar{H}_k.$$ (7.28)

This block-diagonal decomposition, known as the *Dicke decomposition*, follows from spin addition: The Hilbert space of N spin-1/2 particles can be decomposed into irreducible subspaces, each corresponding to an eigenvalue $J(J + 1)$ (with J ranging over $0, 1, \ldots, \frac{N}{2}$ for N even and $\frac{1}{2}, \frac{3}{2}, \ldots \frac{N}{2}$ for N odd). This decomposition of the total angular momentum operator $\hat{\boldsymbol{J}}$ is not unique, since different irreducible subspaces may correspond to the same J.

Each of the Dicke irreducible subspaces has the dimensionality $2J + 1$. For example, two spin-1/2 particles form a triplet ($J = 1$) and a singlet ($J = 0$), whereas three spin-1/2 particles form a quadruplet ($J = \frac{3}{2}$) and two doublets ($J = \frac{1}{2}$).

Every irreducible sub-Hamiltonian $\bar{H}_k$ describes an effective spin-$J(k)$ particle. We may assume that initially the atoms belong to a certain irreducible subspace $\mathcal{H}_k$. Then we effectively have a single (fictitious) particle, with spin $S := J(k)$, and restrict our analysis to the $(2S + 1)$-dimensional Hilbert space $\mathcal{H}_k \subseteq \mathcal{H}$.

If system is initially in a *symmetric state*, such as the collective ground state $|G\rangle$, then the dynamics is described by a single "giant" spin $N/2$, giving rise to Dicke *superradiance*. Namely, the decay rate with maximal cooperative enhancement, γN, is induced by H_{SB} [Eqs. (7.20), (7.22)] from the highest to the lowest symmetric state, instead of the single-spin rate γ.

7.3 Multiatom Cooperative Emission Following Single-Photon Absorption

The Dicke basis discussed above is a powerful systematic means of classifying cooperative effects in multiatom spontaneous emission. The difficulty is the large multiplicity of Dicke states for $N \gg 1$ and their contrasting symmetry properties. Nevertheless, group theory allows us to answer the questions: Which Dicke states become populated during the process of cooperative decay (collective spontaneous emission)? How are their symmetry properties reflected in the spatiotemporal buildup of such emission?

Specifically, we consider *the reemission of a single photon* with the wave vector $\boldsymbol{k}_0$ after its absorption by N *randomly distributed* atoms, initially in their ground state. The absorbed photon frequency ck_0 is at resonance with the atomic transition frequency ω_a. The spatial distribution of the atoms is spherically symmetric, obeying a Gaussian distribution of random atomic positions $\boldsymbol{r}_j$ within a sphere of radius R_0,

$$\mathcal{P}(\boldsymbol{r}) = (\sqrt{\pi} R_0)^{-3} \exp(-r^2/R_0^2). \tag{7.29}$$

Extending the atom–field interaction Hamiltonian in Section 4.1.2 to the case of N atoms at positions $\boldsymbol{r}_j$, we obtain, instead of (4.12),

$$H_{\rm I} = \hbar \sum_{k,j} (\eta_k e^{ik\cdot r_j} \sigma_j^+ a_k + \eta_k^* e^{-ik\cdot r_j} \sigma_j^- a_k^\dagger), \tag{7.30}$$

neglecting the effects of photon polarization. The wave function of the multiatom system combined with the field is expanded as follows:

$$|\Psi(t)\rangle = \sum_{j=1}^{N} \alpha_j(t) e^{-i\omega_{\rm a} t} |e_j\rangle + \sum_k \beta_k(t) e^{-ickt} |k\rangle. \tag{7.31}$$

Here in the first summation $|e_j\rangle$ denotes the excited jth atom, all other atoms are in the ground state, and the electromagnetic field is in the vacuum state, whereas in the second summation $|k\rangle$ denotes all the atoms being in the ground state and one photon in the k mode. The initial amplitudes in (7.31) are

$$\alpha_j(0) = \frac{\exp(ik_0 \cdot r_j)}{\sqrt{N}}, \qquad \beta_k(0) = 0. \tag{7.32}$$

The Schrödinger equation then gives rise to the following coupled equations [analogous to (7.14)]

$$i\dot{\alpha}_j = \sum_k \eta_k e^{ik\cdot r_j} \beta_k, \tag{7.33}$$

$$i\dot{\beta}_k = (ck - \omega_{\rm a})\beta_k + \sum_{j=1}^{N} \eta_k^* e^{-ik\cdot r_j} \alpha_j. \tag{7.34}$$

Upon integrating (7.34) and substituting the result in (7.33), we obtain integro-differential equations for α_j. Then, in the weak-coupling approximation, wherein α_j decays slowly on the scale of the multiatom correlation (cooperation) time, we pull α_j out of the time integral. We thus convert (7.33) into the differential equations,

$$\dot{\alpha}_j = -\sum_{j'=1}^{N} \gamma_{jj'}(t)\alpha_{j'}, \tag{7.35}$$

where

$$\gamma_{jj'}(t) = \sum_k |\eta_k|^2 e^{ik(r_j - r_{j'})} \left[\frac{\sin(\omega_{\rm a} - ck)t}{\omega_{\rm a} - ck} - i\frac{\cos(\omega_{\rm a} - ck)t - 1}{\omega_{\rm a} - ck} \right]. \tag{7.36}$$

The second term in the square brackets in (7.36) represents the atomic-transition Lamb shift. It is cooperative and thus different for each of the N collective excited states formed by the orthogonal linear combinations of $|e_j\rangle$. Since the finite radius R_0 of the sphere corresponds to a wave-vector spread $\Delta k \sim R_0^{-1}$, the ultraviolet divergence of the Lamb shift for a single atom is removed. The cooperative Lamb shift is then $\sim k_0 R_0$ times smaller than the corresponding decay rate, so that the *variance* of the Lamb shift over the space of collective states is negligible. In what follows, we incorporate the Lamb shift into the transition frequency ω_a. Then, only the real $\frac{\sin(\omega_a - ck)t}{\omega_a - ck}$ term in (7.36) is retained. For $t \gg \omega_a^{-1}$ it is approximated by

$$\frac{\sin(\omega_a - ck)t}{\omega_a - ck} \approx \pi \delta(\omega_a - ck), \quad \boldsymbol{k} \approx \hat{\boldsymbol{k}}(k_0 + \delta k), \quad \delta k \ll k_0, \tag{7.37}$$

$\hat{\boldsymbol{k}}$ being a unit vector. Under this long-time approximation, (7.36) is converted into

$$\gamma_{jj'}(t) \approx \int \frac{d\Omega_{\hat{k}}}{4\pi} \pi |\eta_k|^2 \varrho(ck)|_{k=\hat{k}k_0} e^{ik_0\hat{k}\cdot(\boldsymbol{r}_j - \boldsymbol{r}_{j'})} \theta[ct - |\hat{\boldsymbol{k}} \cdot (\boldsymbol{r}_j - \boldsymbol{r}_{j'})|], \tag{7.38}$$

where $\varrho(ck)$ is the density of photon modes and $\theta(x)$ is the Heaviside step function.

If ct is much smaller than the mean interatomic distance, then the decay matrix $\gamma_{jj'}$ is diagonal, all its diagonal elements being equal to the single-atom decay rate,

$$\gamma_1 = \int \frac{d\Omega_{\hat{k}}}{4\pi} \pi |\eta_k|^2 \varrho(ck)|_{k=\hat{k}k_0}. \tag{7.39}$$

At $t \gg R_0/c$, the radiation becomes fully collective, and (7.35), (7.36) yield

$$\dot{\alpha}_j = -\gamma_1 \sum_{j'=1}^{N} \frac{\sin k_0 |\boldsymbol{r}_j - \boldsymbol{r}_{j'}|}{k_0 |\boldsymbol{r}_j - \boldsymbol{r}_{j'}|} \alpha_{j'}, \quad t \gg R_0/c. \tag{7.40}$$

The eigenstates of (7.40) can be approximated by including the phase factors associated with the incident photon in the definition of the excited states,

$$e^{i\boldsymbol{k}_0\cdot\boldsymbol{r}}|e_j\rangle \rightarrow |e_j\rangle, \tag{7.41}$$

thus transforming the probability amplitudes to

$$\tilde{\alpha}_j \equiv e^{-i\boldsymbol{k}_0\cdot\boldsymbol{r}} \alpha_j. \tag{7.42}$$

Then, (7.40) assumes the form

$$\dot{\tilde{\alpha}}_j = -\gamma_1 \sum_{j'=1}^{N} \mathcal{F}(\boldsymbol{r}_j - \boldsymbol{r}_{j'}) \tilde{\alpha}_{j'}, \tag{7.43}$$

where

$$\mathcal{F}(\boldsymbol{r}_j - \boldsymbol{r}_{j'}) = \frac{\sin k_0 |\boldsymbol{r}_j - \boldsymbol{r}_{j'}|}{k_0 |\boldsymbol{r}_j - \boldsymbol{r}_{j'}|} e^{-i\boldsymbol{k}_0\cdot(\boldsymbol{r}_j - \boldsymbol{r}_{j'})}. \tag{7.44}$$

A single excitation acting on the N-atom ground state $|G\rangle$, which is totally symmetric with respect to generalized permutations, yields the totally symmetric state with one atom excited,

$$|\phi^{\{N\}}\rangle = \frac{1}{\sqrt{N}} \sum_{j=1}^{N} |e_j\rangle, \tag{7.45}$$

provided this excitation does not distinguish between the atoms.

The *cooperative decay rate of the fully symmetric state* is then, according to (7.44),

$$\gamma_{\text{col}} = \frac{\gamma_1}{N} \sum_{j=1}^{N} \sum_{j'=1}^{N} \mathcal{F}(\mathbf{r}_j - \mathbf{r}_{j'}). \tag{7.46}$$

Since $N \gg 1$, the double sum in (7.46) amounts to N^2 times the average over the atomic positions $\mathbf{r}_j$, $\mathbf{r}_{j'}$:

$$\gamma_{\text{col}} = \gamma_1 N \int d^3 \mathbf{r}_j \int d^3 \mathbf{r}_{j'} \, \mathcal{P}(\mathbf{r}_j) \mathcal{P}(\mathbf{r}_{j'}) \mathcal{F}(\mathbf{r}_j - \mathbf{r}_{j'}). \tag{7.47}$$

Here we have assumed that there are no correlations between the positions of different atoms, as in gases, so that the integral is over a product of single-atom probability distributions. We evaluate the integral in (7.47), recalling that

$$\frac{\sin k_0 |\mathbf{r}_j - \mathbf{r}_{j'}|}{k_0 |\mathbf{r}_j - \mathbf{r}_{j'}|} = \int \frac{d\Omega_{\hat{k}}}{4\pi} e^{-ik_0 \hat{k} \cdot (\mathbf{r}_j - \mathbf{r}_{j'})}, \tag{7.48}$$

where $\hat{k}$ is uniform over the sphere. Then, assuming that the sample is large enough,

$$k_0 R_0 \gg 1, \tag{7.49}$$

we obtain

$$\gamma_{\text{col}} = \gamma_1 N (k_0 R_0)^{-2}. \tag{7.50}$$

The factor $(k_0 R_0)^{-2}$ is the effective solid angle of the collective forward emission of a photon. This collective process prevails over the incoherent scattering by individual atoms provided $N(k_0 R_0)^{-2} \gg 1$.

7.3.1 Symmetry Considerations

From the N linearly independent states $|e_j\rangle$, we construct, by orthogonalization to $|\phi^{\{N\}}\rangle$, the $N - 1$ states

$$|\phi_l^{\{N-1\}}\rangle = -\frac{1}{\sqrt{N}} |e_N\rangle + \sum_{j=1}^{N-1} \left[\frac{1 + (1/\sqrt{N})}{N - 1} - \delta_{jl} \right] |e_j\rangle$$

$$\equiv \sum_{j=1}^{N} f_j^l |e_j\rangle, \qquad l = 1, 2, \ldots, N - 1. \tag{7.51}$$

The states (7.51) are normalized to 1 and orthogonal to each other and to $|\phi^{\{N\}}\rangle$. They comprise the basis of the irreducible representation characterized by the Young tableau $\{N - 1, 1\}$.

The wave function of the singly excited multiatom state can be expanded into a sum of states comprising the fully (N-fold) symmetric state and the lower ($N-1$)-fold symmetry states,

$$|\psi_{\text{exc}}\rangle = c^{\{N\}}|\phi^{\{N\}}\rangle + \sum_{l=1}^{N-1} c_l^{\{N-1\}}|\phi_l^{\{N-1\}}\rangle, \tag{7.52}$$

yielding the set of equations for their probability amplitudes,

$$\dot{c}^{\{N\}} = -\gamma_{\text{col}} c^{\{N\}} - \sum_{l=1}^{N-1} s_l^* c_l^{\{N-1\}}, \tag{7.53}$$

$$\dot{c}_l^{\{N-1\}} = -s_l c^{\{N\}} - \sum_{l'=1}^{N-1} Q_{ll'} c_{l'}^{\{N-1\}}, \tag{7.54}$$

with the initial conditions

$$\dot{c}^{\{N\}}(0) = 1, \qquad c_l^{\{N-1\}}(0) = 0, \quad l = 1, 2, \ldots N - 1. \tag{7.55}$$

The mixing in (7.53), between the fully symmetric state and the lth state of lower symmetry, occurs at the rate

$$s_l = \frac{\gamma_1}{\sqrt{N}} \sum_{j=1}^{N} \sum_{j'=1}^{N} f_j^l \mathcal{F}(\mathbf{r}_j - \mathbf{r}_{j'}). \tag{7.56}$$

The rates

$$Q_{ll'} = \gamma_1 \sum_{j=1}^{N} \sum_{j'=1}^{N} f_j^l f_{j'}^{l'} \mathcal{F}(\mathbf{r}_j - \mathbf{r}_{j'}) \tag{7.57}$$

in (7.54) characterize the decay of the lth lower-symmetry state if $l = l'$, or the mixing of such states if $l \neq l'$ (Fig. 7.1).

Here, to an accuracy $\sim 1/\sqrt{N}$,

$$f_j^l \approx N^{-1} - \delta_{jl}, \tag{7.58}$$

and

$$s_l = -\gamma_1 \sqrt{N} \left[\int d^3 \mathbf{r}_j \, \mathcal{P}(\mathbf{r}_j) \mathcal{F}(\mathbf{r}_l - \mathbf{r}_j) \right. \\ \left. - \int d^3 \mathbf{r}_l \int d^3 \mathbf{r}_j \, \mathcal{P}(\mathbf{r}_l) \mathcal{P}(\mathbf{r}_j) \mathcal{F}(\mathbf{r}_l - \mathbf{r}_j) \right]. \tag{7.59}$$

Straightforward calculations using (7.48) yield, in the large-sample limit of (7.49), the low-symmetry probability amplitudes,

$$c_l^{\{N-1\}}(t) = \frac{2i\mathbf{k}_0 \cdot \mathbf{r}_l}{\sqrt{N}(k_0 R_0)^2} \left(1 - e^{-\gamma_{\text{col}} t}\right). \tag{7.60}$$

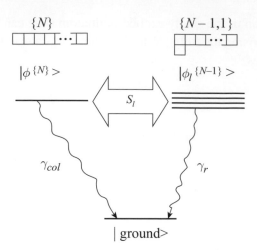

Figure 7.1 *N*-atom cooperative emission following single-photon absorption. The coupled states involved in the process and their respective Young tableaus are shown. (Reproduced with permission from Mazets and Kurizki, 2007. © IOP Publishing. All rights reserved)

These amplitudes are non-negligible only in the mesoscopic regime, where $\sqrt{N}(k_0 R_0)^2$ is not excessively large.

7.4 Discussion

Our analysis has provided insights into the space- and time-cooperativity in spontaneous emission following the absorption of a single photon at $t = 0$. It has revealed the dominance *of the symmetric Dicke state*, in the long-time and large-sample regime. Cooperative states of lower symmetry become mixed with the symmetric state as the decay process unfolds and contribute to slower emission rate ("afterglow") if the sample is mesoscopic, say $k_0 R_0 \lesssim 10$. The entire emission process occurs in the forward preferred direction.

8

Bath-Induced Self-Energy: Cooperative Lamb-Shift and Dipole–Dipole Interactions

Here we are concerned with the interaction (self-energy) that stems from the cooperative coupling of atomic dipoles to a common bath of photonic modes. When the interaction is mediated by *real photons*, it constitutes a dissipative, incoherent process, whereby the generation of entanglement among the atoms is generally probabilistic. Nonradiative (dispersive) interactions are, by contrast, enabled by evanescent fields that are described as an exchange of *virtual (i.e, nonresonant) photons* between the atoms, known as the resonant dipole–dipole interaction (RDDI). In free space, RDDI is dominant over real-photon radiation only at distances shorter than the resonant wavelength. By contrast, photonic cutoffs or bandgaps in field-confining structures (cavities, waveguides or photonic crystals – Chs. 3–5) can suppress this radiation and enhance RDDI or vice versa. In particular, such confining structures may give rise to *coherent* non-dissipative interaction mediated by RDDI at much longer ranges than allowed by free-space RDDI. Such a process would constitute a route towards *high-fidelity long-range deterministic entanglement*. The distance-dependence of the interactions is determined by the geometry-dependent field modes, populated by virtual photons.

8.1 Markovian Theory of Two-Atom Self-Energy

8.1.1 The Model

We consider a pair of identical TLS with energy levels $|g\rangle$ and $|e\rangle$ and transition frequency ω_a. These TLS are coupled to the vacuum field of the structure (photonic bath), and we neglect their coupling to modes outside the structure. The TLS–bath dipole couplings are $\eta_{kj} = -\wp_{eg} \cdot \phi_k(r_j)$, $\wp_{eg}$ denoting [cf. (4.7)] the dipole matrix element of the $|g\rangle \leftrightarrow |e\rangle$ transition (taken to be real), ω_k and $\phi_k(r)$ being, respectively, the kth mode frequency and the spatial function at r_j, the location of the jth TLS ($j = 1, 2$). The interaction Hamiltonian in the dipole approximation has the following form (without the RWA) in the interaction picture,

$$H_I = \hbar \sum_{j=1}^{2} \sum_{k} \left(\eta_{kj} \hat{a}_k e^{-i\omega_k t} + \text{H.c.} \right) \left(\hat{\sigma}_j^- e^{-i\omega_a t} + \text{H.c.} \right), \quad (8.1)$$

where $\hat{a}_k$ and $\hat{\sigma}_j^-$ are the field-mode and the TLS lowering operators, respectively.

The two-atom dissipative linewidth and dispersive energy-shift, $\gamma_{jj'}$ and $\Delta_{jj'}$, respectively, are here calculated in the weak-coupling limit to second order in the coupling strength. To this end, we evaluate the two-atom transition amplitude from the state where atom j is excited to the state where atom j' is excited, which is then given by

$$U_{jj'} = \delta_{jj'} + U_{jj'}^{(2)},$$

$$U_{jj'}^{(2)} = \frac{1}{2\pi i} \int_{-t/2}^{t/2} dt_1 \int_{-t/2}^{t/2} dt_2 \int_{-\infty}^{\infty} d\omega\, e^{i(\omega_a - \omega)(t_2 - t_1)} W_{jj'}(\omega) \quad (8.2)$$

in terms of the two-atom self-energy

$$W_{jj'}(\omega) = \sum_k \frac{\eta_{kj}\eta_{kj'}^*}{\omega - \omega_k + i0^+} + \sum_k \frac{\eta_{kj'}\eta_{kj}^*}{\omega - 2\omega_a - \omega_k + i0^+}. \quad (8.3)$$

This expression can be rewritten as

$$W_{jj'}(\omega) = \int d\omega' \frac{G_{jj'}(\omega')}{\omega - \omega' + i0^+} + \int d\omega' \frac{G_{jj'}(\omega')}{\omega - 2\omega_a - \omega' + i0^+}, \quad (8.4)$$

where

$$G_{jj'}(\omega) = \sum_k \eta_{kj}\eta_{kj'}^* \delta(\omega - \omega_k) \quad (8.5)$$

is the two-point (autocorrelation) spectrum of the bath in the vacuum (zero-temperature) state. In view of the relation $\frac{1}{x+i0^+} = -i\pi\delta(x) + \text{P}\frac{1}{x}$, where P denotes the principal value, the integration in (8.2) yields

$$U_{jj'}^{(2)} = \frac{1}{2\pi i} \int_{-\infty}^{\infty} d\omega\, S_t^2(\omega - \omega_a) \left[-i\frac{1}{2}\gamma_{jj'}(\omega) - i\frac{1}{2}\gamma_{jj'}(\omega - 2\omega_a) \right.$$

$$\left. - \Delta_{jj'}(\omega) - \Delta_{jj'}(\omega - 2\omega_a) \right], \quad (8.6)$$

$S_t(\omega) = \int_{-t/2}^{t/2} dt'\, e^{i\omega t'} = t \operatorname{sinc}(\omega t/2)$ being the sinc function. The imaginary and real parts of the two-atom self-energy represent the cooperative decay rate and Lamb shift, respectively:

$$\gamma_{jj'}(\omega) = 2\pi\, G_{jj'}(\omega), \quad (8.7a)$$

$$\Delta_{jj'}(\omega) = P \int d\omega' \frac{G_{jj'}(\omega')}{\omega' - \omega}, \tag{8.7b}$$

which obey the Kramers–Kronig relations.

In the limit $t \to \infty$, $S_t(\omega) \sim \delta(\omega)$, we recover the Markovian expressions

$$\gamma_{jj'} = \gamma_{jj'}(\omega_a), \quad \Delta_{jj'} = \Delta_{jj'}(\omega_a) + \Delta_{jj'}(-\omega_a), \tag{8.8}$$

upon noting that $G_{jj'}(\omega < 0) = 0$. The Markovian limit is equivalent to the pole approximation, wherein the differences between the field-mode and TLS-resonance frequencies $\omega_\alpha \mp \omega_a$ are neglected.

To assess the validity of the Markovian limit, we take t to be sufficiently large, such that $\Delta_{jj'}(\omega)$ does not change appreciably within the width of $S_t^2(\omega - \omega_a)$ around ω_a. Then, upon expanding $\Delta_{jj'}(\omega)$ around ω_a (and around $-\omega_a$), we obtain

$$\int_{-\infty}^{\infty} d\omega \, S_t^2(\omega - \omega_a) \Delta_{jj'}(\omega) \propto \Delta_{jj'}(\omega_a) + O\left(\frac{\Delta_{jj'}''(\omega_a)}{t^2}\right), \tag{8.9}$$

where $\Delta_{jj'}''$ denotes the second derivative of the cooperative shift with respect to ω. The Markovian approximation demands that the correction to the Markovian result in (8.9) be small,

$$\left| \frac{\Delta_{jj'}''(\omega_a)}{\Delta_{jj'}(\omega_a)} \right| \frac{1}{t^2} \ll 1. \tag{8.10}$$

Over the relevant timescale $t \sim 1/|\Delta_{jj'}|$, (8.10) yields the following conditions for the validity of the Markovian approximation,

$$|\Delta_{jj'}(\omega_a) \Delta_{jj'}''(\omega_a)| \ll 1. \tag{8.11}$$

Under the conditions of (8.11), $\Delta_{jj'}(\omega_a)$ obtained from (8.7) is the Markovian expression for the two-atom cooperative Lamb shift. It corresponds to the resonant dipole–dipole interaction (RDDI) Hamiltonian, induced by the radiation (vacuum-state) bath,

$$H_{DD} = -\frac{\hbar}{2} \sum_{jj'} \Delta_{jj'} \left(\hat{\sigma}_j^+ \hat{\sigma}_{j'}^- + \hat{\sigma}_j^- \hat{\sigma}_{j'}^+ \right),$$

$$\Delta_{jj'} = \Delta_{jj',-} + \Delta_{jj',+},$$

$$\Delta_{jj',\mp} = P \int_0^\infty d\omega \frac{G_{jj'}(\omega)}{\omega \mp \omega_a}. \tag{8.12}$$

The RDDI can be viewed as *virtual-photon* exchange between the atoms.

By analogous considerations, (8.7a) yields the Markovian dissipation rate (linewidth), conforming to Fermi's Golden Rule,

$$\gamma_{jj'} = 2\pi G_{jj'}(\omega_a). \tag{8.13}$$

For $j = j'$ it is the single-atom spontaneous emission rate to the guided modes, and for $j \neq j'$ it is the two-atom (distance-dependent) cooperative emission.

The cooperative emission rate $\gamma_{jj'}$ represents probabilistic, incoherent (dissipative) energy exchange between the atoms. In order to achieve coherent deterministic exchange, $\gamma_{jj'}$ must vanish, leaving intact only the non-radiative dynamics ruled by H_{DD} in (8.12). Then, if initially atom 1 is excited, we obtain a periodic swap of the excitation between the atoms, at a rate $|\Delta_{12}|$. The resulting two-atom state,

$$|\Psi_{12}(t)\rangle = \cos(\Delta_{12}t)|e_1, g_2\rangle + i\sin(\Delta_{12}t)|g_1, e_2\rangle, \qquad (8.14)$$

is thus an oscillatory superposition of singly excited product states of atoms 1 and 2. At odd multiples of the time $t = \pi/(4|\Delta_{12}|)$, this state becomes *maximally entangled*.

8.1.2 Cooperative Self-Energy in Periodic Structures

The $G_{jj'}(\omega)$ two-atom terms are written in free space as sums over wave vectors k of the phase-difference factors $(\wp \cdot \epsilon_\lambda)^2 \exp(ik \cdot R)$, where ϵ_λ is a unit vector of polarization and $R = r_j - r_{j'}$. By contrast, in periodic, dispersive structures, these terms and their contributions to the self-energy can be evaluated in the normal-mode basis of the structure described in Chapters 3 and 4. On separating its real and imaginary parts, we may evaluate in the Markovian limit the principal-value term, $\Delta_{jj'}$, that corresponds to the cooperative Lamb (RDDI) shift, and the δ-function term, $\gamma_{jj'}$, that represents the cooperative contribution to the line width or rate of fluorescence (spontaneous emission):

$$\gamma_{jj'} - i\Delta_{jj'} = \frac{2\pi}{V} \sum_\alpha \omega_\alpha \phi_\alpha^*(r_j)\phi_\alpha(r_{j'})(\wp \cdot \epsilon_\lambda)^2$$

$$\times \left[\delta(\omega_\alpha - \omega_a) - iP\left(\frac{1}{\omega_\alpha - \omega_a} + \frac{1}{\omega_\alpha + \omega_a}\right) \right], \quad (8.15)$$

where (as in Sec. 8.1.1) ω_α and $\phi_\alpha(r_j)$ denote the mode frequency and mode function, respectively, $\alpha = (K, n, \lambda)$, where $\hbar K$ is the mode quasimomentum, n is the Brillouin zone number (Ch. 3), and V is the quantization volume.

8.1.3 Cooperative Self-Energy in Isotropic Structures

In what follows, (8.15) will be evaluated in the mode-continuum limit for isotropic media in which ω depends only on the modulus of K, so that

$$\sum_{\alpha=(K,n,\lambda)} \rightarrow V \sum_{n,\lambda} \int d\Omega_{\hat{K}} \int K^2 dK, \qquad (8.16)$$

where $\int d\Omega_{\hat{K}}$ denotes solid-angle integration. This assumption can be invoked in three-dimensional (3D) periodic structures (photonic crystals), upon approximating the polyhedral Brillouin surface in K space by a sphere, so that $\int K^2 dK$ extends over the nth Brillouin zone.

To evaluate $\Delta_{jj'}$ and $\gamma_{jj'}$ in such media, we must first integrate (8.15) over all angles and sum over the mode polarizations. Then, for plane wave ϕ_α,

$$\sum_{\lambda=1}^{2} \int d\Omega_{\hat{K}} (\hat{\wp} \cdot \epsilon_\lambda)^2 \phi_\alpha^*(r_j) \phi_\alpha(r_{j'}) = \frac{1}{2} \int d\Omega_{\hat{K}} [1 - (\hat{\wp} \cdot K)^2] e^{iK \cdot R} \equiv f(KR),$$

(8.17a)

where $\hat{\wp}$ denotes the unit vector along the atomic dipole $\wp$. This integral may be explicitly performed for two-atom quasi-molecular dimer states, such that $\wp \parallel R$ (dimer Σ states) or $\wp \perp R$ (dimer Π states),

$$f_\Sigma(KR) = 3 \left[\frac{\sin KR}{(KR)^3} - \frac{\cos KR}{(KR)^2} \right],$$

(8.17b)

$$f_\Pi(KR) = -\frac{3}{2} \left[\frac{\sin KR}{(KR)^3} - \frac{\cos KR}{(KR)^2} - \frac{\sin KR}{KR} \right].$$

(8.17c)

In periodic structures conforming to the isotropic approximation, these expressions can be generalized to Bloch waves ϕ_α, yielding, upon averaging over r_1 and the orientations of R,

$$\sum_{\lambda=1}^{2} \int d\Omega_{\hat{K}} (\hat{\wp} \cdot \epsilon_\lambda)^2 \langle \phi_\alpha^*(r_j) \phi_\alpha(r_{j'}) \rangle = \frac{1}{2} f_n(KR) \Phi_n^{jj'}(K),$$

(8.18a)

$$\Phi_n^{jj'}(K) = \sum_g \frac{\sin gR}{gR} \left(|C_{g\alpha}|^2 + \sum_{g' (g' \neq g)} C_{g\alpha}^* C_{g'\alpha} \frac{\sin |g - g'| r_{j'}}{|g - g'| r_{j'}} \right),$$

(8.18b)

where the summation is over g, the reciprocal-lattice vectors of the structure, and $C_{g\alpha}$ are the corresponding coefficients in the mode functions $\phi_\alpha(r) = e^{iK \cdot r} \sum_g C_{g\alpha} e^{ig \cdot r}$. In the isotropic approximation, $C_{g\alpha} = C_{gKn}$. The function $\Phi_n^{jj'}(K)$ that accounts for the lattice diffraction is not necessarily symmetric with respect to $j \leftrightarrow j'$ interchange (i.e., $\Phi_n^{jj'} \neq \Phi_n^{j'j}$), because r_j may not be symmetric with respect to the origin $r = 0$ of the unit cell. Since the choice of the origin determines the $C_{g\alpha}$'s, it is convenient to choose the origin at a maximum or a minimum of the dielectric index $\epsilon(r)$ [Fig. 8.1(a)].

A similar dependence of radiative properties on the location is found for an atom between two mirrors, which define a unit cell of the 1D lattice of image atoms.

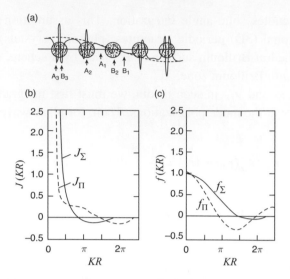

Figure 8.1 (a) Atoms A_1 and B_1 couple via a Bloch-wave photonic mode in a unit cell of a periodic array of spheres. A Bloch wave superimposed on a Mie-resonant variation of the photonic mode in each sphere couples atoms A_2 and B_2 in different spheres or A_3 and B_3 in the same sphere. (b) Separation dependence of the cooperative two-atom (Lamb) shift (RDDI) in the absence of crystal diffraction. (c) Same as (b) for the cooperative two-atom decay rate. (Reprinted from Kurizki, 1990. © 1990 The American Physical Society)

We next convert $K^2 dK$ in (8.16) into the density of modes (DOM),

$$\rho_n(\omega) = \frac{1}{2} \sum_{\lambda=1}^{2} \left[K^2(\omega) \left| \frac{dK}{d\omega} \right| \right]_{n\lambda}, \qquad (8.19)$$

where the sum is over the two polarizations, provided K is a unique-valued function of ω in the nth Brillouin zone. The dissipative and dispersive self-energies, normalized by their free-space counterparts, are then given by

$$\frac{\gamma_{jj'}(\mathbf{R}, \mathbf{r}_{j'})}{\gamma_{jj'}^{\text{free}}(R)} = \frac{c^3 \rho_n(\omega_a) f_n(K_a R) \, \Phi_n^{jj'}(K_a)}{f(\omega_a R/c)}, \qquad (8.20)$$

$$\frac{\Delta_{jj'}(\mathbf{R}, \mathbf{r}_{j'})}{\Delta_{jj'}^{\text{free}}(R)} = \frac{c^3}{\omega_a^3 J(\omega_a R/c)} \sum_n P \int d\omega \frac{\rho_n(\omega) \omega^2 f_n(K R) \, \Phi_n^{jj'}(K)}{\omega^2 - \omega_a^2}, \qquad (8.21)$$

where $K_a = K(\omega_a)$. The summation of principal-value integrals in (8.21) extends over the band(s) adjacent to the ω_a transition. The free-space R-dependence of $\gamma_{jj'}^{\text{free}}(R)$ and $f(\omega_a R/c)$ is given by (8.17b) and (8.17c) and that of $\Delta_{jj'}^{\text{free}}(R)$ is given by

$$J_{\Sigma}(\xi) = -\frac{3}{2}\left(\frac{\cos\xi}{\xi^3} + \frac{\sin\xi}{\xi^2}\right), \tag{8.22}$$

$$J_{\Pi}(\xi) = \frac{3}{4}\left(\frac{\cos\xi}{\xi^3} + \frac{\sin\xi}{\xi^2} - \frac{\cos\xi}{\xi}\right), \tag{8.23}$$

where $\xi = \omega_a R/c$.

In the near zone, $\xi \ll 1$ [Fig. 8.1(c)], we find

$$f_{\Sigma(\Pi)}(\omega_a R/c) = \frac{\gamma_{jj'}^{\mathrm{free}}(R)}{\gamma_1} \to 1, \tag{8.24}$$

γ_1 being the single-atom spontaneous decay rate (Einstein A coefficient). The corresponding $J_{\Sigma(\Pi)}(\xi \ll 1)$ attains the electrostatic R^{-3} RDDI limit [Fig. 8.1(b)],

$$J_{\Sigma(\Pi)}(\omega_a R/c) \to (\omega_a R/c)^{-3}. \tag{8.25}$$

Equations (8.20) and (8.21) show that the enhancement or suppression of $\gamma_{jj'}$ and $\Delta_{jj'}$ by the structure compared to free space is determined by the DOM modification, expressed by the factor $\rho_n(\omega)c^3/\omega^2$ and by the diffraction function $\Phi_n^{jj'}$ (in photonic crystals).

8.1.4 Cooperative Self-Energy in Band Gaps

Let us consider photonic crystals free of dissipation or disorder, where the edge of a band gap is a singularity of the DOM (i.e., $\rho_n(\omega)$ vanishes abruptly). Then, within a bandgap,

$$\gamma_{jj'} \propto \gamma \propto \rho_n(\omega_a) = 0. \tag{8.26}$$

The evaluation of $\Delta_{jj'}$ in a band gap is more involved. The frequency is expanded about the singularity ω_c, the lower or upper edge of the bandgap (Fig. 3.3),

$$\omega = \omega_c + b\kappa^2 + \dots, \tag{8.27}$$

κ being the deviation from K_c, the wave vector of the band edge. The first-derivative term in the expansion vanishes at the cutoff. Just above the cutoff frequency or the lower edge of a band gap ω_c, this expansion corresponds to

$$\rho_n(\omega) \simeq \frac{\omega(\omega^2 - \omega_c^2)^{1/2}}{(2b\omega_c)^{3/2}}\theta(\omega^2 - \omega_c^2), \tag{8.28}$$

θ being the Heaviside step function. An analogous expression is obtained for ω_c at the upper edge of a band gap.

For this form of $\rho_n(\omega)$, $\Delta_{jj'}$ can be evaluated for ω_a just below ω_c upon extending the integral in (8.21) over the domain $\omega_c < \omega < \infty$ and $-\infty < \omega < -\omega_c$. We

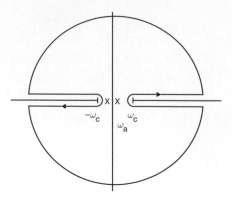

Figure 8.2 Integration contour for the evaluation of RDDI for ω_a below cutoff (ω_c). (Reprinted from Kurizki, 1990. © 1990 The American Physical Society)

consider the integrand to be symmetric in ω, exclude the bandgap and its edges, $-\omega_c \leq \omega \leq \omega_c$, by branch cuts, and close the contour by a circle of infinite radius (Fig. 8.2). The $\pm\omega_a$ residues that contribute to the integral are the product of two factors: (i) the analytic continuation of $\rho_n(\omega)$ [Eq. (8.28)] into the band gap that has the form,

$$\rho_n(\pm\omega_a) = \chi^2 \frac{d\chi}{d\omega} = \pm i \frac{\omega_a(\omega_a^2 - \omega_c^2)^{1/2}}{(2b\omega_c)^{3/2}}, \tag{8.29}$$

where χ is the imaginary part of the wave vector; and (ii) the analytic continuation of $f_n(KR)\,\Phi_n^{jj'}(K)$, which amounts to the replacement of the angular integral in (8.18) by an integral over evanescent modes with complex wave vectors:

$$\int d\Omega_{\hat{K}}(\hat{\wp} \cdot \epsilon_\lambda)^2 \phi_\alpha^*(r_j)\phi_\alpha(r_{j'}) \rightarrow \int d\Omega_{\hat{K}}(\hat{\wp} \cdot \epsilon_\lambda)^2 \exp(iK \cdot R - |\chi \cdot R|). \tag{8.30}$$

In the regime of weak diffraction by a PC structure with period p, we can retain in (8.18) only the reciprocal-lattice vectors 0 and $2\pi/p$, whose coefficients near the band edges are $C_0 \simeq \pm C_1 \simeq \pm 2^{-1/2}$. The diffraction function $\Phi_n^{jj'}(K)$ is also changed, as we move into the band gap, to become [Fig. 8.1(a)]

$$\Phi_n^{jj'} = \left[1 + \frac{\sin(\pi R/p)}{\pi R/p}\right]\left[1 \pm \frac{\sin(\pi r_{j'}/p)}{\pi r_{j'}/p}\right]. \tag{8.31}$$

In the middle of a narrow band gap, $C_0^* C_1 = \pm i/2$.

The analytically continued factors (8.29) and (8.31) inside the band gap drastically change RDDI, as compared to (8.21) that is valid outside a gap:

$$\frac{\Delta_{jj'}(R)}{\Delta_{jj'}^{\text{free}}(R)} \sim \frac{c^3 f_n(K_a R)\,\Phi_n^{jj'}}{(2b\omega_c)^{3/2} J(\omega_a R/c)}|1 - \omega_c^2/\omega_a^2|^{1/2} \exp(-\chi_a R), \tag{8.32}$$

where $\chi_a = \chi(\omega_a)$. Typically, $\chi \ll K \simeq \pi/p$ for ω_a within a narrow band gap, at the first Bragg resonance of the PC. The predominant change in RDDI within a band gap is then that instead of $J(K_a R)$ in an allowed band, we have $f(K_a R)$, because of the $\pi/2$ phase shift and symmetry change in the argument of this function. The analytically continued DOM in (8.29) gives rise to atom–atom coupling via *evanescent* modes that are resonant with a virtual frequency whose imaginary part is $(\omega_a^2 - \omega_c^2)^{1/2}$. This coupling is equivalent to exchange via *allowed* (normal) modes outside the gap that are shifted by $|\omega_a^2 - \omega_c^2|^{1/2}$ from resonance with the atomic frequency ω_a. The $\pi/2$ phase shift caused by this analytical continuation characterizes exchange via the emission of atom j and the *virtual* absorption (i.e., scattering resonance) of atom j', instead of the real absorption of atom j' in an allowed band. Since the resonance is no longer real in the gap, the R^{-3} divergent limit of $J(KR)$ is "smoothed" out, becoming $f(KR)$ [compare Figs. 8.1(b) and 8.1(c)]. For interatomic separations $R \lesssim 1$ nm and optical ω_a, the factor $f(K_a R)/J(\omega_a R/c)$ in (8.32) can be as small as 10^{-9} or 10^{-10} (i.e., *the RDDI in the gap can be nearly completely suppressed*).

8.1.5 Long-Range RDDI near a Waveguide Cutoff

Contrary to RDDI suppression in a band gap of a photonic crystal, the cooperative Lamb shift (RDDI) can be enhanced, while the radiative rate $\gamma_{jj'}$ can be suppressed (compared to free space) when atoms are placed inside a rectangular hollow metallic waveguide (MWG) along its axis z [Fig. 8.3(a)].

These results are determined by two key features of the waveguide structure: (1) below the cutoff ω_{mn} no guided photon modes exist, and (2) the density of states diverges near the cutoff, because it is proportional to

$$\frac{\partial k}{\partial \omega} = \frac{1}{c} \frac{\omega}{\omega_{mn}} \frac{1}{\sqrt{(\omega/\omega_{mn})^2 - 1}}. \tag{8.33}$$

In what follows, feature (1) will be shown to suppress radiative decay and feature (2) to enhance and extend the range of RDDI.

Let the atoms be polarizable in the z direction, $\wp_{eg} = \wp_{eg} e_z$. Since in TE modes the z component of the electric field vanishes, only TM modes contribute to the bath spectrum, which evaluates to [cf. (3.12) and Table 3.1]

$$G_{jj'}(\omega) = \sum_{mn} \frac{\Gamma_{mn}}{2\pi} \frac{\cos\left[k(z_j - z_{j'})\right]}{\sqrt{(\omega/\omega_{mn})^2 - 1}} \theta(\omega - \omega_{mn}). \tag{8.34}$$

Here we have introduced $\Gamma_{mn} \equiv \frac{4\omega_{mn}\tilde{\wp}_{mn,j}\tilde{\wp}_{mn,j'}}{\pi\epsilon_0\hbar cab}$, where

$$\tilde{\wp}_{mn,j} = \wp_{eg}\sin\left(\frac{m\pi}{a}x_j\right)\sin\left(\frac{n\pi}{b}y_j\right), \tag{8.35}$$

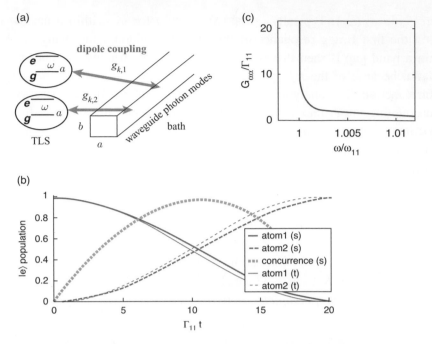

Figure 8.3 (a) Schematic view of two TLS coupled via RDDI in a waveguide structure. (b) Excited-state population for the atoms coupled via the TM_{11} mode below cutoff, with $\omega_{11} = 500\Gamma_{11}$, $\omega_a - \omega_{11} = -100\Gamma_{11}$ and $z_{12} = 0.5\lambda_a$: Markovian theory (t) and simulation (s) results. (c) Divergence of the single-atom spectrum $\bar{G}_{jj}$ for TM_{11} mode near cutoff ω_{11}. (Adapted from Shahmoon and Kurizki, 2013. © 2013 American Physical Society)

(x_j, y_j) being the transverse position of atom j in a waveguide with transverse dimensions a and b.

When the atomic resonance ω_a is below the lowest cutoff frequency, which is $< \omega_{11}$ for TM modes, the atomic dipoles are off-resonant with all field modes, so that the spontaneous emission of radiation is suppressed, according to (8.34), within the Golden-Rule (GR) Markovian approximation (Ch. 4),

$$\gamma_{jj'} = 2\pi G_{jj'}(\omega_a) = 0. \tag{8.36}$$

We are then left only with the nonradiative RDDI (8.7b),

$$\Delta_{12} = \sum_{mn} \frac{\Gamma_{mn}}{2} \frac{1}{\sqrt{1-(\omega_a/\omega_{mn})^2}} e^{-\frac{|z_1-z_2|}{\xi_{mn}}}, \tag{8.37a}$$

where ξ_{mn} is the effective interaction range,

$$\xi_{mn} = \frac{c}{\omega_{mn}} \frac{1}{\sqrt{1-(\omega_a/\omega_{mn})^2}}. \tag{8.37b}$$

Thus, within the GR (Markovian) approximation, radiative dissipation is absent, while the RDDI is mediated by evanescent waves and therefore decays exponentially with interatomic distance, as in (8.32) for a photonic crystal. Most dramatically, (8.37a) and (8.37b) show that as ω_a approaches the lowest cutoff ω_{11} from below, the RDDI contributed by the TM$_{11}$ mode diverges, and so does its range, ξ_{11}. If this prediction held, it would enable deterministic entanglement at large distances.

Yet, as seen from (8.37a) and Figure 8.1(c), in the limit $\omega_a \to \omega_{mn}$, $\Delta_{12}(\omega_a)$ and $\Delta''_{12}(\omega_a)$ diverge and condition (8.11) for the validity of the Markovian approximation is not satisfied. Hence, a non-Markovian theory is required in order to adequately assess the possibility of entanglement via long-distance RDDI in this limit, as detailed in Section 8.2.

8.2 Non-Markovian Theory of RDDI in Waveguides

Here we outline a nonperturbative, non-Markovian, theory for RDDI in a field-confining structure, such as a waveguide. From Hamiltonian (8.1), if atom 1 is initially excited, the state of the combined (atoms+bath) system is, in the RWA,

$$|\psi(t)\rangle = \alpha_1(t)|e_1, g_2, 0\rangle + \alpha_2(t)|g_1, e_2, 0\rangle + \sum_k \beta_k(t)|g_1, g_2, 1_k\rangle.$$

Upon inserting this state into the Schrödinger equation, we obtain dynamical equations for $\alpha_1(t)$, $\alpha_2(t)$, and $\beta_k(t)$. Taking their Laplace transform for the initial conditions, $\alpha_1(0) = 1$, $\alpha_2(0) = \beta_k(0) = 0$, we then obtain the Laplace transform of $\alpha_1(t)$,

$$\tilde{\alpha}_1(s) = \left[s + J_{11}(s) + i\omega_a - \frac{J_{12}(s)J_{21}(s)}{s + J_{22}(s) + i\omega_a}\right]^{-1}, \tag{8.38}$$

where $J_{jj'}(s) = \sum_k \frac{\eta^*_{k,j}\eta_{k,j'}}{s+i\omega_k}$. It can be shown that $J_{jj'}(-i\omega_a) = -i\Delta_{jj',-}$ [cf. (8.12)].

As before, we consider only the MWG transverse mode $m = 1, n = 1$, here for ω_a close to the cutoff ω_{11}, such that the denominator of the spectrum (8.34) is approximated by $\sqrt{(\omega/\omega_{11})^2 - 1} \approx \sqrt{2}\sqrt{\omega/\omega_{11} - 1}$. Upon evaluating the integrals in $J_{jj'}(s)$ for the approximated spectrum, we invert the Laplace transform (8.38) to find

$$\alpha_1(t) = \sqrt{i}e^{-i\omega_{11}t} \sum_{l=1}^{5} c_l \left[\frac{1}{\sqrt{\pi t}} + \sqrt{i}u_l e^{iu_l^2 t}\text{erfc}\left(-u_l\sqrt{it}\right)\right]. \tag{8.39}$$

Although the explicit form of the constants c_l, u_l is complicated, the $1/\sqrt{\pi t}$ dependence of the first term in the square brackets indicates that $\alpha_1(t)$ does not

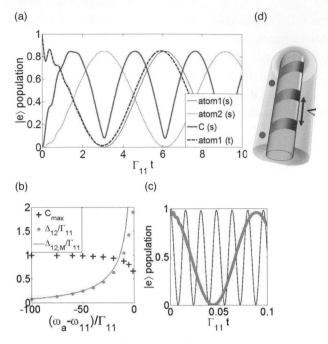

Figure 8.4 (a) Atomic excitation probabilities and concurrence affected by TM_{11} mode as a function of time. We take $|g\rangle$ and $|e\rangle$ to be the two circular states with principal quantum numbers 51 and 50, for which $\omega_a = 2\pi \times 51.1$ GHz and $d \sim 1250ea_0$, respectively, with e the charge of an electron and a_0 the Bohr radius. Near the cutoff, $\Gamma_{11} \simeq \frac{\omega_a^3 d^2}{3\pi\epsilon_0\hbar c^3} \approx 14.7$ Hz and $\lambda_a \sim 6$ mm is the atomic wavelength. Here $\omega_{11} = 500\Gamma_{11}$, $z_{12} = \lambda_a$ and $\omega_a - \omega_{11} = -10\Gamma_{11}$. The simulations (s) are in excellent agreement with the non-Markovian theory (t). (b) Trade-off between RDDI Δ_{12} and maximal achievable concurrence C_{max} as a function of the atomic-resonance detuning from cutoff (simulation for $\omega_{11} = 500\Gamma_{11}$, $z_{12} = 0.5\lambda_a$) in the non-Markovian regime, compared with the Markovian-theory result in Eq. (8.37a) denoted by $\Delta_{12,M}$. (c) The excitation of atom 1 and the concurrence bounded by the maximal value of the plot. In MWG (thin line): $\omega_{11} = 2.17 \times 10^{10}\Gamma_{11}$, $\omega_a - \omega_{11} = -2 \times 10^4\Gamma_{11}$, $z = 100\lambda_a$. In the fiber-Bragg-grating scheme (thick line): $\omega_{11} = 6 \times 10^7\Gamma_{11}$, $\omega_a - \omega_{11} = -1500\Gamma_{11}$, $z = 20\lambda_a$. (d) Atoms are coupled to the guided modes by their transverse evanescent tails in the fiber-Bragg-grating scheme. (Reprinted from Shahmoon and Kurizki, 2013. © 2013 American Physical Society)

decay exponentially at long times, thus deviating drastically from the Markovian (GR) decay.

8.2.1 Long-Distance Entanglement in Waveguides

The theory outlined in Section 8.2 allows us to illustrate the possibility of long-distance entanglement for Rydberg atoms in a cold metallic waveguide (MWG) or a long superconducting cavity (Fig. 8.4). We then find entanglement with concurrence $C = 0.983$, at a distance $z \sim 0.6$ m $= 100\lambda_a$ for interaction time $t \approx 0.2$ ms. For niobium superconducting plates at temperature $T < 1$ K, $\gamma_{loss} = 19.89$ Hz, much slower than the 0.2 ms required for entanglement. Such a temperature ensures that the thermal photon occupancy at ω_a is negligible.

8.3 Cooperative Self-Energy Effects in High-Q Cavities

Cooperative atomic effects in resonant cavities have been studied in Chapter 7, based on the Tavis–Cummings model that assumes many atoms to be identically coupled to a single mode. The limitation of this model is that it ignores the symmetry-breaking dipole–dipole effects, which are important at near-zone separations.

Here we consider a pair of identical TLS (atoms or excitons), sharing a photon with a high-Q mode in a cavity. We treat the two TLS coupled to a bath with an *arbitrary* mode-density spectrum in a *nonperturbative fashion*. The reason for such a treatment is that the standard perturbative theory of two-atom coupling, to second order in the field, consistently with the Markovian approximation, is *inadequate* for a bath whose mode-density spectrum does not vary smoothly, as shown in Sections 5.2 and 8.2. Drastic modifications (as compared to open-space scenarios) of interatomic excitation transfer will be shown to occur at both near-zone and far-zone atomic separations, due to the competition between strong coupling of each atom to a high-Q mode (vacuum Rabi splitting) and interatomic (virtual) photon exchange via all other modes (RDDI).

8.3.1 Model

Here we focus on a system of two atoms that are near resonant with a high-Q cavity mode. We treat a narrow spectral band of the electromagnetic bath, which is associated with the high-Q mode, *separately* from the background mode-density spectrum: The latter bath spectrum is smooth and/or off-resonant and represents bath modes that are weakly coupled to the atoms. The two-atom interaction with the narrow band is evaluated *exactly*, whereas the interaction with the weakly coupled parts of the bath-mode spectrum is treated by second-order perturbation theory. Accordingly, we obtain the effective Hamiltonian in the following second-quantized form:

$$
\begin{aligned}
H = {}& \hbar \sum_{j=1}^{2} \omega_j |e_j\rangle\langle e_j| + \hbar \sum_k \omega_k a_k^\dagger a_k \\
& + \hbar \Delta_{12}(|e_1 g_2\rangle\langle e_2 g_1| + |e_2 g_1\rangle\langle e_1 g_2|) \\
& + \hbar \sum_{j=1}^{2} \sum_k (\eta_{kj} a_k |e_j\rangle\langle g_j| + \text{H.c.}).
\end{aligned}
\tag{8.40}
$$

Here $|e_j\rangle$ and $|g_j\rangle$ are the excited and ground states of the two TLS. The frequency ω_k and annihilation operator a_k pertain to a *near-resonant* bath mode (within the narrow band) whose dipolar coupling to the jth TLS is given by η_{kj}; $\hbar\omega_j$ are the excited-state energies. Those energies include the Lamb shifts caused by single-TLS interactions with the *off-resonant* bath modes (*outside* the near-resonant narrow band). Finally, $\hbar\Delta_{12}$ is the matrix element of the interatom RDDI (excluding the contribution of the near-resonant modes). Both ω_j and Δ_{12} may have imaginary (dissipative) parts. The TLS–bath interaction [the last term in (8.40)] is written in the RWA, since the anti-rotating terms are off-resonant and hence their contributions are assumed to be included in ω_j and Δ_{12}. At near-zone distances, the real part of Δ_{12} reduces to the usual electrostatic RDDI, which varies with the interatom separation R as the inverse cube R^{-3}.

8.3.2 Analysis

The time-dependent wave function for the system of a single photon shared by the field and the atoms (TLS) is written as

$$|\Psi_{j'}(t)\rangle = \sum_{j=1}^{2} V_{jj'}(t)|e_j g_{3-j}, \{0_k\}\rangle + \sum_k \beta_{kj'}(t)|g_1 g_2, 1_k\rangle, \tag{8.41}$$

where j' indicates the atom that is excited initially, $|\{0_k\}\rangle$ and $|1_k\rangle$ being the vacuum and k-mode single-quantum states, respectively. In the subspace $\{|e_1 g_2\rangle, |e_2 g_1\rangle\}$ of the collective atomic states, an initially pure, normalized, state remains pure thereafter (though generally not normalized) and has the form $\alpha_1(t)|e_1 g_2\rangle + \alpha_2(t)|e_2 g_1\rangle$ with the coefficients

$$\alpha_j(t) = \sum_{j'=1}^{2} V_{jj'}(t)\alpha_{j'}(0). \tag{8.42}$$

From the Schrödinger equation we obtain the following *exact* Laplace-transform solution for the excited-state evolution operator

$$\hat{V}(s) = D^{-1}(s)U(s), \tag{8.43}$$

with $D(s)$ being the determinant of the 2×2 matrix $U(s)$. The diagonal and off-diagonal elements of the matrix $U(s)$ represent single-atom and interatomic contributions, respectively,

$$U_{jj}(s) = s + i\omega_j + iJ_{jj}(s),$$
$$U_{jj'}(s) = -i\Delta_{12} - iJ_{jj'}(s) \quad (j \neq j'), \tag{8.44}$$

where the integration over the narrow band of near-resonant modes yields

$$J_{jj'}(s) = \int \frac{G_{jj'}(\omega)d\omega}{is - \omega},$$

$$G_{jj'}(\omega) = \sum_k \eta_{kj}\eta_{kj'}^*\delta(\omega - \omega_k). \tag{8.45}$$

The roots of the determinant equation $D(s) = 0$ correspond to the levels (eigenvalues) of the two-atom system.

The transition frequencies and decay rates of the atoms are taken to be equal (i.e., $\omega_1 = \omega_2 = \omega_a$ and $\gamma_1 = \gamma_2 = \gamma_a$). The transition dipole moments are assumed real and so are the mode functions (standing waves); then $G_{12}(\omega) = G_{21}(\omega)$ is real, $U_{12}(s) = U_{21}(s)$, and $V_{12}(t) = V_{21}(t)$.

Let us first assume that the near-resonant mode is sufficiently broad due to its finite Q-factor, so that the Weisskopf-Wigner approximation for the emission holds. Then the mode-density spectrum is smooth enough, so that $\left|dJ_{jj}/d\omega_a\right| \ll 1$, where $J_{jj'} = J_{jj'}(-i\omega_a + 0)$. The near-resonant contributions $J_{jj'}(s)$ can then be replaced by constants, $J_{jj'}$, and absorbed into the single-atom and two-atom level shifts and the respective decay rates. The resulting eigenvalues and eigenstates of the singly excited system are then

$$\omega_{S(A)} = \text{Re}\tilde{\omega}_a \pm \text{Re}\,\tilde{\Delta}_{12},$$

$$|\Psi_{S(A)}\rangle|\{0_k\}\rangle = 2^{-1/2}(|e_1g_2\rangle \pm |e_2g_1\rangle)|\{0_k\}\rangle, \tag{8.46}$$

where $\tilde{\omega}_a = \omega_a + (J_{11} + J_{22})/2$, $\tilde{\Delta}_{12} = \Delta_{12} + J_{12}$, S (A) denoting the symmetric (antisymmetric) eigenstate and the corresponding eigenvalue. These two states are split by the (cavity-modified) RDDI shift. They give rise to damped sinusoidal oscillations of the excitation-transfer probability between the atoms at a rate of $2|\text{Re}\tilde{\Delta}_{12}|$, which varies as R^{-3} at near-zone separations, $\omega_a R/c \ll 1$.

A very different behavior is obtained in the case of a narrow-linewidth single-mode at ω_0. One can then use in (8.44) the approximation

$$G_{jj'}(\omega) \approx \eta_j\eta_{j'}\delta(\omega - \omega_0). \tag{8.47}$$

This approximation holds if $|\eta_j|$ are much greater than the line width. It amounts to neglecting *all dissipation* (i.e., the decay rates γ_a and $|\text{Im}\Delta_{12}|$), originating from the background density of modes outside the line. These rates are assumed to be much smaller than the oscillation frequencies in the system, since we are interested in times $\ll \gamma_a^{-1}$, $|\text{Im}\Delta_{12}|^{-1}$. We thereby reduce the problem to that of two atoms coupled to a single mode via coupling constants η_j and to each other via a real Δ_{12}. A Hamiltonian describing this problem can be written as a 3×3 matrix in the basis of the states $\{|e_1g_2, \{0\}\rangle, |e_2g_1, \{0\}\rangle, |g_1g_2, 1_0\rangle\}$, where 0 and 1_0 are the

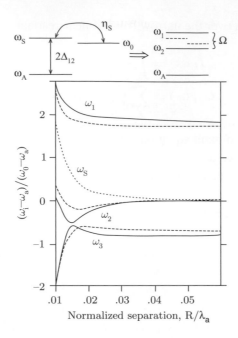

Figure 8.5 Solid curves – exact eigenvalues as a function of the normalized sep-
aration R/λ_a for $\Delta_{12} = (\gamma_a/2)(\omega_a R/c)^{-3}$, γ_a being the free space decay rate of
the atom excited state, $\omega_0 - \omega_a = 10^3 \gamma_a$, $\eta_j = 10^3 \gamma_a \sin(\omega_a r_j/c)$, and $r_1 = 0.3\lambda_a$.
Note the narrow pseudocrossing region. Dashed curves – idem, for $\eta_1 = 1000\gamma_a$
and $\eta_2 = 500\gamma_a$. The pseudocrossing region is now very broad. Inset – dipole–
dipole and vacuum Rabi splittings for two identical atoms almost identically
coupled to a near-resonant mode, $\eta_1 \approx \eta_2$, $\eta_S \approx \sqrt{2}\eta_1$. (Adapted from Kurizki
et al., 1996. © 1996 The American Physical Society)

vacuum and single-photon numbers in the near-resonant mode. The determinant
equation $D(s = -i\omega) = 0$ then becomes

$$
-\left(\omega - \omega_a - \frac{\eta_1^2}{\omega - \omega_0}\right)\left(\omega - \omega_a - \frac{\eta_2^2}{\omega - \omega_0}\right) + \left(\Delta_{12} + \frac{\eta_1 \eta_2}{\omega - \omega_0}\right)^2 = 0. \quad (8.48)
$$

The eigenvalues (Fig. 8.5) and eigenfunctions corresponding to (8.48) will be
illustrated in the following distinct limits.

8.3.3 Oscillating Exchange between Atoms in a Cavity

Let us assume that $\eta_1 \approx \eta_2$. The two atomic couplings to the mode are nearly
equal for identical atoms with parallel dipoles *in the near zone of separations,*
$\omega_a R/c \ll 1$. Under this assumption, the eigenvalues obtained from (8.48) are given
(in descending order) by

$$
\omega_1 \approx \omega_+, \quad \omega_2 \approx \begin{cases} \omega_-, & \text{if } R < R_c, \\ \omega_A, & \text{if } R > R_c, \end{cases} \quad \omega_3 \approx \begin{cases} \omega_A, & \text{if } R < R_c, \\ \omega_-, & \text{if } R > R_c, \end{cases} \quad (8.49)
$$

where $\omega_{S(A)} = \omega_a \pm \Delta_{12}(R)$, R_c is the position of the crossing of $\omega_-(R)$ and $\omega_A(R)$, and

$$\omega_\pm = \frac{1}{2}(\omega_S + \omega_0 \pm \Omega), \quad \Omega = \sqrt{2(\eta_1 + \eta_2)^2 + (\omega_S - \omega_0)^2}. \tag{8.50}$$

Here and below we assume without loss of generality that Δ_{12} is positive in the near zone. These results hold throughout the near-zone (except for the pseudocrossing, $R \approx R_c$, discussed below). For $R \to 0$, the divergence of $\Delta_{12}(R)$ implies that $|\omega_S - \omega_0| \gg |\eta_1 + \eta_2|$, hence $\omega_+ \to \omega_S$ and $\omega_- \to \omega_0$, so that the near-resonant field mode is then decoupled from both RDDI-split (symmetric and antisymmetric) states.

Throughout the range of validity of (8.49), the antisymmetric-state eigenvalue remains uncoupled from the mode, since its coupling is $2^{-1/2}(\eta_1 - \eta_2) \approx 0$ in the near zone. The symmetric state and the single-photon state become increasingly hybridized as R grows, provided the detuning $|\omega_0 - \omega_a|$ is not too large. This hybridization gives rise to two eigenvalues that are split by $\pm\Omega$, the vacuum Rabi frequency of the symmetric state. The trends surveyed above are also exhibited by the dressed-state eigenfunctions: As $R \to 0$, $|\Psi_+\rangle \to |\Psi_S, \{0\}\rangle$ and $|\Psi_-\rangle \to |g_1 g_2, 1_0\rangle$. Otherwise, the symmetric and single-photon states are strongly mixed in $|\Psi_\pm\rangle$. By contrast, $|\Psi_3\rangle \approx |\Psi_A, \{0\}\rangle$ as long as $\eta_1 \approx \eta_2$.

These eigenfunctions can be used to calculate the probability of excitation transfer from the initially excited atom 1 to the initially unexcited atom 2, $P_2(t)$, and the corresponding excitation-trapping probability $P_1(t)$,

$$P_j(t) = \left| \sum_{i=1}^{3} \langle \varphi_j | \Psi_i \rangle \langle \Psi_i | \varphi_1 \rangle e^{-i\omega_i t} \right|^2, \tag{8.51}$$

where Ψ_i are the dressed-state eigenfunctions (labelled by $i = 1, 2, 3$) and $|\varphi_j\rangle = |e_j g_{3-j}, \{0\}\rangle$. We find *three* distinct atomic-state eigenvalues, causing aperiodic oscillations of $P_j(t)$, instead of the sinusoidal oscillations in Chapter 7. The time-averaged probabilities,

$$\bar{P}_j = \sum_{i=1}^{3} |\langle \varphi_j | \Psi_i \rangle \langle \Psi_i | \varphi_1 \rangle|^2, \tag{8.52}$$

are approximately equal, $\bar{P}_1 \approx \bar{P}_2$. They vary from 3/8 at $|\omega_S - \omega_0| \ll |\eta_1 + \eta_2|$ to the free-space value 1/2 at $|\omega_S - \omega_0| \gg |\eta_1 + \eta_2|$. Because of the nonzero probability of the field-mode excitation, we have

$$P_1(t) + P_2(t) < 1 \tag{8.53}$$

for the sum of the excitation probabilities of the two atoms.

8.3.4 Exchange Trapping between Atoms in a Cavity

Let us now consider the effect of a small near-zone difference $\eta_1 - \eta_2$ that scales linearly with the separation R. This effect is most salient at the *pseudocrossing* (near-equality) of two eigenvalues in (8.49) (solid curves, Fig. 8.5), namely for R close to the value R_c such that $\omega_-(R_c) = \omega_A(R_c)$. This equality implies, in view of (8.50), that $\Delta_{12}(R_c) \sim |\eta_1|$ for $|\omega_0 - \omega_a| \lesssim 2|\eta_1|$, or $\Delta_{12}(R_c) \approx 2\eta_1^2/|\omega_0 - \omega_a|$ for $|\omega_0 - \omega_a| \gg 2|\eta_1|$. In both cases the RDDI-induced and cavity-QED level shifts (or splittings) become comparable.

The strong competition of RDDI and Rabi splittings near R_c modifies the eigenvalues in (8.49), replacing them with the more accurate solutions of (8.48),

$$\omega_1 \approx \omega_+, \quad \omega_{2,3} \approx \frac{1}{2}(\omega_- + \omega_A \pm \Omega'), \tag{8.54}$$

where

$$\Omega' = \sqrt{V_0^2 + (\omega_- - \omega_A)^2}, \quad V_0 = \frac{\eta_1^2 - \eta_2^2}{\sqrt{\Omega(\Omega + \omega_0 - \omega_S)}}. \tag{8.55}$$

Here $|V_0|$, the minimal splitting between ω_2 and ω_3, determines the width of the pseudocrossing interval, $|R_1 - R_2|$, where $\omega_a(R_{1,2}) - \omega_-(R_{1,2}) = \pm V_0$. For two atoms far from a node of a sinusoidal mode, $|V_0| \sim |\eta_1(\eta_1 - \eta_2)|/(\sqrt{8}|\eta_1| + |\omega_0 - \omega_S|)$.

Whereas the eigenfunction $|\Psi_1\rangle = |\Psi_+\rangle$ is not affected by the pseudocrossing, $|\Psi_-\rangle$ and $|\Psi_A\rangle$ are strongly mixed near R_c. This mixing signifies the complete breaking of the symmetry [Eq. (8.46)], that characterizes the two-atom system subject to RDDI in open space. At $R = R_c$ and for sufficiently large and positive detuning, such that $\omega_0 - \omega_S \approx \Omega$, we obtain the limit

$$|\Psi_2\rangle \to |e_1 g_2, \{0\}\rangle, \quad |\Psi_3\rangle \to |e_2 g_1, \{0\}\rangle, \tag{8.56}$$

in which the excited eigenstates 2 and 3 become *uncoupled*, due to the interference of $|\Psi_S\rangle$ and $|\Psi_A\rangle$. The corresponding excitation-transfer probability undergoes *strong suppression* in the pseudocrossing interval, as shown by the time-averaged values (solid curves in Fig. 8.6): $\bar{P}_2(R = R_c) = 3c_+^2/8 \ll 1$ and $\bar{P}_1(R = R_c) = 1 - c_+ + (3/8)c_+^2$, tending to 1 with the increase of $\omega_0 - \omega_S$, where $c_+ = [1 + (\omega_S - \omega_0)/\Omega]/2$. *The excitation is then strongly trapped at the initial atom,* owing to the decoupling of the excited eigenstates.

Finally, we estimate R_c for the parameters that maximize $|\eta_j|$: a cavity mode confined to the smallest possible volume $\sim \lambda_a^3$ (where $\lambda_a = 2\pi c/\omega_a$), transition dipole moment $d \sim ea$, a being the excited state radius, and, correspondingly, $\omega_a \sim e^2/(\hbar a)$. Then $|\eta_j| \sim \omega_a(a/\lambda_a)^{3/2} \sim \omega_a[\alpha/(2\pi)]^{3/2}$, where $\alpha = 1/137$, $|\Delta_{12}| \sim (ea)^2/(\hbar R^3) \sim \omega_a(a/R)^3$. As a result,

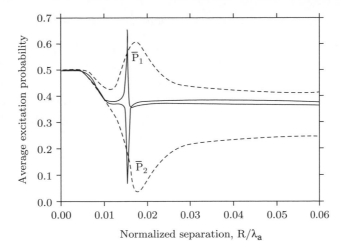

Figure 8.6 Time-averaged excitation probabilities $\bar{P}_1$ and $\bar{P}_2$ [$P_1(0) = 1$] for the parameters in Figure 8.5: solid (dashed) curves – for narrow (broad) pseudocrossing interval. (Adapted from Kurizki et al., 1996. © 1996 The American Physical Society)

$$R_c \sim \sqrt{\alpha/(2\pi)}\lambda_a, \quad \text{if } |\omega_0 - \omega_a| \lesssim 2|\eta_1|;$$
$$R_c \sim (|\omega_0 - \omega_a|/\omega_a)^{1/3}\lambda_a, \quad \text{if } |\omega_0 - \omega_a| \gg 2|\eta_1|. \tag{8.57}$$

In either case, $R_c \ll \lambda_a$, that is, the pseudocrossing occurs *well within the near zone*. Note that for certain dipole orientations, $|\Delta_{12}|$ is further reduced and R_c can become even smaller.

8.4 Macroscopic Quantum-Superposition (MQS) via Cooperative Lamb Shift

Here we present an exactly soluble model, which is an example of collective coupling of a multi-spin system to any thermal bath that can drive the system into a macroscopic quantum-superposition state. This entangled state can spontaneously arise from dispersive interactions (*virtual-quanta exchange*) among spins via the bath.

8.4.1 Model and Dynamics

We here consider N noninteracting spin-1/2 particles or atomic TLS that are identically, linearly coupled to a bosonic (oscillator) bath via σ_z (unlike σ_x in the Dicke model). In the collective basis, the many-body Hamiltonian has the following form, without the RWA,

$$H = H_S + H_B + H_I, \tag{8.58}$$

where

$$H_{\mathrm{S}} = \hbar\omega_x \hat{J}_x, \quad H_{\mathrm{B}} = \hbar\sum_k \omega_k a_k^\dagger a_k, \quad H_{\mathrm{I}} = \hbar\hat{J}_z \sum_k \eta_k(a_k + a_k^\dagger). \quad (8.59)$$

Here the notation is as in Chapter 7, particularly, $a_k^\dagger$ and a_k are the creation and annihilation bosonic operators of the kth bath mode, and the collective spin operators in H_{S} and H_{I} are, as before, $\hat{J}_i = (1/2)\sum_j \sigma_j^i$ $(i = x, y, z)$.

The bath interacts separately with each subspace of the system labeled by the total-spin value J, since H commutes with $\hat{J}^2 = \sum_i \hat{J}_i^2$. It is thus sufficient to study the interaction of the bath with a $(2J + 1)$-dimensional system.

The noncommutativity of $\hat{J}_x$ and $\hat{J}_z$ in (8.59) renders the dynamics of the system insolvable. In order to circumvent this difficulty, we prepare the system in an eigenstate of $\hat{J}_x = (1/2)\sum_k \sigma_k^x$ (a superposition of $\hat{J}_z$ eigenstates) and then switch off $H_{\mathrm{S}} = \omega_x \hat{J}_x$. Equivalently, at time $t = 0$ each spin is prepared in a superposition of its σ_k^z (energy) eigenstates, so that the total system is initially in a product of such superposition states. The individual spins are then uncorrelated (unentangled). The initial state of the system can then be written as

$$|\Psi(0)\rangle \equiv |\theta, \phi\rangle = |\psi_1\rangle \otimes |\psi_2\rangle \cdots |\psi_N\rangle, \quad (8.60)$$

with

$$|\psi_j\rangle = \frac{1}{\sqrt{2}}\left(\cos\frac{\theta}{2}|\uparrow\rangle + \sin\frac{\theta}{2}e^{i\phi}|\downarrow\rangle\right), \quad (8.61)$$

where θ and ϕ are the spherical coordinates of the average particle spin. This state is an *eigenstate* of the collective spin operator $\hat{\boldsymbol{J}} \cdot \hat{\boldsymbol{n}}$, $\hat{\boldsymbol{n}}$ being the unit vector corresponding to the angles θ and ϕ.

The ensuing dynamics, for any *bosonic bath*, is then obtained for a given spin-sector J from the joint time-evolution operator of the system and the bath,

$$U_l(t) = \exp\left\{-it\Delta_{\mathrm{L}}(t)\hat{J}_z^2 + \hat{J}_z \sum_k \left[\alpha_k(t)a_k^\dagger - \alpha_k^*(t)a_k\right]\right\}, \quad (8.62)$$

where

$$\Delta_{\mathrm{L}}(t) = \frac{1}{t}\sum_k \eta_k^2 \frac{\omega_k t - \sin\omega_k t}{\omega_k^2}, \quad \alpha_k(t) = \eta_k \frac{1 - e^{i\omega_k t}}{\omega_k}. \quad (8.63)$$

We thus find from (8.62) that the bath-induced evolution is driven by both $\hat{J}_z$ (linear) and $\hat{J}_z^2$ (*nonlinear*) terms. The linear terms give rise to decoherence by mixing the spin (TLS) and the bath states. The nonlinear $\Delta_{\mathrm{L}}(t)\hat{J}_z^2$ term in Eq. (8.62) coherently shifts the energy of the multi-spin system. Since

$$\hat{J}_z^2 = \frac{1}{4} \left(N\hat{I} + \sum_{j \neq j'} \sigma_j^z \sigma_{j'}^z \right),$$ (8.64)

$\hat{I}$ being the identity operator, this nonlinear term arises only for multiple spins and is absent in the single-spin case where $\hat{J}_z^2 = \sigma_z^2/4 = \hat{I}/4$. It causes the collective (cooperative) Lamb shift of each spin by all others. Its physical origin is virtual-quanta exchange among the spins via the bath.

The basis of the system can be chosen to consist of the states $|Jml\rangle$, where m is an eigenvalue of $\hat{j}_z$ and l enumerates the subspaces with a given J. We consider initial states of the form

$$\rho(0) = \sum_J p_J \rho^{(J)}(0),$$ (8.65)

which is a linear combination of density matrices corresponding to different J with the probabilities pJ. Each component $p^{(j)}$ evolves separately. The state of the system, at any time, obtained upon tracing over the bath, has then the from

$$\rho(t) = \sum_J p_J e^{-it\Delta_L(t)\hat{J}_z^2} \left\{ \sum_{m,m'=-J}^{J} \sum_{l,l'} \rho_{ml,m'l'}^{(J)}(0) e^{-t\gamma(t)(m-m')^2} \right.$$

$$\left. \times |Jml\rangle\langle m'l'| \right\} e^{it\Delta_L(t)\hat{J}_z^2}.$$ (8.66)

Here

$$\Delta_L(t) = \frac{1}{t} \int_0^\infty d\omega \, G_0(\omega) \frac{\omega t - \sin \omega t}{\omega^2},$$

$$\gamma(t) = \frac{1}{t} \int_0^\infty d\omega \, G_s(\omega) \frac{1 - \cos \omega t}{\omega^2},$$ (8.67)

where $G_0(\omega)$ is the zero-temperature coupling (response) spectrum of the bosonic bath,

$$G_0(\omega) = \sum_k \eta_k^2 \delta(\omega - \omega_k),$$ (8.68)

and $G_s(\omega) = G_T(\omega) + G_T(-\omega)$ is the symmetric coupling spectrum of the bath (see also Ch. 11),

$$G_s(\omega) = G_s(-\omega) = G_0(\omega) \coth(\beta\omega/2) \quad (\omega > 0),$$ (8.69)

where β is the inverse temperature of the bath.

It is seen from (8.66) that $\Delta_L(t)$ is the bath-induced Lamb shift (the real part of the bath susceptibility, which is temperature independent), whereas $\gamma(t)$ is the bath-induced decoherence rate (the imaginary part of the susceptibility, which increases with the temperature). The off-diagonal terms in (8.66) decay exponentially at the rate $\gamma(t)(m - m')^2$, so that the multipartite coherence (entanglement) is highly fragile under this decoherence.

If $\gamma(t)$ is negligible, then the effect of the nonlinear term $\Delta_L(t)\hat{J}_z^2$ in (8.62) can be obtained in a simple form for several important special cases. The initial state (8.60) can be written in the form

$$|\theta, \phi\rangle = 2^{-N/2} \sum_{n=0}^{N} c^{N-n} s^n e^{in\phi} |\psi_n\rangle, \qquad (8.70)$$

where $c = \cos(\theta/2)$, $s = \sin(\theta/2)$, and $|\psi_n\rangle$ is the sum over all permutations of $\uparrow$ with $\downarrow$ of the state $|\underbrace{\uparrow \ldots \uparrow}_{N-n} \underbrace{\downarrow \ldots \downarrow}_{n}\rangle$. We have $\hat{J}_z |\psi_n\rangle = (N/2 - n) |\psi_n\rangle$.

Hence, at $t > 0$,

$$|\Psi(t)\rangle = e^{-i\zeta \hat{J}_z^2} |\theta, \phi\rangle = \frac{e^{-i\zeta N^2/4}}{2^{N/2}} \sum_{n=0}^{N} e^{-i\zeta n^2} c^{N-n} s^n e^{in\phi'} |\psi_n\rangle, \qquad (8.71)$$

where $\zeta = t\Delta_L(t)$ and $\phi = \phi + N\zeta$. This expression implies that $|\Psi(t)\rangle$ is periodic (up to a phase factor) as a function of ζ with the period 2π (π) for even (odd) N. For $\zeta = (2k + 1)\pi$ and an even N, we have $|\Psi(t)\rangle = (-1)^{N/2} |\theta, \phi+\pi\rangle$.

It is instructive to consider the evolution at fractions of the period, namely, at $\zeta = 2\pi j/k$, where j and k are positive integers ($j < k$). Since for such ζ, $e^{-i\zeta n^2}$ is periodic in n with the period k, it can represented in the form of the discrete Fourier transform,

$$e^{-i2\pi jn^2/k} = \sum_{q=0}^{k-1} f_q e^{-i2\pi qn/k}, \qquad (8.72)$$

where

$$f_q = \frac{1}{k} \sum_{n=0}^{k-1} e^{-i2\pi jn^2/k + i2\pi qn/k}. \qquad (8.73)$$

Inserting (8.72) in (8.71), we obtain, for $\zeta = 2\pi j/k$,

$$|\psi(t)\rangle = e^{-i\pi jN^2/(2k)} \sum_{q=0}^{k-1} f_q |\theta, \phi' - 2\pi q/k\rangle. \qquad (8.74)$$

Thus, at fractions of the period, the state is a sum of at most k components, which represent the initial state rotated around the z axis at different angles. As a result,

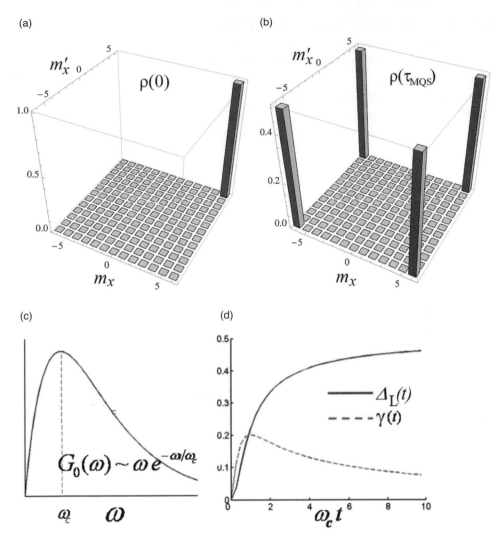

Figure 8.7 Bath-induced macroscopic quantum-superposition (MQS) state. The system is composed of $N = 12$ spins coupled to an Ohmic bath. The cutoff frequency ω_c is chosen such that the collective coupling of the spins to the bath, $\eta = \sqrt{\sum_k \eta_k^2} \sim 0.005\omega_c$. (a)–(b) The absolute values of the density matrix elements, $|\rho_{m_x m_x'}(t)|$, in the $\hat{J}_x$ basis are shown, in the absence of decoherence. (a) The initial state of the system is a spin coherent state, $|\theta = \pi/4, \phi = 0\rangle$. (b) The two off-diagonal peaks signify the formation of an MQS state. (c) The coupling spectrum of an Ohmic bath. (d) The time-dependent functions responsible for the nonlinear Lamb-shift [$\Delta_L(t)$] and decoherence [$\gamma(t)$]. (Adapted from Rao et al., 2011. © 2011 The American Physical Society)

the initial product state (8.60) evolves generally into an entangled macroscopic quantum superposition (MQS) state.

If some of the coefficients f_q vanish, the number of components in (8.74) is less than k. This happens, for example, when k is even: supposing that the fraction j/k is irreducible (i.e., j is odd), the function $f(n) = e^{-i2\pi jn^2/k}$ satisfies $f(n+k/2) = (-1)^{k/2} f(n)$. Using this relation in (8.73), we find that f_q vanish for even (odd) q when $k/2$ is odd (even), whereas the nonzero f_q are given by

$$f_q = \frac{2}{k} \sum_{n=0}^{k/2-1} e^{-i2\pi jn^2/k+i2\pi qn/k} \qquad (q + k/2 \text{ even}). \qquad (8.75)$$

In this case, the function (8.74) contains $k/2$ components. In particular, for $\zeta = \pi/2$ (corresponding to $k = 4$, $j = 1$), we have

$$f0 = f_2^* = e^{-i\pi/4}/\sqrt{2}, \quad f_1 = f_3 = 0, \quad \phi' = \phi + \pi N/2, \qquad (8.76)$$

and the MQS is

$$|\Psi\rangle_{\mathrm{MQS}} = \frac{e^{-i\pi N^2/8}}{\sqrt{2}} \left(e^{-i\pi/4}|\theta, \phi'\rangle + e^{i\pi/4}|\theta, \phi' + \pi\rangle \right). \qquad (8.77)$$

It is a superposition of two equally weighted identical components rotated around the z axis with π-phase difference.

The earliest time when MQS occurs is given by the smallest positive root of the equation

$$\tau_{\mathrm{MQS}} \equiv t = \frac{\pi}{2\Delta_{\mathrm{L}}(t)}, \qquad (8.78)$$

which is independent of the number of spins N.

If, for example, $|\theta, \phi\rangle = |\frac{\pi}{2}, 0\rangle$ and N is even, then (8.77), then is a state with $J = N/2$, which is an equal superposition of the states with the $\hat{J}_x$ eigenvalues $m_x = +N/2$ and $m_x = -N/2$, corresponding to all spins being oriented along $+\hat{x}$ and $-\hat{x}$ directions, respectively [Fig. 8.7(a, b)].

Thus, the interaction with the bath transforms the initially uncorrelated state, $|+x\rangle = |\frac{\pi}{2}, 0\rangle$, under negligible decoherence [$\gamma(t) \approx 0$], into a MQS that is simultaneously oriented along the $+\hat{x}$ and $-\hat{x}$ directions with $\pi/2$ relative phase

$$|+x\rangle \xrightarrow{H_1} \frac{1}{\sqrt{2}} \left(|+x\rangle - (-1)^{N/2} i |-x\rangle \right). \qquad (8.79a)$$

If the spins are rotated by $\pi/2$, then (8.79a) yields

$$\frac{1}{\sqrt{2}} (|\uparrow \cdots \uparrow\rangle + |\downarrow \cdots \downarrow\rangle). \qquad (8.79b)$$

Equations (8.79) represent a macroscopic GHZ-like state in which all the N spins are maximally entangled.

For odd N the results are similar. In particular, (8.79a) is replaced by

$$|+x\rangle \xrightarrow{H_I} \frac{1}{\sqrt{2}} \left(|+y\rangle + (-1)^{(N-1)/2} i |-y\rangle\right) \qquad (8.80)$$

In this case, the MQS is simultaneously oriented along the $+\hat{y}$ and $-\hat{y}$ directions.

Other entangled states ("Schrödinger kittens") will formed at intermediate times. (fractional periods), as discussed above. The same evolution is induced by the nonlinear Kerr Hamiltonian, whereas here it is bath-induced.

The MQS state survives decoherence according to (8.66)–(8.78), if it is induced by $\hat{J}_z^2$ faster than it decays:

$$\tau_{\text{MQS}} \bar{\gamma} N^2 < 1, \qquad (8.81)$$

$\bar{\gamma} N^2$ being the maximal time-averaged decay rate, $\gamma(t)(m - m')^2$, of the off-diagonal (coherence) elements in (8.66).

8.4.2 MQS Formation Conditions

The condition (8.81) for MQS formation is satisfied if $\Delta_L(\tau_{\text{MQS}}) \gg \gamma(\tau_{\text{MQS}})$.

For the satisfaction of condition (8.81), τ_{MQS} should exceed the non-Markovian timescale, as explained in this section. At sufficiently low temperatures, $\gamma(t)$ is drastically reduced in the Markovian limit ($t \gg t_c$) as opposed to its fast initial non-Markovian increase, where t_c, the correlation (memory) time of the bath, is the inverse width of $G_s(\omega)$. Namely,

$$\gamma(t \ll t_c) \gg \gamma(t \to \infty) = \gamma \qquad (8.82)$$

[Figs. 8.7(c, d) and 8.8]. The reason for this trend is that $\gamma(t)$ initially has contributions from all the bath modes, $\int G_s(\omega)d\omega$, but subsequently decreases, as the bath-mode excitations at different frequencies cause the mode states to go out of phase as they approach the Markovian regime. On the other hand, $\Delta_L(t)$ increases in the course of the transition from the non-Markovian to the Markovian regime, so that its long-time value satisfies

$$|\Delta_L(t \to \infty)| \gg |\Delta_L(t \ll t_c)|. \qquad (8.83)$$

Hence, it is beneficial to have τ_{MQS} longer than the bath correlation time t_c, so that MQS formation encounters a much lower γ, and much higher Δ_L, than their non-Markovian counterparts, consistently with (8.81).

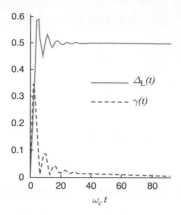

Figure 8.8 As in Figure 8.7, for a spin ensemble that interacts with a photonic cavity bath that has a Lorentzian coupling spectrum with width ω_c centered at $10\omega_c$. The $\Delta_L(t)$ and $\gamma(t)$ functions are shown. (Adapted from Rao et al., 2011. © 2011 The American Physical Society)

As the temperature rises, we are bound to reach the regime $\gamma \geq |\Delta_L|$, where the formation of MQS is prohibited. For an Ohmic bath, the MQS formation condition

$$|\Delta_L| \gg \gamma \tag{8.84}$$

holds as long as the cutoff energy exceeds the thermal energy, $\hbar\omega_c > k_B T$.

These trends are shown in Figure 8.7(c, d) for an Ohmic bath with coupling spectrum $G_0 = \alpha\omega e^{-\omega/\omega_c}$ and in Figure 8.8, for a Lorentzian bath with coupling spectrum $G_0(\omega) = \alpha \frac{\omega_c^2}{\omega_c^2+(\omega-\omega_0)^2}$. Different bath spectra having the same width (i.e., the same inverse correlation time $1/t_c$) may still have different Δ_L and γ values in the long-time Markovian limit and hence yield MQS with different purity values at τ_{MQS}.

A quantitative example of bath-induced MQS formation is provided by a noninteracting atomic TLS ensemble coupled to a single-mode photonic cavity bath. In this case, we can initialize the atomic system governed by $H_S = \hbar\omega_z \hat{J}_z$ in a $\hat{J}_z$ eigenstate and then bring each TLS to degeneracy as $\omega_z \to 0$, by applying Zeeman shifts, so that the TLS attain a degenerate Zeeman state ($H_S = 0$). Then, one can induce $\hat{J}_x$ coupling between the cavity bath and the atomic ensemble by a two-photon (Raman) process. The collective nonlinear evolution can then generate MQS of the atomic ensemble. The cavity, viewed as a Lorentzian bath, induces nonlinear dynamics. The decoherence rate γ is determined by the zero-frequency coupling strength to the cavity $G_s(0)$. This value of γ limits the size of an MQS to $N \sim 100$, for an atom–cavity coupling strength $\eta \sim 1$ MHz according to (8.81).

8.5 Discussion

We have shown that resonant dipole–dipole interactions (RDDI) in multiatom systems can be drastically enhanced or suppressed, as well as extended in range, in appropriately designed field-confining structures. Such modifications allow us to engineer hitherto uncontrollable features of energy, entanglement, or information transfer, as summarized below:

(a) In photonic crystals (PC) or periodic (Bragg-grating) waveguides, RDDI suppression results from self-energy modifications due to a sharp cutoff in the density of modes (DOM), and the analytic continuation of the DOM function beyond the cutoff into a band gap. This modification leads to the elimination of the electrostatic R^{-3}-divergent limit of this interaction. The RDDI is thereby strongly suppressed at interatomic separations that are much smaller than the emission wavelength (Sec. 8.1).

Such a suppression of the RDDI at interatomic separations characterizing quasimolecules (a few A) would have strong implications on their dynamics, spectroscopy and energy transfer properties. Normally, the symmetrized (ungerade) and antisymmetrized (gerade) states of a dimer,

$$|\Psi_{S(A)}\rangle = 2^{-1/2}(|e_1 g_2\rangle \pm |e_2 g_1\rangle), \tag{8.85}$$

where 1 and 2 label the atoms, respectively, are shifted by $\pm \Delta_{12}$ due to RDDI from atomic resonance. These shifts would nearly disappear in a band gap.

In particular, the suppression of donor-acceptor energy transfer by RDDI (known as the Förster–Dexter process) in a band gap would change the fluorescence spectra of donor-acceptor complexes (Fig. 8.9). If the overlap between the emission band of the donor and the absorption band of the acceptor is within the gap, then RDDI suppression would strongly decrease the fluorescence in the emission band of the acceptor, even if the latter band is outside the gap.

(b) We have shown the possibility of long-distance extension of RDDI between dipoles in a waveguide with Bragg grating. Such a nonradiative, deterministic, and coherent process may allow quantum entanglement and information transfer with high fidelity over $\gtrsim 100$ wavelengths (Sec. 8.2).

(c) We have found that two identical atoms interacting via a near-resonant high-Q (narrow-band) cavity mode and an off-resonant bath form a system of *three* mutually coupled excited states (rather than *two* such states in open-space). The competing RDDI and single-atom Rabi splittings interfere, leading to *decoupling* of single-atom excited states at near-zone (quasimolecular) separations. The resulting *suppression* of excitation transfer at quasimolecular separations, may be important in various molecular or excitonic systems within high-Q resonators (Sec. 8.3).

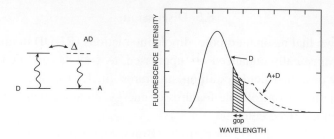

Figure 8.9 Förster–Dexter excitation transfer from a donor (D) to an acceptor (A) via the RDDI. The fluorescence intensity of the donor (D) has a tail caused by the presence of an acceptor (A+D). This tail is expected to be suppressed when the overlap between the emission band of D and the absorption band of A lies within a band gap that suppresses the RDDI. (Adapted from Kurizki, 1990. © 1990 The American Physical Society)

(d) An exactly soluble model has revealed that linear, collective interaction of a multiatom system with a thermal bosonic bath may naturally induce (rather than impede) interatomic entanglement and Schrödinger-cat (MQS) state formation in the system. Only a bosonic bath induces such nonlinear quantum dynamics in the system, whereas a fermionic bath generates also higher-power nonlinearities that may be unfavorable for the generation of high-fidelity MQS states.

9

Quantum Measurements, Pointer Basis, and Decoherence

The consensus regarding quantum measurements rests on two statements: (i) that von Neumann's standard quantum measurement theory leaves undetermined the basis in which observables are measured; (ii) that the environmental decoherence of the measuring device (the "meter") unambiguously determines the measuring ("pointer") basis. The latter statement means that the environment *monitors* (measures) *selected* observables of the meter and (indirectly) of the system. After the decoherence time of the meter, which is typically many orders of magnitude shorter than any other dynamical timescale, a measured quantum state must end up in one of the "pointer states" that persist in the presence of the environment. Here we examine these statements and find that environmental decoherence does not uniquely determine the pointer basis of the meter. By contrast, the measured observable is unique, whether in the presence or in the absence of decoherence.

9.1 Quantum Measurements and Pointer Bases

In the standard (von Neumann) model, a quantum measurement of a system observable is performed *indirectly* by coupling the system S with a "meter" system M and then measuring the latter. Zurek generalized this model by examining what happens when the meter observable (pointer) differs from the "standard pointer," which commutes with the state of the meter. Yet this generalization has given rise to widespread unfounded statements on the fundamentals of quantum measurement theory, as summarized below.

(a) The measured observable of S is uniquely determined by the measured observable of M.
(b) Decoherence "dynamically selects" the pointer basis, whereas other pointer bases cannot be used.

(c) The meter decoherence "dynamically selects" those meter observables that can be measured.

Statement (c) is a consequence of (a) and (b). Below it is shown that statement (a) is not correct, and neither are statements (b) and (c).

9.1.1 Von Neuman's Analysis of Quantum Measurements

Here we consider only the coupling of the quantum system to a quantum meter, both isolated from the environment. The standard (von Neumann) treatment of quantum measurements is then as follows.

Suppose that the observable to be measured is represented in its basis of (orthonormal) eigenstates $|S_n\rangle$, as

$$\hat{S} = \sum_n a_n |S_n\rangle\langle S_n|, \qquad (9.1)$$

with

$$\langle S_m | S_n \rangle = \delta_{mn}, \qquad (9.2)$$

whereas the initial state of the system S is

$$|\psi_S\rangle = \sum_n c_n |S_n\rangle. \qquad (9.3)$$

The system–meter interaction is turned on in the time interval $(0, \tau_M)$. This interaction correlates the initial factorized state of S and M, which then obeys the Schmidt decomposition,

$$|\psi_{SM}(0)\rangle = \sum_n c_n |S_n\rangle \otimes |M\rangle \;\rightarrow\; |\psi_{SM}(\tau_M)\rangle = \sum_n c_n |S_n\rangle \otimes |P_n\rangle, \qquad (9.4)$$

where the meter states also satisfy orthonormality,

$$\langle P_m | P_n \rangle = \delta_{mn}. \qquad (9.5)$$

In the interaction picture, this entangled state is obtained as a result of the action of the system-meter Hamiltonian,

$$H_{SM}^I(t) = e^{i H_0 t} H_{SM}(t) e^{-i H_0 t}, \qquad (9.6)$$

where $H_0 = H_S + H_M$, H_M being the meter Hamiltonian. For a proper measurement of the observable, we impose the back-action evasion condition,

$$[\hat{S}, H_{SM}^I(t)] = 0. \qquad (9.7)$$

We assume that H_{SM} vanishes for $t > \tau_M$, so that the state $|\psi_{SM}(t)\rangle = |\psi_{SM}(\tau_M)\rangle$ remains constant for $t > \tau_M$.

In view of (9.4), a measurement in the basis of the meter states $|P_n\rangle$ yields a system state $|S_n\rangle$ and hence the corresponding eigenvalue a_n of $\hat{S}$ with the probability $|c_n|^2$, whereas the system state "collapses" to $|S_n\rangle$ after the measurement. We can write the meter observable in the form

$$\hat{P} = \sum_n b_n |P_n\rangle\langle P_n|, \tag{9.8}$$

which we dub the *standard pointer* (SP), since it describes ideal (projective) measurements of a quantum system.

If the measurement results are unknown or ignored, we have a *non-selective measurement* of the system observable: it is equivalent to tracing out the meter, which renders the system state diagonal in the measured basis,

$$\rho_S(\tau_M) = \sum_n |c_n|^2 |S_n\rangle\langle S_n|. \tag{9.9}$$

9.1.2 Measurements in Alternative Pointer Bases

Consider, however, another set of basis states of the meter, $\{|R_n\rangle\}$, so that

$$|P_n\rangle = \sum_{n'} |R_{n'}\rangle\langle R_{n'}|P_n\rangle. \tag{9.10}$$

In this basis, the entangled system-meter state in (9.4) changes to

$$|\psi_{SM}(\tau_M)\rangle = \sum_n \sqrt{p_n}\, |\tilde{S}_n\rangle \otimes |R_n\rangle, \tag{9.11}$$

where

$$|\tilde{S}'_n\rangle = \sqrt{p_n}\, |\tilde{S}_n\rangle = \sum_{n'} c_{n'}\langle R_n|P_{n'}\rangle|S_{n'}\rangle, \tag{9.12}$$

$|\tilde{S}_n\rangle$ ($|\tilde{S}'_n\rangle$) being the new normalized (non-normalized) system states, and

$$p_n = \langle \tilde{S}'_m|\tilde{S}'_m\rangle = \sum_{n'} |\langle R_n|P_{n'}\rangle|^2 |c_{n'}|^2. \tag{9.13}$$

The new states are generally not orthogonal,

$$\langle \tilde{S}_m|\tilde{S}_n\rangle = (p_m p_n)^{-1/2} \sum_{n'} |c_{n'}|^2 \langle R_n|P_{n'}\rangle\langle P_{n'}|R_m\rangle \tag{9.14}$$

except in the case where all $|c_n|$ are equal.

Thus, we have replaced the meter observable by

$$\hat{R} = \sum_n r_n |R_n\rangle\langle R_n| \tag{9.15}$$

with some arbitrary eigenvalues r_n. Applying the projection postulate to a measurement of $\hat{R}$ yields, in view of (9.11), the eigenvalue r_n with the probability p_n, and the system is subsequently left in the state $|\tilde{S}_n\rangle$.

This result has been claimed to imply that by measuring $\hat{R}$ we measure an *alternative* observable of the system,

$$\hat{\tilde{S}} = \sum_n \tilde{a}_n |\tilde{S}_n\rangle\langle\tilde{S}_n|, \tag{9.16}$$

which is a Hermitian operator even for non-orthogonal $|\tilde{S}_n\rangle$. However, this claim is incorrect.

In fact, a measurement of a general pointer $\hat{R}$ provides the *same information* as the SP, although in a manner that differs from an ideal (projective) measurement. This can be shown by the formalism of generalized quantum measurements. To this end, consider a projective measurement of the meter resulting in a state $|R_m\rangle$. This measurement leaves the system in the (non-normalized) state

$$|\tilde{S}'_m\rangle = \hat{M}_m |\psi_S\rangle. \tag{9.17}$$

Here $|\psi_S\rangle$ is given by (9.3), and we have introduced the measurement operator $\hat{M}_m$, which equals, in view of (9.12),

$$\hat{M}_m = \sum_n \langle R_m | P_n\rangle |S_n\rangle\langle S_n|. \tag{9.18}$$

The so-called positive-operator valued measure (POVM) operator

$$\hat{E}_m = \hat{M}_m^\dagger \hat{M}_m = \sum_n |\langle R_m | P_n\rangle|^2 |S_n\rangle\langle S_n| \tag{9.19}$$

then yields the probability to obtain the mth outcome [see (9.17)],

$$p_m = \langle\psi_S|\hat{E}_m|\psi_S\rangle = \langle\tilde{S}'_m|\tilde{S}'_m\rangle. \tag{9.20}$$

Thus, the POVM operators are diagonal in the basis $\{|S_n\rangle\}$. This shows that the measurement is performed in the basis $\{|S_n\rangle\}$, *irrespective of the choice of the pointer basis*.

In the special case of a measurement in the SP basis, we have $\langle R_m | P_n\rangle = \langle P_m | P_n\rangle = \delta_{mn}$. Then $\hat{M}_m = \hat{E}_m = |S_m\rangle\langle S_m|$ is a projection operator, and we recover, correspondingly, $p_m = |c_m|^2$ and $|\tilde{S}'_m\rangle \sim |S_m\rangle$. In contrast to projective measurements, generalized (POVM) measurements do not provide the probabilities $|c_n|^2$ directly. Still, in many cases these probabilities can be reconstructed from the measurement results, as we show explicitly.

We can write (9.13) in the form

$$\mathbf{p} = \mathbf{E}\mathbf{c}, \tag{9.21}$$

where $\mathbf{p}$ and $\mathbf{c}$ are column vectors with the components p_n and $|c_n|^2$, respectively, whereas $\mathbf{E}$ is the measurement matrix with the elements $E_{mn} = |\langle R_m | P_n \rangle|^2$. Then $\mathbf{c}$ can be obtained by inverting (9.21),

$$\mathbf{c} = \mathbf{E}^{-1} \mathbf{p}, \tag{9.22}$$

provided $\mathbf{E}^{-1}$ exists, which occurs iff the rows (or, equivalently, columns) of $\mathbf{E}$ are linearly independent. Taking into account that the mth row of $\mathbf{E}$ is the diagonal of the matrix in (9.19), we obtain that *all* $|c_n|^2$ *can be extracted, iff the POVM operators* $\hat{E}_m$ *are linearly independent.*

Typically, this condition is fulfilled, except for special cases. In particular, in the opposite limit where all the POVM operators $\hat{E}_m$ are linear functions (here multiples) of one of them, the condition

$$\sum_m \hat{E}_m = \hat{I}_\mathrm{M}, \tag{9.23}$$

$\hat{I}_\mathrm{M}$ being the identity operator of the meter, yields

$$\hat{E}_m = q_m \hat{I}_\mathrm{M},$$
$$q_m \geq 0, \quad \sum_m q_m = 1. \tag{9.24}$$

In this case, the measurement provides *no information* on the system, since (9.20) then yields the probabilities $p_m = q_m$, which are independent of $|c_n|^2$.

Furthermore, taking into account that

$$\mathrm{Tr}\,\hat{E}_m = \sum_n |\langle R_m | P_n \rangle|^2 = \sum_n \langle R_m | P_n \rangle \langle P_n | R_m \rangle = \langle R_m | R_m \rangle = 1, \tag{9.25}$$

we obtain from (9.24) that $q_m = 1/d$, where d is the system dimensionality, that is,

$$E_{mn} = 1/d, \tag{9.26}$$
$$|\langle R_m | P_n \rangle|^2 = 1/d, \tag{9.27}$$

so that in the present case the actual and standard pointer bases are mutually unbiased. In such cases the measurement results are completely random,

$$p_m = 1/d. \tag{9.28}$$

In the intermediate case, where the number of linearly independent POVM operators is greater than 1 but smaller than d, the results of projective measurements cannot be completely reconstructed, but the measurements still provide some information on the system, since the relations (9.13) yield restrictions on the values of $|c_n|^2$.

9.1.3 Measurements by a Qubit Meter

The foregoing results can be illustrated for a TLS state measured by a qubit meter. For a TLS initially in the general state $|\psi_S\rangle = c_g|g\rangle + c_g|e\rangle$ and a qubit meter in the state $|0\rangle$, a CNOT gate operation on $|\psi_{SM}(0)\rangle = |\psi_S\rangle|0\rangle$ yields the S-M correlated state

$$|\psi_{SM}(\tau_M)\rangle = c_g|g\,0\rangle + c_e|e1\rangle. \tag{9.29}$$

The state (9.29) implies that the SP basis is $\{|0\rangle, |1\rangle\}$, whereas the general pointer basis is

$$|R_0\rangle = a|0\rangle + b|1\rangle, \quad |R_1\rangle = b^*|0\rangle - a^*|1\rangle \tag{9.30}$$

with $|a|^2 + |b|^2 = 1$.

According to (9.17),

$$|\tilde{S}_0'\rangle = a^*c_g|g\rangle + b^*c_e|e\rangle, \quad |\tilde{S}_1'\rangle = bc_g|g\rangle - ac_e|e\rangle. \tag{9.31}$$

Now

$$\mathbf{E} = \begin{pmatrix} |a|^2 & |b|^2 \\ |b|^2 & |a|^2 \end{pmatrix}, \tag{9.32}$$

and its determinant

$$D = |a|^4 - |b|^4 = |a|^2 - |b|^2 = 1 - 2|b|^2. \tag{9.33}$$

When $D \neq 0$, $\mathbf{c}$ can be computed from (9.22), yielding

$$|c_g|^2 = \frac{p_0 - |b|^2}{1 - 2|b|^2}, \quad |c_e|^2 = \frac{p_1 - |b|^2}{1 - 2|b|^2}. \tag{9.34}$$

When $D = 0$, $|a|^2 = |b|^2 = 1/2$, that is, the pointer basis is of the form

$$\left\{ (|0\rangle + e^{i\phi}|1\rangle)/\sqrt{2}, \quad (|0\rangle - e^{i\phi}|1\rangle)/\sqrt{2} \right\}, \tag{9.35}$$

where ϕ is an arbitrary phase. Such pointer bases, lying in the xy-plane of the Bloch sphere, are all the bases that are unbiased with respect to the SP basis. Measurements employing these pointers do not provide information on the system, yielding instead the random results [see (9.28)]

$$p_0 = p_1 = 1/2. \tag{9.36}$$

Result (9.36) disproves the claim that "a Stern-Gerlach magnet with a field gradient in the direction z (which measures the spin component along z) can also measure the spin in the direction y." This claim does not hold since a pointer basis of the form (9.35) does not provide any information on the system, and thus cannot be used for a spin measurement, neither along y nor in any other direction.

Let us consider a measurement in the pointer basis

$$\left\{ (|0\rangle + |1\rangle)/\sqrt{2}, \quad (|0\rangle - |1\rangle)/\sqrt{2} \right\}. \tag{9.37}$$

It yields the (normalized) post-measurement states lying on the x axis,

$$|\tilde{S}_0\rangle = (|g\rangle + |e\rangle)/\sqrt{2}, \quad |\tilde{S}_1\rangle = (|g\rangle - |e\rangle)/\sqrt{2}. \tag{9.38}$$

According to the consensus, one should deduce from (9.38) that the pointer (9.37) leads to a measurement of the system spin along the x-axis, yielding the random result (9.36). However, a projective measurement of the state $|\psi_S\rangle = (|g\rangle + |e\rangle)/\sqrt{2}$ along the x-axis should yield the probabilities 1 and 0, respectively, for the outcomes (9.38), in contradiction to (9.36). Hence, a measurement in the pointer basis (9.37) cannot be considered as a measurement along the x-axis, or any other axis for that matter.

9.2 Decoherence of Entangled System–Meter States

In the preceding section we disproved the statement (a) presented at the beginning of Section 9.1. Below we also disprove statement (b) therein.

Let us now account for the meter–bath (M–B) interaction via a term in the total Hamiltonian written in the interaction representation,

$$H^I(t) = H^I_{SM}(t) + H^I_{MB}(t). \tag{9.39}$$

We assume that there is a nondegenerate meter variable $\hat{Q}$ that satisfies the back-action evasion condition for the M–B interaction,

$$[\hat{Q}, H^I_{MB}(t)] = 0. \tag{9.40}$$

This ensures that the eigenstates $|Q_n\rangle$ of $\hat{Q}$ are invariant under decoherence. The eigenstates $|Q_n\rangle$ are sometimes called "pointer states," although the basis $\{|Q_n\rangle\}$ generally differs from the SP basis.

Still, let us consider first the case where the SP basis coincides with $\{|Q_n\rangle\}$. For simplicity, let us assume that the S–M interaction is much stronger (hence faster) than that of M–B. The whole process then consists of three stages. We may identify as stage 1 the entanglement of S and M in the Schmidt basis,

$$\sum_n c_n |S_n\rangle \otimes |M\rangle \otimes |\phi\rangle \; \rightarrow \; \sum_n c_n |S_n\rangle \otimes |Q_n\rangle \otimes |\phi\rangle, \tag{9.41}$$

where $|\phi\rangle$ is the initial state of the bath B.

On a longer timescale, stage 2, we have the action of the M–B time-evolution operator,

$$U(t) = \hat{I}_S \otimes T_+ \exp\left[-i \int_0^t dt' \sum_n |Q_n\rangle\langle Q_n| \otimes \hat{B}_n(t')\right], \tag{9.42}$$

where $\hat{I}_S$ is the identity operator for the system, T_+ is the chronological (time-ordering) operator, and $\hat{B}_n(t)$ is a bath operator in the interaction representation. On this timescale, an entangled system–meter–bath state ensues,

$$\sum_n c_n|S_n\rangle \otimes |Q_n\rangle \otimes |\phi\rangle \rightarrow \sum_n c_n|S_n\rangle \otimes |Q_n\rangle \otimes |\phi_n(t)\rangle, \tag{9.43}$$

where the time-evolved bath states are

$$|\phi_n(t)\rangle = T_+ \exp\left[-i \int_0^t ds\, \hat{B}_n(s)\right] |\phi\rangle. \tag{9.44}$$

The reduced density matrix of S and M, obtained by tracing out B, is then

$$\rho_{SM}(t) = \sum_{n,m} c_n c_m^* \langle \phi_m(t)|\phi_n(t)\rangle |S_n\rangle |Q_n\rangle \langle S_m|\langle Q_m|. \tag{9.45}$$

The bath-states overlap $\langle \phi_m(t)|\phi_n(t)\rangle$ is describable by

$$|\langle \phi_m(t)|\phi_n(t)\rangle| = \exp[-\Gamma_{mn}(t)], \tag{9.46}$$

with $\Gamma_{mn}(t) \geq 0$ being the decoherence function. This form is exact and holds for any baths and *arbitrary coupling* strengths. Generally, the evaluation of the decoherence function requires detailed knowledge of the M–B interaction *cross-correlation* (time response). At sufficiently long times, the function $\Gamma_{mn}(t)$ becomes linear in time, as in the Markovian case, or nonlinear, and even oscillatory, as long as Markovianity fails (see Ch. 10).

Well beyond the bath correlation time, $t \gg t_c$ (discussed in Chs. 10–12), M decoheres completely,

$$\langle \phi_m(t)|\phi_n(t)\rangle \rightarrow \delta_{mn}, \tag{9.47}$$

and we arrive at the S–M diagonal state,

$$\rho_{SM}(t) \rightarrow \rho_{SM}^\infty = \sum_n |c_n|^2 |S_n\rangle |Q_n\rangle \langle S_n|\langle Q_n|. \tag{9.48}$$

The state ρ_{SM}^∞ is obtained when the meter is measured in the basis of the states $|Q_n\rangle$ nonselectively, without a readout of the measurement results. Decoherence acts as a nonselective projective measurement of the meter by the bath. Since now the standard pointer (SP) coincides with $\hat{Q}$, this means that the bath performs a nonselective projective measurement of the system.

At stage 3, a selective projection measurement of the meter is performed. This measurement is assumed to be fast, so that the evolution of the meter during the measurement may be neglected. In particular, a measurement of $\hat{Q}$ would yield a selective projective measurement of the system.

It is often claimed that the bath selects the meter basis according to its robustness; hence, the states $|Q_n\rangle$ are commonly called "pointer states." This statement, which is equivalent to statement (b) in Section 9.1, is erroneous. It is true that decoherence results in a meter state that is diagonal in the basis $\{|Q_n\rangle\}$, and hence only this basis can yield projective measurements of the system. In other words, decoherence dynamically selects a unique "resilient" basis $\{|Q_n\rangle\}$, whereas any SP basis differing from $\{|Q_n\rangle\}$ cannot yield projective measurements of the system. However, irrespective of the choice of the SP, almost all pointers $\hat{R}$ can be used to obtain the *same information* as that given by projective measurements, as will be now shown.

To this end, consider the general case where the SP $\hat{P}$ does not commute with $\hat{Q}$, since they are determined independently, by the first and second terms in (9.39), respectively. Then, at stage 1 the entangled S-M state is

$$\rho_{\mathrm{SM}}(\tau_{\mathrm{M}}) = |\psi_{\mathrm{SM}}(\tau_{\mathrm{M}})\rangle\langle\psi_{\mathrm{SM}}(\tau_{\mathrm{M}})| = \sum_{l,n} c_n c_l^* |S_n\rangle\langle S_l| \otimes |P_n\rangle\langle P_l|, \qquad (9.49a)$$

where $|\psi_{\mathrm{SM}}(\tau_{\mathrm{M}})\rangle$ is given by (9.4). The meter state traced over S is then

$$\rho_{\mathrm{M}} = \sum_n |c_n|^2 |P_n\rangle\langle P_n|. \qquad (9.49b)$$

Decoherence (at stage 2) renders the state (9.49a) at some time $\tau_{\mathrm{M}}' \gg \tau_{\mathrm{M}}$ diagonal in the basis $\{|Q_n\rangle\}$ of the meter,

$$|P_n\rangle\langle P_l| \;\rightarrow\; \sum_k \langle Q_k|P_n\rangle\langle P_l|Q_k\rangle\, |Q_k\rangle\langle Q_k|, \qquad (9.50)$$

yielding

$$\rho_{\mathrm{SM}}(\tau_{\mathrm{M}}') = \sum_{k,l,n} c_n c_l^* \langle Q_k|P_n\rangle\langle P_l|Q_k\rangle\, |S_n\rangle\langle S_l| \otimes |Q_k\rangle\langle Q_k|. \qquad (9.51)$$

Now the meter state becomes

$$\rho_{\mathrm{M}} = \mathrm{Tr}_{\mathrm{S}}\, \rho_{\mathrm{SM}}(\tau_{\mathrm{M}}') = \sum_n p_n' |Q_n\rangle\langle Q_n|, \qquad (9.52)$$

where the column vector $\boldsymbol{p}' = \{p_n'\}$ is given by

$$\boldsymbol{p}' = \mathbf{E}'\boldsymbol{c}, \qquad (9.53)$$

$\mathbf{E}'$ being a matrix with the elements

$$E'_{mn} = |\langle Q_m | P_n \rangle|^2. \tag{9.54}$$

The matrix $\mathbf{E}'$ is doubly stochastic (as is $\mathbf{E}$ above), that is, it satisfies

$$\sum_m E'_{mn} = \sum_n E'_{mn} = 1. \tag{9.55}$$

A comparison of (9.52) with the state (9.49b) obtained at stage 1 shows that decoherence rotates the eigenbasis of the meter state from $\{|P_n\rangle\}$ to $\{|Q_n\rangle\}$ and changes the eigenvalues from $|c_n|^2$ to p'_n. Since $\mathbf{E}'$ is doubly stochastic, $\boldsymbol{p}'$ is majorized by $\boldsymbol{c}$. As a result, the state (9.52) is more mixed and has a higher von Neumann entropy (i.e., is more randomized) than (9.49b). In other words, decoherence (or a nonselective measurement), in general, *partially erases* the quantum information in a state. The erasure is absent only when $\{|Q_n\rangle\}$ coincides with $\{|P_n\rangle\}$.

At stage 3 ($t \geq \tau'_M$) the meter undergoes a projective measurement in some basis $\{|R_n\rangle\}$. An observation of the m outcome results in the (unnormalized) state of the system,

$$\begin{aligned} \rho'_m &= \text{Tr}_M[\rho_{SM}(\tau'_M)(\hat{I}_S \otimes |R_m\rangle\langle R_m|)] \\ &= \sum_{k,l,n} c_n c_l^* \langle Q_k | P_n \rangle \langle P_l | Q_k \rangle |\langle R_m | Q_k \rangle|^2 |S_n\rangle\langle S_l|. \end{aligned} \tag{9.56}$$

Now, the post-measurement state is generally mixed and can be written in the operator-sum representation as

$$\rho'_m = \sum_k \hat{M}_{mk} |\psi_S\rangle\langle\psi_S| \hat{M}^\dagger_{mk}, \tag{9.57}$$

in terms of the Kraus operators

$$\hat{M}_{mk} = \langle R_m | Q_k \rangle \sum_n \langle Q_k | P_n \rangle |S_n\rangle\langle S_n|. \tag{9.58}$$

The measurement probabilities are given by

$$p_m = \text{Tr}\,\rho'_m = \langle\psi_S|\hat{E}_m|\psi_S\rangle, \tag{9.59}$$

with the POVM operators being

$$\hat{E}_m = \sum_k \hat{M}^\dagger_{mk}\hat{M}_{mk} = \sum_n E_{mn} |S_n\rangle\langle S_n|. \tag{9.60}$$

Here the matrix $\mathbf{E} = \{E_{mn}\}$ is given by

$$\mathbf{E} = \mathbf{E}''\mathbf{E}', \tag{9.61}$$

where

$$\mathbf{E}'' = \{\,|\langle R_m|Q_n\rangle|^2\,\}. \tag{9.62}$$

The POVM operators (9.60) are diagonal in the basis $\{|S_n\rangle\}$, which means that the system is measured in this basis irrespective of the choice of the meter basis $\{|P_n\rangle\}$, $\{|Q_n\rangle\}$, or $\{|R_n\rangle\}$. The probabilities p_m are related to $|c_n|^2$ by (9.21), even though the matrix $\mathbf{E}$, given by (9.61), now differs from $\mathbf{E}$ in Section 9.1.2. Correspondingly, the quantities $|c_n|^2$ can be reconstructed by (9.22), provided $\mathbf{E}^{-1}$ exists (i.e., when all the POVM operators $\hat{E}_m$ are linearly independent). This holds if and only if the determinants of both $\mathbf{E}'$ and $\mathbf{E}''$ are nonzero.

Let us now consider notable special cases:

(i) When the actual pointer basis $\{|R_n\rangle\}$ coincides with $\{|Q_n\rangle\}$, the measurements are performed as described in Section 9.1.2. Namely, the meter–environment interaction and the resulting decoherence, which is equivalent to a projective measurement, does not affect the results of measurements in the basis $\{|Q_n\rangle\}$. Indeed, since now $\langle R_m|Q_k\rangle = \delta_{mk}$, (9.58) yields $\hat{M}_{mk} = \hat{M}_m\delta_{mk}$, which implies that the post-measurement states in (9.57) are pure,

$$\rho'_m = \hat{M}_m|\psi_S\rangle\langle\psi_S|\hat{M}_m^\dagger = |\tilde{S}'_m\rangle\langle\tilde{S}'_m|, \tag{9.63}$$

in agreement with (9.17). This can be understood from the fact that the results of two consecutive projective measurements of the meter should coincide. In this case, the measurement of S is not necessarily projective. It becomes projective only when the bases $\{|P_n\rangle\}$, $\{|Q_n\rangle\}$, and $\{|R_n\rangle\}$ coincide, as mentioned above.

(ii) When the basis $\{|P_n\rangle\}$ coincides with $\{|Q_n\rangle\}$ [see (9.48)], but the actual pointer basis $\{|R_n\rangle\}$ differs from $\{|Q_n\rangle\}$, we have $\langle Q_k|P_n\rangle = \delta_{kn}$ and find that the measurement probabilities and the POVM operators are now the same as in Section 9.1.2. However, the post-measurement states are mixed, being diagonal in the basis $\{|S_n\rangle\}$ [cf. (9.56)],

$$\rho'_m = \sum_n |\langle R_m|Q_n\rangle|^2|c_n|^2\,|S_n\rangle\langle S_n|. \tag{9.64}$$

The diagonal elements of (9.64) coincide with those of the states $|\tilde{S}'_m\rangle\langle\tilde{S}'_m|$ obtained in Section 9.1.2 [see (9.17)].

(iii) The POVM $\hat{E}_m$ in (9.60) have the same properties as in (9.23) and are normalized by $\text{Tr}\,\hat{E}_m = 1$. Therefore, by the same argument as in Section 9.1, such a measurement does not yield any information on the system and provides completely random results (9.28) under the condition (9.26), that is, iff

$$\sum_k |\langle R_m | Q_k \rangle|^2 \, |\langle Q_k | P_n \rangle|^2 = 1/d. \tag{9.65}$$

This condition holds, in particular, when the $\{|Q_n\rangle\}$ basis is mutually unbiased with either the SP $\{|P_n\rangle\}$ or the $\{|R_n\rangle\}$ basis. Hence, decoherence (or a nonselective measurement) in a basis that is mutually unbiased with the SP basis completely erases information on the system.

iv) Decoherence has a remarkable effect on measurements in mutually unbiased bases $\{|P_n\rangle\}$ and $\{|R_n\rangle\}$. In the absence of decoherence (Sec. 9.1), a pair of such bases provides improper measurements that do not yield information on the system. However, when decoherence occurs in the basis $\{|Q_n\rangle\}$, which is not mutually unbiased with any of $\{|P_n\rangle\}$ and $\{|R_n\rangle\}$, condition (9.65) generally *does not hold* (i.e., information is not erased by a measurement in the latter two bases).

This effect is counterintuitive, since decoherence in the meter obliterates information on the system, at least partially. Nevertheless, we see that *decoherence can turn improper measurements into proper ones*. This can be explained by the fact that decoherence rotates the meter-state eigenbasis $\{|P_n\rangle\}$ into the basis $\{|Q_n\rangle\}$, which is not mutually unbiased with $\{|R_n\rangle\}$.

9.3 Qubit Meter of a TLS Coupled to a Bath

As an illustration of the above analysis, let us consider a TLS that is being measured by a qubit meter, with *energy-degenerate states* $|0\rangle$, $|1\rangle$. We may then set the meter Hamiltonian to be zero,

$$H_{\mathrm{M}} = 0. \tag{9.66}$$

The measurement in the TLS energy basis $\{|g\rangle, |e\rangle\}$ is effected via a time-dependent TLS–meter coupling of the form

$$H_{\mathrm{SM}} = h(t)|e\rangle\langle e|(\hat{I}_{\mathrm{M}} - \hat{\sigma}_x^{\mathrm{M}}). \tag{9.67}$$

Here $\hat{\sigma}_x^{\mathrm{M}} = |0\rangle\langle 1| + |1\rangle\langle 0|$ and

$$h(t) = \frac{\pi}{4\tau_{\mathrm{M}} \cosh^2(t/\tau_{\mathrm{M}})} \tag{9.68}$$

is a smooth temporal profile of the TLS coupling to the qubit meter during the measurement that occurs in the interval centered at $t = 0$ with duration τ_{M}.

This form of the measurement Hamiltonian H_{SM} corresponds to

$$e^{-i \int_{-\infty}^{\infty} dt\, H_{\mathrm{SM}}(t)} = U_{\mathrm{CN}}, \tag{9.69}$$

where U_{CN} is the controlled-not (CNOT) operation:

$$\begin{aligned}
|g\rangle|0\rangle &\mapsto |g\rangle|0\rangle, \\
|e\rangle|0\rangle &\mapsto |e\rangle|1\rangle, \\
|g\rangle|1\rangle &\mapsto |g\rangle|1\rangle, \\
|e\rangle|1\rangle &\mapsto |e\rangle|0\rangle.
\end{aligned} \tag{9.70}$$

If the initial state of the meter is $|0\rangle$, only the first two rows are involved in the measurement. If the measurement duration τ_M is much shorter than all other timescales, then only H_{SM} is non-negligible, and the entire action of $H(t)$ during this time is well approximated by the operator U_{CN}. This approximation becomes exact in the impulsive limit $\tau_M \to 0$.

If we allow for the meter–bath interaction via

$$H_{MB} = |1\rangle\langle 1| \otimes \hat{B}_1 + |0\rangle\langle 0| \otimes \hat{B}_0, \tag{9.71}$$

where $\hat{B}_1$ and $\hat{B}_0$ are bath operators that have orthogonal eigenstates, then we may describe the first two stages of the measurement process (Sec. 9.2) as follows:
(a) Stage 1 results in a Schmidt-decomposed S–M correlated state:

$$(c_e|e\rangle + c_g|g\rangle)|0\rangle|\phi\rangle \;\rightarrow\; (c_e|e\rangle|1\rangle + c_g|g\rangle|0\rangle)|\phi\rangle. \tag{9.72}$$

(b) Stage 2 yields the reduced S-M density matrix upon tracing out B:

$$\begin{aligned}
\rho_{SM} = |c_e|^2|e\rangle\langle e||1\rangle\langle 1| + |c_g|^2|g\rangle\langle g||0\rangle\langle 0| \\
+ c_e c_g^* \varepsilon(t)|e\rangle\langle g||1\rangle\langle 0| + c_g c_e^* \varepsilon^*(t)|g\rangle\langle e||0\rangle\langle 1|,
\end{aligned} \tag{9.73}$$

where $\varepsilon(t) = \langle\phi_0(t)|\phi_1(t)\rangle$, the time-evolved bath states $|\phi_n(t)\rangle$ being given by (9.44). This state pertains to the SP basis of M $\{|0\rangle, |1\rangle\}$, which satisfies the back-action evasion condition (9.40), and hence its states are invariant under the decoherence.

At short times, the decoherence function

$$\varepsilon(t) \approx 1 - \frac{t^2}{2}\left\langle(\hat{B}_0 - \hat{B}_1)^2\right\rangle, \tag{9.74}$$

so that the coherences in (9.73) remain nearly intact. They disappear only at long times, such that $\varepsilon(t) \to 0$, rendering (9.73) diagonal,

$$\rho_{SM} \to |c_e|^2|e\rangle\langle e||1\rangle\langle 1| + |c_g|^2|g\rangle\langle g||0\rangle\langle 0|. \tag{9.75}$$

The measurement outcomes may be averaged over by tracing out the meter degrees of freedom, corresponding to nonselective measurements. The total effect of stages 1+2 on the system density-operator is then

$$\rho_S \to \mathrm{Tr}_M[U_{CN}(\rho_S \otimes |0\rangle\langle 0|)] = |e\rangle\langle e|\rho_S|e\rangle\langle e| + |g\rangle\langle g|\rho_S|g\rangle\langle g|, \tag{9.76}$$

(i.e., the diagonal elements are unchanged, whereas the off-diagonal elements are erased). This means that under nonselective measurements, the decoherence of the meter plays no role, since the trace over the meter is independent of the pointer basis.

Consider now *selective generalized measurements*, according to the theory developed in the previous sections. In the present case, the SP basis $\{|P_n\rangle\}$ coincides with $\{|Q_n\rangle\}$. Therefore case iv in Section 9.2 is not applicable here. The remaining cases i–iii merit consideration:

(1) Case i now corresponds to the situation where the actual pointer basis $\{|R_n\rangle\}$ coincides with $\{|P_n\rangle\}$ and $\{|Q_n\rangle\}$. In this case, measurements of the system are projective, and decoherence does not affect them.

(2) In case ii, $\{|R_n\rangle\}$ is arbitrary, being given by (9.30). The measurement results are obtained now as in the absence of decoherence; see (9.34). The post-measurement states are mixed due to the meter decoherence. They are given by (9.64), which yields

$$\rho_0' = |a|^2|c_g|^2|g\rangle\langle g| + |b|^2|c_e|^2|e\rangle\langle e|,$$
$$\rho_1' = |b|^2|c_g|^2|g\rangle\langle g| + |a|^2|c_e|^2|e\rangle\langle e|. \tag{9.77}$$

(3) In case iii, the decoherence does not affect the measurement results. The pointer bases that do not provide information on the system are the same as in (9.35).

9.4 Discussion

Pointer states have been defined by Zurek as the ones that are minimally entangled with the bath following their interaction. To find them for an initial pure state $|\Psi\rangle$, one quantifies the entanglement generated between the system and the bath by the von Neumann entropy

$$S_\Psi(t) = -\mathrm{Tr}[\rho_\Psi(t)\log\rho_\Psi(t)] \tag{9.78}$$

obtained for the reduced density matrix of the system $\rho_\Psi(t)$ [which is initially $\rho_\Psi(0) = |\Psi\rangle\langle\Psi|$]. Pointer states are then obtained by minimizing S_Ψ over $|\Psi\rangle$ and demanding robustness under time variation.

When the dynamics is dominated by the system Hamiltonians, the "pointer" states coincide with the energy eigenstates of this Hamiltonian. This regime conforms with the view that decoherence induced by the bath *"monitors"* the system and selects its "pointer" states. In the absence of the system Hamiltonian, decoherence selects eigenstates of the system–bath interaction Hamiltonian. The same conclusions apply to a meter that is coupled to a bath and measures the system.

However, as we have shown, *a pointer basis is not uniquely selected by decoherence*: There is a broad variety of pointer bases pertaining to a meter under the influence of a bath that still allow us to extract at least partial, and in some cases complete, information on the system. If we allow for measurements that are much faster than the decoherence time of the meter, then the standard pointer (SP) basis determined by the Schmidt decomposition is fully equivalent to almost any other pointer basis, since all pointer bases allow us to extract the same information on the system. In subsequent chapters (Chs. 10–13), we show how to avoid or, at least, minimize information loss due to decoherence, thereby allowing us to restore the complete equivalence of *nearly all pointers*.

10

The Quantum Zeno and Anti-Zeno Effects (QZE and AZE)

According to a prevailing view, quantum mechanics implies that frequently repeated measurements must slow down the evolution of a quantum system and, in particular, the decay of any unstable quantum system. This allegedly general feature, known as the Quantum Zeno Effect (QZE), was discovered by Misra and Sudarshan in their follow-up paper on the early work of Khalfin and Fonda. It has been colloquially described by expressions such as *"a watched arrow never flies,"* or *"a watched pot never boils."* We here show that the QZE is not as general as commonly believed (Fig. 10.1).

In fact, the *opposite* is mostly true for quantum states that decay into open-space continua or baths: The anti-Zeno effect (AZE), that is, *decay acceleration* by frequent measurements, is more ubiquitous than the QZE. This *universal* conclusion, reached by the authors of this book in 2000 (and by Facchi and Pascazio in 2001), has been colloquially referred to as *"a watched pot boils faster"* or *"furtive glances trigger decay."* We explain how the QZE and AZE can be understood, what restricts the QZE universality, and what are the experimental requirements for observing each of these effects.

10.1 The QZE in a Closed System

Let us assume that a system in the initial state $|\psi\rangle$ is subject to impulsive projective measurements of the observable $\hat{X} = |\psi\rangle\langle\psi|$ that are performed at time intervals τ. In between measurements, the state vector of the system, $|\psi(t)\rangle$, is governed by Hamiltonian H and evolves according to the Schrödinger equation

$$i\hbar\frac{\partial}{\partial t}|\psi(t)\rangle = H|\psi(t)\rangle. \tag{10.1}$$

(a)

(b)

Figure 10.1 The Zeno and anti-Zeno effects: The ancient philosopher Zeno of Elea is trying to reassure Wilhelm Tell and his son that the arrow will not fly under observation, in accordance with the Zeno effect. However, they are in for a nasty surprise, since the arrow obeys the anti-Zeno effect and flies faster than in the absence of observation. (Reprinted from Weizmann Magazine, 2000.)

The Schrödinger equation yields for sufficiently small t,

$$|\psi(t)\rangle = \left(\hat{I} - \frac{i}{\hbar}Ht - \frac{1}{2\hbar^2}H^2t^2 + \ldots \right)|\psi\rangle. \qquad (10.2)$$

The probability of populating the state $|\psi\rangle$ following a projection carried out at time $t = \tau$ is here the survival probability of the initial state, given by

$$p(\tau) = |\langle \psi | \psi(\tau) \rangle|^2 = 1 - \frac{1}{\hbar^2}(\Delta E)^2 \tau^2 + O(\tau^4), \tag{10.3}$$

the energy variance in the state $|\psi\rangle$ being

$$(\Delta E)^2 = \langle \psi | H^2 | \psi \rangle - \langle \psi | H | \psi \rangle^2. \tag{10.4}$$

Terms of order τ^4 or higher are neglected here and hereafter.

The probability that the system remains in the initial state after each of the n consecutive measurements is, to the same accuracy,

$$p(t_{\text{tot}}) \approx [1 - (\Delta E)^2 \tau^2 / \hbar^2]^n, \tag{10.5}$$

where $t_{\text{tot}} = n\tau$. Upon keeping t_{tot} fixed and taking the limit of continuous observation, $\tau = t_{\text{tot}}/n \to 0$, (10.5) becomes

$$p(t_{\text{tot}}) \approx \left[1 - \frac{(\Delta E)^2 \tau t_{\text{tot}}}{\hbar^2 n} \right]^n \approx \exp\left[-\frac{1}{\hbar^2}(\Delta E)^2 \tau t_{\text{tot}} \right] \to 1. \tag{10.6}$$

The system thus remains, with near certainty, in the initial state $|\psi\rangle$ if frequent measurements (at intervals $\tau \to 0$) of the observable $\hat{X}$ are performed. Namely, the state reduction induced by frequent measurements strongly inhibits the departure of the system from its initial state. This comes about because the probability of leaving the state $|\psi\rangle$ is

$$1 - p(\tau) \propto \tau^2, \tag{10.7}$$

whereas the number n of measurements increases as τ^{-1}. Hence, the projection onto the initial state by the frequent measurements proceeds at a higher rate than transitions into other states. Misra and Sudarshan named this phenomenon the quantum Zeno effect (QZE), as it evokes the arrow paradox (by Zeno of Elea in the fifth century BC), whose aim was to show that motion is an illusion because an (apparently) flying arrow cannot move in between extremely frequent observations: it is "frozen" when consecutive observations overlap.

We have thus far assumed that the measurement results are known, that is, measurements are selective (Ch. 9), corresponding to the probability $p(t_{\text{tot}})$ that the system is found in the initial state in all measurements. This assumption underlay the early discussions of the QZE. But does the QZE hold for *nonselective* measurements, whose results are unknown or discarded? The answer is affirmative: Even though the probability $p(t_{\text{tot}})$ cannot be measured if the measurements are nonselective, one can observe the slowdown of the density-matrix evolution, since both selective and nonselective measurements cause frequent state reduction, which is key to the QZE, when the measurement rate is extremely high (i.e., in the limit

$\tau \to 0$). As discussed in Chapters 11 and 16, QZE and the AZE can be emulated by purely unitary processes that do not involve measurements.

Instead of projective, strong measurements, one may consider weak measurements that produce only partial state reduction, so that complete state reduction is achieved after several weak measurements. Frequent weak measurements produce results similar to those of strong measurements, provided that the time interval between projective measurements τ is replaced by the characteristic time interval required for complete state reduction.

The QZE induced by nonselective, weak measurements can be illustrated for a two-level system (TLS) that undergoes coherent oscillations at the Rabi frequency Ω between the levels under the influence of the Hamiltonian

$$H = -(\Omega/2)\sigma_x. \tag{10.8}$$

If the system is subject to frequent (or continuous) weak nonselective measurements of the observable σ_z, the mean values of the components of $\boldsymbol{\sigma}$ can be shown to obey the equations of motion, which stem from the appropriate master equation (Ch. 11),

$$\frac{d}{dt}\langle\sigma_x\rangle = -\frac{1}{\tau_d}\langle\sigma_x\rangle,$$

$$\frac{d}{dt}\langle\sigma_y\rangle = -\frac{1}{\tau_d}\langle\sigma_y\rangle + \Omega\langle\sigma_z\rangle,$$

$$\frac{d}{dt}\langle\sigma_z\rangle = -\Omega\langle\sigma_y\rangle. \tag{10.9}$$

Here the continuous weak measurements cause the coherence decay (alias dephasing) at a rate $1/\tau_d$, τ_d being the characteristic time needed to complete a measurement (state reduction).

In the strongly overdamped regime $\Omega\tau \ll 1$, one can set $d\langle\sigma_y\rangle/dt \approx 0$ in (10.9) for $t \gg \tau_d$, then exclude $\langle\sigma_y\rangle$ from the two last equations and obtain the rate equation

$$\frac{d}{dt}\langle\sigma_z\rangle = -2\gamma\langle\sigma_z\rangle \tag{10.10}$$

with the decay rate

$$\gamma = \frac{\Omega^2\tau_d}{2}. \tag{10.11}$$

Taking into account that $\langle\sigma_z\rangle = \rho_{ee} - \rho_{gg}$ and $\rho_{ee} + \rho_{gg} = 1$, we then obtain

$$\dot{\rho}_{ee} = -\dot{\rho}_{gg} = -\gamma\rho_{ee} + \gamma\rho_{gg}. \tag{10.12}$$

In this regime, the system monotonically converges to the stationary state. In the limit $\tau_d \to 0$ we have $\gamma \to 0$, so that the evolution is totally suppressed (see

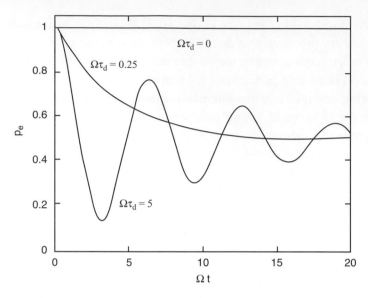

Figure 10.2 The excited $|e\rangle$-state population $\rho_{ee}(t)$ of a TLS with eigenstates $|e\rangle$, $|g\rangle$ as a function of time (in units of the inverse Rabi frequency Ω^{-1}) for different values of the characteristic measurement time τ_d (for the present case of continuous measurements, it is the interval between consecutive state reductions). Each curve corresponds to another value of $\Omega\tau_d$: $\Omega\tau_d = 10$ exhibits underdamped oscillation of $\rho_{ee}(t)$, whereas $\Omega\tau_d = 0.5$ corresponds to overdamped oscillations, and $\Omega\tau_d = 0$ is the QZE limit of evolution freezing.

Fig. 10.2), and the lifetime of the excited state $|e\rangle$ becomes infinite (i.e., the system remains frozen in its initial state). The first experimental observation of the QZE by Itano et al. involved such a system.

Under selective measurements, the probability of no-decay (10.3) tends to zero exponentially at long t, whereas under nonselective measurements [cf. (10.12)] the population of the initial state generally does not vanish at long t (see Fig. 10.2). Nevertheless, the decay rates in both cases obey similar expressions, as discussed below.

10.2 Open-System Decay Modified by Measurements

10.2.1 Short-Time Decay

We next analyze short-time decay in a system governed by the Hamiltonian $H = H_0 + V$, where V causes the decay of a state $|e\rangle$ to other eigenstates of H_0. We shall refer to the set $\{|n\rangle\}$ of these eigenstates as the "bath" (Fig. 10.3), as in an open system.

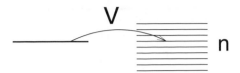

Figure 10.3 The decay of a state into a "bath" consisting of H_0-eigenstates labelled by n via coupling V.

The probability amplitude $\alpha(t)$ of survival in $|e\rangle$, at the energy $\hbar\omega_a$, satisfies the following *exact* integrodifferential equation

$$\dot{\alpha} = -\int_0^t dt' e^{i\omega_a(t-t')} \Phi(t-t')\alpha(t'), \qquad (10.13)$$

where $\alpha(t) = \langle e|\Psi(t)\rangle e^{i\omega_a t}$ and

$$\Phi(t) = \hbar^{-2}\langle e|V^{-iH_0 t/\hbar}V|e\rangle = \hbar^{-2}\sum_n |V_{en}|^2 e^{-i\omega_n t} \qquad (10.14)$$

is the bath autocorrelation function (Ch. 2), expressed by $V_{en} = \langle e|V|n\rangle$, where $|n\rangle \ (\neq |e\rangle)$ are H_0 eigenstates with energies $\hbar\omega_n$, which we treat as the bath.

Equation (10.13) is *exactly* soluble, but here we investigate its short-time behavior, which is obtainable by setting $\alpha(t') \approx \alpha(0) = 1$ in (10.13). This results in the expression

$$\alpha(t) = 1 - \int_0^t dt' e^{i\omega_a t'}(t-t')\Phi(t'), \qquad (10.15)$$

which encompasses *all powers of t* (in the phase factors!) and allows for interference between various decay channels. By contrast, the standard *quadratic* expansion in t yields the population time-dependence [cf. (10.3)],

$$\rho_{ee}(t) = |\alpha(t)|^2 \approx 1 - t^2/\tau_Z^2, \qquad (10.16a)$$

where the Zeno time

$$\tau_Z = \hbar/\Delta E_e \qquad (10.16b)$$

is the inverse variance of the energy in $|e\rangle$ [cf. Eq. (10.4)]. Yet the estimate (10.16) may fail, as discussed below, since (10.15) may not only yield the QZE but also its inverse, the anti-Zeno effect (AZE), depending on the short-time behavior of $|\alpha(t)|^2$.

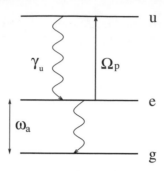

Figure 10.4 The Cook scheme allows to observe the QZE in a decay process from $|e\rangle$ to $|g\rangle$. (Reprinted from Kofman and Kurizki, 2000.)

10.2.2 Decay Modification by Realistic Measurements

Thus far, we have assumed in this section instantaneous measurements of $|e\rangle$. Does a more realistic description of measurements support these foregoing results? The answer is *affirmative* for the two possible types of measurements of $|e\rangle$:

(a) *Impulsive measurements* are realizable, as in Cook's scheme (Fig. 10.4): The decay process is then repeatedly interrupted by short pulses, each pulse transferring the population of $|e\rangle$ to a higher state $|u\rangle$, which then decays back to $|e\rangle$ through photon emission. The basic requirement for impulsive measurements is that the duration of the $|e\rangle \rightarrow |u\rangle$ transfer followed by $|u\rangle \rightarrow |e\rangle$ decay is much shorter than all other timescales in the process, notably the memory time of the bath as discussed below. The same procedure may be carried out by similarly acting on the initially unoccupied state $|g\rangle$.

(b) *Continuous measurements* that act *weakly* on the system, causing only partial state reduction at any given time. Such measurements should not be misconstrued as the limit of successive projections separated by vanishingly short intervals, $\tau \rightarrow 0$: Such infinitely frequent projections are *unphysical*, corresponding to an infinite energy spread $\hbar/\tau$. Continuous measurements are feasible when the state is monitored incessantly, but they still require *a finite time τ* for completing an observation (i.e., they have a *finite effective rate* $1/\tau_d$).

Measurements can be either selective or nonselective, regardless of whether they are impulsive or continuous. Thus, measurements in Cook's scheme are selective if the photons emitted at the $|u\rangle \rightarrow |e\rangle$ transition are detected and the detection results are read out. The measurements are nonselective if the results are unread. The same is true for continuous measurements. Finally, if the photons are not detected, then the process is unitary but has the same effect on the TLS evolution as nonselective measurements. Such a unitary process that emulates nonselective measurements was implemented in the pioneering experiment by Itano et al. (1990).

In what follows we discuss generic examples of frequent impulsive and continuous measurements that can bring about the QZE or AZE.

10.2.3 Impulsive Measurements of an Open TLS

As a generic example of frequent impulsive measurements, we consider an initially excited TLS, here a two-level atom, coupled to a bath with *arbitrary* density-of-modes (DOM) spectrum $\rho(\omega)$ – here a photonic bath in the vacuum state. At time τ a short electromagnetic (EM) pulse is applied. This pulse effects an impulsive quantum measurement of the excited state $|e\rangle$ according to Cook's scheme (Fig. 10.4), since the auxiliary state $|u\rangle$ decays back to $|e\rangle$ incoherently, and *the coherence of the evolution is disrupted.*

To second order in the TLS-bath interaction, $|\alpha(\tau)|^2 \approx 2\mathrm{Re}\,\alpha(\tau) - 1$. Let us assume that the measurements are selective. The probability that the TLS remains in the excited state after each of the k consecutive measurements then has the form

$$p_e(t = k\tau) = |\alpha(\tau)|^{2k} \approx e^{-\gamma t}, \qquad (10.17)$$

where [from (10.15)] the decay rate has the τ-dependent form,

$$\gamma(\tau) = \frac{2}{\tau}\,\mathrm{Re}[1 - \alpha(\tau)] = \frac{2}{\tau}\,\mathrm{Re} \int_0^\tau dt\,(\tau - t)\Phi(t)e^{i\Delta t}. \qquad (10.18)$$

The QZE occurs whenever $\gamma(\tau)$ *decreases with* τ, because $\Phi(t)$ is assumed to fall off slower than the chosen *interruption interval* τ. Equivalently, the correlation (or memory) time of the bath is assumed to be longer than τ.

Equation (10.18) can be rewritten upon replacing the time domain by the spectral domain, yielding

$$\gamma(\tau) = 2\pi \int G(\omega) \left\{ \frac{\tau}{2\pi}\mathrm{sinc}^2 \left[\frac{(\omega - \omega_\mathrm{a})\tau}{2} \right] \right\} d\omega, \qquad (10.19)$$

where $\mathrm{sinc}\,x = \frac{\sin x}{x}$. Here the interruptions sample the response spectrum $G(\omega)$ over the spectral width $\sim 1/\tau$ of the sinc^2-function. The spectral response is identical, according to the Golden Rule, to the emission rate into this bath at frequency ω (Ch. 5), $G(\omega) = |\eta(\omega)|^2 \rho(\omega)$.

A similar result is obtained for nonselective impulsive measurements, if excitation transfer from the bath to the TLS can be neglected. This happens when the bath is at zero temperature and $G(\omega)$ is sufficiently broad so that any excitation transferred to the bath does not return to the system. Then, for an initially excited TLS, the population of the level $|e\rangle$ is

$$\rho_{ee}(t) = e^{-\gamma t}, \qquad (10.20)$$

where γ is given by (10.19).

10.2.4 Open-TLS Continuous Dephasing

As mentioned above, the QZE is obtained via both selective and nonselective measurements, since all measurements yield state reduction, which corresponds to the
destruction of coherence between the initial state and all other states. However, the
coherent TLS evolution may be disrupted not only by measurements. In particular,
effects of nonselective measurements can be emulated by means of coherence dis-
ruption due to *random* fluctuations of the TLS frequency via, for example, *random*
ac-Stark shifts of the level $|e\rangle$ or $|g\rangle$, caused by an off-resonant intensity-fluctuating
field. When the population of the level $|e\rangle$ is averaged over the noise realizations,
it satisfies Eq. (10.20), where γ in (10.19) is now replaced by

$$\gamma = 2\pi \int G(\omega) L(\omega - \omega_{\mathrm{a}}) d\omega. \tag{10.21}$$

In this formula, $L(\omega - \omega_{\mathrm{a}})$ is the Lorentzian-shaped (normalized to 1) relaxation
spectrum of the coherence element $\rho_{eg}(t)$, which is the Fourier transform of the
exponentially decaying $\rho_{eg}(t)$. This behavior represents the common dephasing
model. The width of this Lorentzian relaxation spectrum is $\tau_{\mathrm{d}}^{-1} = \langle \Delta\omega^2 \rangle \tau_c$, which
is the product of the mean-square Stark shift and the noisy-field correlation time.
For (10.21) to be valid, γ should be much less than this spectral width, $\gamma \tau_{\mathrm{d}} \ll 1$.
A *necessary* condition for the QZE is that *the noise-induced width τ_d^{-1} be larger
than the width of the spectral response $G(\omega)$*, as detailed below.

The random ac-Stark shifts both shift and broaden the spectral transition. In
order to avoid the shifting, we may employ a *continuous* driving field that is *reso-
nant* (or *nearly resonant*) with the $|e\rangle \leftrightarrow |u\rangle$ transition. This process is described
by the same scheme as in Figure 10.4, the only difference being that the impulsive
field $\Omega_{\mathrm{d}}(t)$ is replaced by a continuous field. Provided the decay rate of this transition, γ_{u}, is larger than the Rabi frequency Ω_{d} of the driving field, γ can be shown
to be given by (10.21), with a Lorentzian (dephasing) width

$$\frac{1}{\tau_{\mathrm{d}}} = \frac{\Omega_{\mathrm{d}}^2}{2\gamma_{\mathrm{u}}}. \tag{10.22}$$

10.2.5 Universal Formula

The decay rate γ [cf. (10.19), (10.21)] in both of the above schemes is seen to
conform to the same *universal formula* (Fig. 10.5),

$$\gamma = 2\pi \int G(\omega) F(\omega - \omega_{\mathrm{a}}) d\omega, \tag{10.23}$$

where $G(\omega)$ is the spectral bath response, whereas $F(\omega)$ (normalized to 1) is the
spectrum of the coherence fluctuations due to the measurement- or noise-induced

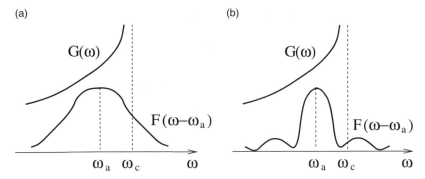

Figure 10.5 Effective decay rate γ as a convolution of the dephasing (relaxation) spectrum $F(\omega - \omega_a)$ and the bath response $G(\omega)$: (a) Lorentzian dephasing spectrum characteristic of noise; (b) sinc-function spectrum (characteristic of repeated impulsive measurements). (Reprinted from Harel et al., 1998. © 1998 OSA)

dephasing: $F(\omega)$ may be, for example, sinc-shaped,

$$F(\omega) = \frac{\tau}{2\pi} \text{sinc}^2 \frac{\omega\tau}{2}, \tag{10.24}$$

or Lorentzian-shaped,

$$F(\omega) = L(\omega) \equiv \frac{1}{\pi} \frac{\tau_d}{\omega^2\tau_d^2 + 1}. \tag{10.25}$$

The (universal) result (10.23) can be rewritten in the form

$$\gamma = \int \gamma_{GR}(\omega) F(\omega - \omega_a) d\omega, \tag{10.26}$$

where $\gamma_{GR}(\omega) = 2\pi G(\omega)$ is the unperturbed ("Golden Rule") decay rate of $|e\rangle$ whose energy is shifted to $\hbar\omega$. Equation (10.26) allows us to interpret the modification of the decay rate as resulting from the energy broadening (uncertainty) ΔE of the level $|e\rangle$, the shape of the level broadening being described by $F(\omega)$. This broadening is caused either by repeated measurements or by the noise (Fig. 10.6), both of which dephase the state $|e\rangle$ at the rate ν, where ν is the characteristic rate of either the measurements or the noise. Such processes are analogous to phase randomization by collisions, which induce a linewidth that scales with the collision rate ν. The trade-off between the spectral broadening of $|e\rangle$ and its dephasing time conforms to the *time-energy uncertainty relation*,

$$\Delta E \Delta t \gtrsim \hbar. \tag{10.27}$$

To obtain the QZE behavior, it is thus *necessary* that the measurement- or noise-induced $F(\omega)$ be *broader* than $G(\omega)$.

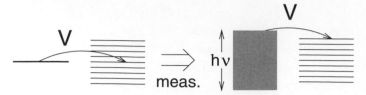

Figure 10.6 Measurements broaden level $|e\rangle$ at rate ν, drastically changing its decay into the reservoir.

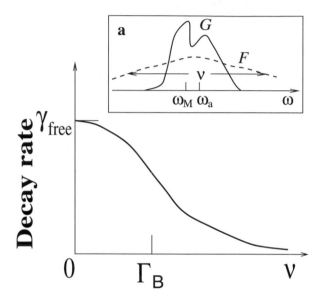

Figure 10.7 Decay rate as a function of the measurement or noise rate ν, in the QZE limit (see inset). (Adapted from Kofman and Kurizki, 2000.)

10.3 QZE and AZE Scaling

A graphical analysis of the universal formula (10.23) yields the conclusions detailed below concerning the occurrence of the QZE or its inverse, the AZE. Each effect expresses a different dependence (scaling) of the level decay rate on the rate ν of measurements or noise for a given spectral bath response $G(\omega)$.

10.3.1 QZE Scaling

The QZE scaling is marked by a *decrease of the decay rate γ with an increase of ν*. It is obtained when the measurement (or dephasing) rate ν is much larger than the spectral width and the detuning of the bath (Fig. 10.7):

$$\nu \gg \Gamma_{\mathrm{B}}, \quad \nu \gg |\omega_{\mathrm{a}} - \omega_{\mathrm{M}}|. \tag{10.28}$$

Here we have assumed that the main part of the integral $\int_0^\infty d\omega G(\omega)$ is concentrated in an interval of the width of order of Γ_B and ω_M is a frequency within this interval. In the special case of a peak-shaped $G(\omega)$, ω_M can be replaced by the position of the maximum. In the limit (10.28), one can approximate the spectrum $G(\omega)$ by a δ-function ($\Gamma_B = 0$).

There is, however, a caveat associated with the limit (10.28). When $G(\omega)$ is too narrow, the evolution (in the absence of measurements or dephasing) is generally non-monotonic and hence cannot be described by means of a positive decay rate. For example, for $G(\omega) \sim \delta(\omega - \omega_a)$, the decay rate is undefined, because the population of state $|e\rangle$ demonstrates resonant Rabi oscillations without decay, periodically exchanging energy with an infinitely narrow band of modes, in accordance with the Jaynes–Cummings model. Namely, condition (10.28) for the QZE presumes the weak-coupling regime of the system and the bath. This regime always holds for a sufficiently broad and smooth $G(\omega)$. Even when $G(\omega)$ is very narrow or has sharp features, so that the weak-coupling regime does not hold in the absence of measurements, this regime is still valid for a sufficiently high rate ν of repeated measurements (or dephasing). This comes about since, in the latter case, the energy level $|e\rangle$ is broadened, acquiring spectral density $F(\omega - \omega_a)$, so that its coupling with the bath is described by the spectrum $G(\omega)$, which is smoothed out by its convolution with $F(\omega - \omega_a)$ [cf. (10.23)].

Assuming that (10.28) holds, the universal formula (10.23) yields the characteristic form of QZE,

$$\gamma \approx C/\nu. \tag{10.29}$$

Here C is the integrated bath-coupling spectrum or, equivalently, the variance of the coupling Hamiltonian in the state $|e\rangle$,

$$C = \int G(\omega)d\omega = \langle e|V^2|e\rangle, \tag{10.30}$$

and we have introduced the general definition,

$$\nu = [2\pi F(0)]^{-1}. \tag{10.31}$$

From (10.24) and (10.25) we then find that $\nu = 1/\tau$ for *instantaneous projections* and $\nu = 1/(2\tau_d)$ for continuous measurements (dephasing). Thus, when $\tau = 2\tau_d$, impulsive and continuous measurements produce the *same decay slowdown* due to the QZE.

Graphically, the origin of the QZE is the *flattening of the spectral peak of $G(\omega)$ by its convolution with a much broader function $F(\omega)$.* Physically, the QZE arises since *the effective decay rate under extremely frequent probing is averaged over all channels*, many of which are negligibly weak. This averaging incurs energy uncertainty that grows with the measurement rate.

The quadratic expansion that yields the QZE may fail since, according to (10.28), the QZE conditions may be more stringent than the requirement $t \sim 1/\nu \ll \tau_Z$. The crucial point stressed below is that (10.28) may be *principally impossible* to satisfy in many scenarios.

The validity of (10.29) presumes that the integral in (10.30) converges (i.e., that $G(\omega)$ falls off faster than $1/\omega$ for $\omega \to \infty$). Yet, it suffices that

$$G(\omega) \to 0 \quad \text{at} \quad \omega \to \infty, \tag{10.32}$$

for the QZE to hold under condition (10.28), except that γ may then decrease with ν more slowly than ν^{-1}, as explained below.

QZE behavior is typically exhibited by frequently interrupted decay into an *asymmetric* peak-shaped bath, which diminishes fast below cutoff, at $\omega < \omega_c$, and slowly falls off above it,

$$G(\omega) = A(\omega - \omega_c)^{-\beta} \quad \text{for} \quad \omega - \omega_c \gg \Gamma_B, \tag{10.33}$$

where $0 < \beta < 1$. For instance, this spectrum holds near a waveguide cutoff, where $\beta = 1/2$. Under condition (10.28), ω_M can be set to ω_c in this case. The QZE scaling of γ for such a bath response, assuming a sufficiently weak coupling ($A \ll \nu^{\beta+1}$), is found to be

$$\gamma = B\nu^{-\beta}. \tag{10.34}$$

Here $\nu = 1/\tau$ and $B = (4\sqrt{2\pi}/3)A$ for $\beta = 1/2$ or, more generally, $B = \{\pi/[\cos(\pi\beta/2)\Gamma(2+\beta)]\}A$, $\Gamma()$ being the gamma-function, for $0 < \beta < 1$. The result (10.34) holds also when the slowly decreasing tail (10.33) is on the low-frequency side of $G(\omega)$, as for $G(\omega) = A(\omega_c - \omega)^{-1/2}$ ($\beta = 1/2$) below the Debye cutoff.

Although condition (10.28) ensures the QZE scaling of γ (for the weak system–bath coupling), it *does not necessarily imply a monotonous decrease* of γ as the rate ν increases (see Fig. 10.7). Monotonic decrease occurs, for instance, when $G(\omega)$ is a peak and ω_a is within its width Γ_B, that is,

$$|\omega_a - \omega_M| \lesssim \Gamma_B. \tag{10.35}$$

Yet, if (10.35) does not hold, the scaling of γ with ν may be highly non-monotonic, as shown in Sections 10.3 and 10.4.

10.3.2 AZE Scaling

The QZE scaling fails whenever ω_a is strongly detuned from the nearest maximum of $G(\omega)$ at ω_m, that is, $G(\omega_a) \ll G(\omega_m)$. Let ν be much smaller than the detuning

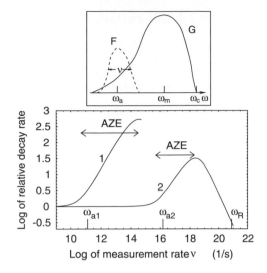

Figure 10.8 Inset: Conditions for the AZE. Graph: the dependence of the log-arithm of the decay rate (normalized to the Golden-Rule decay rate γ_{RG}) $\log_{10}(\gamma/\gamma_{RG})$ on $\log_{10}\nu$ for a spontaneously emitting hydrogen-atom state. The atomic transition frequencies corresponding to curves 1, 2 are $\omega_{a1} = 1.32 \times 10^{11}$ s^{-1}, $\omega_{a2} = 1.55 \times 10^{16}$ s^{-1}. The corresponding Bohr frequencies are $\omega_{B1} = 1.22 \times 10^{15}$ s^{-1}, $\omega_{B2} = 8.50 \times 10^{18}$ s^{-1}. The relativistic cutoff $\omega_R = 7.76 \times 10^{20}$ s^{-1}. The AZE ranges are marked. (Adapted from Kofman and Kurizki, 2000.)

of ω_a from ω_m, yet larger than δ_a, the interval around ω_a over which $G(\omega)$ changes appreciably (Fig. 10.8 - inset):

$$\delta_a \lesssim \nu \ll |\omega_m - \omega_a|. \tag{10.36a}$$

In this range, the decay rate γ *grows* with ν, since the dephasing function $F(\omega)$ then probes more of the *rising* part of $G(\omega)$ in the convolution. This yields the scaling

$$\gamma \propto \nu^\beta \quad (\beta > 0). \tag{10.36b}$$

This scaling is the signature of the *anti-Zeno effect* (AZE) of decay acceleration by frequent measurements. Physically, this effect arises whenever, as the rate ν increases, *the state of the probed system decays into more and more channels, whose weight $G(\omega)$ is progressively larger.*

Remarkably, we may impose condition (10.36a) in *any bath that is not spectrally flat*, which attests to the *universality* of the AZE.

10.3.3 Golden-Rule Limit

In the limit $\nu \ll \delta_a$, (10.23) yields *the Golden-Rule (GR) decay rate* in the absence of measurements,

$$\gamma_{GR} = 2\pi G(\omega_a). \tag{10.37}$$

This rate is the Markovian long-time limit of the decay rate, the limit in which the system relaxes only into the *resonant* modes of the bath.

10.3.4 Intermediate Scaling

More subtle behavior occurs in the intermediate range between the QZE and AZE regimes. Let us assume, for simplicity, that $G(\omega)$ is single-peaked and satisfies condition (10.32). When ν increases up to the range $\nu \gg |\omega_m - \omega_a|$, then condition (10.28) implies the QZE scaling of (10.29) or (10.34). Yet γ *remains larger than the Golden Rule rate* γ_{GR} (10.37), up to much higher ν, according to the following condition for *"genuine QZE,"*

$$\gamma < \gamma_{GR} \quad \text{for} \quad \nu > \nu_{QZE}, \tag{10.38}$$

where in the case of a finite C

$$\nu_{QZE} = \frac{C}{\gamma_{GR}} = \frac{C}{2\pi G(\omega_a)}, \tag{10.39a}$$

or in the case of (10.33)

$$\nu_{QZE} = \left(\frac{B}{\gamma_{GR}}\right)^{1/\beta} = \left[\frac{B}{2\pi G(\omega_a)}\right]^{1/\beta}. \tag{10.39b}$$

The rate ν_{QZE} may be much greater than the minimal ν value of the QZE-scaling regime, as shown in Section 10.4.3.

The value of ν_{QZE} given by (10.39a) was termed the reciprocal "jump time," (i.e., the longest time interval between measurements for which the decay rate is appreciably changed). However, the reciprocal jump time is in fact δ_a in (10.36a), which may be smaller by many orders of magnitude than ν_{QZE}.

10.4 QZE and AZE Scaling in Various Baths

10.4.1 Decay in Lorentzian Bath

The above analysis pertains to the case of a TLS coupled to a near-resonant Lorentzian bath centered at ω_s, as, for example, in a high-Q cavity mode. In this case (Sec. 5.1),

$$G_s(\omega) = L(\omega - \omega_s) = \frac{g_s^2 \Gamma_s}{\pi[\Gamma_s^2 + (\omega - \omega_s)^2]}, \tag{10.40}$$

where g_s is the resonant coupling strength and Γ_s is the line width. Here $G_s(\omega)$ represents the sharply varying part of the DOM distribution, associated with the narrow cavity mode. The broad portion of the DOM distribution $G_b(\omega)$ represents the TLS coupling to the unconfined (background) free-space modes. The *exponential* decay factor in the excited state probability is then additive,

$$\gamma = \gamma_s + \gamma_b, \tag{10.41}$$

where γ_s is the contribution to γ from the sharply varying modes and $\gamma_b = 2\pi G_b(\omega_a)$ is the rate of spontaneous emission into the background modes.

Since the Fourier transform of the Lorentzian $G_s(\omega)$ is $\Phi_s(t) = g_s^2 e^{-\Gamma_s t}$, (10.15) becomes at short times (without the background-modes contribution)

$$\alpha(\tau) \approx 1 - \frac{g_s^2}{\Gamma_s - i\delta}\left[\tau + \frac{e^{(i\delta - \Gamma_s)\tau} - 1}{\Gamma_s - i\delta}\right], \tag{10.42}$$

where $\delta = \omega_a - \omega_s$. The QZE condition is then expressed by

$$\tau \ll (\Gamma_s + |\delta|)^{-1}, g_s^{-1}. \tag{10.43}$$

On resonance ($\delta = 0$), (10.18) and (10.42) yield

$$\gamma_s = g_s^2 \tau. \tag{10.44}$$

Whereas the background-DOM contribution cannot be changed by the QZE, the sharply varying DOM contribution γ_s may allow for the QZE. Only the γ_s term decreases with τ due to the QZE. However, since Γ_s has dropped out of (10.44), the decay rate γ is the *same* for both strong-coupling ($g_s > \Gamma_s$) and weak-coupling ($g_s \ll \Gamma_s$) regimes. Physically, this comes about since the energy uncertainty of the emitted quantum for $\tau \ll g_s^{-1}$ is too large to allow for distinction between reversible and irreversible evolutions.

The evolution inhibition has a different meaning for the two regimes. In the weak-coupling regime, the excited-state population decays *nearly exponentially* at the rate $g_s^2/\Gamma_s + \gamma_b$ (at $\delta = 0$), so that irreversible decay is inhibited, in the spirit of the original QZE prediction. By contrast, in the strong-coupling regime, the damped Rabi oscillations at the frequency $2g_s$ of the excited-state population are destroyed by repeated measurements. Yet, in both cases the QZE slows down the evolution, resulting in the same exponential law, with the rate (10.44).

10.4.2 QZE and AZE for Intracavity Radiative Decay

Consider atoms within an open cavity that repeatedly interact with a pump laser, which is resonant with the $|e\rangle \rightarrow |u\rangle$ transition frequency. The resulting $|e\rangle \rightarrow |g\rangle$ spontaneous-emission rate is monitored as a function of the laser-pulse repetition rate $1/\tau$. Each short pump pulse of duration t_p and Rabi frequency Ω_p is followed

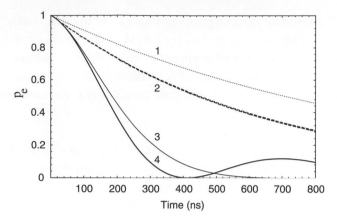

Figure 10.9 Excited-state population p_e in a TLS coupled to cavity mode with Lorentzian lineshape on resonance ($\delta = 0$) as a function of time: 1 – decay to background-mode continuum at rate $\gamma_b = 10^6$ s^{-1}; 3 – uninterrupted decay in a cavity with finesse $F = 10^5$, L_c=15 cm, and f_c=0.02 ($\Gamma_s = 6.3 \times 10^6$ s^{-1}, $g_s = 4.5 \times 10^6$ s^{-1}); 4 – idem, but for $F = 10^6$ ($\Gamma_s = 2 \times 10^6$ s^{-1}; damped Rabi oscillations); 2 – evolution replacing *both* curves 3 and 4, under measurements at intervals $\tau - 3 \times 10^{-8}$ s. Here $\Gamma_s = (1 - r_m)c/L$ and $g_s = \sqrt{cf_c\gamma_b/(2L_c)}$, where r_m is the mirror reflectivity, f_c is the fractional solid angle (normalized to 4π) subtended by the confocal cavity, and L_c is the cavity length. (Adapted from Kofman, A. G., and Kurizki, G. 1996. *Phys. Rev. A*, **54**, R3750. © 1996 The American Physical Society)

by spontaneous decay from $|u\rangle$ back to $|e\rangle$, at a rate γ_u. This *destroys the coherence* of the atomic system, as well as *reshuffles the population* from $|e\rangle$ to $|u\rangle$ and back (Fig. 10.4). Since the interval between measurements must significantly exceed the measurement time, we impose the inequality $\tau \gg t_p$. This inequality can be reduced to the requirement $\tau \gg \gamma_u^{-1}$ if the "measurements" are performed by π pulses, such that $\Omega_p t_p = \pi$, $t_p \ll \gamma_u^{-1}$. This implies choosing a $|u\rangle \to |e\rangle$ transition with a much shorter radiative lifetime than that of $|e\rangle \to |g\rangle$.

Figure 10.9, describing the QZE for a Lorentzian line on resonance ($\delta = 0$), assuming feasible cavity parameters, shows that the population of $|e\rangle$ decays nearly exponentially at times well within the interruption intervals τ, but when those intervals become too short, the decay is strongly inhibited.

Figure 10.10 shows that the detuning $\delta = \omega_a - \omega_s$ renders the decay oscillatory. The interruptions by measurements now *enhance* the decay, in accordance with the AZE.

10.4.3 QZE and AZE for Free-Space Radiative Decay

The spectral response for hydrogenic-atom radiative decay via the $\boldsymbol{p} \cdot \boldsymbol{A}$ free-space interaction (4.1) is given by the exact expression (without the dipole approximation),

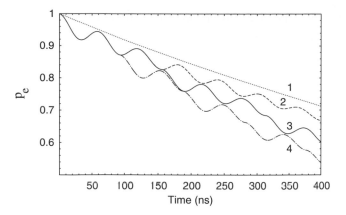

Figure 10.10 As in Figure 10.9, for detuning $\delta = 10^8 \text{s}^{-1}$ and $F = 10^6$: 1 – decay to background-mode continuum; 2 – uninterrupted free evolution; 3 – interrupted evolution at intervals $\tau = 5\pi \times 10^{-8}$ s ($\delta\tau = 5\pi$); 4 – idem, for $\tau = 3\pi \times 10^{-8}$ s ($\delta\tau = 3\pi$). (Adapted from Kofman, A. G., and Kurizki, G. 1996. *Phys. Rev. A*, **54**, R3750. © 1996 The American Physical Society)

$$G(\omega) = \frac{g_e \omega}{[1 + (\omega/\omega_c)^2]^4}, \tag{10.45}$$

where g_e is the effective field–atom coupling constant. The cutoff frequency is the Bohr frequency

$$\omega_c \approx 10^{19} \text{ s}^{-1} \sim \frac{c}{a_B}, \tag{10.46}$$

where a_B is the Bohr radius: Physically, ω_c is the inverse time of photon escape from the atom. Using measurement control that causes Lorentzian broadening [(10.21), (10.25)] we obtain

$$\gamma = \frac{g_e \omega_c}{3} \text{Re} \left[\frac{f(2f^4 - 7f^2 + 11)}{2(f^2 - 1)^3} - \frac{6f \ln f}{(f^2 - 1)^4} - \frac{3i\pi(f^2 + 4f + 5)}{16(f + 1)^4} \right], \tag{10.47}$$

where

$$f = \frac{2\nu - i\omega_a}{\omega_c}, \qquad \nu = (2\tau_d)^{-1}. \tag{10.48}$$

For

$$\nu \ll \omega_c, \tag{10.49}$$

we obtain from (10.47) the *AZE* of accelerated decay, caused by the *rising* of the spectral response $G(\omega) \approx g_e \omega$ as a function of frequency (for $\omega \ll \omega_c$).

The QZE can occur only for $\nu \gtrsim \omega_c \sim 10^{19}$ s^{-1}. Yet, this frequency range is well beyond the limit of validity of the present analysis, since $\Delta E \sim \hbar\nu \gtrsim \hbar\omega_c$

is then close to the relativistic cutoff $\hbar\omega_R$ (Fig. 10.8) where pair creation becomes possible, giving rise to decay channels other than spontaneous transitions to $|g\rangle$. By contrast, (10.47) may yield the following AZE signature for open-space radiative decay,

$$\gamma \approx 2g_e\nu[\ln(\omega_c/\nu) + C_1], \qquad C_1 = -0.339, \qquad (10.50)$$

provided

$$\omega_a \ll \nu \ll \omega_c. \qquad (10.51)$$

The AZE should be observable (Fig. 10.8) for $\nu \gtrsim \omega_a$. For Rydberg transitions at microwave frequencies this requires τ on a picosecond scale, while for optical transitions τ must be on the attosecond scale.

The transition from the AZE to the QZE range occurs at $\nu \sim \omega_c$, so that for $\nu \gg \omega_c$ (10.29) holds, where now $C = g_e\omega_c^2/6$. The genuine-QZE condition (10.38) corresponds to

$$\nu > \nu_{QZE} = \omega_c^2/(12\pi\omega_a) \qquad (10.52)$$

[cf. (10.39a)]. Only this condition ensures that $\gamma < \gamma_{GR} = 2\pi g_e\omega_a$.

According to this analysis, the "genuine QZE" range $\nu > \nu_{QZE}$ is *principally unattainable* for spontaneous emission (open-space radiative decay) at an optical transition, where ω_c conforms to (10.46). The reason is that this QZE range then requires measurement rate ν above the relativistic cutoff ω_R. Such rates may, however, be *detrimental* to the atom, capable of electron–positron pair creation. A similar situation may arise for nuclear decay. By contrast, the AZE is *accessible* in spontaneous emission or β-decay. The required rate ν can essentially always be chosen to induce the AZE regime which corresponds to pulse rates that exceed the atomic transition frequency ω_a as per (10.51).

10.4.4 QZE and AZE for Photon Depolarization

We next consider the decay of photon polarization into a noise bath induced by a polarization-randomizing element (polarization rotator) that is controlled by a noisy field (see Fig. 10.11). Although the photon evolution is unitary, it emulates the effects of different types of measurements: selective or nonselective, as well as strong or weak measurements.

We consider the photon polarization evolution at the time instants that are multiples of the round-trip time τ_r in the cavity. Let us denote by P_h (P_v) the probability to find a photon with horizontal (vertical) polarization. We first assume that the polarization rotator rotates the polarization by angle $\Delta\varphi$ in each passage of the

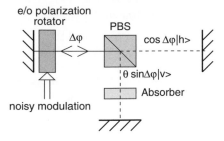

Figure 10.11 Control scheme for the polarization decay of photons bouncing between the mirrors of a cavity. Measurements are performed by a polarization beam-splitter (PBS) and an absorber with transparency θ that varies between 0 and 1. The bath into which the polarization decays is realized by a noisy field that modulates a Pockels cell which rotates the polarization by a random angle $\Delta\varphi$. (Reprinted from Kurizki et al., 2001. © 2001 The American Physical Society)

photon. Coherent, uninterrupted evolution without absorption (transparency $\theta = 1$) then corresponds to Rabi oscillations of the horizontal polarization probability:

$$P_h(n) = \cos^2(n\Delta\varphi), \tag{10.53}$$

where n is the number of photon round trips. In this case, the polarization beam splitter (PBS) in Figure 10.11 has no effect whatsoever on the polarization evolution.

The opposite limit of a perfect absorber, $\theta = 0$, corresponds to decay of the horizontal polarization probability via perfect (projective) measurements resulting in

$$P_h(n) = \cos^{2n}(\Delta\varphi), \tag{10.54}$$

which for small rotation angles, $|\Delta\varphi| \ll 1$, becomes

$$P_h(n) = e^{-n(\Delta\varphi)^2}. \tag{10.55}$$

This decay is slower than Rabi oscillations and thus signifies the QZE. This case corresponds to selective measurements, since each round trip results in a projection of the polarization state to the horizontal polarization.

The case $0 < \theta < 1$ corresponds to nonselective, imperfect (weak) measurements, for which the evolution of P_h can be obtained analytically. For small rotation angles and sufficiently strong absorption such that $|\Delta\varphi| \ll 1 - \theta$, the evolution is an exponential decay expressed by

$$P_h(t = n\tau_r) = \exp\left[-\frac{(\Delta\varphi)^2}{\tau_r^2 \nu}t\right]. \tag{10.56}$$

Here, as before, τ_r is the round-trip time and

$$\nu = \frac{1-\theta}{1+\theta}\frac{1}{\tau_r} \tag{10.57}$$

is the measurement rate that decreases with the increase of the transparency θ. According to (10.56), the decay rate *diminishes with* ν. This regime conforms with the QZE. In the extremal case $\theta = 0$, (10.56) reduces to (10.55), as expected.

We next consider random rotation angles $\Delta\varphi_n$, caused by noisy modulation of the polarization rotator. For sufficiently small angles, a master equation (ME) can be derived for the evolution of the polarization probabilities averaged over the realizations of the random process and smoothed over the evolution time. This ME yields the following rate equations for probabilities of the two orthogonal polarizations,

$$\dot{P}_h = -\gamma\,P_h + \gamma\,P_v,$$
$$\dot{P}_v = \gamma\,P_h - (\gamma + 2\Gamma_0)\,P_v. \tag{10.58}$$

Here

$$2\Gamma_0 = -2\ln\theta/\tau_r \tag{10.59}$$

is the time-averaged photon absorption rate. The vertical-polarization decay rate in (10.58) is given by

$$\gamma = \frac{1+C\theta}{1-C\theta}\frac{\delta\varphi^2}{\tau_r}. \tag{10.60}$$

This rate depends on $\delta\varphi^2 = \langle\Delta\varphi_n^2\rangle$, the variance of the random rotation angles $\Delta\varphi_n$, whereas the mean $\langle\Delta\varphi_n\rangle$ is assumed to be zero; C is the correlation parameter of consecutive polarization-angle jumps $\Delta\varphi_n$ and $\Delta\varphi_{n+1}$ which quantifies their degree of correlation as per the relation

$$\langle\Delta\varphi_{n+1}\Delta\varphi_n\rangle = C\langle\Delta\varphi_n^2\rangle \quad (-1 \le C \le 1). \tag{10.61}$$

The validity condition for (10.58)–(10.60) is

$$\delta\varphi^2 \ll (1-C)(1-C\theta), \tag{10.62}$$

that is, the rotation-angle variance should be sufficiently small, and C should not be too close to 1.

The general solution of (10.58) is

$$P_h(t) = e^{-(\gamma+\Gamma_0)t}\left(\cosh St + \frac{\Gamma_0}{S}\sinh St\right), \tag{10.63}$$

where $S = \sqrt{\gamma^2 + \Gamma_0^2}$. This solution is physically meaningful only at the discrete moments that are multiples of τ_r. Let us consider the special cases of this solution:

(a) In the absence of measurements ($\theta = 1$), the solution of (10.58) is

$$P_{\mathrm{h}}(t) = \frac{1 + e^{-2\gamma_0 t}}{2}, \tag{10.64}$$

with

$$\gamma_0 = \frac{1 + \mathcal{C}}{1 - \mathcal{C}} \frac{\delta\varphi^2}{\tau_{\mathrm{r}}}. \tag{10.65}$$

Thus, random polarization fluctuations lead to complete depolarization, $P_{\mathrm{h}}(t), P_{\mathrm{v}}(t) \to 1/2$ as $t \to \infty$.

(b) For strong selective measurements ($\theta = 0$), the polarization probability exponentially decays to zero,

$$P_{\mathrm{h}}(t) = e^{-\gamma t}, \tag{10.66}$$

as one would expect from (10.17).

(c) For weak measurements ($\theta > 0$), the exponential dependence (10.66) still holds, as long as $\Gamma_0 \gg \gamma$.

We next investigate the polarization decay rate (10.60), which obeys the universal formula (10.23),

$$\gamma = 2\pi \int_{-\pi/\tau_{\mathrm{r}}}^{\pi/\tau_{\mathrm{r}}} d\omega\, G(\omega) F(\omega). \tag{10.67}$$

Here

$$G(\omega) = \frac{\delta\varphi^2}{2\pi\tau_{\mathrm{r}}} \frac{1 - \mathcal{C}^2}{1 + \mathcal{C}^2 - 2\mathcal{C}\cos\omega\tau_{\mathrm{r}}} \tag{10.68}$$

is the spectral density of the rotation-angle fluctuations, whereas the measurement function

$$F(\omega) = \frac{\tau_{\mathrm{r}}}{2\pi} \frac{1 - \theta^2}{1 + \theta^2 - 2\theta\cos\omega\tau_{\mathrm{r}}} \tag{10.69}$$

depends on the measurement parameter θ. In the present setup, the frequency domain for $G(\omega)$ and $F(\omega)$ is restricted to the interval $(-\pi/\tau_{\mathrm{r}}, \pi/\tau_{\mathrm{r}})$, since the time evolution is discrete.

For projective (strong) measurements ($\theta = 0$), $F(\omega)$ acquires the constant value $\tau_{\mathrm{r}}/(2\pi)$, whereas for $\theta \neq 0$ $F(\omega)$ has a peak with the maximum at $\omega = 0$ and the characteristic width $\nu = [2\pi F(0)]^{-1}$ [cf. (10.31)] given by (10.57).

For highly correlated jumps ($\mathcal{C} \approx 1$) we find that

$$G(\omega) \approx \frac{\delta\varphi^2}{\pi\tau_{\mathrm{r}}^2} \frac{\Gamma_{\mathrm{B}}}{\Gamma_{\mathrm{B}}^2 + \omega^2}, \tag{10.70}$$

that is, it is a narrow Lorentzian (Fig. 10.12) of width $\Gamma_{\mathrm{B}} = (1 - \mathcal{C})/\tau_{\mathrm{r}}$.

In the *correlated* case $0 < \mathcal{C} < 1$, $G(\omega)$ is peaked at $\omega = 0$. Then, *perfect measurements* ($\theta = 0$), corresponding to a flat $F(\omega)$, *reduce* γ, which is a signature

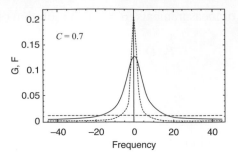

Figure 10.12 The overlap of F (dashed or dotted lines) and G (solid line) for correlated polarization-angle (phase) jumps (see Fig. 10.11) with correlation parameter $\mathcal{C} = 0.7$. Dashed line: $F(\omega)$ for $\theta = 0$ (perfect projections). Dotted line: $F(\omega)$ for $\theta = 0.9$ (weak measurements). Here $\delta\varphi = 0.1$, $\tau_r = 0.07$. (Reprinted from Kurizki et al., 2001. © 2001 The American Physical Society)

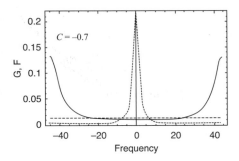

Figure 10.13 Idem, for anticorrelated polarization-angle (phase) jumps, $\mathcal{C} = -0.7$. (Reprinted from Kurizki et al., 2001. © 2001 The American Physical Society)

of the QZE (Fig. 10.12). This QZE signature disappears for the uncorrelated case $\mathcal{C} = 0$, where $G(\omega)$ is flat, $G(\omega) = \delta\varphi^2/(2\pi\tau_r)$.

For highly *anticorrelated* jumps ($\mathcal{C} \approx -1$) the bath spectrum is a sum of two shifted narrow Lorentzians,

$$G(\omega) \approx \sum_{j=\pm 1} \frac{\delta\varphi^2}{\pi\tau_r^2} \frac{\Gamma'_{\mathrm{B}}}{\Gamma'_{\mathrm{B}}{}^2 + (\pi/\tau_r + j\omega)^2} \tag{10.71}$$

of width $\Gamma'_{\mathrm{B}} = (1 + \mathcal{C})/\tau_r$.

In general, $G(\omega)$ is peaked at $\omega = \pm\pi/\tau_r$ for anti-correlated consecutive polarization-angle jumps, $-1 < \mathcal{C} < 0$ (Fig. 10.13). Such anti-correlated phase jumps correspond to the AZE trend whereby γ increases with the measurement rate ν, as opposed to the QZE trend, obtained for correlated jumps, whereby γ decreases with ν. These QZE and AZE trends are compared in Figure 10.14.

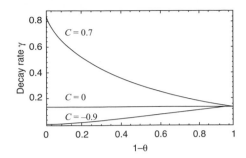

Figure 10.14 Polarization decay rate in the scheme of Figure 10.11 as a function of the measurement weakness $1 - \theta$ (which is equivalent to the dimensionless measurement rate $\nu \tau_r$). The three curves show correlated ($\mathcal{C} = 0.7$), Markovian ($\mathcal{C} = 0$) and anticorrelated ($\mathcal{C} = -0.9$) rotation-angle fluctuations. (Reprinted from Kurizki et al., 2001. © 2001 The American Physical Society)

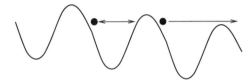

Figure 10.15 Atoms escape out of a "washboard" potential. (Reprinted from Kofman and Kurizki, 2001. © Verlag der Zeitschrift für Naturforschung, Tübingen)

10.4.5 Atom Escape from Accelerated Potential

Experimental studies by Raizen's group considered cold atoms trapped in a "washboard"-shaped potential that is subject to alternating periodic tilting and levelling-off, that is, to periodic acceleration (Fig. 10.15). The finding was that the trapped atoms escape from this periodically accelerated potential non-exponentially, at times $t \lesssim \omega_g/(k_L a)$, where $\hbar \omega_g$ is the band gap, which is roughly half the potential well depth, a the acceleration and k_L the wave number of the laser creating this potential. The present analysis, confirmed by experiment, shows that AZE should arise for measurement (tilting) intervals $\tau \gg 1/\omega_g$ and QZE for $\tau \ll 1/\omega_g$. The two trends are plotted in Figures 10.16 and 10.17, respectively. The two curves in Figure 10.17 intersect at $\tau_{\mathrm{QZE}} = 1/\nu_{\mathrm{QZE}}$, and the genuine QZE occurs for $\tau < \tau_{\mathrm{QZE}}$: In this case, τ_{QZE} ($\approx 10^{-6} \tau_0$) is much shorter than the boundary of the QZE-scaling region $\tau \sim 1/\omega_g$ (Fig. 10.16).

10.4.6 Experimental Accessibility

Table 10.1 lists processes wherein the AZE may be observed at accessible measurement intervals $t \sim \tau_{\mathrm{AZE}}$. The QZE ($\tau_{\mathrm{QZE}} \ll \tau_{\mathrm{AZE}}$) may be *principally unobservable in open systems* or, as a rule, is much less accessible than the AZE ($\tau_{\mathrm{QZE}} \ll \tau_{\mathrm{AZE}}$),

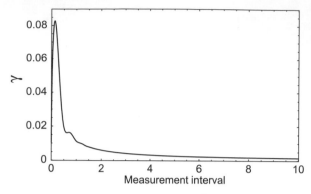

Figure 10.16 The escape rate γ (in units of τ_0^{-1}) of atoms from a periodically accelerated (tilted) "washboard" potential (Fig. 10.15) as a function of the scaled tilting interval τ/τ_0 ($\omega_g = 10/\tau_0$, $M^2 a/8\hbar^2 k_L^3 = 0.01$, where M is the atomic mass). (Adapted from Kofman and Kurizki, 2001. © Verlag der Zeitschrift für Naturforschung, Tübingen)

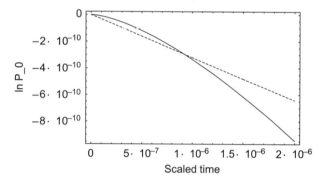

Figure 10.17 The escape law atoms from a "washboard" potential (solid line) as a function of the scaled time t/τ_0. Dashed line: exponential decay at the rate $\gamma_{GR} = 3.2 \times 10^{-4}\tau_0^{-1}$. The short-time portion of the solid line signifies the QZE. Same parameters as in Figure 10.16. (Adapted from Kofman and Kurizki, 2001. © Verlag der Zeitschrift für Naturforschung, Tübingen)

as argued in Section 10.4.3. The AZE and QZE are equally accessible in open systems only when the system resonance ω_a is near the peak of the bath spectrum, and this spectrum is narrow, as in the cases of radiative or photon-polarization decay in a cavity.

10.5 Discussion

Effects of frequent measurements on the evolution of a quantum system discussed in this chapter are commonly explained as resulting from state reduction (collapse of the wave function), that is, an irreversible state change, which marks any quantum measurement. Yet, all evolution changes obtained from either

Table 10.1 *Accessible processes for QZE and AZE.*

Process	References	τ_{QZE}	τ_{AZE}
Resonant Rabi oscillations	Itano et al. (1990)	microsecond (μs)	not applicable
Radiative decay in a cavity	Kofman & Kurizki (1996)	nanosecond (ns)	10 ns
Radiative decay in open space	Kofman & Kurizki (2000)	10^{-19} s	picosecond (ps) (Rydberg transition), femtosecond (fs) (optical transition)
Photon polarization decay via random modulation in a cavity	Kofman, Kurizki & Opatrny (2001)	ns	10 ns
Transmission of tunneling emitting atoms	Japha & Kurizki (1996); Barone et al. (1996)	ps	ns
Atomic escape from "washboard" accelerated potential	Wilkinson et al. (1997)	0.01 μs	μs
Macroscopic tunneling under current bias in SQUID ("washboard" potential)	Silvestrini et al. (1997)	?	10 ns
Nuclear β-decay	Kofman & Kurizki (2000)	too short	10^{-20} s
Near-threshold photodetachment	Lewenstein & Rzążewski (2000)	fs	millisecond (ms)

nonselective or selective measurements can also be produced by suitable unitary processes, as discussed in what follows.

Under frequent nonselective measurements of an observable $\hat{X} = |\psi\rangle\langle\psi|$, the QZE or the AZE are caused by repeated state reduction, marked by the destruction of coherences between $|\psi\rangle$ and all the states orthogonal to it. This effect can also be achieved unitarily, for example, by entangling the system at short time intervals with auxiliary systems acting as meters (Ch. 9) or by frequently changing the phase or amplitude of the system–bath coupling and thereby the joint (entangled) system–bath state (Chs. 11 and 16). A random field can also mimic the QZE or the AZE. Thus, frequent weak measurements described in (10.9) by the τ-dependent terms can be reproduced by random Stark shifts of the TLS resonance frequency (Sec. 10.2). The notion of wave function collapse is thus superfluous in this analysis, since nonselective measurements are describable as dephasing (coherence-breaking) processes of finite (albeit short) duration.

Under selective measurements, the QZE or the AZE are obtained upon selecting a subensemble of those systems found in the state $|\psi\rangle$ following each measurement. This procedure can be replaced by a unitary process, caused by multiple repeated interactions, followed by a single selective measurement that collapses the wave function only at the end of the process. For example, in the setup described in Section 10.4 for photons passing through a polarization-rotating medium, multiple polarization beam splitters consecutively placed along the optical path cause unitary changes as they transmit the original polarization and reflect the orthogonal one. The QZE or the AZE are then revealed by measuring the decrease of the light intensity that passed through the entire medium.

The QZE prediction by Misra and Sudarshan has been that *irreversible decay* of an excited state can be inhibited by repeated interruption of the system evolution via measurements (as in the case of an unstable particle on its flight through a bubble chamber). However, this prediction has not been experimentally verified as yet! Instead, the QZE has been mostly investigated for interrupted Rabi oscillations and analogous forms of *nearly reversible* evolution in closed systems, starting from Itano's pioneering experiment.

The inhibition of *nearly exponential* excited-state decay by the QZE and the converse effect of decay acceleration (AZE) have been demonstrated for atoms leaking out of a "washboard" potential by the landmark experiment of Raizen's group (2000), followed by others. Two-level emitters in cavities are also adequate for radiative decay control by measurements of the excited state.

Our analysis of the rate of TLS excitation decay to a bath with or without interruptions (by measurements or noise) has revealed the conditions for observing the QZE and the AZE in various structures (cavities, phonon and photonic band structures or atomic trapping potentials) as compared to open space.

Our simple universal formula (10.23) results in the following general criteria:
(i) The QZE can only occur in open systems coupled to baths whose spectral width is less than the resonance energy of the system. If the bath response has a much higher cutoff, as in open-space decay, *nondestructive* frequent measurements are much more likely to cause the AZE.
(ii) The QZE is *principally unattainable* for radiative or nuclear β-decay in open space, since the required measurement rates would be likely to cause the creation of new particles.
It follows from the discussion above that frequent measurements can be chosen to *accelerate* essentially any decay process. Hence, the AZE is inherently *more ubiquitous than the QZE in open systems.*

11

Dynamical Control of Open Systems

In this chapter, we derive a generalized non-Markovian master equation (ME) and its Markovian limit for a quantum system in a thermal bath under dynamical control of their interaction. We develop the framework for universal dynamical control by modulating fields, aimed at suppressing or preventing their decoherence or relaxation. A general master equation (ME) is obtained for a multilevel system, weakly coupled to an arbitrary bath and subject to arbitrary temporal driving or modulation. This ME does not invoke the RWA and therefore applies to arbitrarily fast modulations. The transition from non-Markovian to Markovian evolution is analyzed. The two types of decoherence that require suppression by dynamical control are amplitude and phase noise (AN and PN). AN changes the populations of energy states of the system via their coupling to the thermal bath. PN or proper dephasing randomizes the phases and the energies of these states. The characteristic AN time, T_1, is typically much slower than its PN counterpart, T_2. The QZE and the AZE (Ch. 10) are analogous to the slowdown or speedup, respectively, of T_2 or T_1 by driving or inter-level modulations, respectively, that rely on non-Markovian behavior. Ultrafast modulations also yield the same effect as the TLS coupling to a squeezed bath, on account of the breakdown of the RWA and the secular approximation.

11.1 Non-Markovian Master Equation for Dynamically Controlled Systems in Thermal Baths

11.1.1 Driven-System Master Equation by the Nakajima-Zwanzig Method

We here analyze the general evolution of a perturbed or driven quantum system coupled to a thermal bath. The Liouville equation for ρ_{tot}, the joint density matrix of the system plus the bath, reads (setting here and henceforth $\hbar = 1$):

$$\dot{\rho}_{\text{tot}}(t) = -i[H(t), \rho_{\text{tot}}(t)] \equiv -i\mathcal{L}(t)\rho_{\text{tot}}(t), \tag{11.1}$$

where

$$H(t) = H_{\text{S}}(t) + H_{\text{B}} + H_{\text{I}}(t) \tag{11.2}$$

and

$$\mathcal{L}(t) = \mathcal{L}_{\text{S}}(t) + \mathcal{L}_{\text{B}} + \mathcal{L}_{\text{I}}(t). \tag{11.3}$$

Here H_{S}, H_{B}, and $H_{\text{I}} \equiv H_{\text{SB}}$ are the Hamiltonians of the system, bath, and their coupling, respectively. The action of each operator H_i on the right-hand side of (11.2) is represented in (11.3) by the corresponding Liouville superoperator $\mathcal{L}_i$ that acts linearly on *operators* in our Hilbert space. We restrict our treatment to the regime of weak system–bath coupling, so that

$$\mathcal{L}_0(t) = \mathcal{L}_{\text{S}}(t) + \mathcal{L}_{\text{B}} \tag{11.4}$$

describes, to lowest order, the evolution of the system and the bath, whereas $\mathcal{L}_{\text{I}}(t)$ acts as a perturbation.

The master equation (ME) for the reduced density matrix of the system $\rho \equiv \text{Tr}_{\text{B}}\rho_{\text{tot}}$ should allow for *arbitrary* time dependence of $H(t)$, hence it must not invoke the RWA. We shall derive the ME for the system evolution by the Nakajima–Zwanzig projection-operator technique for a given bath state ρ_{B}. The projection operator $\mathcal{P}(\cdot) \equiv \text{Tr}_{\text{B}}(\cdot) \otimes \rho_{\text{B}}$ (that satisfies $\mathcal{P}^2 = \mathcal{P}$) and the complementary projection operator $\mathcal{Q} \equiv 1 - \mathcal{P}$ are invoked to rewrite (11.1) as:

$$\mathcal{P}\dot{\rho}_{\text{tot}}(t) = -i\mathcal{P}\mathcal{L}(t)\mathcal{P}\rho_{\text{tot}}(t) - i\mathcal{P}\mathcal{L}(t)\mathcal{Q}\rho_{\text{tot}}(t), \tag{11.5a}$$

$$\mathcal{Q}\dot{\rho}_{\text{tot}}(t) = -i\mathcal{Q}\mathcal{L}(t)\mathcal{P}\rho_{\text{tot}}(t) - i\mathcal{Q}\mathcal{L}(t)\mathcal{Q}\rho_{\text{tot}}(t). \tag{11.5b}$$

Upon integrating (11.5b), we obtain

$$\mathcal{Q}\rho_{\text{tot}}(t) = -i\int_0^t \mathcal{K}_+(t, \tau)\mathcal{Q}\mathcal{L}(\tau)\mathcal{P}\rho_{\text{tot}}(\tau)d\tau + \mathcal{K}_+(t, 0)\mathcal{Q}\rho_{\text{tot}}(0), \tag{11.6}$$

where

$$\mathcal{K}_+(t, \tau) = \text{T}_+ e^{-i\mathcal{Q}\int_\tau^t \mathcal{L}(s)ds}, \tag{11.7}$$

T_+ being the chronological (time-ordering) operator.

By plugging (11.6) into (11.5a), one obtains the non-Markovian ME for $\mathcal{P}\rho_{\text{tot}}$ in the form

$$\mathcal{P}\dot{\rho}_{\text{tot}}(t) = -i\mathcal{P}\mathcal{L}(t)\mathcal{P}\rho_{\text{tot}}(t) - \int_0^t \mathcal{P}\mathcal{L}(t)\mathcal{K}_+(t, \tau)\mathcal{Q}\mathcal{L}(\tau)\mathcal{P}\rho_{\text{tot}}(\tau)d\tau$$
$$-i\mathcal{P}\mathcal{L}\mathcal{K}_+(t, 0)\mathcal{Q}\rho_{\text{tot}}(0). \tag{11.8}$$

The corresponding ME for the system state $\rho(t)$ is obtained by tracing out the bath.

The time dependence of ρ_{tot} is given by

$$\rho_{\text{tot}}(\tau) = \mathcal{G}_-(t, \tau)\rho_{\text{tot}}(t). \tag{11.9}$$

Here the retarded Green function is

$$\mathcal{G}_-(t, \tau) \equiv T_- e^{+i \int_\tau^t \mathcal{L}(s)ds}, \tag{11.10}$$

where T_- denotes the anti-chronological ordering. Upon substituting (11.9) into (11.6), we then obtain:

$$\mathcal{Q}\rho_{\text{tot}}(t) = -\int_0^t \mathcal{K}_+(t, \tau)i\,\mathcal{Q}\mathcal{L}(\tau)\mathcal{P}\mathcal{G}_-(t, \tau)d\tau\,(\mathcal{P} + \mathcal{Q})\,\rho_{\text{tot}}(t)$$
$$+ \mathcal{K}_+(t, 0)\mathcal{Q}\rho_{\text{tot}}(0). \tag{11.11a}$$

We can rewrite this equation as

$$\mathcal{F}(t)\mathcal{Q}\rho_{\text{tot}}(t) = [1 - \mathcal{F}(t)]\,\mathcal{P}\rho_{\text{tot}}(t) + \mathcal{K}_+(t, 0)\mathcal{Q}\rho_{\text{tot}}(0), \tag{11.11b}$$

where

$$\mathcal{F}(t) = 1 + \int_0^t \mathcal{K}_+(t, \tau)i\,\mathcal{Q}\mathcal{L}(\tau)\mathcal{P}\mathcal{G}_-(t, \tau)d\tau \equiv 1 + \Sigma(t). \tag{11.11c}$$

Assuming that $\mathcal{F}(t)$ is invertible, we obtain

$$\mathcal{Q}\rho_{\text{tot}}(t) = [\mathcal{F}^{-1}(t) - 1]\,\mathcal{P}\rho_{\text{tot}}(t) + \mathcal{F}^{-1}(t)\mathcal{K}_+(t, 0)\mathcal{Q}\rho_{\text{tot}}(0). \tag{11.11d}$$

Upon plugging (11.11d) into (11.5a), the ME assumes the "time-convolutionless" form,

$$\mathcal{P}\dot\rho_{\text{tot}}(t) = -i\mathcal{P}\mathcal{L}(t)\mathcal{F}^{-1}(t)\mathcal{P}\rho_{\text{tot}}(t) - i\mathcal{P}\mathcal{L}(t)\mathcal{F}^{-1}(t)\mathcal{K}_+(t, 0)\mathcal{Q}\rho_{\text{tot}}(0). \tag{11.12}$$

We next substitute (11.3) for $\mathcal{L}(t)$ and take into account that $\mathcal{L}_S(t)$ commutes with $\mathcal{P}$, whereas $\mathcal{P}\mathcal{L}_B = 0$, $\mathcal{P}\mathcal{Q} = 0$, and

$$\mathcal{P}\mathcal{K}_+(t, \tau) = \mathcal{P}, \quad \mathcal{P}\mathcal{F}^{-1}(t) = \mathcal{P}, \tag{11.13}$$

as follows from (11.7) and (11.11c). Making use of these results, the ME (11.12) can be recast as

$$\mathcal{P}\dot\rho_{\text{tot}} = -i[\mathcal{L}_S(t) + \mathcal{P}\mathcal{L}_I(t)\mathcal{F}^{-1}(t)]\mathcal{P}\rho_{\text{tot}} - i\mathcal{P}\mathcal{L}_I(t)\mathcal{F}^{-1}(t)\mathcal{K}_+(t, 0)\mathcal{Q}\rho_{\text{tot}}(0). \tag{11.14}$$

We have thus transformed (11.8) to a differential or "time-convolutionless" equation, in which $\rho(t)$ has been pulled out of the integral.

We henceforth assume *initial factorization* of the system state ρ and the bath state,

$$\rho_{\text{tot}}(0) = \rho(0) \otimes \rho_B. \tag{11.15}$$

Under this assumption, $\mathcal{Q}\rho_{\text{tot}}(0) = 0$, the last term in (11.14) vanishes, and all memory effects are contained in $\mathcal{F}^{-1}(t)$. The ME for the reduced density matrix of the system then becomes, in the notation $\langle \cdot \rangle_{\text{B}} \equiv \text{Tr}_{\text{B}}(\cdot \rho_{\text{B}})$,

$$\dot{\rho} = -i \left[\mathcal{L}_S(t) + \langle \mathcal{L}_I(t)\mathcal{F}^{-1}(t)\rangle_{\text{B}} \right] \rho. \tag{11.16}$$

This differential ME is *exact* (under the assumption of initial factorization of the state of S and B). In order to render it soluble, one must treat the effects of $\mathcal{L}_I$ *perturbatively*, as detailed below.

11.1.2 Born Approximation

The ME (11.16) is perturbatively expanded in $\mathcal{L}_I$ upon noting from (11.11c) that

$$\langle \mathcal{L}_I(t)\mathcal{F}^{-1}(t)\rangle_{\text{B}} = \langle \mathcal{L}_I(t)\left[1 - \Sigma(t) + \mathcal{O}(\Sigma^2)\right]\rangle_{\text{B}}, \quad \Sigma(t) = \mathcal{O}(\mathcal{L}_I). \tag{11.17}$$

In order to expand this expression to *second order* in $\mathcal{L}_I$, consistently with the weak-coupling assumption, we evaluate $\mathcal{K}_+$ and $\mathcal{G}_-$ only *to zeroth order in $\mathcal{L}_I$*. In particular, we obtain from (11.7)

$$\mathcal{K}_+(t, \tau) \simeq \mathcal{V}_0(t, \tau) \equiv \text{T}_+ e^{-i\mathcal{Q}\int_\tau^t \mathcal{L}_0(s)ds}. \tag{11.18}$$

We shall employ the relation

$$\mathcal{V}_0\mathcal{Q}(t, \tau) = \mathcal{Q}\mathcal{U}_0(t, \tau)\mathcal{Q} = \mathcal{Q}\mathcal{U}_0(t, \tau), \tag{11.19a}$$

where

$$\mathcal{U}_0(t, \tau) = \text{T}_+ e^{-i\int_\tau^t \mathcal{L}_0(s)ds}, \tag{11.19b}$$

and we have made use of the commutation of $\mathcal{U}_0$ with $\mathcal{Q}$ in (11.19a). The retarded Green function (11.10) can be written, to zeroth order in $\mathcal{L}_I$, as

$$\mathcal{G}_-(t, \tau) = [1 + \mathcal{O}(\mathcal{L}_I)]\,\text{T}_- e^{+i\int_\tau^t \mathcal{L}_0(s)ds} = \mathcal{U}_0^{-1}(t, \tau) + \mathcal{O}(\mathcal{L}_I). \tag{11.20}$$

We thus have, under the weak-coupling assumption,

$$\Sigma(t) = i \int_0^t \mathcal{V}_0(t, \tau)\mathcal{Q}\mathcal{L}_I(\tau)\mathcal{P}\mathcal{U}_0^{-1}(t, \tau)d\tau. \tag{11.21}$$

Upon introducing (11.21) into (11.17) and using (11.16), we then obtain the following ME for $\rho(t)$,

$$\dot{\rho} = -\left[i\mathcal{L}_S(t) + i\langle \mathcal{L}_I(t)\rangle_{\text{B}} + \int_0^t d\tau \langle \mathcal{L}_I(t)\mathcal{Q}\mathcal{U}_0(t, \tau)\mathcal{L}_I(\tau)\mathcal{P}\mathcal{U}_0^{-1}(t, \tau)\rangle_{\text{B}} \right]\rho, \tag{11.22}$$

where (11.19a) has been allowed for.

Equation (11.22) can be further simplified as follows. Equations (11.19b) and (11.4) imply that

$$\mathcal{U}_0(t, \tau) = \mathcal{U}_S(t, \tau)\mathcal{U}_B(t - \tau), \tag{11.23}$$

where

$$\mathcal{U}_S(t, \tau) = T_+ e^{-i\int_\tau^t \mathcal{L}_S(s)ds}, \quad \mathcal{U}_B(t) = e^{-i\mathcal{L}_B t}. \tag{11.24}$$

Hence $\mathcal{U}_0^{-1}(t, \tau) = \mathcal{U}_S^{-1}(t, \tau)\mathcal{U}_B^{-1}(t - \tau)$. Since $\mathcal{P}\mathcal{U}_B^{-1}(t - \tau)\rho_B = \mathcal{P}\rho_B = \rho_B$, we then obtain the equation

$$\dot{\rho} = -\left[i\mathcal{L}_S(t) + i\langle\mathcal{L}_I(t)\rangle_B + \int_0^t d\tau \langle\mathcal{L}_I(t)\mathcal{Q}\mathcal{U}_0(t, \tau)\mathcal{L}_I(\tau)\rangle_B \, \mathcal{U}_S^{-1}(t, \tau) \right] \rho. \tag{11.25}$$

We have thus obtained a differential (time-convolutionless) equation, valid to the second order in the system–bath coupling. In this approach, called the Born approximation, the back-effect of the system on the bath is absent due to the weakness of the system–bath coupling.

Equation (11.25) is not the only ME that can be obtained in the second-order (Born) approximation. Alternatively, an integrodifferential ME (involving time convolution) can be obtained in the same approximation. Both equations have generally the same accuracy, since the difference between them is only revealed beyond second order. In this book, we use the differential ME, which is more convenient for our purposes than the integrodifferential ME.

11.1.3 Factorizable Interaction Hamiltonians

We shall explicitly write the foregoing ME for the factorizable interaction Hamiltonian,

$$H_I(t) = S(t)B, \tag{11.26}$$

where H_I is the product of operators S and B that act on the system and the bath, respectively. We further assume that $\langle B\rangle_B = 0$, so that $\langle\mathcal{L}_I\rangle_B = 0$ and $\mathcal{P}\mathcal{L}_I\mathcal{P} = 0$. The ME (11.25) then simplifies to

$$\dot{\rho} = -i\mathcal{L}_S(t)\rho - \int_0^t d\tau \langle\mathcal{L}_I(t)\mathcal{U}_0(t, \tau)\mathcal{L}_I(\tau)\rangle_B \, \mathcal{U}_S^{-1}(t, \tau)\rho. \tag{11.27}$$

Any operator A satisfies the relation

$$\mathcal{U}_0(t, \tau)A = U_S(t, \tau)U_B(t - \tau)AU_S^\dagger(t, \tau)U_B^\dagger(t - \tau), \tag{11.28a}$$

where

$$U_S(t, \tau) = T_+ e^{-i\int_\tau^t H_S(t')dt'}, \tag{11.28b}$$

$$U_B(t) = e^{-iH_Bt}. \tag{11.28c}$$

The integrand in (11.27) can therefore be written as follows,

$$I(t,\tau) = \text{Tr}_B\left[S(t)B, \mathcal{U}_0(t,\tau)\left[S(\tau)B, \mathcal{U}_S^{-1}(t,\tau)\rho(t)\rho_B\right]\right]$$
$$= \text{Tr}_B\left[S(t)B, \left[\tilde{S}(\tau,t)\tilde{B}(\tau-t), \rho(t)\rho_B\right]\right], \tag{11.29}$$

where $\tilde{S}(\tau,t)$ and $\tilde{B}(t)$ are the system and bath operators $S(\tau)$ and B in the interaction representation,

$$\tilde{S}(\tau,t) = U_S^\dagger(\tau,t)S(\tau)U_S(\tau,t), \tag{11.30}$$
$$\tilde{B}(t) = U_B^\dagger(t)BU_B(t). \tag{11.31}$$

The commutativity of S and B, and that of ρ_B and H_B, using the cyclic property of the trace, yield:

$$I(t,\tau) = \Phi_T(t-\tau)\left[S(t), \tilde{S}(\tau,t)\rho(t)\right] + \text{H.c.}, \tag{11.32}$$

where

$$\Phi_T(t) = \langle\tilde{B}(t)B\rangle_B \tag{11.33}$$

is the bath autocorrelation function. Here and henceforth we take ρ_B to be the thermal density operator,

$$\rho_B = Z^{-1}e^{-H_B/T}, \tag{11.34}$$

where Z is the unit-trace normalization constant and the temperature T is given, from here on, in energy units, with the Boltzmann constant $k_B = 1$.

In the derivation of (11.32), we used the equality

$$\Phi_T(t,\tau) \equiv \langle\tilde{B}(t)\tilde{B}(\tau)\rangle_B = \Phi_T(t-\tau). \tag{11.35}$$

This relation relies on (11.28c), (11.31), and (11.34), which yield

$$\Phi_T(t,\tau) = \text{Tr}[U_B^\dagger(t)BU_B(t)U_B^\dagger(\tau)BU_B(\tau)\rho_B]$$
$$= \text{Tr}[U_B(\tau)U_B^\dagger(t)BU_B(t)U_B^\dagger(\tau)B\rho_B] = \Phi_T(t-\tau). \tag{11.36}$$

In the second equality, we used the fact that $U_B(\tau)$ commutes with ρ_B as well as the cyclic property of the trace, whereas in the last equality we used the relation $U_B(t)U_B^\dagger(\tau) = U_B(t-\tau)$. From (11.35) it follows that

$$\Phi_T^*(t) = \langle\tilde{B}(0)\tilde{B}(t)\rangle_B = \Phi_T(-t). \tag{11.37}$$

For a bath in a thermal state (or, more generally, in a state commuting with the Hamiltonian), one can show that, the Fourier transform of $\Phi_T(t)$,

$$G_T(\omega) = \frac{1}{2\pi}\int_{-\infty}^{\infty} dt\, \Phi_T(t)e^{i\omega t}, \tag{11.38}$$

is a real, nonnegative function. It represents the spectral response of the bath. The bath correlation function is related to $G_T(\omega)$ by the inverse Fourier transform,

$$\Phi_T(t) = \int_{-\infty}^{\infty} d\omega \, G_T(\omega) e^{-i\omega t}. \tag{11.39}$$

Upon performing the one-sided Fourier transform of the both sides of this equality and using the formula (Ch. 8)

$$\int_0^{\infty} dt \, e^{i\omega t} = \pi \delta(\omega) + i P \frac{1}{\omega}, \tag{11.40}$$

P being the principal value, we obtain

$$\int_0^{\infty} dt \, \Phi_T(t) e^{i\omega t} = \pi G_T(\omega) + i \Delta_T(\omega), \tag{11.41}$$

where the function

$$\Delta_T(\omega) = P \int_{-\infty}^{\infty} d\omega' \frac{G_T(\omega')}{\omega - \omega'} \tag{11.42}$$

determines the bath-induced system-energy (Lamb) shifts. Thus, we have the relations,

$$G_T(\omega) = \frac{1}{\pi} \mathrm{Re} \int_0^{\infty} dt \, \Phi_T(t) e^{i\omega t}, \tag{11.43}$$

$$\Delta_T(\omega) = \mathrm{Im} \int_0^{\infty} dt \, \Phi_T(t) e^{i\omega t}. \tag{11.44}$$

The ME for $\rho(t)$ beyond the Markovian approximation, but within the Born approximation, can finally be written in the form

$$\dot{\rho} = -i \, [H_S(t), \rho] + \int_0^t d\tau \left\{ \Phi_T(t - \tau) \left[\tilde{S}(\tau, t)\rho, S(t) \right] + \text{H.c.} \right\}. \tag{11.45}$$

This ME is *convolutionless*, yet it retains non-Markovian effects to second order in H_I.

In the special case, where H_S in (11.2) and S in (11.26) are time independent, the operator in the interaction picture $\tilde{S}(\tau, t)$, (11.30), depends on the time difference,

$$\tilde{S}(\tau, t) = \tilde{S}(\tau - t), \tag{11.46}$$

where

$$\tilde{S}(t) = e^{i H_S t} S e^{-i H_S t}. \tag{11.47}$$

In general, the system-interaction operator $\tilde{S}(\tau, t)$ in the interaction picture, where it is affected by the time dependence of $S(t)$ and/or $H_S(t)$, cannot be assumed to satisfy the stationarity condition $\tilde{S}(\tau, t) = \tilde{S}(\tau - t)$. Stationarity is, however, not required for the validity of this convolutionless (differential) ME.

11.1.4 Master Equation for an Oscillator Bath

The above results hold for an arbitrary bath. Below we consider, for definiteness, the case of a bosonic oscillator bath (Ch. 3) in thermal equilibrium. In this case, the bath Hamiltonian in (11.2) reads

$$H_B = \sum_k \omega_k a_k^\dagger a_k, \qquad (11.48)$$

where k labels the bath-mode frequencies ω_k, the bosonic creation and annihilation operators $a_k^\dagger$ and a_k, respectively. The S-B interaction Hamiltonian (11.26) for a TLS coupled to such a bath reads as in (4.10) with antiresonant terms included or as in (4.12) in the RWA. In the interaction picture [as per (11.31)], the bath operator has then the form

$$\tilde{B}(t) = \sum_k (\eta_k a_k e^{-i\omega_k t} + \eta_k^* a_k^\dagger e^{i\omega_k t}), \qquad (11.49)$$

where η_k are the dipolar k-mode coupling strengths. The bath autocorrelation function (11.33) is then given by

$$\Phi_T(t) = \sum_k \sum_{k'} (\eta_k \eta_{k'}^* e^{-i\omega_k t} \langle a_k a_{k'}^\dagger \rangle_B + \eta_k^* \eta_{k'} e^{i\omega_k t} \langle a_k^\dagger a_{k'} \rangle_B)$$

$$= \sum_k |\eta_k|^2 \{ e^{-i\omega_k t} [\bar{n}_T(\omega_k) + 1] + e^{i\omega_k t} \bar{n}_T(\omega_k) \}, \qquad (11.50)$$

where we used the equilibrium bosonic bath properties $\langle a_k a_{k'} \rangle_B = \langle a_k^\dagger a_{k'}^\dagger \rangle_B = 0$, $\langle a_k a_{k'}^\dagger \rangle_B = \delta_{kk'} [\bar{n}_T(\omega_k) + 1]$, $\langle a_k^\dagger a_{k'} \rangle_B = \delta_{kk'} \bar{n}_T(\omega_k)$, $\bar{n}_T(\omega)$ being the average quanta number corresponding to the temperature-dependent Planck distribution at frequency ω,

$$\bar{n}_T(\omega) = \left[\exp\left(\frac{\omega}{T} \right) - 1 \right]^{-1}. \qquad (11.51)$$

From (11.43) and (11.40), we then obtain the thermal-bath response spectrum,

$$G_T(\omega) = G_0(\omega)[1 + \bar{n}_T(\omega)] + G_0(-\omega) n_T(-\omega), \qquad (11.52)$$

where the bath response spectrum at $T = 0$ is

$$G_0(\omega) = \sum_k |\eta_k|^2 \delta(\omega - \omega_k). \qquad (11.53)$$

On taking the continuum limit of bath modes, $\sum_k \to \int d\omega \rho(\omega)$, (11.53) acquires the form

$$G_0(\omega) = \rho(\omega) |\eta(\omega)|^2. \qquad (11.54)$$

11.1.5 Markovian Limit of the Master Equation

Here we consider the case of time-independent H_S and S, so that (11.46) holds, and assume sufficiently long times, $t \gg t_c$, where the bath correlation (response) time t_c is the characteristic decay or memory time of $\Phi_T(t)$ (Sec. 11.3). Then the integral in (11.45) tends to its $t \to \infty$ limit, and (11.45) becomes a Markovian ME (MME),

$$\dot{\rho} = -i\,[H_S, \rho] + \int_0^\infty d\tau \left\{ \Phi_T(\tau) \left[\tilde{S}(-\tau)\rho, S \right] + \text{H.c.} \right\}. \tag{11.55}$$

We next perform the secular simplification, as follows. We write H_S in the form

$$H_S = \sum_E E\Pi(E), \tag{11.56}$$

where $\{E\}$ is the complete set of those Hamiltonian eigenvalues, which differ from each other (i.e., if some E is degenerate, it appears in $\{E\}$ only once), $\Pi(E)$ is an orthogonal projector onto the eigenspace corresponding to E, such that $\sum_E \Pi(E) = I_S$, and I_S is the identity operator for the system. Then we can decompose the system-interaction operator,

$$S = \sum_{E,E'} \Pi(E)S\Pi(E') = \sum_\alpha S_\alpha. \tag{11.57}$$

The S_α are dubbed the "jump" operators. They are related to S by the equality

$$S_\alpha = \sum_{E'-E=\omega_\alpha} \Pi(E)S\Pi(E'), \tag{11.58}$$

where the sum is over all E' and E with a given difference ω_α.

The operator (11.57) in the interaction picture results from (11.47) in the form

$$\tilde{S}(t) = \sum_\alpha e^{-i\omega_\alpha t} S_\alpha. \tag{11.59}$$

Upon inserting this expression and its $\tau = 0$ value (11.57) (rewritten in the form $S = \sum_{\alpha'} S_{\alpha'}^\dagger$) into (11.55) and using (11.41), the MME acquires the form

$$\dot{\rho} = -i\,[H_S, \rho] + \sum_{\alpha,\alpha'} \left\{ (\gamma_\alpha/2 + i\Delta_\alpha) \left[S_\alpha\rho, S_{\alpha'}^\dagger \right] + \text{H.c.} \right\}, \tag{11.60}$$

where the meaning of γ_α and Δ_α is explained below.

The transformation of (11.60) to the interaction picture introduces in the double sum $\sum_{\alpha,\alpha'}$ phase factors that oscillate at frequencies $\omega_\alpha - \omega_{\alpha'}$. The "non-secular" terms, for which these frequencies are nonzero, oscillate fast and therefore self-average after a few oscillation periods. Hence, they do not appreciably contribute to the solution of (11.60). The secular simplification consists in neglecting these oscillatory terms, thus retaining only the terms with $\alpha = \alpha'$. This approximation yields the Lindblad equation,

$$\dot{\rho} = -i\,[H_S + H_{LS}, \rho] + \frac{1}{2}\sum_{\alpha} \gamma_{\alpha}(2S_{\alpha}\rho S_{\alpha}^{\dagger} - S_{\alpha}^{\dagger}S_{\alpha}\rho - \rho S_{\alpha}^{\dagger}S_{\alpha}), \qquad (11.61)$$

where H_{LS} is the "Lamb-shift" Hamiltonian,

$$H_{LS} = \sum_{\alpha} \Delta_{\alpha} S_{\alpha}^{\dagger} S_{\alpha}. \qquad (11.62)$$

In (11.61) and (11.62), the parameters γ_{α} are proportional to the relaxation (or dephasing) rates, whereas Δ_{α} are proportional to the bath-induced energy (Lamb) shifts (Chs. 5, 8):

$$\gamma_{\alpha} := 2\pi\,G_T(\omega_{\alpha}), \quad \Delta_{\alpha} := \Delta_T(\omega_{\alpha}). \qquad (11.63)$$

11.2 Non-Markovian Master Equation for Periodically Modulated TLS in a Thermal Bath

In this section, we restrict the non-Markovian ME derived above to the case of a TLS that is coupled to a thermal bath, under periodic modulations of the TLS frequency. We dwell on unconventional *squeezing effects* incurred by a modulation that is fast enough to violate the RWA.

In the joint system-plus-bath Hamiltonian (11.2), now the modulated TLS Hamiltonian is given by

$$H_S(t) = \frac{1}{2}\sigma_z\,[\omega_a + \delta_a(t)]\,, \qquad (11.64)$$

where $\delta_a(t)$ is the frequency-modulation function. We take in (11.26) the dipolar system-interaction operator that couples it to the bath to be

$$S(t) = \tilde{\epsilon}(t)\sigma_x, \qquad (11.65)$$

where $\tilde{\epsilon}(t)$ is the real amplitude describing the interaction-strength modulation. The operator (11.30) is given by

$$\tilde{S}(\tau, t) = \tilde{\epsilon}(\tau)[\varepsilon^*(t)\varepsilon(\tau)e^{-i\omega_a(t-\tau)}\sigma_+ + \varepsilon(t)\varepsilon^*(\tau)e^{i\omega_a(t-\tau)}\sigma_-], \qquad (11.66)$$

where the time-dependent phase factor follows from (11.64) to be

$$\varepsilon(t) = e^{i\int_0^t dt'\delta_a(t')}. \qquad (11.67)$$

On inserting (11.65) and (11.66) into (11.45) and performing algebraic calculations, we obtain the non-Markovian ME,

$$\dot{\rho} = -i[\tilde{H}_S(t), \rho] + \frac{\gamma_g(t)}{2}(2\sigma_+\rho\sigma_- - \rho\sigma_-\sigma_+ - \sigma_-\sigma_+\rho)$$

$$+ \frac{\gamma_e(t)}{2}(2\sigma_-\rho\sigma_+ - \rho\sigma_+\sigma_- - \sigma_+\sigma_-\rho)$$

$$+ \tilde{\gamma}(t)\sigma_+\rho\sigma_+ + \tilde{\gamma}^*(t)\sigma_-\rho\sigma_-. \tag{11.68}$$

Here $\tilde{H}_S(t)$ differs from $H_S(t)$ by the substitution

$$\omega_a \rightarrow \tilde{\omega}_a(t) = \omega_a + \Delta_a(t), \tag{11.69a}$$

the frequency (energy) shift

$$\Delta_a(t) = \Delta_e(t) - \Delta_g(t) \tag{11.69b}$$

being expressed through the Lamb shifts $\Delta_{e,g}(t)$ of the energy levels e and g, and

$$\tilde{\gamma}(t) = \gamma(t) - i\Delta_a(t), \tag{11.70}$$

where

$$\gamma(t) = \frac{\gamma_e(t) + \gamma_g(t)}{2}. \tag{11.71}$$

The relaxation rates $\gamma_{e(g)}(t)$ of the levels e and g and the Lamb shifts $\Delta_{e(g)}(t)$ are given by the real and imaginary parts of the expressions

$$\gamma_e(t)/2 + i\Delta_e(t) = \int_0^t dt'\epsilon(t)\epsilon^*(t')\Phi_T(t - t')e^{i\omega_a(t-t')}, \tag{11.72a}$$

$$\gamma_g(t)/2 + i\Delta_g(t) = \int_0^t dt'\epsilon^*(t)\epsilon(t')\Phi_T(t - t')e^{-i\omega_a(t-t')}, \tag{11.72b}$$

where $\epsilon(t)$ is the dipole-modulation function allowing for both amplitude and phase modulations,

$$\epsilon(t) = \tilde{\epsilon}(t)\varepsilon(t). \tag{11.73}$$

It is remarkable that both types of modulations allow for a unified description by a single function, but commonly only one type of modulation (amplitude or phase) is used. Below we focus on phase modulation, as the amplitude modulation can be considered very similarly. Thus, we set $\tilde{\epsilon}(t) = 1$, and hence, in view of (11.73), $\epsilon(t) = \varepsilon(t)$, where $\varepsilon(t)$ is given in (11.67).

We assume that the modulation function $\delta_a(t)$ is periodic with the period T_m. This implies that

$$\int_0^t dt'\delta_a(t') = \delta t + f(t), \tag{11.74}$$

where

$$\delta = \frac{1}{T_{\mathrm{m}}} \int_0^{T_{\mathrm{m}}} dt \, \delta_{\mathrm{a}}(t) \tag{11.75}$$

is the average of $\delta_{\mathrm{a}}(t)$ over the period and $f(t)$ is a periodic function of t with the period T_{m}. The phase factor can then be expressed as a Fourier series,

$$\varepsilon(t) \equiv e^{i \int_0^t dt' \delta_{\mathrm{a}}(t')} = \sum_q \epsilon_q e^{i \nu_q t}. \tag{11.76}$$

Here the sum is over all integers, $q = 0, \pm 1, \ldots$,

$$\nu_q = \delta + q \omega_{\mathrm{m}}, \tag{11.77}$$

where $\omega_{\mathrm{m}} = 2\pi/T_{\mathrm{m}}$, and

$$\epsilon_q = \frac{1}{T_{\mathrm{m}}} \int_0^{T_{\mathrm{m}}} dt \, e^{i \int_0^t dt' \delta_{\mathrm{a}}(t') - i \nu_q t}. \tag{11.78}$$

The ME (11.68) yields the following equations for the density matrix elements,

$$\dot{\rho}_{ee} = -\dot{\rho}_{gg} = -\gamma_e(t) \rho_{ee} + \gamma_g(t) \rho_{gg}, \tag{11.79a}$$

$$\dot{\rho}_{eg} = \dot{\rho}_{ge}^* = -\{\gamma(t) + i[\bar{\omega}_{\mathrm{a}}(t) + \delta_{\mathrm{a}}(t)]\} \rho_{eg} + \tilde{\gamma}(t) \rho_{ge}. \tag{11.79b}$$

The fact that the coefficients here are time dependent may seem to complicate these equations. However, in the present weak coupling approximation, this time dependence is important only for short times, where the density matrix is still close to its initial value. This short-time behavior can be found perturbatively, as follows.

We first transform to the interaction picture (equivalently, to the rotating frame), in which the populations do not change, whereas the coherences are transformed as $\rho_{eg(ge)} \to \tilde{\rho}_{eg(ge)}$ via

$$\rho_{eg}(t) = e^{-i\bar{\omega}_{\mathrm{a}}(t)t} \epsilon^*(t) \tilde{\rho}_{eg}(t), \quad \rho_{ge}(t) = e^{i\bar{\omega}_{\mathrm{a}}(t)t} \epsilon(t) \tilde{\rho}_{ge}(t), \tag{11.80}$$

where

$$\bar{\omega}_{\mathrm{a}}(t) = \omega_{\mathrm{a}} + \bar{\Delta}_{\mathrm{a}}(t), \qquad \bar{\Delta}_{\mathrm{a}}(t) = \bar{\Delta}_e(t) - \bar{\Delta}_g(t) \tag{11.81}$$

and the overbar denotes averaging over the interval $[0, t]$,

$$\bar{f}(t) = \frac{1}{t} \int_0^t d\tau f(\tau). \tag{11.82}$$

Then (11.79b) becomes

$$\dot{\tilde{\rho}}_{eg} = \dot{\tilde{\rho}}_{ge}^* = -\gamma(t) \tilde{\rho}_{eg} + e^{2i\bar{\omega}_{\mathrm{a}}(t)t} \epsilon^2(t) \tilde{\gamma}(t) \tilde{\rho}_{ge}. \tag{11.83}$$

The last term on the right-hand sides of (11.79b) and (11.83) is commonly omitted upon adopting the secular approximation (Sec. 11.1.5), as, in the absence of modulation, this term oscillates fast in (11.83) for a nondegenerate TLS and hence negligibly contributes to the solution. By contrast, under fast modulation of the level distance, the secular approximation may break down as discussed below. As a result of the secular approximation, (11.79b) and (11.83) become

$$\dot{\rho}_{eg} = \dot{\rho}_{ge}^* = -\{\gamma(t) + i[\tilde{\omega}_a(t) + \delta_a(t)]\}\rho_{eg} \tag{11.84}$$

and, in the interaction picture,

$$\dot{\tilde{\rho}}_{eg} = \dot{\tilde{\rho}}_{ge}^* = -\gamma(t)\tilde{\rho}_{eg}. \tag{11.85}$$

At short times, the solution of (11.79a) and (11.85), to first order in the relaxation parameters, is

$$w(t) = \{1 - [\bar{\gamma}_g(t) + \bar{\gamma}_e(t)]t\}w(0) + [\bar{\gamma}_g(t) - \bar{\gamma}_e(t)]t, \tag{11.86a}$$

$$\tilde{\rho}_{eg}(t) = \tilde{\rho}_{eg}^*(t) = [1 - \bar{\gamma}(t)t]\tilde{\rho}_{eg}(0), \tag{11.86b}$$

where $w = \rho_{ee} - \rho_{gg}$ is the population inversion, so that $\rho_{ee(gg)} = (1 \pm w)/2$.

The expressions for the average rates and Lamb shifts can be recast as integrals in the frequency domain, as follows. Integrating both sides of (11.72a) and dividing them by t yields

$$\frac{1}{2}\bar{\gamma}_e(t) + i\bar{\Delta}_e(t) = \frac{1}{t}J(t). \tag{11.87}$$

Here,

$$J(t) = \int_0^t dt' \epsilon(t') e^{i\omega_a t'} I_1(t', t''), \tag{11.88}$$

where

$$I_1(t', t'') = \int_0^{t'} dt'' \epsilon^*(t'') \Phi_T(t' - t'') e^{-i\omega_a t''}. \tag{11.89}$$

It is useful to recast $I_1(t', t'')$ in the form

$$I_1(t', t'') = \int_{-\infty}^{t'} dt'' \epsilon^*(t, t'') \Phi_T(t' - t'') e^{-i\omega_a t''}$$

$$= \frac{1}{\sqrt{2\pi}} \int_{-\infty}^{t'} dt'' \Phi_T(t' - t'') e^{-i\omega_a t''} \int_{-\infty}^{\infty} d\omega \epsilon_t^*(\omega) e^{-i\omega t''}. \tag{11.90}$$

Here,

$$\epsilon(t, \tau) = \epsilon(\tau)\theta(t - \tau)\theta(\tau) = \frac{1}{\sqrt{2\pi}} \int_{-\infty}^{\infty} d\omega \epsilon_t(\omega) e^{i\omega \tau}, \tag{11.91}$$

where $\theta(t)$ is the Heaviside step function and $\epsilon_t(\omega)$ is the Fourier transform of $\epsilon(t, \tau)$ as a function of τ,

$$\epsilon_t(\omega) = \frac{1}{\sqrt{2\pi}} \int_{-\infty}^{\infty} d\tau \epsilon(t, \tau) e^{-i\omega\tau} = \frac{1}{\sqrt{2\pi}} \int_0^t d\tau \epsilon(\tau) e^{-i\omega\tau}. \tag{11.92}$$

The last formula shows that $\epsilon_t(\omega)$ is the finite-time Fourier transform of the modulation function $\epsilon(t)$ in the "window" $[0, t]$. Inserting (11.90) into (11.88), using the substitution $\tau = t' - t''$ for t'', and changing the order of integration yields

$$\begin{aligned} J(t) &= \frac{1}{\sqrt{2\pi}} \int_{-\infty}^{\infty} d\omega \epsilon_t^*(\omega) \int_0^t dt' \epsilon(t') e^{-i\omega t'} \int_0^{\infty} d\tau \Phi_T(\tau) e^{i(\omega+\omega_a)\tau} \\ &= \int_{-\infty}^{\infty} d\omega \, |\epsilon_t(\omega)|^2 \int_0^{\infty} d\tau \, \Phi_T(\tau) e^{i(\omega+\omega_a)\tau} \\ &= \int_{-\infty}^{\infty} d\omega \, |\epsilon_t(\omega - \omega_a)|^2 \int_0^{\infty} d\tau \, \Phi_T(\tau) e^{i\omega\tau}, \end{aligned} \tag{11.93}$$

where, in the second equality, we have used (11.92).

Finally, inserting (11.41) into (11.93), we obtain from (11.87) the modulation-modified average over $[0, t]$ of the e-level relaxation rate and energy shift, respectively, are

$$\bar{\gamma}_e(t) = 2\pi \int_{-\infty}^{\infty} d\omega \, G_T(\omega) F_t(\omega - \omega_a), \tag{11.94a}$$

$$\bar{\Delta}_e(t) = \int_{-\infty}^{\infty} d\omega \, \Delta_T(\omega) F_t(\omega - \omega_a). \tag{11.94b}$$

Here,

$$F_t(\omega) = \frac{|\epsilon_t(\omega)|^2}{t} = \frac{1}{2\pi t} \left| \int_0^t d\tau \epsilon(\tau) e^{-i\omega\tau} \right|^2 \tag{11.95}$$

is the spectrum of the modulation function $\epsilon(t)$ in the "window" $[0, t]$, the spectrum being normalized to one,

$$\int_{-\infty}^{\infty} d\omega \, F_t(\omega) = 1. \tag{11.96}$$

This spectrum is dubbed the spectral filter function, as detailed in Sections 11.3 and 11.4. By repeating the above calculations with minor changes, starting from (11.72b), we obtain the g-level counterparts of (11.94),

$$\bar{\gamma}_g(t) = 2\pi \int_{-\infty}^{\infty} d\omega \, G_T(\omega) F_t(-\omega - \omega_a), \tag{11.97a}$$

$$\bar{\Delta}_g(t) = \int_{-\infty}^{\infty} d\omega \, \Delta_T(\omega) F_t(-\omega - \omega_a). \tag{11.97b}$$

Let us consider $F_t(\omega)$ in more detail. In the case of phase modulation with a periodic $\delta_a(t)$, upon inserting $\epsilon(t) = \varepsilon(t)$ [given by (11.76)] into (11.92), we have

$$\epsilon_t(\omega) = \frac{1}{\sqrt{2\pi}} \sum_q \epsilon_q \frac{e^{i(\nu_q - \omega)t} - 1}{i(\nu_q - \omega)}$$

$$= \frac{t}{\sqrt{2\pi}} \sum_q \epsilon_q \, \mathrm{sinc}\left(\frac{(\omega - \nu_q)t}{2}\right) e^{i(\nu_q - \omega)t/2}, \qquad (11.98)$$

and hence,

$$|\epsilon_t(\omega)|^2 = \frac{t^2}{2\pi} \sum_{q,q'} \epsilon_q \epsilon_{q'}^* \, \mathrm{sinc}\left(\frac{(\omega - \nu_q)t}{2}\right) \mathrm{sinc}\left(\frac{(\omega - \nu_{q'})t}{2}\right) e^{i(\nu_q - \nu_{q'})t/2}.$$
$$(11.99)$$

This expression can be simplified for $t \gg 1/\omega_m$. The terms with $q = q'$ are then peaked at $\omega = \nu_q$, whose spacing ω_m is much greater than the peak width $\sim 1/t$. The other terms are fast oscillating at such times and, although they also form peaks at $\omega = \nu_q$, the ratio of their height to that of the $q = q'$ peaks is small, $\sim (\omega_m t)^{-1} \ll 1$. Thus, at times $t \gg 1/\omega_m$, we can neglect the terms with $q \neq q'$ and reduce (11.95) to

$$F_t(\omega) = \sum_q |\epsilon_q|^2 S_t(\omega - \nu_q), \qquad (11.100)$$

where

$$S_t(\omega) = \frac{t}{2\pi} \, \mathrm{sinc}^2(\omega t/2) = \frac{2 \sin^2(\omega t/2)}{\pi \omega^2 t} \qquad (11.101)$$

is a positive function of ω normalized to 1.

In the absence of modulation, $\epsilon(\tau) = 1$ in (11.92), we obtain

$$F_t(\omega) = S_t(\omega). \qquad (11.102)$$

The function $S_t(\omega)$ is a narrow peak with the width $\sim 1/t$ tending to 0 with time. Thus, for a sufficiently long time one can use the approximation $S_t(\omega - \nu_q) \approx \delta(\omega - \nu_q)$ in (11.100), yielding

$$F_t(\omega) \to F(\omega) = \sum_q |\epsilon_q|^2 \delta(\omega - \nu_q) \quad \forall \, t \to \infty, \qquad (11.103)$$

and the parameters (11.94) and (11.97) reach the long-time limits,

$$\gamma_{e(g)} = 2\pi \sum_q |\epsilon_q|^2 G_T(\pm\omega_q), \qquad (11.104a)$$

$$\Delta_{e(g)} = \sum_q |\epsilon_q|^2 \Delta_T(\pm\omega_q), \qquad (11.104b)$$

where $\omega_q = \omega_a + \nu_q$. The foregoing long-time approximation holds even when $1/t$ is much less than the scales of change of $G_T(\pm\omega)$ and $\Delta_T(\pm\omega)$ in the vicinities of $\pm\omega_q$. Denoting the characteristic value of these frequency scales by $1/t_c$, where t_c is the correlation (memory) time of the bath, we find that (11.104) holds for

$$t \gg t_c. \tag{11.105}$$

In the weak bath-coupling regime, the limits (11.104) are achieved when the perturbative solution (11.86) still holds. In this regime, for the times (11.105), the non-Markovian ME (11.68) can be replaced by its Markovian limit under the same assumptions as (11.61):

$$\dot{\rho} = -i[H_S'(t), \rho] + \frac{\gamma_g}{2}(2\sigma_+\rho\sigma_- - \rho\sigma_-\sigma_+ - \sigma_-\sigma_+\rho)$$
$$+ \frac{\gamma_e}{2}(2\sigma_-\rho\sigma_+ - \rho\sigma_+\sigma_- - \sigma_+\sigma_-\rho)$$
$$+ \tilde{\gamma}\sigma_+\rho\sigma_+ + \tilde{\gamma}^*\sigma_-\rho\sigma_-. \tag{11.106}$$

Here, all the time-dependent bath-induced parameters in (11.68) have been replaced by the corresponding constants (11.104), and $H_S'(t)$ differs from $H_S(t)$ in (11.64) by the substitution [cf. (11.69)]

$$\omega_a \to \tilde{\omega}_a = \omega_a + \Delta_a. \tag{11.107}$$

The first term in the Markovian ME (11.106) represents unitary evolution of the TLS. The second and third terms correspond to absorption and emission from and to the bath, respectively. The last two terms are non-secular. They are commonly neglected in the secular approximation, as in (11.61), but are important in the special case discussed below.

For the validity of (11.106), the bath-induced non-Markovian evolution should negligibly affect the solutions at $t \gg t_c$. This is satisfied under the conditions

$$\gamma_{e(g)}t_c \ll 1, \quad |\Delta_a|t_c \ll 1, \tag{11.108}$$

consistently with the weak bath-coupling regime. The Markovian conditions (11.108) may hold even in the presence of modulation, which, according to (11.104a), changes the effective coupling to the bath (Sec. 11.3). Remarkably, in cases where, in the absence of modulation this ME approach is not valid owing to a strong system–bath coupling, modulation can bring about their weak coupling and thus validate the ME derived above.

11.2.1 Engineering Squeezing-like Effects by Phase Modulation

Let us consider the situation where the non-secular terms in (11.106) become important. To this end, we transform (11.106) to the interaction picture, where

$\rho \to \tilde{\rho}$ [see (11.80)] and omit, for simplicity, the tilde over $\tilde{\rho}$, obtaining

$$\dot{\rho} = \frac{\gamma_g}{2}(2\sigma_+\rho\sigma_- - \rho\sigma_-\sigma_+ - \sigma_-\sigma_+\rho)$$
$$+ \frac{\gamma_e}{2}(2\sigma_-\rho\sigma_+ - \rho\sigma_+\sigma_- - \sigma_+\sigma_-\rho)$$
$$+ e^{2i\tilde{\omega}_a t}\epsilon^2(t)\tilde{\gamma}\sigma_+\rho\sigma_+ + e^{-2i\tilde{\omega}_a t}\epsilon^{*2}(t)\tilde{\gamma}^*\sigma_-\rho\sigma_-. \tag{11.109}$$

The time-dependent factor here can be written similarly to (11.76)–(11.78),

$$e^{2i\tilde{\omega}_a t}\epsilon^2(t) = \sum_q \epsilon'_q e^{i\omega'_q t}, \tag{11.110}$$

where the qth harmonic frequency and amplitude are

$$\omega'_q = 2\tilde{\omega}_a + 2\delta + q\omega_m, \tag{11.111}$$

$$\epsilon'_q = \frac{1}{T_m}\int_0^{T_m} dt\, e^{2i\int_0^t dt'\delta_a(t') - 2i\delta t - iq\omega_m t}. \tag{11.112}$$

If the frequencies ω'_q are not small, then the factor (11.110) is fast-oscillating and the last two terms in (11.109) can be neglected (i.e., the secular approximation is applicable). However, if for some harmonic q_0, the frequency

$$2\tilde{\omega}_a + 2\delta + q_0\omega_m = 0, \tag{11.113}$$

then the corresponding term in (11.110) does not oscillate and hence cannot be neglected. The secular approximation is then applicable only to the terms with $q \neq q_0$. Then, (11.109) becomes

$$\dot{\rho} = \frac{\gamma_g}{2}(2\sigma_+\rho\sigma_- - \rho\sigma_-\sigma_+ - \sigma_-\sigma_+\rho)$$
$$+ \frac{\gamma_e}{2}(2\sigma_-\rho\sigma_+ - \rho\sigma_+\sigma_- - \sigma_+\sigma_-\rho)$$
$$+ \gamma_s\sigma_+\rho\sigma_+ + \gamma_s^*\sigma_-\rho\sigma_-, \tag{11.114}$$

where [cf. (11.69)–(11.71)]

$$\gamma_s = \epsilon'_{q_0}\tilde{\gamma} = \epsilon'_{q_0}(\gamma - i\Delta_a). \tag{11.115}$$

Equation (11.114) is the quantum optical ME for a non-modulated TLS in a squeezed (thermal or vaccum) photonic bath. For such a bath, the complex relaxation parameter γ_s results from squeezing and vanishes in the absence of squeezing.

The value of ω_m, as follows from (11.113), is

$$\omega_m = \frac{2(\tilde{\omega}_a + \delta)}{-q_0}, \tag{11.116}$$

where q_0 is any negative integer. According to this relation, the condition to have nonzero "squeezing" terms in the ME is for ω_m to equal twice the effective TLS frequency, $\tilde{\omega}_a + \delta$, divided by an integer. Since typically $|\epsilon'_{q_0}|$ is maximized for $q_0 = -1$, the "squeezing" effect is maximal when

$$\omega_m = 2(\tilde{\omega}_a + \delta). \tag{11.117}$$

Thus, effects similar to those of a squeezed photonic bath can be engineered with the help of phase modulation for a TLS coupled to an arbitrary bath. However, there is no complete equivalence between a modulated TLS in a thermal bath and a non-modulated TLS in a squeezed photonic bath. In particular, for a non-modulated TLS in a squeezed photonic bath, the population relaxation rates are related by

$$\gamma_e - \gamma_g = \gamma_a, \tag{11.118}$$

where $\gamma_a = 2\pi G_0(\omega_a)$ is the (Golden-Rule), decay rate of the TLS in the vacuum (Ch. 5). For a thermal photonic bath, (11.118) is obtained by noting that $\gamma_{e(g)} = 2\pi G_T(\pm\omega_a)$ and taking into account (11.52). By contrast, for a modulated TLS, (11.118) generally does not hold, since (11.104a) and (11.52) yield, instead of (11.118),

$$\gamma_e - \gamma_g = \gamma_{e0} - \gamma_{g0}, \tag{11.119}$$

$\gamma_{e(g)0}$ being the values of $\gamma_{e(g)}$ at $T = 0$. Under modulation, γ_{e0} generally differs from γ_a and we may have $\gamma_{g0} \neq 0$.

11.2.2 TLS in a Cavity Bath

To illustrate the effect of the ME (11.114), we consider a two-level atom that resonantly interacts with a single cavity-mode in the "bad-cavity" limit. The bath is a cavity mode with the frequency ω_a coupled to the modes outside the cavity. The system–bath interaction is given by (11.26) with $S = \sigma_x$, whereas the bath operator in the interaction representation (11.31) is then

$$\tilde{B}(t) = \eta[\tilde{a}(t)e^{-i\omega_a t} + \tilde{a}^\dagger(t)e^{i\omega_a t}], \tag{11.120}$$

with dipolar coupling η. Here, the time dependence of $\tilde{a}(t)$ is caused by the coupling of the cavity mode to the outside modes. Taking the bath to be in a thermal state, the correlation function (11.33) acquires the form

$$\Phi_T(t) = \langle \tilde{B}(t)\tilde{B}(0)\rangle_B = \eta^2[\langle \tilde{a}(t)\tilde{a}^\dagger(0)\rangle_B e^{-i\omega_a t} + \langle \tilde{a}^\dagger(t)\tilde{a}(0)\rangle_B e^{i\omega_a t}]. \tag{11.121}$$

The atom dynamics caused by its coupling to the cavity mode can be treated beyond the Markovian approximation. Yet, in the bad-cavity limit, the cavity-mode damping by its coupling to the continuum of free-space modes outside the cavity

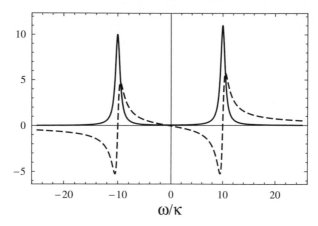

Figure 11.1 Real and imaginary parts of the response spectrum (11.41) of a Lorentzian-cavity bath at temperature T. Here $\pi G_T(\omega)$ (solid line) and $\Delta_T(\omega)$ (dashed line), respectively, are normalized to $\pi G_0(\omega_a) = 2\eta^2/\kappa$, $\omega_a = 10\kappa$, $\bar{n}_a = 10$.

is effectively Markovian. Then, as described in textbooks on quantum optics, the cavity-mode correlation functions decay exponentially,

$$\langle \tilde{a}(t)\tilde{a}^\dagger(0)\rangle = (\bar{n}_a + 1)e^{-\frac{\kappa}{2}|\tau|},$$
$$\langle \tilde{a}^\dagger(t)\tilde{a}(0)\rangle = \bar{n}_a e^{-\frac{\kappa}{2}|\tau|}, \qquad (11.122)$$

where $\bar{n}_a = \bar{n}_T(\omega_a)$ and κ, the damping rate or line width of the cavity mode due to its coupling to outside modes, satisfies $\kappa \gg \eta$ in the bad-cavity limit. Realistically, $\kappa \ll \omega_a$ in any existing cavity.

For such an exponentially decaying correlation function (11.121)–(11.122), the temperature-dependent dissipative and dispersive parts of the bath response follow from (11.43) and (11.44) to be

$$G_T(\omega) = \frac{\gamma_a}{2\pi}\left[(\bar{n}_a + 1)\frac{(\frac{\kappa}{2})^2\theta(\omega)}{(\frac{\kappa}{2})^2 + (\omega - \omega_a)^2} + \bar{n}_a\frac{(\frac{\kappa}{2})^2\theta(-\omega)}{(\frac{\kappa}{2})^2 + (\omega + \omega_a)^2}\right], \quad (11.123)$$

$$\Delta_T(\omega) = \frac{\gamma_a}{2}\left[(\bar{n}_a + 1)\frac{\frac{\kappa}{2}(\omega - \omega_a)}{(\frac{\kappa}{2})^2 + (\omega - \omega_a)^2} + \bar{n}_a\frac{\frac{\kappa}{2}(\omega_a + \omega)}{(\frac{\kappa}{2})^2 + (\omega + \omega_a)^2}\right], \quad (11.124)$$

where $\theta(\omega)$ (as before) is the Heaviside step function and

$$\gamma_a \equiv 2\pi G_0(\omega_a) = \frac{4\eta^2}{\kappa} \qquad (11.125)$$

is the atomic (Golden-Rule) damping rate by the cavity at $T = 0$. The bath-response spectrum (11.123) is a Lorentzian at $T = 0$ and a sum of two Lorentzians at $T > 0$. The real (dissipative) and imaginary (dispersive) response functions $G_T(\omega)$ and $\Delta_T(\omega)$ are connected by the Kramers–Kronig relations (Fig. 11.1).

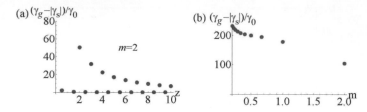

Figure 11.2 Effective bath squeezing (the difference between γ_g and $|\gamma_s|$) as a function of the modulation parameters z and $m \equiv \omega_m/\omega_a$ (ω_a and $\bar{n}_a$ being fixed). The squeezing-induced rate γ_s modifies the TLS dephasing rate, compared to that caused by a thermal bath, (a) as a function of z for fixed $m = 2$ and (b) as a function of m for $z = 1$ ($\omega_a = 10\kappa$, $\bar{n}_a = 10^3$). The difference between γ_g and $|\gamma_s|$ cannot violate the "classical squeezing" bound $|\gamma_s| \leq \gamma_g$. (Adapted from Shahmoon and Kurizki, 2013. © 2013 American Physical Society)

11.2.3 Sinusoidal Modulation

A thermal bath may act as if it were squeezed on a TLS under sinusoidal modulation of the TLS level spacing,

$$\delta_a(t) = \delta[1 - \sin(\omega_m t)], \tag{11.126}$$

where $\delta > 0$. Upon introducing this modulation in (11.76) and using the identity $e^{iz\cos\phi} = \sum_{q=-\infty}^{\infty} i^q J_q(z) e^{iq\phi}$ with integer q, we obtain

$$\epsilon_q = i^q e^{-iz} J_q(z), \quad \epsilon'_q = i^q e^{-2iz} J_q(2z), \quad \nu_q = \delta + q\omega_m, \tag{11.127}$$

where $z = \delta/\omega_m$ and $J_q(z)$ is the Bessel function of the first kind of order q. Since $J_{-q}(z) = (-1)^q J_q(z)$, we have $\epsilon_q = \epsilon_{-q}$ and $\epsilon'_q = \epsilon'_{-q}$. In the expressions for the coefficients (11.104), we now obtain $|\epsilon_q|^2 = |\epsilon_{-q}|^2 = J_q^2(z)$ and ω_m is given by (11.116).

Below we set, for simplicity, $\tilde{\omega}_a \approx \omega_a$, which is often a good approximation. The Lorentzian bath-response described by (11.123) with the parameters given in Figure 11.2 yields an appreciable squeezing effect under modulations, as $|\gamma_s| \sim \gamma_g/2$ in (11.114). For $z = 0$ we retrieve the secular-approximation results without modulation, $\gamma_g = \gamma_a \bar{n}_a$, $\gamma_e = \gamma_a(\bar{n}_a + 1)$, and $\gamma_s = 0$. Thus, $|\gamma_s|$ does not exceed γ_g, so that we only obtain "classical squeezing" due to the secular-approximation breakdown.

11.2.4 Modified Resonance Fluorescence Spectrum

The resonance fluorescence spectrum of a two-level atom driven by a resonant strong (classical) field with Rabi frequency $\Omega e^{i\theta}$ is well known to be given by the

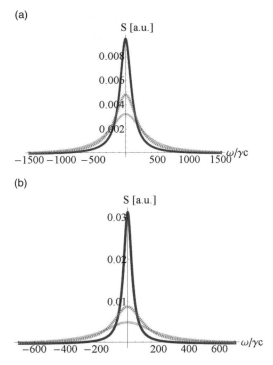

Figure 11.3 Resonance fluorescence spectrum (only the central peak is shown) modified by modulation-induced effective squeezing. The spectrum is plotted for several values of the relative driving field phase ϕ; $\phi = \pi$ (solid line), $\phi = \pi/2$ (dashed line), and $\phi = 0$ (dotted line), with modulation parameters: (a) $z = 1$, $m = 2$, and (b) $z = 3$, $m = 2$. Here $\omega_a = 10\kappa$ and $\bar{n}_a = 10^3$, as in Figure 11.2. (Reprinted from Shahmoon and Kurizki, 2013. © 2013 American Physical Society)

Fourier transform of the bath autocorrelation function,

$$S(\omega) \propto \frac{\gamma_\phi}{(\omega - \omega_a)^2 + \gamma_\phi^2} + \left[\frac{\Gamma_\phi}{(\omega - \omega_a + \Omega)^2 + \Gamma_\phi^2} + \frac{\Gamma_\phi}{(\omega - \omega_a - \Omega)^2 + \Gamma_\phi^2} \right].$$

(11.128)

In the present case,

$$\gamma_\phi = \gamma + |\gamma_s| \cos\phi, \quad \Gamma_\phi = \frac{3}{2}\gamma - \frac{1}{2}|\gamma_s| \cos\phi, \quad \phi = 2(\theta - \varphi), \quad (11.129)$$

where φ is the squeezing phase defined by $\gamma_s = |\gamma_s| e^{2i\varphi}$. The spectrum consists of three Lorentzian peaks centered at ω_a and $\omega_a \pm \Omega$, as in the case of the vacuum state of the bath, but these peaks are modified by effective squeezing: their widths, γ_ϕ and Γ_ϕ, change as the phase of the strong field is varied. This *phase dependence* of the fluorescence line widths is a signature of the effective squeezing induced by the modulation (Fig. 11.3). When $\phi = \pi/2$, the only effect of the modulation is a modification of γ relative to the unmodulated case, resulting in a narrower peak. When $\phi = 0$, the line width is enhanced by the squeezing term $\gamma_\phi = \gamma + |\gamma_s|$. The narrowest peak $\gamma_\phi = \gamma - |\gamma_s|$ arises for $\phi = \pi$.

The two quadrature rates $\gamma_{x,y}$ become different when squeezing arises. In the cases discussed above, these rates are, respectively, $\sim 0.1\gamma_T$, $\sim \gamma_T$. Here the TLS damping rate in the cavity, in the absence of modulation, is $\gamma_T = 2\pi G_T(\omega_a) = \gamma_a(\bar{n}_a + 1)$ according to (11.123) and (11.125). In the limit of high bath temperature, $T \gg \omega_a$ (where T is given in energy units and $\hbar = 1$ as before), $\gamma_T \approx \gamma_a T/\omega_a$.

According to (11.125), $\gamma_T \approx \frac{4\eta^2}{\kappa}\frac{T}{\omega_a}$, with $\eta = d\sqrt{\frac{2\pi\omega_a}{V}}$, d being the dipole matrix element of the TLS transition and $V \sim \lambda_a^3$ the cavity mode volume, the transition wavelength cubed. We then obtain

$$\gamma_T \approx \frac{3T}{4\pi^2\kappa}\gamma_a, \tag{11.130}$$

where $\gamma_a = \frac{4d^2\omega_a^3}{3c^3}$ is the atomic-transition spontaneous emission rate in free space. Thus, the effective squeezing becomes more pronounced with the growth of temperature divided by the cavity line width in the bad-cavity and high-temperature limits.

11.3 Finite-Temperature TLS Decoherence Control

We here focus on two scenarios: a TLS subject to either an amplitude- or phase-noise (AN or PN) in a thermal bath (Fig. 11.4). Unlike in Section 11.2, we here adopt the secular approximation, neglecting the nonsecular terms in the universal ME (11.45). The Hamiltonian (in either scenario) is given by (11.2), (11.48), and the S-B interaction Hamiltonian (11.26), $S(t)$ being the control-modified Pauli-matrix operator and B the bosonic bath operator, having the form (11.49) in the interaction picture.

11.3.1 Dynamically Modified Bloch Equations for TLS Noise Control

We can adapt the universal non-Markovian ME (11.45) to a dynamically controlled TLS in both AN and PN scenarios.

In the AN scenario, $H_S(t)$ is given by (11.64), which is designed to counter AN by phase modulation (PM) due to the time-dependent dynamical AC Stark shift $\delta_a(t)$ and by $\tilde{\epsilon}(t)$, the amplitude of $S(t)$ in (11.65).

In the PN scenario, $H_S(t)$ in the Hamiltonian (11.2) is given by

$$H_S(t) = \frac{1}{2}\omega_a\sigma_z + V(t)\sigma_x, \tag{11.131}$$

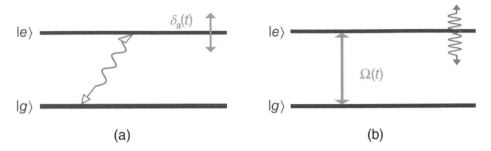

Figure 11.4 (a) Amplitude noise (AN) (wavy arrow) countered by AC-Stark shift modulation (vertical arrow). (b) Phase noise (PN) (wavy arrow) countered by resonant-field modulation (vertical arrow). (Adapted from Kurizki, 2013.)

where

$$V(t) = \Omega(t)\cos(\omega_a t),\qquad(11.132)$$

and the interaction Hamiltonian (11.26) involves

$$S(t) = \tilde{\epsilon}(t)\sigma_z.\qquad(11.133)$$

Here, PN is countered by Rabi oscillations caused by a resonant field $V(t)$ with the time-dependent Rabi frequency $\Omega(t)$ and by $\tilde{\epsilon}(t)$, the time-dependent modulation of the interaction strength. In this scenario, it is convenient to transform the Hamiltonian (11.2) to the interaction picture with respect to the Hamiltonian $H_0 + H_B = \omega_a \sigma_z/2 + H_B$. In the RWA with respect to $V(t)$ (which means the neglect of the counterrotating component of the control field), this transformation yields the Hamiltonian

$$H(t) = \tilde{\epsilon}(t)\tilde{B}(t)\sigma_z + \frac{1}{2}\Omega(t)\sigma_x,\qquad(11.134)$$

where the transformed bath operator $\tilde{B}(t)$ is defined in (11.31).

We next recast the Hamiltonian (11.134) in a form similar to that of the AN-scenario Hamiltonian by performing a rotation on the Bloch sphere through the Hadamard transformation

$$U_H = \frac{\sigma_x + \sigma_z}{\sqrt{2}} = \frac{1}{\sqrt{2}}\begin{pmatrix} 1 & 1 \\ 1 & -1 \end{pmatrix}.\qquad(11.135)$$

Under this transformation, (11.134) becomes

$$\tilde{H}(t) = \frac{1}{2}\Omega(t)\sigma_z + \tilde{\epsilon}(t)\tilde{B}(t)\sigma_x,\qquad(11.136)$$

the eigenvectors of σ_z being

$$|\uparrow, \downarrow\rangle = \frac{1}{\sqrt{2}}(|e\rangle \pm |g\rangle), \tag{11.137}$$

as compared to the σ_z eigenbasis in the AN case,

$$|\uparrow, \downarrow\rangle = |e, g\rangle. \tag{11.138}$$

The TLS interaction (dipole) operator under control via *any* modulation *in both AN and PN scenarios* can be explicitly written in the general form

$$\tilde{S}(\tau, t) = \tilde{\varepsilon}(\tau)\left[\varepsilon^*(t)\varepsilon(\tau)e^{-i\omega_a'(t-\tau)} |\uparrow\rangle \langle\downarrow| + \varepsilon(t)\varepsilon^*(\tau)e^{i\omega_a'(t-\tau)} |\downarrow\rangle \langle\uparrow|\right]. \tag{11.139}$$

Here,

$$\omega_a' = \begin{cases} \omega_a & \text{for AN,} \\ 0 & \text{for PN} \end{cases} \tag{11.140}$$

and

$$\varepsilon(t) = e^{i\phi(t)} \tag{11.141}$$

is the modulation phase factor with

$$\phi(t) = \begin{cases} \int_0^t dt'\delta_a(t') & \text{for AN,} \\ \int_0^t dt'\Omega(t') & \text{for PN.} \end{cases} \tag{11.142}$$

In order to counter AN effects, we can apply in (11.142) a time-integrated AC Stark shift,

$$\delta_a(t) = \int_0^t dt' \frac{|\Omega_c(t')|^2}{\delta_c(t')}, \tag{11.143}$$

where $\Omega_c(t)$ is the Rabi frequency of the control field and $\delta_c(t')$ is its detuning. The corresponding phase factor for PN is the integral (pulse area) of the Rabi frequency $\Omega(t)$.

The TLS evolution operator in the interaction picture is then given by

$$U_S(t, \tau) = U_S(t, 0)U_S^\dagger(\tau, 0), \tag{11.144}$$

where

$$U_S(t, 0) = e^{-i\phi(t)-\omega_a't} |\uparrow\rangle \langle\uparrow| + e^{i\phi(t)+\omega_a't} |\downarrow\rangle \langle\downarrow|. \tag{11.145}$$

The dynamically modified Bloch equations can now be obtained from the ME (11.45) in the suitable diagonalizing basis [(11.138) for AN or (11.137) for PN]. They have the following form in the secuar approximation,

$$\dot{\rho}_{\uparrow\uparrow} = -\dot{\rho}_{\downarrow\downarrow} = -\gamma_\uparrow(t)\rho_{\uparrow\uparrow} + \gamma_\downarrow(t)\rho_{\downarrow\downarrow}, \tag{11.146a}$$

$$\dot{\rho}_{\uparrow\downarrow} = \dot{\rho}^*_{\downarrow\uparrow} = -[i\dot{\phi}(t) + \gamma(t) + i\tilde{\omega}'_a(t)]\rho_{\uparrow\downarrow}. \tag{11.146b}$$

Here,

$$\gamma(t) = \frac{\gamma_{\uparrow}(t) + \gamma_{\downarrow}(t)}{2} \tag{11.147}$$

is the mean of the time-dependent relaxation rates of the upper and lower levels, $\tilde{\omega}'_a = \omega'_a + \Delta_a(t)$, and the Lamb shift

$$\Delta_a(t) = \Delta_{\uparrow}(t) - \Delta_{\downarrow}(t) \tag{11.148}$$

is the difference between the shifts of the two levels. The $\gamma_{\uparrow(\downarrow)}(t)$ and $\Delta_{\uparrow(\downarrow)}(t)$ are determined by the real and imaginary parts, respectively, of the following expression,

$$\gamma_{\uparrow(\downarrow)}(t)/2 + i\Delta_{\uparrow(\downarrow)}(t) = \int_0^t dt' \Phi_T(t - t') K_{\pm}(t, t'). \tag{11.149}$$

Here $\Phi_T(t)$ is determined by (11.33) and

$$K_+(t, t') = K^*_-(t, t') = \epsilon(t)\epsilon^*(t')e^{i\omega'_a(t-t')}, \tag{11.150}$$

where $\epsilon(t)$ is given by (11.73).

The difference between $\gamma_{\uparrow}$ and $\gamma_{\downarrow}$ or between $\Delta_{\uparrow}$ and $\Delta_{\downarrow}$ in (11.149) is due to the mutually conjugated factors $K_{\pm}(t, t')$. This difference exists only for a complex bath correlation function $\Phi_T(t)$.

The dynamically modified Bloch equations (11.146) provide a unified description of AN and PN control by modulation. In the case of AN, (11.146) coincides with (11.79), if the nonsecular terms in the latter are neglected.

As in Section 11.2, the *time-averaged* relaxation rates and energy shifts in the interval $(0, t)$ can be expressed, respectively, by

$$\bar{\gamma}_{\uparrow(\downarrow)}(t) = 2\pi \int_{-\infty}^{\infty} d\omega\, G_T(\pm\omega) F_t(\omega - \omega'_a),$$

$$\bar{\Delta}_{\uparrow(\downarrow)}(t) = \int_{-\infty}^{\infty} d\omega\, \Delta_T(\pm\omega) F_t(\omega - \omega'_a). \tag{11.151}$$

Here, for both AN [Eq. (11.64)] and PN [Eq. (11.131)], we employ the temperature-dependent bath-response function $G_T(\omega)$, (11.52), which is the Fourier transform of $\Phi_T(t)$, (11.43). Equations (11.151) show that $\bar{\gamma}_{\uparrow}(t)$ [$\bar{\gamma}_{\downarrow}(t)$] *are proportional to the overlap of the modulation spectrum* $F_t(\omega - \omega'_a)$ *with the bath-response spectrum* $G_T(\omega)$ [$G_T(-\omega)$], and $\bar{\Delta}_{\uparrow(\downarrow)}(t)$ are similarly related to $\Delta_T(\pm\omega)$.

For simplicity, we assume for the rest of this section that there is no amplitude modulation, $\tilde{\epsilon}(t) = 1$, so that $\epsilon(t) = \varepsilon(t)$. Then the modulation spectrum (normalized to 1) for both AN and PN is expressed by [cf. (11.95) and (11.92)]

$$F_t(\omega) = \frac{1}{2\pi t} \left| \int_0^t dt' e^{i\phi(t')-i\omega t'} \right|^2, \qquad (11.152)$$

in terms of the modulation phase given in (11.142).

We may conclude that under the appropriate substitutions, the Bloch equations (11.146) assume the same universal form for either relaxation (AN) or decoherence (PN) control, by off-resonant or near-resonant modulation, respectively.

At long times compared to the bath memory time, $t \gg t_c$, $\bar{\gamma}_{\uparrow(\downarrow)}$ tend to the constant limits $\bar{\gamma}_{\uparrow(\downarrow)} \equiv \gamma_{\uparrow(\downarrow)}$, $\gamma_{\uparrow(\downarrow)}(t)$ coincide with these limits (up to negligible oscillations), and we can then omit the bar over $\gamma_{\uparrow(\downarrow)}$. Likewise, at such long times, $\Delta_a(t)$ tends to a constant Δ_a. Then, (11.146) become the Markovian Bloch equations,

$$\dot{\rho}_{\uparrow\uparrow} = -\dot{\rho}_{\downarrow\downarrow} = -\gamma_\uparrow \rho_{\uparrow\uparrow} + \gamma_\downarrow \rho_{\downarrow\downarrow}, \qquad (11.153a)$$

$$\dot{\rho}_{\uparrow\downarrow} = \dot{\rho}_{\downarrow\uparrow}^* = -[i\dot{\phi}(t) + \gamma + i\tilde{\omega}_a']\rho_{\uparrow\downarrow}, \qquad (11.153b)$$

where $\tilde{\omega}_a' = \omega_a' + \Delta_a$.

The validity conditions for both the non-Markovian and the Markovian Bloch equations, (11.146) and (11.153), are similar to (11.108),

$$\gamma_{\uparrow(\downarrow)} t_c \ll 1, \quad |\Delta_a| t_c \ll 1, \qquad (11.154)$$

corresponding to sufficiently weak system–bath coupling. The correlation time generally depends on the modulation, as described below [(11.207)].

11.3.2 Dynamically Modified TLS Relaxation (AN) beyond RWA

The dynamically controlled relaxation or dissipation (AN) rates for the $|e\rangle \to |g\rangle$ and $|g\rangle \to |e\rangle$ transitions, given by (11.94a) and (11.97a), can be compactly rewritten as

$$\bar{\gamma}_{e(g)}(t) = 2\pi \int_{-\infty}^{\infty} d\omega F_t(\omega - \omega_a) G_T(\pm\omega), \qquad (11.155)$$

where the upper (lower) sign corresponds to the subscript e (g) and the "filter function" $F_t(\omega)$ is the spectral density of phase modulation (PM). These expressions have been obtained using (11.26), the full TLS–bath interaction Hamiltonian (4.10), with the antiresonant terms included.

Had we used the RWA interaction Hamiltonian (4.12) without the antiresonant terms, we would have obtained instead the transition rates

$$\bar{\gamma}_{e(g)}^{\text{RWA}}(t) = 2\pi \int_0^{\infty} d\omega F_t(\omega - \omega_a) G_T(\pm\omega), \qquad (11.156)$$

where the integration over ω extends from 0 to ∞, rather than from $-\infty$ to ∞ as in (11.155). These RWA transition rates are valid for a modulation that is not excessively fast, such that $F_t(\omega - \omega_a) \simeq 0$ at $\omega < 0$, being peaked near ω_a. Yet, if the task of suppressing the transition rates $\gamma_{e(g)}$ requires modulation rates comparable to ω_a, then the RWA is inadequate. In particular, (11.52) and (11.156) imply that, at zero temperature ($T = 0$), the rate $\bar{\gamma}_g^{\text{RWA}}(t)$ vanishes, irrespective of how fast the modulation, whereas the true upward-transition rate $\bar{\gamma}_g(t)$ in (11.155) may be comparable to $\bar{\gamma}_e(t)$ under ultrafast modulation. The RWA only describes a downward (upward) transition that corresponds to the emission (absorption) of a bath quantum. By contrast, the antiresonant (negative-frequency) contribution to $\bar{\gamma}_{e(g)}(t)$ in (11.155) accounts for downward (upward) transitions that are accompanied by absorption (emission) of a bath quantum. Such antiresonant (non-RWA) processes are possible since ultrafast modulation may shift and broaden levels $|e\rangle$ and $|g\rangle$ to the extent that they can no longer be identified as the upper and lower levels (Ch. 16).

We further assume that the effective transition rates $\gamma_{e(g)j}$ from level e (g) to *any* other level j, are suppressed by the modulation, to the extent that the TLS model remains valid as long as the control is on (Ch. 12).

A notable class of AN control is that of phase modulation (PM) that consists in periodic jumps of the modulation phase by an amount ϕ at times $\tau, 2\tau, \ldots$. Such modulation can be effected by a train of identical, equidistant, narrow pulses of nonresonant radiation, which produce pulsed AC Stark shifts of ω_a. At sufficiently long times [Eq. (11.105)] one can then use (11.104a), with

$$\omega_q = \omega_a + \frac{\phi + 2q\pi}{\tau}, \quad |\epsilon_q|^2 = \frac{4\sin^2(\phi/2)}{(\phi + 2q\pi)^2}. \tag{11.157}$$

For *small phase shifts*, $\phi \ll 1$, the $q = 0$ harmonic dominates, with weight

$$|\epsilon_0|^2 \approx 1 - \frac{\phi^2}{12}, \tag{11.158}$$

whereas the other harmonics have much smaller weights,

$$|\epsilon_q|^2 \approx \frac{\phi^2}{4\pi^2 q^2} \quad (q \neq 0). \tag{11.159}$$

In this case, one can retain only the $q = 0$ term in (11.104a), unless $G_T(\omega)$ is changing very rapidly with frequency ω. The $q = 0$ modulation acts as a constant spectral shift [Fig. 11.5(d)],

$$\delta_a = \phi/\tau. \tag{11.160}$$

As $|\phi|$ increases, the difference between the $q = 0$ and $q = -1$ (for positive ϕ) or $q = 1$ (for negative ϕ) harmonic weights diminishes, until for $\phi = \pm\pi$ they become equal. Thus, for $\phi = \pi$, $F_t(\omega)$ has two identical peaks, oppositely shifted

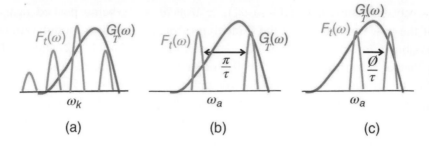

Figure 11.5 Overlap of the bath coupling spectrum, $G_T(\omega)$, and the modulation "filter" spectrum, $F_t(\omega)$. (a) Quasiperiodic modulation with spectral peaks at ω_q. (b) Impulsive PM (π-pulses), $\phi = \pi$. (c) Impulsive PM, with small phase shifts, $\phi \ll 1$, and $1/\tau$ repetition rate. (Adapted from Kurizki, 2013.)

in frequency [Fig. 11.5(c)], with the weights

$$|\epsilon_0|^2 = |\epsilon_{-1}|^2 = 4/\pi^2, \tag{11.161}$$

whereas the weights of other harmonics $|\epsilon_q|^2$ decrease with q as $(2q+1)^{-2}$, totaling $\sum_{q \neq 0, -1} |\epsilon_q|^2 = 0.19$].

The foregoing features allow one to adjust the modulation parameters for a given AN (bath-coupling) spectrum to obtain an optimal decrease of the relaxation rate γ. Thus, PM with small ϕ is preferable near a cutoff of $G_T(\omega)$ [Fig. 11.5(c),(d)], since it yields a spectral shift in the required direction [upshift, i.e., frequency shift toward the upper cutoff of $G_T(\omega)$]. The adverse effect of the higher harmonics ($q \neq 0$) in $F_t(\omega)$ then scales as ϕ^2 and hence can be suppressed by decreasing $|\phi|$. If, however, ω_a is near a symmetric peak of $G_T(\omega)$, then γ is reduced more effectively for $\phi \simeq \pi$, since the main peaks of $F_t(\omega)$ at ω_0 and ω_1 then shift stronger with $1/\tau$ than the peak at $\omega_0 = -\phi/\tau$ for $\phi \ll 1$.

11.3.3 Dynamically Modified Dephasing due to Classical Noise

The case of PN, also known as proper dephasing, is phenomenologically ascribed to random frequency fluctuations $\delta_r(t)$, commonly characterized by the system–bath interaction Hamiltonian

$$H_I \equiv H_{SB} = \frac{1}{2}\sigma_z \delta_r(t), \tag{11.162}$$

with a (single) correlation time t_c and ensemble mean $\bar{\delta}_d = 0$. The time-dependent bath operator can then be described by the classical random (noise) function $\tilde{B}(t) = \delta_r(t)/2$ that yields the phenomenological bath correlation function

$$\Phi_r(t - t') = \frac{1}{4}\overline{\delta_r(t)\delta_r(t')}, \tag{11.163}$$

where the overbar denotes the ensemble average over the noise fluctuations. The correlation function $\Phi_r(t)$ is real and time-symmetric, $\Phi_r(t) = \Phi_r(-t)$. From (11.41), we have

$$\int_0^\infty dt\, \Phi_r(t)e^{i\omega t} = \pi G_r(\omega) + i\Delta_r(\omega), \qquad (11.164)$$

where the bath spectral response functions $G_r(\omega)$ and $\Delta_r(\omega)$ have the symmetry properties,

$$G_r(-\omega) = G_r(\omega),$$
$$\Delta_r(-\omega) = -\Delta_r(\omega). \qquad (11.165)$$

Let us first consider free decay due to proper dephasing, in the absence of modulation, $V(t) = 0$, $\tilde{\epsilon}(t) = 1$. The Bloch equations (11.146) cannot be employed in this case, since they involve the secular approximation, which holds only under a sufficiently large $V(t)$ [cf. (11.177) below]. Instead, we then use the universal ME (11.45), where, now $H_S = \omega_a \sigma_z/2$, $\Phi_T(t) \rightarrow \Phi_r(t)$, and $\tilde{S}(\tau, t) = S(t) = \sigma_z$, resulting in the ME

$$\dot{\rho} = -i[H_S, \rho] + \frac{1}{2}\gamma_{d0}(t)(\sigma_z \rho \sigma_z - \rho), \qquad (11.166)$$

with the time-dependent dephasing rate

$$\gamma_{d0}(t) = 4\int_0^t \Phi_r(t')dt'. \qquad (11.167)$$

In the $(|e\rangle, |g\rangle)$ basis, (11.166) yields

$$\dot{\rho}_{ee} = \dot{\rho}_{gg} = 0,$$
$$\dot{\rho}_{eg} = \dot{\rho}_{ge}^* = -[i\omega_a + \gamma_{d0}(t)]\rho_{eg}. \qquad (11.168)$$

Let us consider the effect of the control Hamiltonian (11.131) on a TLS that undergoes such noise-induced dephasing. A near-resonant driving field $V(t)$ (11.132) may dynamically modify the dephasing rate, as per (11.149) if one substitutes $\Phi_T(t) \rightarrow \Phi_r(t)$. The Bloch equations (11.146) then yield in the basis (11.137)

$$\dot{\tilde{w}} = -\gamma_d(t)\tilde{w},$$
$$\dot{\rho}_{\uparrow\downarrow} = \dot{\rho}_{\downarrow\uparrow}^* = -[i\Omega(t) + \gamma_d(t)/2 + i\Delta_d(t)]\rho_{\uparrow\downarrow}, \qquad (11.169)$$

where $\tilde{w} = \rho_{\uparrow\uparrow} - \rho_{\downarrow\downarrow}$ and the rate $\gamma_d(t)$ extends (11.167) to account for modulation by $V(t)$. Since $\Phi_r(t)$ in (11.163) is real, (11.149) yields

$$\gamma_\uparrow(t) = \gamma_\downarrow(t) \equiv \gamma_d(t)/2, \qquad (11.170)$$
$$\Delta_\uparrow(t) = -\Delta_\downarrow(t) \equiv \Delta_d(t)/2. \qquad (11.171)$$

The average dynamically modified decay or decoherence rates and frequency shifts are, according to (11.151) and (11.165),

$$\bar{\gamma}_\uparrow(t) = \bar{\gamma}_\downarrow(t) \equiv \bar{\gamma}_d(t)/2 = 2\pi \int_{-\infty}^{\infty} d\omega F_t(\omega)G_r(\omega), \qquad (11.172)$$

$$\bar{\Delta}_\uparrow(t) = -\bar{\Delta}_\downarrow(t) \equiv \bar{\Delta}_d(t)/2 = \int_{-\infty}^{\infty} d\omega F_t(\omega)\Delta_r(\omega), \qquad (11.173)$$

where $F_t(\omega)$ is given by (11.152).

If the time-dependent Rabi frequency $\Omega(t)$ changes periodically with the period $2\pi/\Omega_0$, then at long times, such that $\Omega_0 t \gg 1$ and $t \gg t_c$, the decoherence rate and shift approach their infinite-time asymptotic values,

$$\gamma_d = \lim_{t\to\infty} \bar{\gamma}_d(t) = 4\pi \int_{-\infty}^{\infty} d\omega F(\omega)G_r(\omega), \qquad (11.174)$$

$$\Delta_d = \lim_{t\to\infty} \Delta_d(t). \qquad (11.175)$$

Correspondingly, at $t \gg t_c$, the Bloch equations (11.169) become Markovian,

$$\dot{\tilde{w}} = -\gamma_d \tilde{w},$$

$$\dot{\rho}_{\uparrow\downarrow} = \dot{\rho}_{\downarrow\uparrow}^* = -[i\Omega(t) + \gamma_d/2 + i\Delta_d]\rho_{\uparrow\downarrow}. \qquad (11.176)$$

In the ME (11.169) [as in (11.146)], we have made the secular approximation, which here holds when

$$\Omega(t) \gg \gamma_d, |\Delta_d|. \qquad (11.177)$$

Namely, $\Omega(t)$ should not vanish, so that $\min_t \Omega(t)$ be sufficiently large. As in Section 11.2, the relaxation parameters must satisfy

$$\gamma_d, |\Delta_d| \ll 1/t_c \qquad (11.178)$$

for dephasing to be considered a weak perturbation.

In the presence of a continuous-wave (CW) field $V(t)$, Ω is constant, $F(\omega) = \delta(\omega - \Omega)$, and the asymptotic rate becomes

$$\gamma_d = 4\pi G_r(\Omega). \qquad (11.179)$$

Consistently with (11.179), in the absence of a driving field, the dephasing rate (11.167) tends at $t \gg t_c$ to the Markovian (stationary) limit

$$\gamma_{d0}(t) \rightarrow \gamma_{d0} = 4\pi G_r(0). \qquad (11.180)$$

For an exponentially decaying correlation function,

$$\Phi_r(t) = \frac{A}{4} e^{-t/t_c}, \qquad (11.181)$$

where $A = \overline{\delta_{\mathrm{r}}^2}$ is the random-phase dispersion, we have

$$G_{\mathrm{r}}(\omega) = \frac{1}{4\pi} \frac{A t_{\mathrm{c}}}{\omega^2 t_{\mathrm{c}}^2 + 1}. \qquad (11.182)$$

Then the stationary *proper-dephasing* rate (11.180) becomes

$$\gamma_{\mathrm{d}0} = A t_{\mathrm{c}}. \qquad (11.183)$$

In the presence of a CW field, according to (11.179) and (11.182), the dephasing rate (11.183) is changed to

$$\gamma_{\mathrm{d}} = \frac{A t_{\mathrm{c}}}{\Omega^2 t_{\mathrm{c}}^2 + 1}. \qquad (11.184)$$

For a sufficiently strong CW field, the dephasing rate γ_{d} is suppressed by the factor $(\Omega t_{\mathrm{c}})^{-2} \ll 1$, which signifies that the TLS frequency is shifted out of the noise bandwidth, or that the noise effects are averaged out.

11.3.4 Dynamically Modified Dephasing due to a Quantum Bath

The foregoing results, obtained for dephasing due to classical noise (frequency fluctuations), remain valid, under proper substitutions, for σ_z-coupling to a quantum bath. Namely, (11.26) is then

$$H_{\mathrm{I}} = \sigma_z B. \qquad (11.185)$$

Upon performing calculations similar to those made in Section 11.3.3, we obtain that in the absence of modulations, (11.166) and (11.168) remain valid if we substitute

$$\Phi_{\mathrm{r}}(t) \rightarrow \Phi_{\mathrm{s}}(t)/2, \qquad (11.186)$$

where

$$\Phi_{\mathrm{s}}(t) = 2\,\mathrm{Re}\,\Phi_T(t) = \langle \tilde{B}(t)\tilde{B}(0) + \tilde{B}(0)\tilde{B}(t)\rangle_{\mathrm{B}} \qquad (11.187)$$

is the symmetrized bath correlation function. Then, in (11.166) and (11.168) [cf. (11.167) and (11.180)]

$$\gamma_{\mathrm{d}0}(t) = 2\int_0^t \Phi_{\mathrm{s}}(t')dt'. \qquad (11.188)$$

The spectrum of $\Phi_{\mathrm{s}}(t)$ [cf. (11.38)] is the symmetrized bath-response spectrum,

$$G_{\mathrm{s}}(\omega) = \frac{1}{2\pi}\int_{-\infty}^{\infty} dt\,\Phi_{\mathrm{s}}(t)e^{i\omega t} = G_T(\omega) + G_T(-\omega). \qquad (11.189)$$

If the KMS condition holds, so that $G_T(-\omega) = e^{-\beta\omega}G_T(\omega)$, then $G_T(\omega)$ and $G_s(\omega)$ are related by

$$G_T(\omega) = \frac{G_s(\omega)}{1 + e^{-\beta\omega}}. \tag{11.190}$$

In the presence of modulations, the above results [(11.169)–(11.179)] can be used for a quantum bath under the substitutions

$$\Phi_r(t) \to \Phi_T(t), \qquad G_r(\omega) \to G_T(\omega). \tag{11.191}$$

11.3.5 Dynamical Decoupling

The dynamical control of dephasing considered above requires $\Omega(t)$ never to become too small [see (11.177)]. An alternative control method, called dynamical decoupling (DD), employs short π-pulses. Originally, DD was suggested for dephasing due to bosonic baths or a classical Gaussian noise, where an exact solution can be obtained. Here, we use our general approach to show that the same theory holds for *arbitrary baths*, if only the system–bath coupling is sufficiently weak.

We assume the interaction-picture Hamiltonian (11.134) to have $\tilde{\epsilon}(t) = 1$ and

$$\Omega(t) = \pi \sum_{i=1,2,\dots} \delta(t - t_i) \qquad (0 < t_1 < t_2 < \dots), \tag{11.192}$$

corresponding to a sequence of impulsive (infinitely short) π-pulses that are not necessarily equidistant. We next transform this Hamiltonian to a doubly rotating frame by the unitary transformation,

$$U_{\text{dr}}(t) = \exp\left[-\frac{i}{2}\int_0^t dt'\Omega(t')\sigma_x\right] = \begin{cases} (-1)^k, & n(t) = 2k, \\ (-1)^k i\sigma_x, & n(t) = 2k - 1, \end{cases} \tag{11.193}$$

where k is an integer and $n(t) = \sum_i \theta(t - t_i)$ is the number of the pulses in the interval $(0, t)$. Then, the Hamiltonian becomes

$$\tilde{H}(t) = \epsilon(t)\sigma_z\tilde{B}(t), \tag{11.194}$$

where

$$\epsilon(t) = e^{i\phi(t)} = (-1)^{n(t)}, \tag{11.195}$$

$\phi(t)$ being given by the second line in (11.142).

We can now employ the universal ME (11.45), with $H_S = 0$ and $\tilde{S}(\tau, t) = S(t) = \epsilon(t)\sigma_z$. It has then the form

$$\dot{\rho} = \frac{1}{2}\gamma_d'(t)(\sigma_z\rho\sigma_z - \rho) \tag{11.196}$$

with the dephasing rate

$$\gamma_d'(t) = 2 \int_0^t d\tau \, \Phi_s(t - \tau)\epsilon(t)\epsilon(\tau). \tag{11.197}$$

This rate is a DD modification of the rate (11.188), which holds in the absence of modulation. The density-matrix elements governed by (11.196) obey

$$\dot{\rho}_{ee} = \dot{\rho}_{gg} = 0,$$
$$\dot{\rho}_{eg} = \dot{\rho}_{ge}^* = -\gamma_d'(t)\rho_{eg}. \tag{11.198}$$

The solution of these equations yields

$$\rho_{ee}(t) = \rho_{ee}(0), \quad \rho_{gg}(t) = \rho_{gg}(0), \tag{11.199}$$
$$\rho_{eg}(t) = \rho_{eg}(0)e^{-J_d'(t)}, \tag{11.200}$$

where

$$J_d'(t) = \int_0^t dt' \gamma_d'(t'). \tag{11.201}$$

The transformation back to the interaction picture is given, according to (11.193), by

$$\rho(t) \rightarrow U_{dr}^\dagger(t)\rho(t)U_{dr}(t) = \rho(t) \qquad \forall \, n(t) = 2k. \tag{11.202}$$

Thus, the foregoing results for $\rho(t)$ in the doubly rotating frame remain valid in the interaction picture when the number of pulses is even [i.e., for t such that $n(t) = 2k$]. The results [(11.196)–(11.201)] hold not only for dephasing caused by quantum baths but also for dephasing due to classical noise, provided we substitute $\Phi_s(t) \rightarrow 2\Phi_r(t)$.

These results are exact for bosonic baths or for a classical Gaussian noise. They are also applicable to arbitrary baths or non-Gaussian classical noise, when the system–bath coupling is sufficiently weak, so that the Born approximation used for the derivation of these results holds. Even if the ME (11.168) is not applicable in the absence of modulation due to a strong coupling of the system with the bath or noise, the modulation may suppress the coupling sufficiently to render the MEs (11.169) and (11.198) applicable.

To recast the relaxation parameter (11.201) as a frequency-domain integral, we note that (11.197) coincides with $\gamma_e(t)$ in (11.72a), if in (11.72a) $\omega_a = 0$, $\Phi_T(t)$ is replaced by $\Phi_s(t)$, and $\epsilon(t)$ is real. Therefore, we can use (11.94a), under the same substitutions, to obtain

$$\bar{\gamma}_d'(t) \equiv \frac{J_d'(t)}{t} = 2\pi \int_{-\infty}^\infty d\omega \, G_s(\omega) F_t(\omega) = 4\pi \int_0^\infty d\omega \, G_s(\omega) F_t(\omega). \tag{11.203}$$

Here, the last equality follows from the fact that, since now $\epsilon(t)$ is real [see (11.195)], $F_t(\omega)$ in (11.95) is a symmetric function, $F_t(\omega) = F_t(-\omega)$.

11.4 Dynamical "Filter Function" Control

In what follows, we consider several generic phase-modulation spectra that may
act as spectral "filter functions," which modify decay or relaxation rates. All modulations are taken to be *quasiperiodic*, namely, the quantity $\epsilon(t) = \varepsilon(t)$ in (11.141)
[with $\tilde{\epsilon}(t) = 1$] has the form,

$$\epsilon(t) = e^{i\phi(t)} = \sum_q \epsilon_q e^{i\nu_q t}, \qquad (11.204)$$

ν_q ($q = 0, \pm 1, \dots$) being arbitrary discrete frequencies such that

$$|\nu_q - \nu_{q'}| \geq \Omega_0 \quad \forall q \neq q', \qquad (11.205)$$

with the minimal spectral interval Ω_0. The complex coefficients ϵ_q satisfy
$\sum_q |\epsilon_q|^2 = 1$.

Since it is not necessarily periodic, (11.204) is more general than (11.76). Nevertheless, the derivation leading to (11.104) still holds. Thus, the relaxation rates
are dynamically modified at long times as

$$\gamma_{e(g)} = 2\pi \sum_q |\epsilon_q|^2 G_T(\pm\omega_q). \qquad (11.206)$$

These long-time limits are approached when

$$\Omega_0 t \gg 1, \quad t \gg t_c \equiv \max_q\{1/\xi_q\}. \qquad (11.207)$$

Here, the bath–memory (correlation) time t_c is defined as the inverse of the narrowest spectral interval ξ_q over which $G_T(\omega)$ changes appreciably near the relevant
frequencies ω_q [Fig. 11.5(a)]. Thus, the correlation time t_c depends on the modulation frequencies ω_q. The quasiperiodic modulation (11.204) is thus a filter function
that shifts or splits the TLS resonant response at frequency ω_a or 0 for AN or PN,
respectively, converting it into peaks at the frequencies ω_q: The relaxation rate
(11.206) is "filtered" by the modulation spectrum that assigns the desired weight
$|\epsilon_q|^2$ to the coupling spectrum value at the frequency ω_q (Fig. 11.5).

Similarly, if (11.204) holds for the DD function (11.195), then for sufficiently
long times ($t \gg t_c, \Omega_0^{-1}$), we have

$$\bar{\gamma}_d'(t) \rightarrow \gamma_d' = 2\pi \sum_q |\epsilon_q|^2 G_s(\omega_q). \qquad (11.208)$$

This means that for a sufficiently weak system–bath interaction (the case considered here), the decay in (11.200) is approximately exponential,

$$\rho_{eg}(t) \approx \rho_{eg}(0)e^{-\gamma_d' t}. \qquad (11.209)$$

A necessary validity condition for such decay is

$$\gamma_d' \ll 1/t_c, \Omega_0. \qquad (11.210)$$

11.4.1 Phase Modulation (PM) of the Coupling for AN Control

A monochromatic (CW) modulation control of AN yields a constant frequency shift δ_a, so that

$$\epsilon(t) = e^{-i(\omega_a + \delta_a)t}, \qquad (11.211)$$

resulting in the upper state decay rate

$$\gamma_e = 2\pi G_T(\omega_a + \delta_a). \qquad (11.212)$$

Such a shift is induced by the AC Stark effect of the control field (for atoms) or by the Zeeman effect (for spins). This shift may either enhance or suppress the Golden-Rule decay rate,

$$\gamma_{GR} = 2\pi G_T(\omega_a). \qquad (11.213)$$

Equation (11.212) yields the maximal change of the decay rate achievable by external control, since it does not involve any smoothing of the bath response (coupling spectrum) $G_T(\omega)$ incurred by the width of the filter function $F_t(\omega)$. The modified γ can vanish, if the shifted frequency $\omega_a + \delta_a$ is beyond the cutoff frequency of the bath coupling spectrum, where $G_T(\omega) = 0$ [Fig. 11.5(d)]. Conversely, the increase of γ due to a shift δ_a can exceed that achievable by repeated measurements, that is, the anti-Zeno effect (AZE) (Ch. 10). However, AC Stark shifts are usually small for CW perturbations. Typically, only pulsed perturbations may result in multiple shifted frequencies ω_q of the coupling spectrum as per (11.206).

11.5 Discussion

In this chapter, we have expounded a universal approach to the dynamical control of qubits subject to AN and PN, by either off- or on-resonant modulating fields, respectively. This approach is based on a general non-Markovian master equation valid for weak system–bath coupling and arbitrary control by modulations, since it does not invoke the RWA. The universal convolution formulae provide intuitive clues as to the tailoring of the modulation and the noise (bath-coupling) spectra.

AN characterizes a variety of decoherence or relaxation processes (e.g., spontaneous emission of photons by excited atoms, vibrational and collisional relaxation of trapped ions, and the relaxation of current-biased Josephson junctions).

Here, we have presented a *universal form of the AN decay rate* of unstable states into *any* bath (continuum), dynamically modified by perturbations with arbitrary time dependence, focusing on non-Markovian timescales. An analogous form is obtained for the dynamically modified rate of PN (proper dephasing).

The theory applies to finite temperatures and to systems driven by an *arbitrary* time-dependent field, which may cause the failure of the RWA. It allows for decoherence in *multilevel systems*, where quantum interference between the levels may either inhibit or accelerate the relaxation/decay.

When the general non-Markovian ME derived here is applied to either AN or PN, the resulting dynamically controlled relaxation or decoherence rates obey *analogous formulae*, provided the corresponding density-matrix (generalized Bloch) equations are written in the appropriate basis. This underscores the universality of our treatment. The choice of an appropriate time-dependent basis allows here to simplify the AN treatment. More importantly, it allows us to present a PN treatment that does not describe noise phenomenologically, but rather dynamically, starting from the ubiquitous spin-boson Hamiltonian.

A thoroughly studied approach to the suppression of decoherence is the "dynamical decoupling" (DD) of the system from the bath. DD is effected by a series of pulses for *stroboscopic* suppression of PN (proper dephasing): π-phase flips of the system–bath coupling via strong and sufficiently fast resonant pulses applied to the system.

DD is based on the assumption that the phase-modulation (PM) control fields have the form of short and strong (stroboscopic) pulses, to the extent that the free evolution can be neglected during these pulses. Consistently with this assumption, the propagator is written as the product of the free propagator, followed by alternating with the control-field propagator, and this is repeated for as many pulses as there are in the sequence. Each control pulse has an area of π, so that a periodic sequence of control pulses corresponds to a periodic accumulation of π-phase shifts of the TLS. As in the spin-echo technique, the free propagator after a control π-pulse negates the dephasing effects of the free propagator prior to the control pulse, to first order in the Magnus–Baker–Hausdorf expansion.

Since the DD formalism requires stroboscopic pulses, it is different from the formalism presented here. ME (11.169) is written in the secular approximation, which requires $\Omega(t)$ to be nonzero at all times. This requirement changes the time dependence of the density-matrix components in comparison to DD, but the decay or decoherence rates are modified similarly by both methods under similar pulse sequences.

However, two major features make the present universal approach to dynamical control advantageous compared to the DD approach. The first is that it bypasses the DD requirement that the control fields must consist of very short and very strong pulses, since in the universal dynamical control approach, unlike DD, control fields are considered *concurrently* with the coupling to the bath, thus allowing a much greater variety of control fields, ranging from continuous modulation all the way to DD sequences. In particular, the present approach requires generally much weaker field intensities than DD to achieve a comparable PN suppression. The second advantage is that, whereas DD prescribes the same pulse sequence regardless of the shape of the bath spectrum, universal dynamical control explicitly considers the bath spectrum and allows optimal tailoring of the modulation to a given bath spectrum. On the other hand, the DD formalism holds for an arbitrary TLS–bath coupling strength, whereas the universal dynamical control approach holds only for the weak-coupling regime.

This universal approach bears analogy to DD in the case of AN control by PM consisting of periodic π-pulses, but only for TLS coupling to *spectrally symmetric* (e.g., Lorentzian or Gaussian) noise baths with limited spectral width. The type of PM here advocated for the suppression of coupling to *spectrally asymmetric* baths (e.g., phonon or photon baths with frequency cutoff), is, however, drastically different from the DD periodic π-pulse sequences. Other situations to which the universal approach applies, but not the DD method, include *amplitude modulation* of the coupling to the continuum, as in the case of decay from quasibound states of a periodically tilted washboard potential: such modulation has been experimentally shown to give rise to either slowdown of the decay (Zeno-like behavior) or its speedup (anti-Zeno-like behavior), depending on the modulation rate (Ch. 10).

The equivalence between counterrotating terms in the TLS–bath interaction and the relaxation of a TLS coupled to a squeezed bath has been shown within the universal dynamical control approach to allow engineering a thermal bath so that it emulates a squeezed bath. This goal is achievable through the RWA breakdown even at long times, by fast and strong TLS level-distance modulation. To this end, we have derived a general master equation that does not invoke the Markovian and rotating-wave approximations.

Possible measurable effects are foreseen in a system or a TLS comprised of two atomic Rydberg levels in a Lorentzian cavity bath under sinusoidal level-distance modulations. Squeezed-bath effects on atomic-dipole dephasing and fluorescence rates are then anticipated.

12

Optimal Dynamical Control of Open Systems

Standard dynamical decoupling (DD) sequences are not optimal, in general, with respect to decoherence suppression, since they do not depend on the bath coupling spectrum. Here, we apply variational principles to the universal dynamical control of decoherence (expounded in Ch. 11) in order to find the optimal modulation for the suppression of decoherence caused by a given bath. We derive an equation for the optimal, energy-constrained control by modulation that minimizes decoherence for a given bath-coupling spectrum and compare the effects of optimal modulation to those of energy-constrained DD pulses.

12.1 Euler–Lagrange Optimization

12.1.1 Control, Score, and Constraint

Here, we consider the general question: How to find a system dynamics that optimizes a desired pertinent quantity? One can define a control problem in terms of control parameters, a score P that measures the success of control optimization, and a constraint E (e.g., the energy allotted for control), as detailed below.

The real control parameters f_l ($1 \leq l \leq N$) form an N-dimensional vector $\boldsymbol{f}$. In the case of time-dependent control, the $f_l(\tau)$ parameterize the system Hamiltonian or the unitary evolution operator as

$$\hat{H}_S = \hat{H}_S[\boldsymbol{f}(\tau)], \quad \hat{U}(\tau) = \mathrm{T}e^{-(i/\hbar)\int_0^\tau d\tau' \hat{H}_S(\tau')} \equiv \hat{U}[\boldsymbol{f}(\tau)]. \tag{12.1}$$

Such *parameterization of $\hat{U}$ circumvents the complication of time-ordered integration of its exponent.* The evolution operator $\hat{U}(\tau)$ thus obtained can then be used to calculate the system Hamiltonian

$$\hat{H}_S(\tau) = i\hbar \left[\frac{\partial}{\partial \tau} \hat{U}(\tau) \right] \hat{U}^\dagger(\tau). \tag{12.2}$$

The score P that measures the success of controlling the quantity of interest can be written as a real-valued functional of the system state $\hat{\varrho}(t)$ at the time t_f when the control ends. This score may be, for example, the maximal fidelity of a given pure state $|\Psi\rangle$ under bath-induced decoherence, $F_{|\Psi\rangle} = \langle\Psi|\hat{\varrho}|\Psi\rangle$. Alternatively, the score may be the maximal concurrence Co

$$Co_{|\Psi_{AB}\rangle} = [2(1 - \text{Tr}\hat{\varrho}_A^2)]^{1/2}, \quad \hat{\varrho}_A = \text{Tr}_B|\Psi_{AB}\rangle\langle\Psi_{AB}|, \tag{12.3}$$

which is a measure of the entanglement of a bipartite state $|\Psi_{AB}\rangle$. More generally, the score may be the maximum value of any real-valued function $P[\hat{\varrho}(t)]$ in the time interval $[0, t_f]$,

$$P = \max_{t\in[0,t_f]} P[\hat{\varrho}(t)]. \tag{12.4}$$

The choice of a *constraint* that is generally required to ensure the existence of a physical solution for the control is dictated by the most critical source of error. A possible constraint is the average speed with which the controls change,

$$E = \hbar \int_0^{t_f} d\tau \, \dot{f}^2(\tau), \tag{12.5}$$

which depends on the control spectral bandwidth. Another choice is the mean-squared modulation energy,

$$E = \hbar^{-1} \int_0^{t_f} d\tau \langle(\Delta\hat{H})^2(\tau)\rangle_{id}, \tag{12.6}$$

where $\langle\dots\rangle_{id} = d^{-1}\text{Tr}(\dots)$ refers to the average over the maximally mixed state of a d-dimensional system, $\Delta\hat{H}$ being the difference between the dynamically controlled and the uncontrolled system Hamiltonians.

A projected gradient search can be performed in parameter space (Fig. 12.1) to find the set of controls f that optimize a score $P(f)$ subject to a constraint $E(f)$. To this end, we start at the initial point f_0 at which $E(f_0)$ is the constraint value. If we follow the gradient δP, we may maximize or minimize P, but this may also change E. Instead, in order to optimize P while keeping E constant, we move along the projection of δP orthogonal to δE, that is,

$$\delta P_\perp = \delta P - \frac{\delta P \cdot \delta E}{(\delta E)^2}\delta E. \tag{12.7}$$

The gradients depend on f, hence the iteration consists of small steps

$$f_{n+1} = f_n \pm \epsilon\delta P_\perp(f_n), \quad \epsilon \ll 1. \tag{12.8}$$

If neither δP nor δE vanish, the iteration stops where $\delta P_\perp$ vanishes because the gradients are parallel, as determined by the condition

$$\delta P = \lambda\delta E, \tag{12.9}$$

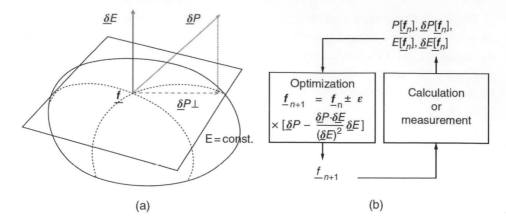

(a)   (b)

Figure 12.1 (a) Optimization of the score $P(f)$ subject to a constraint $E(f)$ in control-parameter space $\{f\}$. The optimization is obtained by walking along the component $\delta P_\perp$ of the gradient δP orthogonal to δE. (b) Iteration steps of f differing by a small parameter ϵ. The iteration yields a local solution depending on the starting point f_0. (Reprinted from Clausen et al., 2012. © 2012 American Physical Society)

which constitutes the Euler–Lagrange (EL) equation, the proportionality constant λ being the Lagrange multiplier. The solutions of the EL equation correspond to *local optima* of the constrained P. We may repeatedly perform the search with randomly chosen initial control-parameter sets f_0 and select the best solution. The gradients at each point f_n may be obtained from prior knowledge of the bath-coupling spectrum. A discretization of the time interval $0 \leq \tau \leq t_f$ then replaces the variational δ by a finite-dimensional vector gradient ∇.

12.2 Bath-Optimized Task-Oriented Control (BOTOC)

12.2.1 Fixed Time Approach

Our goal is to achieve, by means of classical control fields, a time dependence of the system Hamiltonian within the interval $[0, t_f]$ that sets the score $P(t_f) = P[\hat{\varrho}(t_f)]$ at a desired value in the presence of the bath. This should be the optimal (maximal or minimal) value of the score $P(t_f)$. If the initial system state $\hat{\varrho}(0)$ is given, then the *change* in the score $\Delta P = P(t_f) - P(0)$ due to the effects of control and the bath can be used instead of $P(t_f)$ as the score. To first order in the Taylor expansion of the score change as a function of the state change $\Delta\hat{\varrho} = \hat{\varrho}(t_f) - \hat{\varrho}(0)$ in a chosen basis, we have

$$\Delta P \approx \sum_{m,n} \frac{\partial P}{\partial \varrho_{mn}} \Delta \varrho_{mn} = \text{Tr}(\hat{P}\Delta\hat{\varrho}). \qquad (12.10)$$

Here, the coefficients

$$\left. \frac{\partial P}{\partial \varrho_{mn}} \right|_{t=0} \equiv (\hat{P})_{nm} \tag{12.11a}$$

are the matrix elements (in the chosen basis) of a Hermitian operator $\hat{P}$, which is the gradient of $P[\hat{\varrho}(t)]$ with respect to $\hat{\varrho}$ at $t = 0$:

$$\hat{P} = (\nabla_{\hat{\varrho}} P)|_{t=0} = (\partial P / \partial \hat{\varrho})|_{t=0} = 0. \tag{12.11b}$$

The operator $\hat{P}$ contains the complete information on the controlled variable. The transposition in (12.11a) allows us to express the sum over the entries (the Hadamard matrix product) in (12.10) as a trace of the operator product $\hat{P} \Delta \hat{\varrho}$. Equation (12.10) holds when $\Delta \hat{\varrho}$ and ΔP are small, which implies weak system–bath coupling. The score change ΔP then quantifies the bath effects and not the internal system dynamics.

If P is the state purity, $P = \mathrm{Tr}(\hat{\varrho}^2)$, then (12.11a) is proportional to the state, $\hat{P} = 2\hat{\varrho}(0)$.

12.2.2 Averaged Interaction Energy

Equation (12.10) expresses the score ΔP as an overlap between the gradient $\hat{P}$ and the bath-induced change of system state $\Delta \hat{\varrho}$. In what follows, we express ΔP in terms of physically pertinent quantities. To this end, we decompose the total Hamiltonian into system (S), bath (B), and interaction (I) parts,

$$\hat{H}(t) = \hat{H}_{\mathrm{S}}(t) + \hat{H}_{\mathrm{B}} + \hat{H}_{\mathrm{I}}, \tag{12.12}$$

and employ the Liouville–von Neumann equation of the total (system plus bath) state in the interaction picture,

$$\frac{\partial}{\partial t} \hat{\varrho}_{\mathrm{tot}}(t) = -\frac{i}{\hbar}[\hat{H}_{\mathrm{I}}(t), \hat{\varrho}_{\mathrm{tot}}(t)]. \tag{12.13}$$

Here, the interaction Hamiltonian is transformed from the Schrödinger picture to the interaction picture by

$$\hat{H}_{\mathrm{I}}(t) = \hat{U}_{\mathrm{F}}^{\dagger}(t) \hat{H}_{\mathrm{I}}^{(\mathrm{S})}(t) \hat{U}_{\mathrm{F}}(t), \tag{12.14}$$

the free-evolution operator $\hat{U}_{\mathrm{F}}(t)$ being given by the chronologically ordered expression

$$\hat{U}_{\mathrm{F}}(t) = \mathrm{T} \exp\left[-\frac{i}{\hbar} \int_0^t d\tau\, \hat{H}_{\mathrm{S}}(\tau) - \frac{i}{\hbar} \hat{H}_{\mathrm{B}} t \right]. \tag{12.15}$$

The solution of (12.13) can be written as the Dyson (state) expansion

$$\hat{\varrho}_{\mathrm{tot}}(t) = \hat{\varrho}_{\mathrm{tot}}(0) + \frac{-i}{\hbar} \int_0^t dt_1 [\hat{H}_{\mathrm{I}}(t_1), \hat{\varrho}_{\mathrm{tot}}(0)]$$

$$+ \left(\frac{-i}{\hbar} \right)^2 \int_0^t dt_1 \int_0^{t_1} dt_2 [\hat{H}_{\mathrm{I}}(t_1), [\hat{H}_{\mathrm{I}}(t_2), \hat{\varrho}_{\mathrm{tot}}(0)]] + \dots, \tag{12.16}$$

which is obtained either by an iterated integration of (12.13) or from its formal
solution,

$$\hat{\varrho}_{\text{tot}}(t) = \hat{U}_{\text{I}}(t)\hat{\varrho}_{\text{tot}}(0)\hat{U}_{\text{I}}^{\dagger}(t), \quad \hat{U}_{\text{I}}(t) = \text{T}e^{-(i/\hbar)\int_0^t dt' \hat{H}_{\text{I}}(t')}, \tag{12.17}$$

by applying the Magnus–Baker–Hausdorf expansion

$$\hat{U}_{\text{I}}(t) = e^{-(it/\hbar)\hat{H}_{\text{eff}}(t)}, \tag{12.18}$$

where

$$\hat{H}_{\text{eff}}(t) = \frac{1}{t}\int_0^t dt_1 \hat{H}_{\text{I}}(t_1)$$
$$- \frac{i}{2\hbar t}\int_0^t dt_1 \int_0^{t_1} dt_2 [\hat{H}_{\text{I}}(t_1), \hat{H}_{\text{I}}(t_2)] + \dots \tag{12.19}$$

is obtained upon sorting the terms according to their order in $\hat{H}_{\text{I}}$.

We assume that the system is initially brought in contact with the bath (rather
than being in equilibrium with it). This scenario corresponds to the factorizing
initial conditions

$$\hat{\varrho}_{\text{tot}}(0) = \hat{\varrho}(0) \otimes \hat{\varrho}_{\text{B}}. \tag{12.20}$$

Tracing out the bath in (12.16) then yields $\Delta\hat{\varrho}$ over time t, which we insert
into (12.10). We further assume that the expectation value of the interaction
Hamiltonian with respect to the bath state vanishes,

$$\langle\hat{H}_{\text{I}}\rangle_{\text{B}} \equiv \text{Tr}_{\text{B}}(\hat{\varrho}_{\text{B}}\hat{H}_{\text{I}}) = 0. \tag{12.21}$$

Then, the "drift" term corresponding to the first order in (12.16) vanishes, and
the second-order term is the lowest-order nonvanishing contribution. We finally
assume that the initial system state commutes with the score-gradient operator $\hat{P}$
(12.11b),

$$[\hat{\varrho}(0), \hat{P}] = 0. \tag{12.22}$$

Then, the mean-value evolution of $\hat{P}$ is given by

$$\text{Tr}[e^{(i/\hbar)\hat{H}t}\hat{\varrho}(0)e^{-(i/\hbar)\hat{H}t}\hat{P}] = \text{Tr}[\hat{\varrho}(0)\hat{P}] + \frac{it}{\hbar}\text{Tr}(\hat{H}[\hat{\varrho}(0), \hat{P}]) + O(\hat{H}^2 t^2/\hbar^2) \tag{12.23}$$

for a system Hamiltonian $\hat{H}$ that causes weak effects relative to $\text{Tr}[\hat{\varrho}(0)\hat{P}]$, known
as the "kinematic critical point."

To second order, we then evaluate for the score change (12.10) as

$$\Delta P = \frac{t_{\text{f}}^2}{\hbar^2}\langle[\hat{H}, \hat{P}]\hat{H}\rangle, \tag{12.24}$$

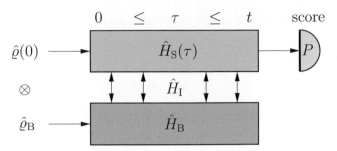

Figure 12.2 In the considered control approach, a system is brought in contact with a bath over a time interval $0 \leq \tau \leq t_f$. During this time, the system Hamiltonian $\hat{H}_S(\tau)$ is chosen to be such that a system variable (score) P attains the desired value at t_f. (Reprinted from Clausen et al., 2012. © 2012 American Physical Society)

where $\langle \cdot \rangle = \mathrm{Tr}[\hat{\varrho}_{\mathrm{tot}}(0)(\cdot)]$ and

$$\hat{H} = \frac{1}{t_f} \int_0^{t_f} d\tau \, \hat{H}_I(\tau). \tag{12.25}$$

The score change is thus expressed in terms of the interaction Hamiltonian, averaged in the interaction picture over the time interval of interest (Fig. 12.2).

12.2.3 Spectral Overlap

Equation (12.24) may be rewritten in the form of an overlap of the system and the bath-response spectral matrices. This can be done by expanding the interaction Hamiltonian in a d-dimensional Hilbert space as a sum of products of the system and bath operators,

$$\hat{H}_I = \hbar \sum_{j=1}^{d^2-1} \hat{S}_j \otimes \hat{B}_j. \tag{12.26}$$

If the mean values vanish, $\langle \hat{B}_j \rangle = 0$, then (12.21) is satisfied. Otherwise, we may include these mean values in the system Hamiltonian,

$$\hat{B}_j \rightarrow \hat{B}'_j = \hat{B}_j - \langle \hat{B}_j \rangle \hat{I}, \quad \hat{H}'_S = \hat{H}_S + \sum_j \langle \hat{B}_j \rangle \hat{S}_j. \tag{12.27}$$

We then transform (12.26) to the interaction picture and expand

$$\hat{S}_j(t) = \sum_k \epsilon_{jk}(t) \hat{S}_k \tag{12.28}$$

in the basis of operators $\hat{S}_j$ that are Hermitian, traceless, and orthonormalized to $\mathrm{Tr}(\hat{S}_j \hat{S}_k) = d\delta_{jk}$. This expansion defines a (real orthogonal) rotation matrix $\boldsymbol{\epsilon}(t)$ in the Hilbert space of the system, with elements

$$\epsilon_{jk}(t) = \langle \hat{S}_j(t) \hat{S}_k \rangle_{\mathrm{id}}, \tag{12.29}$$

in the notation of (12.6). These matrix elements are the dynamical correlation functions of the system basis operators at infinite temperature. We similarly define the bath correlation (response) matrix $\boldsymbol{\Phi}(t)$ at a finite temperature with elements

$$\Phi_{jk}(t) = \langle \hat{B}_j(t) \hat{B}_k \rangle_{\mathrm{B}} \tag{12.30}$$

and a Hermitian matrix $\boldsymbol{\Xi}$ with elements

$$\Xi_{kj} = \langle [\hat{S}_j, \hat{P}] \hat{S}_k \rangle, \tag{12.31}$$

where $\langle \ldots \rangle = \mathrm{Tr}[\hat{\varrho}(0) \ldots]$. This matrix represents the gradient operator $\hat{P}$ [cf. (12.11b)] in the basis of operators $\hat{S}_j$. Using (12.29) and (12.30), we evaluate the bath and (finite-time) system spectra to be

$$\boldsymbol{G}(\omega) = \int_{-\infty}^{\infty} dt e^{i\omega t} \boldsymbol{\Phi}(t), \tag{12.32}$$

$$\boldsymbol{\epsilon}_t(\omega) = \frac{1}{\sqrt{2\pi}} \int_0^t d\tau e^{i\omega\tau} \boldsymbol{\epsilon}(\tau). \tag{12.33}$$

We may now express the score, (12.24), as the overlap of the matrices defined in (12.29)–(12.31),

$$\Delta P = \iint_0^{t_{\mathrm{f}}} dt_1 dt_2 \mathrm{Tr}[\boldsymbol{\epsilon}^{\mathrm{T}}(t_1) \boldsymbol{\Phi}(t_1 - t_2) \boldsymbol{\epsilon}(t_2) \boldsymbol{\Xi}] \tag{12.34}$$

$$= \int_{-\infty}^{\infty} d\omega \mathrm{Tr}[\boldsymbol{\epsilon}_{t_{\mathrm{f}}}^{\dagger}(\omega) \boldsymbol{G}(\omega) \boldsymbol{\epsilon}_t(\omega) \boldsymbol{\Xi}] \tag{12.35}$$

$$= t_{\mathrm{f}} \int_{-\infty}^{\infty} d\omega \mathrm{Tr}[\boldsymbol{F}_{t_{\mathrm{f}}}(\omega) \boldsymbol{G}(\omega)]. \tag{12.36}$$

Here, the superscript T denotes the transpose operation; we have employed the cyclic property of the trace and have combined the rotation matrix spectra $\boldsymbol{\epsilon}_t(\omega)$ and the gradient representation $\boldsymbol{\Xi}$ into a spectral matrix $\boldsymbol{F}_t(\omega)$ that characterizes the controlled system,

$$\boldsymbol{F}_t(\omega) = \frac{1}{t} \boldsymbol{\epsilon}_t(\omega) \boldsymbol{\Xi} \boldsymbol{\epsilon}_t^{\dagger}(\omega). \tag{12.37}$$

A derivation similar to that of (11.37) shows that

$$\Phi_{jk}^*(t) = \Phi_{kj}(-t). \tag{12.38}$$

This equality implies, on account of (12.32), that the bath-response spectral matrix $\boldsymbol{G}(\omega)$ is Hermitian for all ω,

$$G^*_{jk}(\omega) = G_{kj}(\omega). \tag{12.39}$$

It is also *positive semidefinite* for all ω. We assume that the same holds for the control matrix $\boldsymbol{F}_t(\omega)$, so that P is always positive. For example, P may represent a gate error, the goal being to minimize it.

The matrix spectral overlap (12.36) can be made as small as desired by sufficiently fast (high-frequency) control modulation of the system, such that the entire weight of the system spectrum is shifted beyond that of the bath, under the common assumption that the bath spectrum has an upper cutoff (Chs. 3, 4), that is, it vanishes at sufficiently high frequencies. Such high-frequency modulation may, however, result in unbounded growth of the system energy. A meaningful characterization of the control therefore requires a constraint, which is commonly chosen to be the total energy invested in control. The maximization or minimization of the score P, as the case may be, is then constrained by the minimization of the control energy.

In general, $\boldsymbol{F}_t(\omega)$ in (12.37) is Hermitian but need not necessarily be positive semidefinite: the score may either increase or decrease over t, so that our goal may be to maximize ΔP with either positive or negative sign. Equation (12.36) shows that the score change ΔP over time t is determined by the overlap between the controlled system-variable and bath-response spectra, analogously to dynamical control by modulation (Ch. 11) or to measurement-induced quantum Zeno and anti-Zeno control of open systems (Ch. 10). Whenever $\boldsymbol{F}_t(\omega)$ is positive semidefinite, its effect on ΔP in (12.36) may be viewed as a *spectral filter* that acts to suppress certain frequencies in the bath-response spectral matrix $\boldsymbol{G}(\omega)$, while enhancing others.

12.2.4 Optimal Universal Control of Qubit Decoherence

Let us seek to minimize the decoherence rate, $\bar{\gamma}_d(t_f)$, averaged over the control time-interval $(0, t_f)$, of a TLS that interacts with a bath characterized by the temperature-dependent bath-coupling spectrum, $G_T(\omega)$. To this end, we look for the optimal PM (Sec. 11.4),

$$\epsilon(t) = e^{i\phi(t)}, \tag{12.40}$$

under the energy constraint,

$$\hbar \int_0^{t_f} dt |\dot{\phi}(t)|^2 = E, \tag{12.41}$$

with the boundary conditions for the accumulated phase $\phi(0) = \dot{\phi}(0) = 0$. Upon eliminating the Lagrange multiplier λ, we find that the optimal control PM obeys the following equation:

$$\ddot{\phi}(t) = \frac{-\sqrt{E/\hbar} Z[t, \phi(t)]}{\left[\int_0^{t_f} dt_1 \left| \int_0^{t_1} dt_2 Z[t_2, \phi(t_2)] \right|^2 \right]^{1/2}}, \tag{12.42}$$

where

$$Z[t, \phi(t)] = \frac{1}{t_f} \int_0^{t_f} dt_1 \tilde{\Phi}(|t - t_1|) \sin[\phi(t) - \phi(t_1)]. \tag{12.43}$$

Equations (12.42) and (12.43) will be used in Chapter 14 to optimize the control spectrum for minimizing the TLS–bath interaction.

12.3 Comparison of BOTOC and DD Control

It is appropriate to compare BOTOC with the optimal dephasing rate calculated by the bath-optimized task-oriented control and by dynamical decoupling (DD) control (Ch. 11). For a meaningful comparison, we impose the *same energy constraint* on both control methods. Finite-duration periodic DD consists of n equidistant π-pulses with Rabi frequency

$$\Omega(t) = \begin{cases} \pi/t_w; & j\tau \leq t < j\tau + t_w; \quad j = 0, \ldots, n-1; \\ 0; & \text{otherwise}; \end{cases} \tag{12.44}$$

where $t_w \ll \tau$ is the temporal width of each pulse and τ is the interval between pulses. The energy constraint E and the total modulation duration (t_f) are related as follows:

$$t_f = n\tau + t_w, \quad n = t_w E/(\pi^2 \hbar). \tag{12.45}$$

The corresponding modulation spectrum consists of a series of peaks, where the two main peaks are at the frequencies $\pm\pi/\tau$. Thus, the peaks are shifted proportionally to the energy invested in the modulation.

Whereas the DD sequence consists of equidistant π-pulses, a modified DD sequence devised by Uhrig (UDD) assigns unequal temporal spacings to the pulses, in order to minimize dephasing. The UDD sequence is derived by requiring the vanishing of the first n derivatives of the modulation power spectrum (filter function) $F_t(\omega)$ with respect to ω at $\omega = 0$. This condition corresponds to a modulation spectrum with enhanced high-frequency components and suppressed low-frequency peaks, as compared to the standard DD. The resulting spectral filter function for n π-pulses distributed over time interval t_f is

$$F_{t_f}(\omega) \approx \frac{8(n+1)^2 J_{n+1}^2(\omega t_f/2)}{\pi \omega^2 t_f}, \tag{12.46}$$

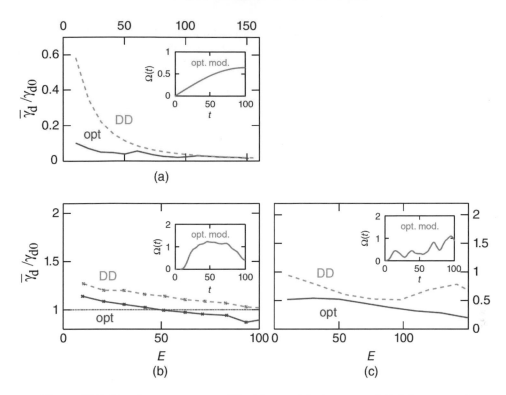

Figure 12.3 Decoherence rate modified by modulation control $\bar{\gamma}_d(t_f)$, averaged over time interval t_f and divided by the unmodulated rate γ_{d0} as a function of energy constraint. DD – dashed, optimal (BOTOC) modulation – solid. Insets: optimal Rabi-frequency modulation $\Omega(t)$ under different energy constraints. (a) Single-peak resonant dephasing spectrum (inset: $E = 20$). (b) $1/f$ spectrum (inset: $E = 30$). (c) Multipeaked spectrum (inset: $E = 30$). (Adapted from Gordon et al., 2008. © 2008 The American Physical Society)

where $J_{n+1}()$ is the $(n+1)$-order Bessel function. This UDD optimization procedure does not depend on the temperature-dependent bath-coupling spectrum $G_T(\omega)$. Therefore, the UDD can be optimal only if the coupling to higher-frequency modes of the bath is weaker than to lower-frequency modes. This can be seen from (11.155), where $G_T(\omega)$ must fall off at higher frequencies for $F_{t_f}(\omega)$ to be effective. By contrast, the UDD optimization modulation based on the universal BOTOC formula is *adapted* to the bath-coupling spectrum. This difference may render BOTOC advantageous compared to DD or UDD, as shown in what follows for generic bath (dephasing) spectra.

(a) Single-peak resonant dephasing spectrum. This dephasing spectrum corresponds to the common model $\Phi(t) = e^{-t/t_c}\gamma_{d0}/(4t_c)$, where γ_{d0} is the long-time dephasing rate $\gamma_{d0}(t \to \infty)$ in the absence of modulation and t_c is the noise

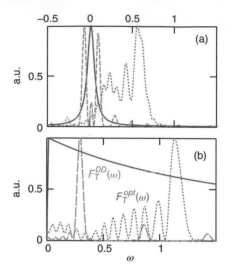

Figure 12.4 Dephasing spectra $G_T(\omega)$ (solid) and modulation spectra (filter functions) $F_{t_f}(\omega)$: optimal (dotted), and DD (dashed), in arbitrary units (a.u.). Same parameters as in Figure 12.3(a), (b), respectively. (Adapted from Gordon et al., 2008. © 2008 The American Physical Society)

correlation time [cf. (11.181) and (11.183)]. Figure 12.3(a) shows the modulation-modified average dephasing rate $\bar{\gamma}_d(t_f)$ as a function of the energy constraint E. The more energy is invested in the modulation, the lower is the dephasing rate. For low E, the optimal BOTOC modulation outperforms DD, whereas at higher E the two coincide. These results can be understood from Figure 12.4(a), in which the two main DD spectral peaks significantly overlap with $G_T(\omega)$ at low E, but as E is increased at fixed t_f, these peaks are pushed farther apart, and therefore have less overlap with $G_T(\omega)$, thus suppressing the dephasing more strongly. This superiority of BOTOC compared to DD [Fig. 12.4(a)] stems from the fact that when high frequencies have lower coupling strength in $G_T(\omega)$, the BOTOC optimal control filters by maximizing its weight at high frequencies, to the extent permitted by the energy constraint. The modulation Rabi frequency can be approximated by

$$\Omega(t) = \Omega_0[1 + e^{-t/t_f}(t/t_f - 1)], \qquad (12.47)$$

where Ω_0 is determined by the energy constraint E that fits the inset in Figure 12.3(a). As the number of pulses increases, DD shifts the weight of $\boldsymbol{F}(\omega)$ toward higher frequencies, until the overlap (12.36) becomes sufficiently small. This is illustrated in Figure 12.5(a), (b) for periodic dynamical decoupling (PDD) of a single TLS, with two different numbers of pulses.

(b) 1/f dephasing spectrum. The $1/f$ dephasing spectrum (for example, charge noise in superconducting qubits) has the form $G_T(\omega) \propto 1/\omega$, with $\omega_{\min}$ and $\omega_{\max}$ cutoffs. Figure 12.3(b) shows that the more energy is invested in

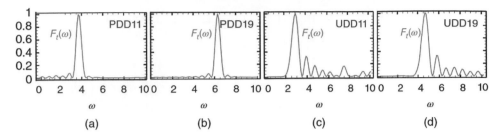

Figure 12.5 DD modulation spectra $F_t(\omega)$ (scaled to their maximum value) as a function of ω (in units of the inverse bath correlation time) generated by two methods: (a,b) Periodic DD with n π-pulses (PDDn); (c,d) Uhrig DD with the same number of π-pulses (UDDn). The spectra are drawn for $n = 11$ (a,c) and $n = 19$ (b,d) pulses. (Adapted from Kurizki, 2013.)

modulation, the lower the dephasing rate. Due to the cutoff ω_{min}, a low-energy modulation enhances dephasing compared to having no modulation at all. The higher the frequency, the lower is the coupling strength to the bath. Thus, BOTOC assigns the largest weight to the highest-frequency range allowed by the energy constraint [Fig. 12.4(b)]. The corresponding modulation strength [Fig. 12.3(b) – inset] first increases with time ($t < 50$), then starts decreasing, because at long times, corresponding to the vicinity of the lower cutoff, the optimal modulation benefits from the lower modulation amplitude. The $1/f$ [Fig. 12.4(b)] and the Lorentzian spectra [Fig. 12.4(a)] require different optimal long-time control because of the lower cutoff in the $1/f$ case.

The foregoing analysis may be generalized to any dephasing spectrum with a monotonically decreasing $G_T(\omega)$ as a function of ω. The optimal modulation for any such spectrum is an energy-constrained chirped modulation, modified by the spectra cutoffs.

(c) Multipeaked dephasing spectrum. In general, $G_T(\omega)$ may have several resonances (peaks) and is characterized by multiple noise correlation times. Since DD does not adapt itself to the dephasing spectrum, its performance may be much worse in this case than that of the optimal BOTOC filter function, whose peaks are predominantly *anticorrelated* with the peaks of $G_T(\omega)$, so as to minimize $\bar{\gamma}_d(t_f)$.

The plots in Figure 12.5 indicate that the suppression of low-frequency components in UDD is achieved in return for a smaller shift of the main peak. Therefore, shifting the main peak beyond a given cutoff frequency requires more pulses in UDD than in PDD.

In contrast to all DD sequences whose aim is only to fight dephasing, BOTOC [shown in Fig. 12.6 for a bath whose coupling spectrum $G_T(\omega)$ has both a Lorentzian peak and a low-frequency tail] optimizes the TLS evolution $\hat{U}(\tau)$ *simultaneously for all 3D Pauli matrix couplings* to the bath (Z, Y, and X). The resulting modulation spectra $F_t(\omega)$ are plotted in Figure 12.6 for different

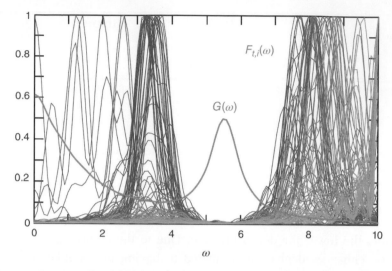

ω

Figure 12.6 BOTOC-minimization of the error for a TLS π-flip. The error is caused by pure dephasing with a given bath spectrum $G_T(\omega)$ (bold line), scaled to 0.61 with respect to its maximum value; scaling of ω as in Figure 12.5. The Z-component of the Pauli matrix undergoes modulation with spectra $F_{t,i}(\omega)$ that are shown for energy constraints $E_i = 0.1 + 4(i-1)$, $i = 1, 2, \ldots, 101$ (thin lines), each spectrum scaled to 1. (Adapted from Kurizki, 2013.)

energy constraints E. For low E, $F_t(\omega)$ has a single peak below the bath spectral peak. Upon increasing E, another peak of $F_t(\omega)$ emerges above the bath spectral peak. The higher-frequency peak of $F_t(\omega)$ grows, while the lower-frequency peak diminishes, until, at high E, only the higher-frequency peak remains. Thus, $F_t(\omega)$ generated by BOTOC avoids overlap with the main peak of $G_T(\omega)$ irrespective of E. BOTOC can therefore outperform all forms of DD (including UDD), if the bath coupling spectrum has high cutoff but bandgaps at low frequencies.

12.4 Discussion

The general BOTOC approach to dynamical control of TLS coupled to an arbitrary bath or continuum has reaffirmed the anticipation that, in order to suppress their relaxation or dephasing, we must modulate the system–bath coupling at a rate exceeding the spectral range over which this coupling is significant. BOTOC can serve as a general recipe for *optimized* design of the modulation aimed at an effective use of the fields for decay and decoherence suppression.

Under a fixed control-energy constrains, BOTOC may well outperform all DD control methods because the latter are not adapted to the dephasing spectrum. Furthemore, BOTOC can simultaneously fight noise acting along all three axes (Z, Y, and X) of the Bloch sphere, whereas DD is restricted to countering dephasing.

13

Dynamical Control of Quantum Information Processing

Quantum computations, which promise to be much faster than their classical analogs, can be performed via single-qubit and two-qubit operations only. However, their experimental implementation has been proven to be difficult due to dephasing and relaxation effects that hamper quantum information. Namely, entangling two-qubit gate operations, which are the cornerstone of quantum computation, are corrupted resulting in the loss of computational fidelity.

Here, we resort to the universal dynamical-control approach described in Chapters 11 and 12 with the aim of suppressing dephasing or relaxation during all stages of quantum-information processing: (a) information storage, (b) manipulation by single-qubit gates, or (c) manipulation by entangling two-qubit gates. We show that in order to suppress these adverse effects, it is advantageous to exert addressable dynamical control on all the qubits at once, whether or not they are manipulated by quantum gates. Conventional approaches, whereby one tries to either reduce the gate duration or increase its coherence time, are shown not to be necessarily the best options. Instead, one can increase the gate duration and simultaneously reduce the effects of dephasing or relaxation, resulting in higher gate fidelity.

13.1 Decoherence Control during Quantum Computation

13.1.1 Principles of Control Scheme

We here extend the universal dynamical decoherence control approach of Chapters 11 and 12 to multiqubit systems where storage and single- or two-qubit gate operations occur intermittently.

We assume a system of N qubits, with ground and excited states $|g\rangle_j$, $|e\rangle_j$, respectively, and identical transition frequency ω_0. Each qubit (labeled by j)

undergoes different random fluctuations, $\delta_j(t)$, of the transition frequency causing random dephasing. The total Hamiltonian

$$\hat{H} = \hat{H}_S(t) + \hat{H}^{(1)}(t) + \hat{H}^{(2)}(t) \tag{13.1}$$

consists of three terms:

$$\hat{H}_S(t) = \hbar \sum_{j=1}^{N} \left[\omega_0 + \delta_j(t) \right] |e\rangle_{jj} \langle e| \bigotimes_{k \neq j} I_k, \tag{13.2}$$

$$\hat{H}^{(1)}(t) = \hbar \sum_{j=1}^{N} \left[V_j^{(1)}(t) |e\rangle_{jj} \langle g| + \text{H.c.} \right] \bigotimes_{k \neq j} I_k, \tag{13.3}$$

$$\hat{H}^{(2)}(t) = \hbar \sum_{j=1}^{N} \sum_{k=j+1}^{N} \left[V_{jk}^{(2)\Psi}(t) |ge\rangle_{jk} \langle eg| \right.$$
$$\left. + V_{jk}^{(2)\Phi}(t) |ee\rangle_{jk} \langle gg| + \text{H.c.} \right] \bigotimes_{l \neq j,k} I_l. \tag{13.4}$$

Here, I is the identity matrix and H.c. is the Hermitian conjugate. The superscript (1) or (2) distinguishes between one- and two-qubit manipulations, respectively. The subscript labels the manipulated qubits: thus, $V_{jk}^{(2)\Psi}(t)$, $V_{jk}^{(2)\Phi}(t)$ are the possible time-dependent two-qubit gates, acting on qubits j and k, in the Bell-states basis,

$$|\Psi_\pm\rangle = \frac{1}{\sqrt{2}} e^{-i\omega_0 t} (|eg\rangle \pm |ge\rangle), \quad |\Phi_\pm\rangle = \frac{1}{\sqrt{2}} (e^{-i2\omega_0 t} |ee\rangle \pm |gg\rangle). \tag{13.5}$$

Let us assume, as in Chapter 11, that the random dephasing of each qubit is caused by a stochastic process. We take it to be a Gaussian process with the first and second ensemble-average moments $\overline{\delta_j(t)} = 0$, $\Phi_{jk}(t) = \overline{\delta_j(t)\delta_k(0)}$. We take the driving fields of the gates to be resonant on their transition, with a time-dependent real envelope, as in Chapter 11,

$$V_j^{(1)}(t) = \Omega_j^{(1)}(t)e^{-i\omega_0 t} + \text{c.c.},$$
$$V_{jk}^{(2)\Psi}(t) = \Omega_{jk}^{(2)\Psi}(t) + \text{c.c.},$$
$$V_{jk}^{(2)\Phi}(t) = \Omega_{jk}^{(2)\Phi}(t)e^{-i2\omega_0 t} + \text{c.c.}, \tag{13.6}$$

where the notation of (13.3) and (13.4) is used and the RWA is adopted.

Three generic cases will be considered: (i) single-qubit gates are applied, (ii) two-qubit gates are applied on different qubit pairs, and (iii) each qubit is manipulated by either a single- or a two-qubit gate, but never by both concurrently.

The one- ($q=1$) and two- ($q=2$) qubit cases can be analyzed upon transforming to the frame rotating at the frequency ω_0, performing the rotating-wave approximation, and diagonalizing $\hat{H}^{(q)}(t)$. The entire Hamiltonian has 2^N states in the diagonalizing basis, $|\Psi_{l_q}^{(q)}\rangle = \bigotimes_{j=1}^{N} |b_j^{l_q}\rangle_j$, where $l_1 = 0, \ldots, 2^N - 1$; $\{b_j^{l_1}\}$ is the binary representation of $l_1 = b_1^{l_1} b_2^{l_1} \ldots b_N^{l_1}$, and $b_j^l = 0, 1$ corresponds to

$$|\pm\rangle_j = \frac{1}{\sqrt{2}} \left(e^{-i\omega_0 t} |e\rangle_j \pm |g\rangle_j\right), \tag{13.7}$$

respectively, $l_2 = 0, \ldots, 2^N - 1$; $\{c_j^{l_2}\}$ is the quartary representation of $l_2 = c_1^{l_2} c_2^{l_2} \ldots c_N^{l_2}$, with $c_j^{l_2} = 0, 1, 2, 3$ corresponding to $|\Psi_+, \Psi_-, \Phi_+, \Phi_-\rangle_{kk'}$, respectively. The density matrix of the ensemble is then obtained in the diagonalized basis to second order in $\delta_j(t)$,

$$\bar{\rho}(t) = \overline{|\psi\rangle \langle \psi|} = \rho(0) - \int_0^t dt' \int_0^{t'} dt'' \overline{[\hat{W}^{(q)}(t'), [\hat{W}^{(q)}(t''), \rho(0)]]}, \tag{13.8}$$

where

$$\hat{W}^{(q)}(t) = \sum_{l,m=1}^{2^N} w_{lm}^{(q)}(t) |\Psi_l^{(q)}\rangle \langle \Psi_m^{(q)}| \tag{13.9}$$

is written in terms of

$$w_{lm}^{(q)}(t) = w_{ml}^{(q)*}(t) = \frac{1}{2} \begin{cases} \delta_j(t)\epsilon_j^{(1)}(t), & b_j^{l_1} = 0, \ b_j^{m_1} = 1, \\ & b_k^{l_1} = b_k^{m_1} \ \forall k \neq j, \\ \delta_{j-}\epsilon_j^{(2)\Psi}, & b_j^{l_2} = 0, \ b_j^{m_2} = 1, \\ & b_k^{l_2} = c_k^{m_2} \ \forall k \neq j, \\ \delta_{j+}\epsilon_j^{(2)\Phi}, & b_j^{l_2} = 2, \ b_j^{m_2} = 3, \\ & b_k^{l_2} = c_k^{m_2} \ \forall k \neq j, \\ 0 & \text{otherwise} \end{cases} \tag{13.10}$$

and

$$\delta_{j\pm}(t) = \delta_k(t) \pm \delta_{k'}(t), \quad \epsilon_j^{(1)}(t) = e^{i\phi_j^{(1)}(t)}, \quad \epsilon_j^{(2)\Psi,(2)\Phi}(t) = e^{i\phi_{pj}^{(2)\Psi,(2)\Phi}(t)}, \tag{13.11}$$

the accumulated phases being

$$\phi_j^{(1)}(t) = \int_0^t dt' \Omega_j^{(1)}(t'), \quad \phi_{pj}^{(2)\Psi,(2)\Phi}(t) = \int_0^t dt' \Omega_{pj}^{(2)\Psi,(2)\Phi}(t'). \tag{13.12}$$

This general scheme of dephasing control satisfies all the requirements for universal quantum computation, which is performed via single- and two-qubit gate operations on different qubits. Thus, quantum computation requires the evaluation of the interaction operator $\hat{W}(t)$ as a combination of the expressions (13.9) with $q = 1, 2$.

The three stages of quantum computation are defined by the conditions imposed on the overall phase accumulated by the state under the gate fields after each stage, at time t_f: (a) Storage requires $\phi_j^{(q)}(t_f) = 2\pi M_j$ ($M_j = 0, \pm 1, \ldots$). (b) A Hadamard gate applied to the jth qubit is restricted by $\phi_j^{(1)}(t_f) = \pi/4$, all other qubits being in the storage regime. (c) A SWAP gate that acts on qubits k and k' is restricted by $\phi_{kk'}^{(2)\psi}(t_f) = \pi/4$, all other qubits being in the storage regime.

The success score of dephasing control schemes (Ch. 12) is measured by the fidelity,

$$f(t_f) = \mathrm{Tr}[\rho_{\text{target}}^{1/2} \overline{\rho}(t_f) \rho_{\text{target}}^{1/2}], \tag{13.13}$$

where ρ_{target} is the target density matrix after the quantum computation, but during the storage stage $\rho^{\text{target}} = \rho(0)$. Since quantum computation requires lack of knowledge of the initial qubits' state, it is appropriate to average the fidelity over all possible initial pure states,

$$\bar{f}(t_f) = \langle f(t_f) \rangle . \tag{13.14}$$

13.1.2 Applications to One- and Two-Qubit Gates

We next analyze quantum computations by two qubits that undergo random dephasing. For single-qubit gates on each qubit, the average fidelity is obtained as

$$\bar{f}(t_f) = 1 - \frac{5}{12} \left[J_{11}^{(1)}(t_f) + J_{22}^{(1)}(t_f) \right], \tag{13.15}$$

where

$$J_{jk}^{(q)}(t) = \int_0^t dt' \int_0^{t'} dt'' \Phi_{jk}(t' - t'') \epsilon_j^{(q)}(t') \epsilon_k^{(q)*}(t''), \tag{13.16}$$

which implies

$$\mathrm{Re} J_{jk}^{(q)}(t) = \pi \int_{-\infty}^{\infty} d\omega \, G_{jk}(\omega) \epsilon_{j,t}^{(q)}(\omega) \epsilon_{k,t}^{(q)*}(\omega). \tag{13.17}$$

Here, $J_{jk}^{(q)}(t)$ is the dephasing function modified by the fields with Rabi frequencies $\Omega_{j,k}^{(q)}$ ($q = 1, 2_\Psi, 2_\Phi$), $G_{jk}(\omega) = (2\pi)^{-1} \int_{-\infty}^{\infty} dt \Phi_{jk}(t) e^{i\omega t}$ is the dephasing spectrum, and $\epsilon_{j,t}^{(q)}(\omega) = (2\pi)^{-1/2} \int_0^t dt' \epsilon_j^{(q)}(t') e^{i\omega t'}$ is the finite-time Fourier transform of the modulation.

Equations (13.15)–(13.17) show that the fidelity is maximized by *reducing the spectral overlap* of the dephasing and modulation (control) spectra. Single-qubit gate fields do not cause cross-dephasing. Since (13.15) depends only on single-qubit dephasing, $\Phi_{jj}(t)$, because of the averaging over all initial qubits. For each initial entangled state that "suffers" from cross-dephasing (e.g., triplet, $|\Phi_-\rangle$), there is another entangled state that "benefits" from cross-dephasing (e.g., the singlet,

$|\Psi_-\rangle$). Equation (13.15) also shows that if one applies a gate field on one qubit, one can still benefit from applying a control field on the other, stored, qubit.

For two-qubit gate operations, the average fidelity at t_f is found to be:

$$\bar{f}(t_f) = 1 - \frac{5}{24} \sum_{j,k=1,2} \left[J_{jk}^{(2)\Phi}(t_f) + (-1)^{j+k} J_{jk}^{(2)\Psi}(t_f) \right]. \tag{13.18}$$

In this expression, cross-dephasing corresponding to $j \neq k$ does not cancel out despite averaging. Yet, the cross-terms have opposite signs for the different two-gate fields acting on the Ψ and Φ Bell states. Thus, remarkably, the SWAP-gate fidelity may benefit from cross-dephasing.

It is noteworthy that by applying together both two-qubit gate fields, one can reduce dephasing, even if only one field is needed for the actual gate operation. For example, a two-qubit storage-gate field, with $\phi_{1,2}^{(2)\Phi}(t_f) = 2\pi M$ $(M = 1, 2, \ldots)$, concurrently applied with, for example, a SWAP gate, can reduce dephasing. This approach may result in longer gate durations, as the maximal peak power in the gate fields may be limited. Nevertheless, according to (13.15) and (13.18), this approach may be beneficial if the reduction in the dephasing due to the applied fields outweighs the increase in dephasing due to longer gate duration.

13.1.3 Optimal Gate Protection

The protection of a given quantum operation from decoherence is most effective under bath-optimized task-oriented control (BOTOC), expounded in Chapter 12. Here, we consider the implementation of a quantum-gate unitary operation within a given "gate time" t for a pure input state $|\Psi\rangle$. In the interaction picture with respect to the desired gate operation, the projector $\hat{P} = \hat{\varrho}(0) = |\Psi\rangle\langle\Psi|$ is then used as the gradient operator, so that (12.22) is satisfied. Then, (12.24) yields the fidelity change as the score

$$P = \langle\Psi|\Delta\hat{\varrho}|\Psi\rangle = -t^2(\langle\Psi|\hat{H}^2|\Psi\rangle - \langle\Psi|\hat{H}|\Psi\rangle^2)_B. \tag{13.19}$$

To eliminate the dependence on $|\Psi\rangle$, we uniformly average over all $|\Psi\rangle$, whereby for any two operators $\hat{A}$ and $\hat{B}$:

$$\overline{\langle\Psi|\hat{A}|\Psi\rangle\langle\Psi|\hat{B}|\Psi\rangle} = \frac{\text{Tr}\hat{A}\hat{B} + \text{Tr}\hat{A}\text{Tr}\hat{B}}{d(d+1)}, \tag{13.20}$$

d being the Hilbert-space dimensionality of the system. The average score is then

$$\bar{P} = -t^2 \frac{d}{d+1} \langle\hat{H}^2\rangle_{\text{id}}, \tag{13.21}$$

where [in the notation of (12.6)] $\langle\ldots\rangle_{\text{id}} = \text{Tr}(d^{-1}\hat{I} \otimes \hat{\varrho}_B \ldots)$. Here we have used $\text{Tr}_S\hat{H} = 0$, corresponding to $\text{Tr}\,\hat{S}_j = 0$. On account of (12.21), $\langle\hat{H}\rangle_B = \langle\hat{H}_I\rangle_B = 0$,

$\langle \hat{H} \rangle_{id} = 0$, so that (13.21) is proportional to the variance of the Hamiltonian: $\text{Var}(\hat{H}) = \langle \hat{H}^2 \rangle_{id} - \langle \hat{H} \rangle_{id}^2$. The *gate error* $\mathcal{E}$, which is the average fidelity decline (or the infidelity), then satisfies

$$\mathcal{E} \equiv -\bar{P} = t^2 \frac{d}{d+1} \text{Var}(\hat{H}). \tag{13.22}$$

The average over the initial states in the matrix Ξ, defined in (12.31), yields $\bar{\Xi} = -\frac{d}{d+1} I$, upon using $\text{Tr}(\hat{S}_j \hat{S}_k) = d\delta_{jk}$ and $\text{Tr}\hat{S}_j = 0$. Hence

$$\mathcal{E} = \frac{d}{d+1} \int_{-\infty}^{\infty} d\omega \ \text{Tr}[\epsilon_t(\omega)\epsilon_t^\dagger(\omega)G(\omega)], \tag{13.23}$$

where $G(\omega)$ and $\epsilon_t(\omega)$ are given by (12.32) and (12.33), respectively. Because of the requirement that $\mathcal{E} \geq 0$, $G(\omega)$ must be a positive semidefinite matrix for any ω. BOTOC then aims at finding the evolution operator of the system, $\hat{U}(t)$ (as in the control examples in Sec. 12.2), that minimizes $\mathcal{E}$, subject to the condition that the desired gate is executed over time interval t_f.

We may therefore conclude that, whereas each gate operation should be as fast as possible, the optimal overall pulse sequence may take longer than the gate time because of the storage control duration. This general principle is unparalleled by other approaches.

13.1.4 Implementation in Trapped-Ion Systems

A possible implementation of this approach may involve a string of ions in a linear trap. The qubits, encoded by two internal states of each ion [$|g(e)\rangle_j$], are manipulated by laser beams. An additional qubit, encoded by the ground and first excited common vibrational levels ($|0(1)\rangle_N$), acts as the "bus mode." The qubit gates are executed by applying laser pulses on the "carrier" [$\Omega_j^{(1)}(t)$, $|g\rangle \leftrightarrow |e\rangle$], the "blue sideband" [$\Omega_{jN}^{(2)\Phi}(t)$, $|g\rangle |0\rangle \leftrightarrow |e\rangle |1\rangle$], and the "red sideband" [$\Omega_{jN}^{(2)\Psi}(t)$, $|g\rangle |1\rangle \leftrightarrow |e\rangle |0\rangle$] of the electronic quadrupole transition [Fig. 13.1(a)]. In a harmonic trap, the blue sideband also couples to higher levels, for example, $|g\rangle |1\rangle \leftrightarrow |e\rangle |2\rangle$, and the red one to $|e\rangle |1\rangle \leftrightarrow |g\rangle |2\rangle$. Such unwarranted excitations complicate and hamper the fidelity of the concurrent application of both two-qubit gates. These excitations may be suppressed by resorting to trap anharmonicity. Dephasing in the trapped-ion system arises due to magnetic-field fluctuations that give rise to random Zeeman shifts of the qubit levels. Simulations of a SWAP gate involving the lowest two common vibrational levels (in an anharmonic trap) in the presence of dephasing [Fig. 13.1(b)] show that the gate fidelity may be raised by means of the optimized pulse sequence compared to its standard counterpart, despite the longer duration of the former.

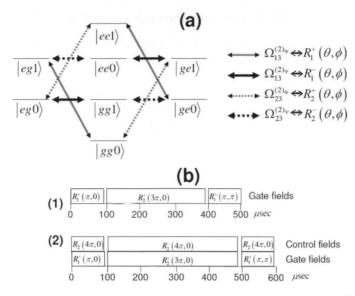

Figure 13.1 (a) Energy levels and two-qubit gate fields [cf. (13.6)] acting on two internal states of two ions $|g(e)\rangle$ and their lowest two common vibrational levels $|0(1)\rangle$. (b) Conventional pulse sequence (1) compared to sequence (2) outlined above, in the presence of dephasing. Here $\gamma^{-1} = 1$ ms, $t_c = 300\ \mu$s. The conventional pulse sequence, resulting in $\bar{f}(t = 500\ \mu s) = 0.93$, whereas the optimized sequence yields $\bar{f}(t = 600\ \mu s) = 0.97$. Here $R_j^+(\theta, \phi)$ and $R_j^-(\theta, \phi)$ denote pulse-induced rotations on blue and red sidebands, respectively, for the jth ion, θ denoting the angle of the rotation and ϕ the phase of the field. (Adapted from Gordon and Kurizki, 2007. © 2007 The American Physical Society)

13.2 Multipartite Decoherence Control

Control of multipartite systems, aimed at suppressing their decoherence that is caused by thermal baths, is a far more challenging task than its single-particle counterpart, for two reasons: (i) Bath-induced multipartite disentanglement is much faster than single-particle decoherence, as shown below. (ii) Cross-decoherence, which arises when many particles interact with the baths, is shown to complicate the control. Nevertheless, extension of the decoherence control approach surveyed in Chapters 11 and 12 to multiparticle systems that undergo either relaxation or dephasing is shown to tackle the challenges indicated above.

To this end, we here address the basic questions: How does decoherence scale with the number of particles and is it state dependent? Can dynamical control combat decoherence and its scaling with the number of particles? What timescales are required for such control?

13.2.1 Multipartite Decoherence Control at $T = 0$

We here consider a multipartite system relaxing into a common $T = 0$ bath. Each particle (labeled by j) has the ground state $|g\rangle_j$ and M_j nondegenerate excited states $|n\rangle_j$ (see Ch. 6). The particle–bath interaction couples the excited states of the particle with its ground state but does not mix different excited states. The total Hamiltonian is the sum of the system (S), bath (B), and interaction (I) Hamiltonians of the form

$$H_S(t) = \hbar \sum_{j=1}^{N} \sum_{n=1}^{M_j} [\omega_{j,n} + \Delta_{j,n}(t)] \, |n\rangle_{jj} \langle n| \,, \tag{13.24}$$

$$H_B = \hbar \sum_k \omega_k \, |1_k\rangle \langle 1_k| \,, \tag{13.25}$$

$$H_I(t) = \hbar \sum_{j=1}^{N} \sum_{n=1}^{M_j} \sum_k \left[\tilde{\epsilon}_{j,n}(t) \eta_{k,j,n} \, |n\rangle_j \, |\{0_k\}\rangle_{\,j} \langle g| \, \langle 1_k| \bigotimes_{j' \neq j} I_{j'} + \text{H.c.} \right]. \tag{13.26}$$

Here $\hbar\omega_{j,n}$ is the energy of $|n\rangle_j$, $\hbar\Delta_{j,n}(t)$ is the time-dependent AC Stark shift modulating the energy of $|n\rangle_j$, $|\{0_k\}\rangle$ is the bath vacuum state, $|1_k\rangle$ is the bath state with single-quantum excitation in the kth mode, and I_j is the identity operator for the jth particle. The strength $\eta_{k,j,n}$ of the coupling of $|n\rangle_j$ to $|1_k\rangle$ undergoes a time-dependent modulation described by the function $\tilde{\epsilon}_{j,n}(t)$.

Although the same control strategy applies to any number of excitations, simple closed-form solutions are only obtainable for an (initially) *singly* excited multipartite system, for which the combined system–bath wave function

$$|\Psi(t)\rangle = \sum_{j=1}^{N} \sum_{n=1}^{M_j} \tilde{\alpha}_{j,n}(t) \, |\{0_k\}\rangle \, |n\rangle_j \bigotimes_{j' \neq j} |g\rangle_{j'} + \sum_k \beta_k(t) \, |1_k\rangle \bigotimes_{j=1}^{N} |g\rangle_j. \tag{13.27}$$

In order to analyze the time evolution of this wave function, we evaluate the column vector of the multipartite excited-state probability amplitudes, $\boldsymbol{\alpha}(t) = \{\alpha_{j,n}(t)\}$, which is expressed in the interaction picture as

$$\alpha_{j,n}(t) = e^{i\omega_{j,n}t + i\int_0^t d\tau \, \Delta_{j,n}(\tau)} \tilde{\alpha}_{j,n}(t). \tag{13.28}$$

The Schrödinger equation for the coupled $\{\alpha_{j,n}(t)\}$ and $\{\beta_k(t)\}$ amplitudes may be reduced, upon eliminating the ground-state $\{\beta_k(t)\}$ amplitudes, to the following *exact* integrodifferential equations, which are the multipartite analogs of the single-TLS equation (5.6) and (10.13),

$$\dot{\alpha}_{j,n}(t) = \int_0^t dt' \sum_{j',n'} \Phi_{jj',nn'}(t-t')\epsilon_{j,n}(t)\epsilon^*_{j',n'}(t')e^{i\omega_{j,n}t - i\omega_{j',n'}t'}\alpha_{j',n'}(t'). \quad (13.29)$$

Here $\boldsymbol{\Phi}(t) = \{\Phi_{jj',nn'}(t)\}$ is the bath-response matrix with elements

$$\Phi_{jj',nn'}(t) = \sum_k \eta_{k,j,n}\eta^*_{k,j',n'}e^{-i\omega_k t}, \quad (13.30)$$

and

$$\epsilon_{j,n}(t) = \tilde{\epsilon}_{j,n}(t)e^{i\int_0^t d\tau \Delta_{j,n}(\tau)} \quad (13.31)$$

represent the dynamical-control field amplitudes.

Assuming that the amplitudes are slowly varying on the timescale of the bath response, (13.29) can be solved to the same accuracy as its single-TLS counterpart in Chapter 10. In matrix notation, we then obtain the following approximate solution *for dynamically modified relaxation of singly excited multipartite entangled states caused by a zero-temperature bath,*

$$\boldsymbol{\alpha}(t) = \mathrm{T}_+ e^{-\tilde{\boldsymbol{J}}(t)}\boldsymbol{\alpha}(0). \quad (13.32)$$

Here, T_+ is the chronological (time-ordering) operator, and $\tilde{\boldsymbol{J}}(t) = \{\tilde{J}_{jj',nn'}(t)\}$ is the dynamically modified decoherence matrix, with elements given by the convolution

$$\tilde{J}_{jj',nn'}(t) = 2\pi \int_{-\infty}^{\infty} d\omega \, \tilde{G}_{jj',nn'}(\omega)\tilde{F}_{t,jj',nn'}(\omega) \quad (13.33)$$

of the bath-coupling spectral matrix

$$\tilde{G}_{jj',nn'}(\omega) = \pi^{-1}\int_0^{\infty} dt \, \Phi_{jj',nn'}(t)e^{i\omega t}, \quad (13.34)$$

with the following modulation spectral matrix evaluated over the $[0, t]$ time interval, designed to suppress the decoherence,

$$\tilde{F}_{t,jj',nn'}(\omega) = \epsilon_{t,j,n}(\omega - \omega_{j,n})\epsilon^*_{t,j',n'}(\omega - \omega_{j'n'}), \quad (13.35)$$

where

$$\epsilon_{t,j,n}(\omega) = \int_0^t d\tau \epsilon_{j,n}(\tau)e^{i\omega\tau}. \quad (13.36)$$

A prominent class of multilevel dynamical modulations are impulsive phase modulations: sequences of pulsed AC Stark shifts effected by fields with amplitudes that satisfy

$$\epsilon_{j,n}(t) = e^{i[t/\tau_{j,n}]\theta_{j,n}}, \quad (13.37)$$

so that

$$\epsilon_{t,j,n}(\omega) = \frac{(e^{i\omega\tau_{j,n}} - 1)(e^{i(\theta_{j,n} + \omega\tau_{j,n})[t/\tau_{j,n}]} - 1)}{i\omega(e^{i(\theta_{j,n} + \omega\tau_{j,n})} - 1)}. \tag{13.38}$$

Here $[\dots]$ is the integer part of a number, and $\tau_{j,n}$ and $\theta_{j,n}$ are the pulse duration and the phase change for level n of particle j, respectively. For weak pulses with area $|\theta_{j,n}| \ll \pi$, (13.38) yields

$$\epsilon_{t,j,n}(\omega) \simeq \tilde{\epsilon}_{t,j,n}\delta(\omega - \Delta_{j,n}), \tag{13.39}$$

where

$$\Delta_{j,n} = \theta_{j,n}/\tau_{j,n} \tag{13.40}$$

is the effective frequency shift caused by the applied pulses.

These spectral (AC Stark) shifts are generally time-dependent, so that by the end of the gate operation at time $t = t_f$, the AC Stark shifts $\Delta_{j,n}$ satisfy

$$\int_0^{t_f} d\tau\, \Delta_{j,n}(\tau) = 2\pi m, \quad m = 0, \pm 1, \dots. \tag{13.41}$$

This requirement ensures that modulations only affect the decoherence matrix (13.33), but do not change the relative phases of the entangled particles when their state is manipulated by logic operations at $t = t_f$.

It is expedient to rewrite the fidelity of the evolving state that is corrupted by the bath as a product of two factors,

$$f(t) \equiv |\langle\Psi(0)|\Psi(t)\rangle|^2 = f_p(t)f_c(t). \tag{13.42}$$

Here,

$$f_p(t) = |A(t)|^2 = \sum_{j=1}^{N}\sum_{n=1}^{M_j} |\alpha_{j,n}(t)|^2 \tag{13.43}$$

is the total excitation probability and

$$f_c(t) = \frac{|\sum_{j=1}^{N}\sum_{n=1}^{M_j} \alpha_{j,n}^*(0)\alpha_{j,n}(t)|^2}{|A(t)|^2} \tag{13.44}$$

is the autocorrelation function normalized by the total excitation probability $|A(t)|^2$. Hence, $1 - f_p(t)$ measures population loss from any energy eigenstate $|n\rangle_j$, whereas $1 - f_c(t)$ measures multipartite decorrelation.

For mixed states, the fidelity, $f(t)$, the excitation probability, $f_p(t)$, and their ratio, $f_c(t)$, can be expressed, in the multiqubit density-matrix notation, as

$$f(t) = \text{Tr}\,[\rho(0)\rho(t)], \tag{13.45}$$

$$f_p(t) = \text{Tr}\,[(|e\rangle_{jj}\langle e|)\rho(t)], \tag{13.46}$$

$$f_{\rm c}(t) = f(t)/f_{\rm p}(t). \tag{13.47}$$

In the absence of dynamical control, $f_{\rm c}(t)$ decays much faster than $f_{\rm p}(t)$ and is much more affected by the asymmetry between local particle–bath couplings. As shown below, population and correlation loss can be *independently controlled*.

13.2.2 Decoherence-Free Subspace for Three-Level Multipartite System

In the absence of modulations, decoherence has, in general, no inherent symmetry. Yet, one can dynamically *symmetrize the decoherence* by appropriate modulations. A "global" dynamical modulation that has the same form, (13.35), for all N particles and levels cannot satisfy $N \gg 1$ symmetrizing requirements at all times. Only different, "local," modulations applied to the individual particles and levels can cause *controlled interference* and/or spectral shifts between the couplings of different particles to the bath. The dynamical control matrix elements (13.35) (for N qubits) can then be made to satisfy $2N$ requirements at all times and be tailored to impose the symmetries described below.

The most desirable symmetry is that of *identical coupled particles* (ICP), whereby all the modulated particles (and all transitions, in multilevel particles) undergo the *same* dynamically modified decoherence and cross-decoherence. Then, the following $N \times N$ fully symmetrized decoherence matrix would ensue:

$$J_{jj'}^{\rm ICP}(t) = J(t) \ \forall j, j'. \tag{13.48}$$

In the case of N qubits that share a single excitation, ICP symmetry would yield to an $(N - 1)$-dimensional decoherence-free subspace (DFS), the entire single-excitation sector, excluding the totally symmetric entangled state. Any initial state in this DFS would keep perfect fidelity 1 at all times.

Yet, it is generally impossible to impose the ICP symmetry, since it must satisfy too many conditions: $N(N - 1)/2$ conditions for N modulating fields in the case of N qubits [in the case of multilevel particles, there are even more conditions, for each pair of levels (n, n')]. Even if we accidently succeeded with N particles, the success would not be scalable to $N+1$ or more particles. Moreover, such symmetry fails completely if not all particles are coupled to all others through the bath, that is, if some $G_{jj'}(\omega)$ elements vanish. Nevertheless, in what follows we consider a system of three-level particles where local modulation may impose a DFS.

In order to impose a DFS, we make use of an auxiliary level (here, a third level) and local modulations to form *a decoherence-free singly excited N-qubit* system. Specifically, we here invoke three-level particles with excited states $|1\rangle$ and $|2\rangle$. In a single three-level particle, an external field can impose an intraparticle DFS if $|1\rangle$ and $|2\rangle$ decay at the same rate to the ground state $|0\rangle$; this DFS consists of $|0\rangle$

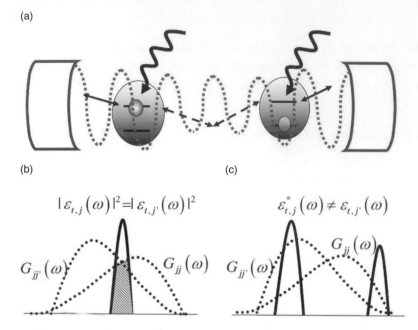

(a)

(b)   (c)

$$|\varepsilon_{t,j}(\omega)|^2 = |\varepsilon_{t,j'}(\omega)|^2 \qquad\qquad \varepsilon_{t,j}^*(\omega) \neq \varepsilon_{t,j'}(\omega)$$

$G_{jj'}(\omega)$ $G_{jj}(\omega)$ $G_{jj'}(\omega)$ $G_{jj}(\omega)$

Figure 13.2 (a) Two qubits in a cavity, coupled to the cavity-modes bath (dashed) and driven by local control fields (thick lines). (b),(c) Spectral overlap of the bath-coupling spectrum (dotted) and the modulation matrix elements (solid) for this case, resulting in modified decoherence matrix elements (shaded): (b) global modulation under ICP symmetry, (c) cross-decoherence elimination under IIP symmetry. (Reproduced with permission from Gordon et al., 2007. © IOP Publishing. All rights reserved.)

and the "dark" (trapping) state, which is the antisymmetric superposition of the two excited states $(1/\sqrt{2})(|1\rangle - |2\rangle)$. However, this intraparticle DFS would not survive in a system of N three-level particles that are coupled to the bath and undergo cross-decoherence, as detailed below in Section 13.2.2. The remedy consists in local modulations acting *differently on the two excited levels* within each particle.

Such modulations can be tailored to impose *"independent identical trapping"* (IIT) symmetry, under which all particles acquire identical trapping states and become (effectively) independent, *without cross-decoherence*. As shown below, N spectral shifts Δ_j determined by $|\epsilon_{t,j}(\omega)|^2$ suffice to eliminate cross-decoherence. Thus, N destructively interfering pairs of local fields can cause trapping in each particle, resulting in an N-dimensional DFS, sharing a single excitation with anti-symmetrically superposed excited states (Fig. 13.2). Under the IIT symmetry, the three-level, N-particle decoherence matrix acquires the block-diagonal form,

$$J_{jj',nn'}^{\text{IIT}}(t) = \delta_{jj'} J_{j,nn'}(t), \tag{13.49}$$

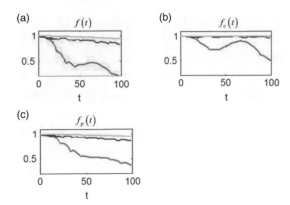

Figure 13.3 Fidelity as a function of time [in units of $(10\gamma)^{-1}$]: (a) overall fidelity, (b) correlation preservation, and (c) population preservation. Global π-phase flips impose no symmetry [bottom curves in (a)–(c)] ($\tau_{j,n} = 1.1$, $\theta_{j,n}/\pi = 1.0$). Individual phase shifts can impose independent identical particles symmetry [middle curves in (a)–(c), $\tau_{j,n} = (0.75, 0.85, 0.95, 1.05)$, $\theta_{j,n}/\pi = (0.834, 0.806, 0.836, 0.82)$] or independent identical-trapping symmetry [top curves in (a)–(c), $\tau_{j,n} = (0.85, 0.85, 1.05, 1.05)$, $\theta_{j,n}/\pi = (0.924, 0.9, 0.945, 0.91)$]. The parameters are $\omega_1 = 0.5$, $\omega_2 = 0.6$, $\gamma = 0.1$, $k_0 x_{min} = 1.0$, $t_{j,n} = (0.7, 1.0, 1.06, 1.1)$, $x_j = (0.0, 0.1)$, and $\eta n/\pi = (0.0, 0.1)$. (Reproduced with permission from Gordon, 2009. © IOP Publishing. All rights reserved.)

where the excited levels 1 and 2 are assumed to decay at the same rate to the ground state. Then, both disentanglement and population loss are nearly completely eliminated within the multipartite subspace, allowing $f(t)$ to be close to 1 (Fig. 13.3).

An example of the IIT recipe will now be given for two three-level particles, where each level of each particle has *different* coupling to the bath, and cross-decoherence exists. Let the initial state of particles 1 and 2 be

$$|\Psi_S(0)\rangle = (|-\rangle_1|0\rangle_2 \pm |0\rangle_1|-\rangle_2)/\sqrt{2}, \tag{13.50}$$

where

$$|-\rangle_j = (|1\rangle_j - |2\rangle_j)/\sqrt{2} \quad (j = 1, 2). \tag{13.51}$$

To keep the initial state intact, $|\Psi_S(t)\rangle \simeq |\Psi_S(0)\rangle$, we should prevent cross decoherence and impose intraparticle destructive interference. The bath response matrix can be

$$\Phi_{jj',nn'}(t) = \gamma \frac{e^{-t^2/(4t_{j,n}^2)} e^{-t^2/(4t_{j',n'}^2)}}{k_0(x_{min} + \delta x_{jj'})}. \tag{13.52}$$

Here, γ is the single-particle unperturbed decay rate (according to Fermi's Golden Rule), $t_{j,n}$ is the correlation time of level $n = 1, 2$ of particle j, k_0 is the wave vector of the resonant transition $|0\rangle \leftrightarrow |1(2)\rangle$, x_{min} is the distance below which the particles are identically coupled to the bath, and $\delta x_{jj'} = |x_j - x_{j'}|$, where x_j is the position of particle j. This model describes the *distance-dependent cross-decoherence* of either radiatively or vibrationally relaxing atoms or ions at different sites in a trap, lattice, or cavity. The parameters of the pulse sequence in (13.37) for creating the IIT symmetry can be chosen such that the faster-decaying particle experiences the stronger and/or faster pulses. Namely, $\theta_{j,n}/\tau_{j,n}$ can be adapted to the bath response parameters of each particle. The pulse sequences can be chosen to be different enough for each particle in order to eliminate the desired cross-decoherence terms [see (13.54) below]. Individual pulses can be selected to obey the other IIT requirements described above (Fig. 13.3).

13.2.3 Multiqubit Dynamical Symmetry Control

Below, we mostly assume that the particles are qubits [Fig. 13.4(a)], so that they have only one excited state $|1\rangle_j \equiv |e\rangle_j$. The bath-coupling off-diagonal matrix element $G_{jj'}(\omega)$ is then associated with the interaction of qubits j and j' via the bath, and $\tilde{F}_{t,jj'}(\omega)$ is the dynamical modulation matrix, which we design at will to suppress the decoherence or the relaxation. The relaxation or decoherence of the system of qubits is described by the decoherence matrix (13.33) in the simplified notation $J_{jj'}(t)$.

The diagonal elements, $J_{jj}(t)$, are determined by the coupling of individual qubits (labeled by j) to their respective baths. On the other hand, the off-diagonal elements, $J_{jj'}(t)$ with $j \neq j'$, represent the results of *cross-decoherence* induced on the jth qubit by the j'th qubit. Such cross-decoherence exists only for qubits coupled to a common bath mode or sharing a correlated dephasing-noise mode [Fig. 13.4(a)]. Specifically, cross-dephasing stems from the cross correlation of the stochastic noise functions of qubits j and j'. Thus, for example, ions subject to the same fluctuating magnetic field undergo cross-dephasing. This off-diagonal cross-decoherence may be viewed as either virtual off-resonant or real (resonant) emission of a quantum into the bath by qubit j and its reabsorption by qubit j' [Fig. 13.4(a)].

In the relaxation scenario, the modulation spectral function $\tilde{F}_{t,jj'}(\omega)$ represents the effective change in qubit j's energy splitting or AC Stark shift [Fig. 13.4(b)]. In the dephasing scenario, it determines the time-dependent Rabi splitting of the $|\pm\rangle$ states [Fig. 13.4(c)]. In both scenarios, the spectral function in the integrand in (13.33) represents the effective coupling between the qubits via the bath, filtered by the dynamical control (modulation). This filter determines which spectral bath modes the qubits are coupled to.

Importantly, the off-diagonal terms of both the coupling-spectrum matrix, $G_{jj'}(\omega)$, and the dynamical modulation matrix, $\tilde{F}_{t,jj'}(\omega)$, are generally complex. In particular, the matrix elements $\tilde{F}_{t,jj'}(\omega)$ are complex, because they depend on

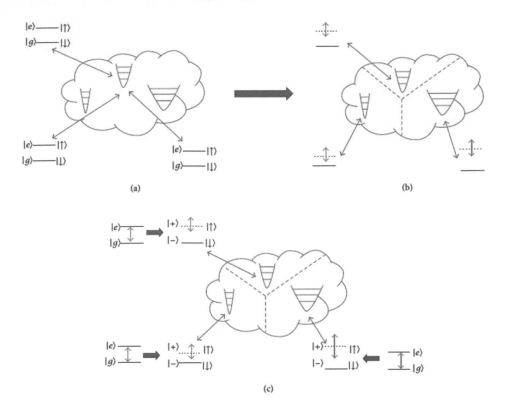

Figure 13.4 Multipartite systems coupled to their environments. (a) Relaxation of identical qubits via coupling to the same bath modes causes cross-decoherence. (b) *Different* AC Stark shifts on individual qubits couple them to different bath modes, eliminating cross-decoherence. (c) Different Rabi driving of dephasing qubits is analogous to (b). (Reprinted from Kurizki, 2013.)

phase difference between modulations of the relevant particles. The real and imaginary parts of these matrices are inseparable. By contrast, the diagonal elements of the decoherence matrix are separable into real parts corresponding to decay and imaginary parts associated with self-energy, alias Lamb shifts (Ch. 8).

A more restricted symmetry than ICP without DFS may be ensured for N qubits: namely, the symmetry of *independent identical particles* (IIP). This symmetry arises when spectral shifts or interferences imposed by N modulations cause the N different particles (qubits) to be subject to the *same* single-particle decoherence at a rate

$$\gamma(t) = \dot{J}_{jj}(t) \tag{13.53}$$

and undergo no cross-decoherence. To this end, we may choose fields as in (13.37) that realize AC Stark (spectral) shifts Δ_j that are different for each j, while all unmodulated frequencies are identical, $\omega_j = \omega_0$. The spectral shifts Δ_j can be

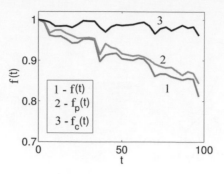

Figure 13.5 Fidelity of the IIP symmetry for two initially entangled qubits, each coupled to a distinct zero-temperature bath. (Adapted with permission from Gordon et al., 2007. © IOP Publishing. All rights reserved.)

made different enough to couple each particle to a different spectral range of bath modes so that their cross-coupling vanishes,

$$J_{jj'}(t) = \epsilon_{t,j}^* \epsilon_{t,j'} \int d\omega G(\omega) \delta(\omega - \Delta_j) \delta(\omega - \Delta_{j'}) \rightarrow 0. \tag{13.54}$$

As opposed to the ICP symmetry discussed in Section 13.2.2, the vanishing of $G_{jj'}(\omega)$ for some pairs of particles (j, j') is not a limitation. The N single-particle decoherence factors can be equated by an appropriate choice of N parameters $\{\Delta_j\}$ [Fig. 13.2(c)],

$$J_{jj'}^{\text{IIP}}(t) = |\epsilon_{t,j}|^2 G_{jj}(\Delta_j) \delta_{jj'} = \delta_{jj'} J(t). \tag{13.55}$$

The IIP symmetry completely preserves quantum correlations (entanglement), that is $f_c(t) = 1$, but still permits population decay, $f(t) = f_p(t) = e^{-2\mathrm{Re}J(t)}$ (Fig. 13.5). If the single-particle decoherence factor $J(t)$ is dynamically suppressed, that is, if the spectrally shifted bath response $G_{jj}(\omega_j + \Delta_j)$ is small enough, this $f(t)$ can be kept close to 1 (Fig. 13.3). Under IIP symmetry, the particles become effectively independent in terms of their coupling to the bath. Hence, collective coupling to the bath, which is a prerequisite for a DFS, *does not occur* in this case.

13.2.4 Multiqubit Decoherence Control at $T \neq 0$

Our system is comprised of N identical qubits labeled $j = 1, ..., N$ with ground and excited states $|g\rangle_j$ and $|e\rangle_j$, respectively, and energy separation $\hbar\omega_a$ (Fig. 13.6). Each qubit may be differently (weakly) coupled to a bath, the Hamiltonian being (henceforth we take $\hbar = 1$)

$$H(t) = H_S + H_B + H_I + H_C(t),$$

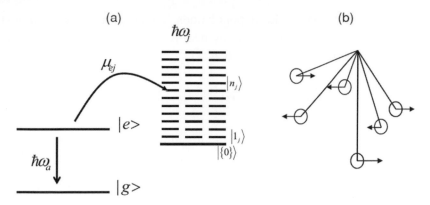

Figure 13.6 (a) A TLS with off-diagonal coupling to a continuum or a bath. (b) A bath comprised of many oscillators with different frequencies depicted by pendula of different lengths. Their getting out of phase (dephasing) after correlation time t_c renders the system–bath interaction practically irreversible. (Reproduced with permission from Gordon, 2009. © IOP Publishing. All rights reserved.)

$$H_S = \omega_a \sum_{j=1}^{N} |e\rangle_{jj}\langle e|, \quad H_I = \sum_{j=1}^{N} \hat{S}_j \hat{B}_j. \tag{13.56}$$

Here H_S, H_B, and H_I are the system, bath, and interaction Hamiltonians, respectively, $H_C(t)$ is the time-dependent control Hamiltonian, $\hat{B}_j$ being the jth qubit bath operator, and $\hat{S}_j$ the system–bath coupling operator of qubit j.

Two kinds of decoherence can be similarly treated: (i) *Relaxation*, caused by an off-diagonal system–bath coupling Hamiltonian (in the energy basis), corresponding to the X-Pauli matrix of the jth qubit,

$$\hat{S}_j = \sigma_{xj} = |e\rangle_{jj}\langle g| + |g\rangle_{jj}\langle e|. \tag{13.57}$$

Relaxation involves temperature-dependent level-population change, as well as the erasure (randomization) of the phase relations. Such a process is controllable if each qubit is subject to a strong but off-resonant driving field that causes an *AC Stark shift* [Fig. 13.4(b)], that is, energy modulation of the system Hamiltonian,

$$H_C(t) = \sum_{j=1}^{N} \Delta_j(t)|e\rangle_{jj}\langle e|. \tag{13.58}$$

(ii) *Dephasing* is caused by a system–bath coupling Hamiltonian that is diagonal in the energy basis, via the Z-Pauli matrix of the jth qubit,

$$\hat{S}_j = \sigma_{zj} = |e\rangle_{jj}\langle e| - |g\rangle_{jj}\langle g|. \tag{13.59}$$

Dephasing does not change level populations but only erases (scrambles) the phases. Such a process is controllable by near-resonant driving fields, $V_j(t)$, that cause transitions between qubit levels (Ch. 11),

$$H_{\rm C}(t) = \hbar \sum_{j=1}^{N} [V_j(t) e^{-i\omega_{\rm a}t} |e\rangle_{jj} \langle g| + {\rm H.c.}]. \qquad (13.60)$$

In this scenario, the ME becomes the multiqubit counterpart of the single-qubit non-Markovian ME in Chapter 11, assuming the qubits do not interact directly,

$$\dot{\rho} = \sum_{j,j'=1}^{N} [\gamma_{jj'}(t)\sigma_{xj}\rho\sigma_{xj'} - \gamma_{jj'}(t)\sigma_{xj'}\sigma_{xj}\rho - \gamma_{jj'}^{*}(t)\rho\sigma_{xj'}\sigma_{xj} + \gamma_{jj'}^{*}(t)\sigma_{xj'}\rho\sigma_{xj}],$$

$$(13.61)$$

where $\gamma_{jj'}(t)$ are the instantaneous dynamically controlled decoherence rates.

To second order in the system–bath coupling, we can obtain $\rho(t)$ from this ME for an initially pure state $|\psi\rangle$ by formally integrating (13.61), upon setting $\rho(t) = \rho(0) = |\psi\rangle\langle\psi|$ on its right-hand side. As a result, we have, at sufficiently short times,

$$f_{\psi}(t) \simeq 1 - t\gamma_{\psi}(t; N), \qquad (13.62)$$

where the infidelity is given by

$$t\gamma_{\psi}(t; N) = \sum_{j,j'=1}^{N} J_{jj'}(t)(\langle\psi|\sigma_{xj}\sigma_{xj'}|\psi\rangle - \langle\psi|\sigma_{xj}|\psi\rangle\langle\psi|\sigma_{xj'}|\psi\rangle). \qquad (13.63)$$

This infidelity is determined by the elements of the (time-averaged) decoherence-rate matrix

$$J_{jj'}(t) = 2 \int_{0}^{t} dt' \gamma_{jj'}(t'), \qquad (13.64)$$

which are given by (13.33), without the n, n' indices in the case of qubits.

There is an important distinction between the controllability of single-qubit decoherence factors, $J_{jj}(t)$, and that of their cross-decoherence counterpart, $J_{jj'}(t)$ with $j \neq j'$. The single-qubit dynamically modified $J_{jj}(t)$ can only be reduced by decreasing the overlap between the modulation spectra $|\epsilon_{t,j}(\omega)|^2$ and the bath-coupling spectra, $G_{jj}(\omega)$, both positive. No matter how fast the modulation, $|\epsilon_{t,j}(\omega)|^2$ cannot suppress the single-qubit decoherence factor if the bath-coupling spectrum is spectrally flat. By contrast, cross-decoherence factors $J_{jj'}(t)$ can be *completely eliminated* by choosing spectrally distinct local modulations whose spectral overlap $\epsilon_{t,j}(\omega)\epsilon_{t,j'}^{*}(\omega)$ vanishes, irrespective of the bath-coupling spectrum, $G_{jj'}(\omega)$, even if it is spectrally flat. The qubits effectively couple to orthogonal bath modes, so that, by modulating each qubit differently, one can effectively decouple the qubits, as if they interacted with separate baths.

13.3 Decoherence-Control Scalability

13.3.1 Decoherence of Entangled States

In the following, we consider the scaling of decoherence with the number of qubits N for several generic multipartite entangled states, such as:

(i) GHZ states,

$$|\text{GHZ}\rangle = \frac{1}{\sqrt{2}} \left(\bigotimes_j |e\rangle_j + \bigotimes_j |g\rangle_j \right). \tag{13.65}$$

(ii) W-states,

$$|W\rangle = \frac{1}{\sqrt{N}} \sum_{j=1}^{N} |e\rangle_j \bigotimes_{j \neq j'} |g\rangle_{j'}. \tag{13.66}$$

(iii) Bright (symmetrized, maximally cooperative) Dicke states,

$$|B\rangle = |J = N/2, m = 0\rangle, \tag{13.67}$$

where the Dicke states $|J, m\rangle$ are labeled by J, the total angular momentum, and m, the excitation number (Ch. 7).

(iv) Dark (antisymmetrized) states for even N,

$$|D\rangle = \bigotimes_{j=2,4,\dots,2N} |\Psi_-\rangle_{j-1,j}, \tag{13.68}$$

where

$$|\Psi_-\rangle_{j,j+1} = \frac{1}{\sqrt{2}}(|g\rangle_j |e\rangle_{j+1} - |e\rangle_j |g\rangle_{j+1}). \tag{13.69}$$

It follows from (13.63) that the infidelity rate depends on the multipartite state in question and on the decoherence rates $\gamma_{jj'}(t)$ whose time dependence plays out on non-Markovian timescales (Ch. 11).

In the absence of cross-decoherence, when each qubit is locally coupled to an individual bath $J_{jj'}(t) = 0$, for all $j \neq j'$, the first state-dependent term of the infidelity rate in (13.63) scales linearly with N, for all states. The second term then arises only for certain states (e.g., W-states) and can only lower the N scalability (Table 13.1), down to (at most) linear dependence on N.

Table 13.1 *N-scalability of entangled states decoherence for relaxation and dephasing scenarios, with and without cross-(X) decoherence.*

	Relaxation		Dephasing	
	No X-deco.	X-deco.	No X-deco.	X-deco.
GHZ	N	N	N	N^2
W-state	N	$3N - 2$	$4(N - 1)/N$	0
Bright state	N	$N + N^2/2$	N	0
Dark state	N	0	N	0

The infidelity rate is much more *state dependent* in systems with cross-decoherence, where additional decoherence channels may interfere, either constructively or destructively, depending on the initial entangled state and the phases of $\langle \psi | \sigma_{xj} \sigma_{xj'} | \psi \rangle$. Consequently, the N scalability under cross-decoherence can either rise to N^2 or vanish altogether.

Specifically, under global dephasing, the GHZ state becomes a bright state in the $|\pm\rangle$ basis and hence experiences constructive interference of all decoherence channels, causing its dephasing to scale quadratically with N. The W-state and bright state, on the other hand, do not undergo global dephasing since they are symmetric superpositions of identically excited states, so that the dephasing only induces a global phase factor and does not affect decoherence. The dark state $|\Psi_-\rangle$ is decoupled from global relaxation and dephasing, due to its invariance under the basis rotation from $|e(g)\rangle$ to $|\pm\rangle$.

13.3.2 Coherence Time Scaling in the Zeno Regime

The coherence time may be defined as the time at which the fidelity (13.62) or the infidelity, $1 - f(t)$, reaches a certain threshold:

$$J_\psi(t_{\text{coh}}; N) = 1 - f_{\text{threshold}} \equiv e_{\text{threshold}}, \qquad (13.70)$$

where $J_\psi(t)$ is the state-dependent decoherence (relaxation) factor for $|\psi\rangle$. Coherence time is important, since its ratio to the duration of a single quantum operation (e.g., a two-qubit gate)] determines how many such operations can be successfully performed. Here, we consider the unmodulated (uncontrolled) scenario where all qubits have identical coupling to the bath. Then the decoherence factor is

$$J_\psi(t; N) = c_\psi(N) J(t), \qquad (13.71)$$

where $c_\psi(N)$ is the state-dependent scalability and $J(t)$ is the single-qubit decoherence factor.

In the Markovian domain, the decoherence rate is constant, $J(t) = t\gamma$, and, correspondingly, the coherence time satisfies

$$t_{\mathrm{coh}}(N)|_{\mathrm{Markov}} = \frac{e_{\mathrm{threshold}}}{c_\psi(N)\gamma}. \tag{13.72}$$

For large N, the fidelity threshold (13.70) is reached in accordance with (13.62), after much shorter time than the correlation (memory) time of the bath, t_c, so that the non-Markov dependence of $J(t)$ starts playing a role.

For still larger N, this threshold will be reached within the extremely short Zeno timescale, t_{Zeno}, in which the decoherence factor scales quadratically with t, that is,

$$J_{\mathrm{Zeno}}(t) = At^2, \tag{13.73}$$

and the decoherence rate scales linearly with t. For sufficiently large N, such that

$$c_\psi(N) \geq \frac{e_{\mathrm{threshold}}}{t_{\mathrm{Zeno}}^2 A}, \tag{13.74}$$

the coherence time invariably falls within the Zeno time regime,

$$t_{\mathrm{coh}}(N)|_{\mathrm{Zeno}} = \frac{e_{\mathrm{threshold}}}{\sqrt{c_\psi(N)A}}. \tag{13.75}$$

For an N-qubit GHZ state wherein the qubits decohere independently, without cross-decoherence, $c_\psi(N) = N$. The Zeno coherence time then scales as the *inverse square root* of N. This paradoxical result expresses the *inevitable improvement* of decoherence scaling for sufficiently large systems, as a result of non-Markovian effects.

13.3.3 Dynamically Modified Decoherence Scaling

To understand how modulations affect decoherence scaling with N, let us consider two limits of the general scenario:

(A) For *locally decohering* qubits, in the absence of a common bath, modulations cannot induce cross decoherence. Therefore, they cannot change the scaling law $\gamma_\psi(t; N) \propto N$, which applies to many multipartite entangled states (Table 13.1). Nevertheless, modulations can still improve the infidelity rate by decreasing the overlap of the bath-coupling spectrum and the modulation spectrum for each qubit.

(B) Under collective decoherence of qubits coupled to a common bath, the spectral distinction of N different local modulations can completely eliminate cross-decoherence and effectively render the system similar to individually (locally) decohering qubits, for which decoherence scales linearly with N.

Such local modulations may be advantageous for suppressing the relaxation of a bright state, or the dephasing of a GHZ state, by reducing their infidelity-rate scaling by a factor of N.

Enhanced cross-decoherence can also drastically improve the coherence time by rendering certain states completely dark: without modulation, partially dark states, because of unequal coupling strengths of individual qubits to the bath, have infidelity rate that scales as N. Then the modulations can be chosen to compensate their unequal coupling strengths. Such modulations can impose destructive interference of the decoherence channels, thereby rendering the states completely dark.

To assess the effectiveness of local dynamic modulations, the relevant timescales must be considered. The longest one is the coherence time of a single-qubit without modulation, $t_{\mathrm{coh}}(1) \simeq 1/\gamma_\psi$. Another is the correlation (memory) time of the bath, t_{c}. In the N-qubit case, the fundamental weak-coupling assumption amounts to $t_{\mathrm{c}} \ll t_{\mathrm{coh}}(N)$. Even if the weak-coupling assumption does not hold for a large (unmodulated) system, modulations that drastically increase the coherence time by eliminating cross decoherence can make the system obey the weak-coupling regime for which the analysis holds.

Only modulations faster than the bath (noise) correlation time can affect the single-qubit decoherence factors $J_{jj}(t)$. Yet a crucial insight transpires: If each qubit decoheres individually (cross-decoherence is eliminated), then the (unmodulated) N-particle coherence time is irrelevant to the effectiveness of control and so is the infidelity rate (13.63). The longest modulation period, τ, is unaffected by scaling. It must only be shorter than t_{c}, but can be much longer than $t_{\mathrm{coh}}(1)/N$.

The conditions for an increase of the coherence time of entangled multivariate systems by modulations with a period τ are thus:

$$t_{\mathrm{coh}}(N) \underbrace{<}_{a} \tau \underbrace{<}_{b} t_{\mathrm{c}} \underbrace{<}_{c} t_{\mathrm{coh}}(1) \underbrace{<}_{d} t_{\mathrm{coh}}^{\mathrm{mod}}(N). \qquad (13.76)$$

Here, (a) is the striking result that modulation can be slower than any coherence time of the system, (b) is the main requirement for effective modulation, (c) is the single-qubit weak-coupling assumption, and (d) determines the possibility of attaining (via control modulations) multipartite coherence time $t_{\mathrm{coh}}^{\mathrm{mod}}(N)$ that *is longer than the single-qubit coherence time*.

13.3.4 Application to an Ion Trap

The foregoing analysis is applicable to a variety of experimental setups. For example, consider a string of ions in a linear trap, in which the qubits are encoded by two

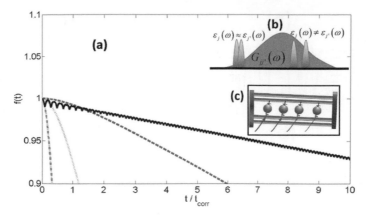

Figure 13.7 (a) Fidelity as a function of time for entangled states undergoing collective relaxation under different modulations in comparison to that of a single qubit without modulation (dashed): bright-state, $N = 10$ without modulation (dot-dashed); GHZ with $N = 10$ without modulation (dotted); bright-state with $N = 10$ under asymmetric modulation (solid). (b) The overlap of the bath-coupling spectrum $G_{jj'}(\omega)$ (broad peak) and the modulation spectra $\epsilon_{t,j}(\omega)$ and $\epsilon_{t,j'}(\omega)$ (narrow peaks) determines the decoherence rate for qubits j and j'. If the qubits undergo asymmetric modulations such that $\epsilon_j(\omega) \neq \epsilon_{j'}(\omega)$, their lack of overlap eliminates cross-decoherence. (c) Individual ions in a trap driven by control fields with different amplitudes. Their asymmetric Rabi flippings eliminate cross-decoherence. (Adapted from Kurizki, 2013.)

internal states of each ion ($|g(e)\rangle_j$) (Fig. 13.7). Fluctuating magnetic fields give rise to a collective fluctuation of qubit energies and thereby to cross-dephasing. Modulations that locally address each ion ($j = 1, \ldots, N$) may be realized by laser fields whose amplitude varies spatially and differently affects the jth "carrier" frequency $\omega_j(t)$ of the transition $|g\rangle_j \leftrightarrow |e\rangle_j$: a gradient of the laser field amplitude along the ion chain would suffice to achieve this.

A local modulation may reduce the decoherence rate of each qubit by at least an order of magnitude, depending on the noise spectrum and the available modulation frequencies. Local modulations may also eliminate cross-dephasing between ions, thereby increasing the coherence time of bright or GHZ states by a factor of N.

Time-dependent decoherence rates may exhibit oscillations whose frequency and amplitude are controllable via the *local modulations* applied on the individual qubits and thus *resuscitate their entanglement* at specific desired times.

To demonstrate these effects, we consider two limiting cases: (a) distant particles coupled to separate baths [$\Phi_{T,jj'}(t) = 0$, $j \neq j'$], such that an initially entangled state does not experience cross-decoherence [Fig. 13.8(a)] and (b) adjacent particles coupled to the same bath, where *local modulations* may control their time-dependent cross-decoherence [Fig. 13.8(c)].

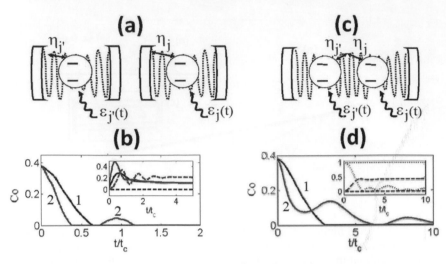

Figure 13.8 (a) Two locally modulated atoms in separated cavities, coupled to the cavity modes. (b) Concurrence as a function of time. Unmodulated (curve 1) and modulated (curve 2) cases, with no cross decoherence. Inset: decoherence factor magnitude $J(t)$ (solid) and asymmetry $\delta J(t)$ (dashed) as a function of time. Here η_j and Ξ_j are taken to be equal for the two TLS, thus having no asymmetry in the unmodulated case. The results are renormalized such that the long-time limit of $J(t)$ is equal in all cases to that of the unmodulated case. This is achieved by adjusting the coupling strength η for each scenario. Parameters: $a = 0.75$, $\omega_a = 20/t_c$, $\omega_0 = 0.8\omega_a$, $\beta = 1/\omega_a$, $\gamma_{unmod} = 0.032/t_c$, $\gamma_{mod} = 2.65\gamma_{unmod}$, $\tau_{1,2}/t_c = \{0.175, 0.1\}$, and $\phi_{1,2} = \pi/10$. (c) Two locally modulated atoms in the same cavity, coupled to the cavity modes. (d) Concurrence as a function of time. Unmodulated (curve 1) and modulated (curve 2) cases, with cross decoherence. Inset: cross-decoherence (dotted) and asymmetry (dashed) as a function of time. Werner-state concurrence dies faster due to this cross-decoherence. Different modulations on each TLS result in the suppression of cross decoherence and the increase of asymmetry, both of which are beneficial for entanglement preservation. As the two modulation frequencies are close to one another, we may periodically kill and resuscitate the concurrence even on Markovian timescales ($t \gg t_c$). Parameters: $a = 0.75$, $\omega_a = 20/t_c$, $\omega_0 = \omega_a$, $\beta = 10/\omega_a$, $\gamma_{unmod} = 0.032/t_c$, $\gamma_{mod} = 1.1\gamma_{unmod}$, $\tau_{1,2}/t_c = 0.175, 0.1$, $\phi_{1,2} = \pi/10$, and $J(t \gg t_c) = 0.28/t_c$. (Adapted from Kurizki, 2013.)

Both the decoherence rate magnitude and asymmetry exhibit controlled oscillations on non-Markovian timescales [Fig. 13.8(b)–inset]. This can be observed by assuming the most general quasiperiodic modulation,

$$\epsilon_j(t) = \sum_k \epsilon_{j,k} e^{-i\omega_{j,k}t},
\tag{13.77}$$

where $\omega_{j,k}$ are arbitrary discrete frequencies with minimal spectral distance $\Omega = \min\{\omega_{j,k} - \omega_{j',k'}\}$ for all $j \neq j'$, $k \neq k'$. In this case, the modulation spectrum,

$\tilde{F}_{t,jj}(\omega)$, is a sum of terms peaked at $\omega = \omega_{j,k}$ that oscillate on non-Markovian timescales.

In the limit of weak pulses, whose area $|\phi_j| \ll \pi$, (13.38) yields $\epsilon_{t,j}(\omega) \cong \epsilon_{t,j}\delta(\omega - \delta_j)$, where $\delta_j = \phi_j/\tau_j$ is the effective spectral shift caused by the pulses.

13.4 Bell-State Entanglement and Decoherence Control

13.4.1 Concurrence and Fidelity

We here analyze the case of bipartite (two-qubit) systems, $N = 2$. In this case, entanglement is quantified by the concurrence Co,

$$Co = \max\{0, \sqrt{\lambda_1} - \sqrt{\lambda_2} - \sqrt{\lambda_3} - \sqrt{\lambda_4}\}, \tag{13.78}$$

where λ_i are the positive eigenvalues, in descending order, of the matrix

$$\boldsymbol{\mu} = \rho(\sigma_y \otimes \sigma_y)\rho^*(\sigma_y \otimes \sigma_y). \tag{13.79}$$

Here σ_y is the Pauli matrix Y, and the asterisk denotes the complex conjugation in the energy basis $\{|gg\rangle, |ge\rangle, |eg\rangle, |ee\rangle\}$.

Dynamical modulations may affect the following decoherence parameters.

(i) *Decoherence magnitude*

$$J_{e(g)}(t) = \sum_{j=1,2} |J_{jj}^{e(g)}(t)|. \tag{13.80}$$

The case of $J_{e(g)}(t) = 0$ corresponds to the absence of decoherence, with complete preservation of the quantum information. The decoherence rate magnitude, being the sum of all individual qubit decoherence rates, is best suppressed by single-qubit decoherence control, that is, by reducing the overlap of the system–bath coupling and modulation spectra.

(ii) *Decoherence asymmetry*

$$\delta J(t) = \frac{|J_{11}^{e}(t) - J_{22}^{e}(t)|}{J_e(t)} \tag{13.81}$$

expresses the difference between the decoherence factors of two particles. This asymmetry is best controlled by differently modulating the (two) TLS. One can increase the asymmetry by enhancing the decoherence of one qubit and reducing that of the other, or decrease this asymmetry by equating the spectral overlap of the bath coupling and modulation spectra for both qubits.

(iii) *Cross-decoherence*

$$J_x(t) = \frac{|J_{12}^{e}(t) + J_{21}^{e}(t)|}{J_e(t)} \tag{13.82}$$

expresses the "crosstalk" between the two particles via the bath modes. Here $J_x(t) = 1$ denotes maximal cross-decoherence and $J_x(t) = 0$ denotes completely separable, uncorrelated systems. Cross-decoherence can be eliminated by choosing for each qubit such modulations that the two qubits are effectively coupled to different bath modes.

(iv) The effective dimensionless temperature is defined as

$$\tilde{T}(t) = 1/\ln[J_e(t)/J_g(t)], \tag{13.83}$$

where, for unmodulated memoryless Markovian identical baths, $\tilde{T} = k_B T/(\hbar \omega_a)$. For initial Bell singlet and triplet states,

$$|\Psi(0)\rangle = \frac{1}{\sqrt{2}}(|g\rangle_1|n\rangle_2 \pm |n\rangle_1|g\rangle_2), \tag{13.84}$$

which do not experience cross-decoherence but only different local decoherence rates,

$$\alpha_{1(2)}(t) = \frac{1}{\sqrt{2}}e^{-J_{1(2)}(t)}. \tag{13.85}$$

We find

$$f_p(t) = [e^{-2J_1(t)} + e^{-2J_2(t)}]/2,$$
$$f_c(t) = [1 + Co(t)]/2, \tag{13.86}$$

where the concurrence $Co(t) = \mathrm{sech}(J_1(t) - J_2(t))$.

The effective temperature is mainly governed by the bath temperature. However, when ultrafast modulations have very high spectral components, they can strongly affect the ratio of upward (governed by J_{jj}^g) and downward (governed by J_{jj}^e) transitions. This effect arises due to the RWA breakdown, which renders negative-frequency components, $\omega_{j,k} + \omega_a < 0$, accessible by modulations.

For an initial Bell-state $\bar{\rho}_l(0) = |\Psi_l\rangle \langle \Psi_l|$, where $l = 1 \ldots 4$, one finds the fidelity as:

$$f_l(t) = \langle \Psi_l| \bar{\rho}_l(t) |\Psi_l\rangle = \cos(\phi_{\pm}(t))\mathrm{Re}\left[e^{i\phi_{\pm}(t)} \left(1 - \frac{1}{2}\sum_{jj'} J_{jj',l}(t) \right) \right]$$
$$(j, j' = 1, 2), \tag{13.87}$$

expressed in terms of

$$\phi_{\pm}(t) = [\phi_1(t) \pm \phi_2(t)]/2, \tag{13.88}$$

where ϕ_+ corresponds to Bell states with $l = 1, 3$ and ϕ_- to $l = 2, 4$,

$$\phi_j(t) = 2\int_0^t d\tau V_{0,j}(\tau), \tag{13.89}$$

$V_{0,j}(t)$ is the amplitude of the resonant field applied to qubit j, and

$$J_{jj',l}(t) = 2\pi \int_{-\infty}^{\infty} d\omega \, G_{jj'}(\omega) \tilde{F}_{t,jj',l}(\omega). \qquad (13.90)$$

The dephasing matrix elements (13.90) are, as before, the convolution of the cross-dephasing correlation spectra,

$$G_{jj'}(\omega) = (2\pi)^{-1} \int_{-\infty}^{\infty} dt \, \Phi_{jj'}(t) e^{i\omega t}, \qquad (13.91)$$

and the driving (modulating) field spectral densities,

$$\tilde{F}_{t,jj,l}(\omega) = |\epsilon_{t,j}(\omega)|^2, \qquad (13.92)$$

$$\tilde{F}_{t,jj',3}(\omega) = -\tilde{F}_{t,jj',1}(\omega) = \epsilon_{t,j}^*(\omega)\epsilon_{t,j'}^*(\omega), \qquad (13.93)$$

$$\tilde{F}_{t,jj',4}(\omega) = -\tilde{F}_{t,jj',2}(\omega) = \epsilon_{t,j}(\omega)\epsilon_{t,j'}^*(\omega). \qquad (13.94)$$

Expressions (13.87)–(13.94) *minimize the Bell-state fidelity losses*, for *any* dephasing time-correlations and *arbitrary* driving fields.

A driving (modulation) scheme, aimed at preserving an initial quantum state, requires us to equate the modified dephasing and cross-dephasing rates of all qubits, $J_{jj',l}(t) = J(t)$. However, only the singlet Bell state is then fully preserved (i.e., $f_2(t) = 1$ for all t), whereas the fidelity of the triplet Bell state $f_1(t)$ is lowered.

13.4.2 Bell-State Disentanglement and Resuscitation

We here analyze the disentanglement of the initial bipartite fully entangled Bell states,

$$|\Phi_\pm\rangle = (1/\sqrt{2})(|e\rangle_1|e\rangle_2 \pm |g\rangle_1|g\rangle_2),$$
$$|\Psi_\pm\rangle = (1/\sqrt{2})(|e\rangle_1|g\rangle_2 \pm |g\rangle_1|e\rangle_2). \qquad (13.95)$$

For initial Bell singlet and triplet states $|\Psi_\pm(0)\rangle$, provided they do not undergo cross-decoherence but only different local decoherence rates, we obtain:

$$f_p(t) = \frac{e^{-2J_1(t)} + e^{-2J_2(t)}}{2}, \qquad (13.96)$$

$$f_c(t) = \frac{1 + Co(t)}{2} = \frac{1}{2} + \frac{e^{-\delta J(t)}}{1 + e^{-2\delta J(t)}}, \qquad (13.97)$$

where $Co(t)$ is the concurrence, $J_{1(2)}$ are the local decay factors of the $j = 1, 2$ qubit excited states, and $\delta J(t) = J_1(t) - J_2(t)$. Decoherence is then the fidelity loss (infidelity), caused by $\delta J(t)$, the decay-factor *asymmetry*.

The long-time concurrences of the Bell states for zero effective temperature are found to be

$$Co(|\Phi_\pm\rangle)|_{\mu=0} = e^{-\delta J}\left[1 - \sqrt{(1 - e^{-\delta J(1-J_x)})(1 - e^{-\delta J(1+J_x)})}\right],$$

$$Co(|\Psi_\pm\rangle)|_{J_x=0} = e^{-\delta J(1\pm J_x)}. \tag{13.98}$$

Thus, an initial (pure) Bell state disentangles exponentially in time in zero-temperature baths but does not vanish abruptly. By contrast, the coupling of the two qubits to a finite-temperature bath results in *entanglement sudden death* (ESD). Namely, the concurrence vanishes at a finite time, which we shall dub the time of death (TOD).

Equation (13.98) implies that the TOD can be postponed by modulations that (i) reduce the decoherence magnitude, (ii) increase the asymmetry for $|\Phi_\pm\rangle$, (iii) reduce (increase) the cross decoherence for $|\Psi_+\rangle$ ($|\Psi_-\rangle$), or (iv) decrease the effective temperature.

Disentanglement is best avoided by the Bell singlet state, which, however, is not always experimentally accessible. In such a case, we may resort to the $|\Phi_\pm\rangle$ states and dynamically modulate each qubit so as to eliminate the cross-decoherence and maximize the decoherence asymmetry, thereby achieving maximal entanglement preservation.

Local modulation may control not only the magnitude of disentanglement but also its period of oscillation, and thus may give rise to entanglement sudden death with controlled partial resuscitation (ESD with CPR). This can be shown, for example, for the initial partially entangled Werner states

$$\rho(0) = \frac{1}{3}(2|\Psi_+\rangle\langle\Psi_+| + (1 - a)|gg\rangle\langle gg| + a|ee\rangle\langle ee|), \tag{13.99}$$

where $0 \le a \le 1$. The numerical solutions of the modified Bloch equations on non-Markovian short timescales, and local (individual-qubit) impulsive-phase modulation,

$$\epsilon_j(t) = e^{\phi_j[t/\tau_j]}, \tag{13.100}$$

are shown in Figure 13.8 for finite-temperature Lorentzian bath-coupling spectra

$$G_{0,jj}(\omega) = \eta_j \frac{\Gamma_{B,j}^2}{\Gamma_{B,j}^2 + (\omega - \omega_0)^2}, \tag{13.101}$$

where η_j is the coupling strength of the jth TLS to the bath, ω_0 is the center of the Lorentzian, $\Gamma_{B,j} = 1/t_{c,j}$, and $G_{0,jj'}(\omega) = 0 \ \forall j \ne j'$.

Figure 13.8(a) presents the case of two atoms in remote cavities with no cross-decoherence. While the concurrence in the unmodulated case with zero asymmetry rapidly dies out, the non-Markovian oscillation of the asymmetry caused by the modulation results in ESD with CPR [Fig. 13.8(b)]. The "rise of concurrence from the dead" can only occur at a time comparable to the correlation time of the bath t_c, allowing the entanglement to be "transferred" to the bath modes and back with partial fidelity. The oscillations of the asymmetry facilitate the return of entanglement back to the two TLS. Notably, ESD with CPR is attainable by *local modulations*; although the two TLS are neither coupled through the control fields nor through the bath modes. This control of entanglement oscillations via local modulations is only enabled by *non-Markovian* multipartite decoherence control.

The case of systems experiencing cross-decoherence due to coupling to the same bath is shown in Figure 13.8(c). This case demonstrates long-time controlled oscillations that result in ESD with CPR in the Markovian limit. To investigate this case, we set the bath coupling responsible for cross-decoherence to be

$$G_{0,jj'}(\omega) = \sqrt{G_{0,jj}(\omega)G_{0,j'j'}(\omega)}. \tag{13.102}$$

The cross-decoherence (13.82) may vanish under quasiperiodic modulation in the long-time limit, $\Omega t \gg 1$, where the cross-modulation spectrum, $\tilde{F}_{t,jj'}(\omega)$, becomes a series of $\delta(\omega_{j,k} - \omega_{j',k'})$. This long-time limit is reached only after the inverse of the minimal spectral distance of the two modulations at $t \gg 1/\min|\omega_{j,k} - \omega_{j'\neq j,k'}|$, thus enabling a long oscillatory behavior of the cross-decoherence [Fig. 13.8(d)–inset].

13.5 Discussion

We have formulated a universal protocol for dynamical decoherence (dephasing) control during all stages of quantum-information processing, namely, storage and single- and two-qubit gate operations. This protocol amounts to controlling continuously all the qubits all the time, whether they are involved in the computation or not, and tailoring specific gate and control fields that optimally reduce the dephasing. This counterintuitive protocol has an advantage over others in that it increases the fidelity of the operation required, whether storage, manipulation, or computation, despite the fact that it requires longer duration.

The protocol presented here is aimed at protecting gate operations in multiqubit systems via dynamical modulation. Hence, our requirements are more complex than those concerned only with storage, which requires operating a single qubit at a fixed optimum point.

Relaxation or decoherence of sparsely distributed quantum systems (spins, atoms, excitons, quantum dots, etc.) that are in contact with a thermal bath are usually viewed as local, single-body processes. The opposite limit is the *collective* relaxation or decoherence of a (*localized*) many-body system, which typically is more rapid than their one-body counterparts. Yet, the possibility of *dynamically* creating a *decoherence-free subspace* has been shown to exist for three-level particles coupled to a common bath.

In general, *nonlocal* disentanglement or decoherence of many-body (multipartite) systems may occur on a timescale much shorter than the time for either body to undergo local decoherence, but much longer than the time each particle takes to disentangle from its environment. As shown here, disentanglement of multipartite systems can be dynamically controlled on the *same non-Markovian timescales, as a single particle*, upon eliminating cross-decoherence.

Our analysis of multiple, field-driven, qubits has resulted in universal formulae (13.33) for their coupling to a zero-temperature bath, and (13.87) for Bell-state preservation under local dephasing. This analysis allows one to come up with an optimal choice between *global and local control*, based on the consideration that *the maximal suppression of decoherence is not necessarily the best one*, for the following reasons:

- Local modulation can effectively *decorrelate* the dephasing of multiple qubits, resulting in equal dephasing rates for all states. For two qubits, the singlet and triplet Bell states acquire the same dynamically modified dephasing rate. Such local modulation is beneficial compared to the standard global dynamical decoupling (DD) by π-phase flips if both states are used (intermittently) for information transmission or storage.
- For different couplings of the qubits to a zero-temperature bath, it is preferable to preserve any initial state by local modulation that can reduce not only the decay but also the mixing with other states, a feature that is unavailable under global modulation. Local modulation that eliminates cross-decoherence can increase the fidelity more than its global modulation counterpart. Thus, for two qubits local modulation has been shown to better preserve an initial Bell state, whether a singlet or a triplet, compared to global π-phase "parity kicks."

In the context of *one-way quantum computing*, fidelity protection is more important than preventing disentanglement, since maintaining the *specific initial state* is essential for the computation.

Fidelity in such systems can be protected by fields applied to the system that can maintain its coherence, as long as their effects are faster than the correlation (memory) time of the bath, that is, they act on non-Markovian timescales. We have shown that by locally (specifically) modulating each qubit, one can drastically suppress the decoherence and its scaling with the number of qubits through modifying the strength and number of effective decoherence channels. The surprising result is that a modulation that acts slower than the multipartite coherence time can still restore coherence and entanglement, if it is within the memory time of the bath, irrespective of the number of qubits involved.

The entanglement sudden death with controlled partial resuscitation (ESD-with-CPR) scenario discussed here has several aspects: (a) non-Markovian timescale behavior, (b) asymmetry between the two decoherence channels, (c) cross-decoherence of the two qubits. The universal dynamical control of disentanglement uses all these aspects, by locally modulating entangled systems to control frequencies and amplitudes of entanglement oscillations.

14

Dynamical Control of Quantum State Transfer in Hybrid Systems

Quantum information (QI) manipulations are schematically divided into the following stages: writing-in, processing, storage, and reading-out. Some systems are better suited for writing-in or reading-out than for processing or storage. This has prompted the suggestion of *hybrid*, composite quantum systems: quantum processing operations are rapidly performed and efficiently written in qubits susceptible to decoherence (e.g., superconducting qubits); then the QI is transferred to storage qubits resilient to decoherence (e.g., ultracold atoms); then, on demand, transferred back and read out from the same fragile (say, superconducting) qubit. We here examine a strategy for *maximizing* the average fidelity of quantum state transfer in such hybrid systems, from a subspace that is susceptible to decoherence to a robust subspace, by means of appropriate dynamical control fields.

14.1 Optimized Control of Transfer between Multipartite Open-System Subspaces

We resort to the general BOTOC approach (detailed in Ch. 12) to the control of *arbitrary* quantum operations in multidimensional open systems. The desired operation is disturbed via operators $\hat{S} \otimes \hat{B}$, where $\hat{S}$ is the traceless operator of the system, which is hereafter assumed to consist of qubits, and $\hat{B}$ is the bath operator. One can choose controls to maximize the operation fidelity by acting on the multiqubit system according to the following protocol, which holds to second order in the system–bath coupling:

(i) The control (modulation) transforms the system–bath coupling operators to the time-dependent form $\hat{S}(t) \otimes \hat{B}(t)$ in the interaction picture, via the set of time-dependent coefficients $\epsilon_i(t)$ that define a notation in the Pauli basis $\hat{\sigma}_i$,

$$\hat{S}(t) = \sum_i \epsilon_i(t)\hat{\sigma}_i. \tag{14.1}$$

(ii) We next write the time-independent *gradient-control matrix* (12.31), describing how the fidelity score changes for each pair of the Paui basis operators,

$$\Xi_{ij} \equiv \overline{\langle\psi|[\sigma_i, \sigma_j|\psi\rangle\langle\psi|]|\psi\rangle}, \tag{14.2}$$

the overline being an average over all possible initial states.

(iii) Using the matrix Ξ whose elements are Ξ_{ij}, one arrives at the following expression for the average fidelity of the desired operation (to second-order accuracy in the system–bath coupling):

$$\bar{f}(t) = 1 - t\int_{-\infty}^{\infty} d\omega \, \mathrm{Tr}[\boldsymbol{G}(\omega)\boldsymbol{F}_t(\omega)], \tag{14.3}$$

where $\boldsymbol{G}(\omega)$ is the bath-coupling (-response) spectral matrix defined in (12.32) and the modulation (control) spectral matrix $\boldsymbol{F}_t(\omega)$ is defined in (12.37) according to the operation, via the gradient-control matrix Ξ_{ij}.

(iv) The fidelity is maximized by the variational Euler–Lagrange method described in Chapter 12 that minimizes the overlap between $\boldsymbol{G}(\omega)$ and $\boldsymbol{F}_t(\omega)$ under the constraint of a given control energy.

14.2 Optimized State Transfer from Noisy to Quiet Qubits

The general approach outlined above will next be used to optimize the fidelity-versus-speed trade-off in the transfer of quantum states from a fragile (noisy) qubit to a robust (quiescent) qubit. We focus here on the case of two resonant qubits with temporally controlled coupling strength. The free Hamiltonian is then

$$\hat{H}_{\mathrm{S}}(t) = \frac{\hbar\omega_0}{2}(\hat{\sigma}_{z1} + \hat{\sigma}_{z2}) + \hat{H}_{\mathrm{c}}(t),$$
$$\hat{H}_{\mathrm{c}}(t) = \hbar V(t)\hat{\sigma}_{x1} \otimes \hat{\sigma}_{x2}, \tag{14.4}$$

where $\hat{H}_{\mathrm{c}}(t)$ is the two-qubit controlled-interaction Hamiltonian, $V(t)$ being the interaction amplitude (see Fig. 14.1–inset), where the interaction amplitude is adjustable by an external laser field. The system–bath interaction Hamiltonian is taken to represent proper dephasing in the noisy source qubit 1 due to the bath operator $\hat{B}$, whereas the target qubit 2 is quiescent, that is, robust against decoherence,

$$\hat{H}_{\mathrm{I}} = \hbar\hat{S} \otimes \hat{B}(t) = \hbar\hat{\sigma}_{z1} \otimes \hat{B}(t), \tag{14.5}$$

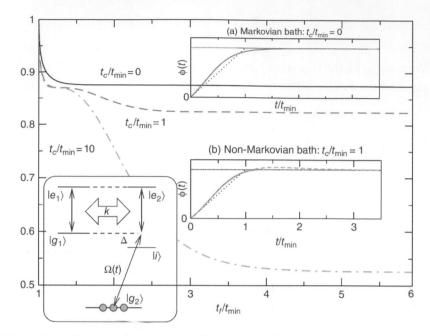

Figure 14.1 Inset – scheme of coupling a "noisy" (source) qubit 1 to a quiescent (target) qubit 2: off-resonant two-photon transfer through $|i\rangle$ corresponds to effective $\sigma_{x1} \otimes \sigma_{x2}$ coupling with a strength $V(t) = \frac{\kappa\Omega(t)}{\Delta}$, $\Omega(t)$ being the Rabi frequency of an external laser field. Main panel: the lowest achievable average infidelity as a function of the transfer time t_f in units of the fastest transfer time $t_{\min}$ at a given transfer energy for various bath memory times: (solid) $t_c = 0$ (Markovian); (dashed) $t_c = t_{\min}$; (dash-dot) $t_c = 10 t_{\min}$. The best solution is not the fastest one even for Markovian baths. For non-Markovian baths ($t_c \gtrsim t_{\min}$), two regions of insensitivity to t_f (plateaux) can be seen. The first plateau is independent of the memory time, whereas the second plateau depends on the memory of the bath. (a) and (b): the transfer phase $\phi(t)$ plotted versus $t/t_{\min}$. The fastest modulation (dotted) with the Markovian optimal modulation (solid) and the non-Markovian optimal modulation (dashed) for Markovian (a) and non-Markovian (b) baths. For the Markovian bath, the optimal modulation transfer "overtakes" the "fastest" transfer (when the information is still "fresh") and slows down. For the non-Markovian bath, optimal modulation attains full transfer [$\phi(t) = \pi/2$] well within the modulation time, but then "overshoots" [$\phi(t) > \pi/2$] and eventually resumes the value $\phi(t) = \pi/2$. (Adapted with permission from Escher et al., 2011. © IOP Publishing. All rights reserved.)

where $\hat{B}(t)$ is the bath operator $\hat{B}$ that evolves under the action of the free bath Hamiltonian $\hat{H}_B$. This model can be generalized to allow for *any asymmetry* between the dephasing (decoherence) of the two qubits.

Since (14.4) and (14.5) conserve the parity of the number of excitations, the two-qubit system can be split into two subsystems

$$\mathcal{O} = \text{span}\{|g_1 e_2\rangle, |e_1 g_2\rangle\}, \quad \mathcal{EV} = \text{span}\{|g_1 g_2\rangle, |e_1 e_2\rangle\}, \tag{14.6}$$

where $\mathcal{O}$ and $\mathcal{E}V$ stand for *odd* and *even* excitation numbers, respectively:

$$
\begin{aligned}
\hat{H}_S + \hat{H}_I &= \hat{H}_\mathcal{O} + \hat{H}_{\mathcal{E}V}, \\
\hat{H}_\mathcal{O} &= \hbar V(t)\hat{\sigma}_x^\mathcal{O} + \hbar\hat{\sigma}_z^\mathcal{O} \otimes \hat{B}(t), \\
\hat{H}_{\mathcal{E}V} &= \hbar\omega_0\hat{\sigma}_z^{\mathcal{E}V} + \hbar V(t)\hat{\sigma}_x^{\mathcal{E}V} + \hbar\hat{\sigma}_z^{\mathcal{E}V} \otimes \hat{B}(t),
\end{aligned} \tag{14.7}
$$

the appropriate Pauli matrices in the $\mathcal{O}$ ($\mathcal{E}V$) subsystems being

$$
\begin{aligned}
\hat{\sigma}_x^\mathcal{O} &= |g_1 e_2\rangle\langle e_1 g_2| + \text{H.c.}, \quad \hat{\sigma}_z^\mathcal{O} = |e_1 g_2\rangle\langle e_1 g_2| - |e_1 g_2\rangle\langle g_1 e_2|, \\
\hat{\sigma}_x^{\mathcal{E}V} &= |g_1 g_2\rangle\langle e_1 e_2| + \text{H.c.}, \quad \hat{\sigma}_z^{\mathcal{E}V} = |e_1 e_2\rangle\langle e_1 e_2| - |g_1 g_2\rangle\langle g_1 g_2|. \tag{14.8}
\end{aligned}
$$

Here, $|g_1\rangle$ ($|g_2\rangle$) and $|e_1\rangle$ ($|e_2\rangle$) are, respectively, the ground and the excited states of the source qubit 1 and the target qubit 2. Thus, we have a resonant and a nonresonant TLS, both coupled to the same dephasing bath, which renders them inseparable. The $V(t)$ control determines the transfer fidelity via the $\mathcal{O}$ subsystem while keeping the $\mathcal{E}V$ subsystem unchanged.

The control function is the accumulated phase

$$
\phi(t) = \int_0^t V(t')dt'. \tag{14.9}
$$

In the absence of decoherence, state transfer from qubit 1 to qubit 2 can be perfectly realized if at the final time, t_f, the phase $\phi(t)$ satisfies $\phi(t_f) = \pi/2$. Then, any initial state of qubit 1 is mapped onto that of qubit 2 (initially in the ground state)

$$
(\alpha|g_1\rangle + \beta|e_1\rangle)|g_2\rangle \rightarrow |g_1\rangle(\alpha|g_2\rangle - i\beta|e_2\rangle), \tag{14.10}
$$

for any normalized probability amplitudes α, β.

Two conflicting error considerations affect the transfer: (i) Because of the interaction between the source qubit 1 and the bath, the longer the information is stored in qubit 1, the lower the transfer fidelity as determined by the evolution of subsystem $\mathcal{O}$. (ii) If, however, the transfer is extremely fast, it may result in population leakage from $|g_1\rangle$ ($|g_2\rangle$) to $|e_1\rangle$ ($|e_2\rangle$), thereby lowering the transfer fidelity. Such leakage occurs in subsystem $\mathcal{E}V$ and signifies the RWA violation: fast driving (modulation) $V(t)$ may incur unwarranted off-resonant transitions under the transfer rate comparable to the level distance of the qubits, ω_0.

Let us first assume that there is negligible leakage due to RWA violation, since the transfer time is much slower than $1/\omega_0$. The control Hamiltonian then acquires the RWA form:

$$
\hat{H}_c(t) \equiv \hbar V(t)(|e_1 g_2\rangle\langle g_1 e_2| + |g_1 e_2\rangle\langle e_1 g_2|). \tag{14.11}
$$

The general expression (14.3) for the average transfer fidelity, completed at t_f, has then the form (Ch. 11, 12)

$$\overline{f(t_f)} = 1 - \int_{-\infty}^{\infty} d\omega\, G(\omega) F_{t_f}(\omega), \tag{14.12}$$

where the driving (modulation) control spectrum is

$$F_t(\omega) = \frac{2}{3} \left| \int_0^t d\tau \cos^2[\phi(\tau)] e^{-i\omega\tau} \right|^2 + \frac{1}{2} \left| \int_0^t d\tau \sin[2\phi(\tau)] e^{-i\omega\tau} \right|^2. \tag{14.13}$$

The problem at hand is to find the transfer that minimizes the average infidelity at time t_f f, $1 - \overline{f}(t_f)$. Obviously, infinitely fast (zero-time) transfer has zero infidelity, but this would require unphysical infinitely strong control. A constraint on the total energy E invested in the transfer process,

$$\hbar \int_0^{t_f} dt [V(t)]^2 = \hbar \int_0^{t_f} dt \left[\frac{d\phi(t)}{dt}\right]^2 = E, \tag{14.14}$$

defines the minimum possible transfer time $t_{\min} = \frac{\pi^2}{4E}$.

The general expressions (14.12)–(14.14) can be illustrated for a non-Markovian Lorentzian bath spectrum, which is the Fourier transform of an exponentially decaying autocorrelation function $\Phi(t) = \frac{\gamma}{t_c} e^{-|t|/t_c}$, t_c being the correlation (memory) time. One might expect the best strategy to be the fastest possible transfer under the energy constraint, which would imply $V(0 \le t \le t_{\min}) = 2E/\pi$. Surprisingly, *slower* transfer ($t_f > t_{\min}$) with an appropriate modulation $\phi(t)$ is shown below to improve the average fidelity even for a *purely Markovian* bath, with negligible correlation (memory) time $t_c/t_{\min} \to 0$, but for baths with long memory times, $t_c \gtrsim t_{\min}$, the improvement turns out to be much larger [Fig. 14.1(a) and (b)].

The analytical solution for the optimal modulation phase $\phi(t)$ is obtainable by the Euler–Lagrange variational method (Ch. 12). For a Markovian bath, the optimal phase ϕ_M evolves according to

$$\frac{d\phi_M(x)}{dx} = \left\{ \frac{\sin^2[2\phi_M(x)]}{2} + \frac{2\cos^4 \phi_M(x)}{3} \right\}^{1/2}, \tag{14.15}$$

where $\phi_M(0) = 0$. Equation (14.15) determines the shape of $\phi_M(x)$ and its "energy" $e_M = \int_0^{\infty} |d\phi_M(x)/dx|^2 dx = 1.038\ldots$ (x and e_M being dimensionless). The Markovian optimal modulation for energy E is then $\phi(t) = \phi_M\left(\frac{Et}{\hbar e_M}\right)$, with an infidelity of $\hbar\gamma e_M^2/E \simeq \hbar\gamma/E$, γ being the dephasing rate of the source qubit 1. The fastest modulation with energy E has an infidelity of $\hbar\gamma \frac{\pi^2}{8E} \simeq 1.2\hbar\gamma/E$. The optimal modulation has typically $\sim 10\%$ less infidelity than the fastest modulation for the same energy.

One can see that the optimal solution starts off faster and then slows down, being overtaken by the "fastest" solution only at $t \approx 0.9 t_{\min}$. The reason for this fidelity increase is that the optimal solution must start off faster, so as to transfer more of the information while it is still nearly untainted by the bath. Eventually it must slow down so as to comply with the energy constraint, the result being that the total transfer time t_f is longer than the fastest time t_{min} for the given energy.

When the memory time of the bath t_c exceeds the minimal transfer time $t_c \gtrsim t_{min}$, a much larger improvement can be achieved. Remarkably, the optimal solution attains full transfer, $\phi(t) = \pi/2$, well within the modulation time. Then, rather than stop at $\phi = \pi/2$, it "overshoots" the transfer, so that $\phi(t) > \pi/2$, and subsequently returns slowly to $\pi/2$. During the "overshooting," the information partially returns from the target (storage) qubit to the source (noisy) qubit but with a negative sign. Thus, similarly to the spin-echo method, the noise now acts in the reverse direction, correcting itself, so that the non-Markov bath effect is undone. The required transfer times are *significantly* larger than t_{min}, yet the result is a substantial *fidelity increase* (up to 50% in Fig. 14.1).

Let us now account for the RWA breakdown. A coupling of the form $\sigma_{x1} \otimes \sigma_{x1}$, as in (14.4), has non-RWA terms of the form $|g_1 g_2\rangle\langle e_1 e_2|$, which do not conserve excitation. Such terms are commonly discarded or considered as a form of leakage. However, there may be a drastic difference between the effect of non-RWA terms in (14.4) and leakage to higher levels: the TLS level separation ω_0 is often orders of magnitude smaller than the separation to a higher level. Hence, leakage to higher levels requires timescales that are orders of magnitude shorter than those that break the RWA.

The RWA in (14.4) breaks down when $t_{min} \lesssim \omega_0^{-1}$. For TLS whose resonance frequency is in the microwave (GHz) or radio frequency (RF; MHz) range, the optimization must account for both the dephasing as in (14.12)–(14.14) and the non-RWA error, when minimizing the infidelity.

If one of these errors dominates, it is the only one to consider. However, if the dephasing and the non-RWA errors are comparable, we find that dephasing of the doubly excited state caused by $\hat{\sigma}_x^{\mathcal{E}V} \otimes \hat{B}$ in (14.7) is a fourth-order effect and hence can be ignored in the present second-order approximation. In this approximation, the system Hamiltonian can be separated into two parts pertaining to odd ($\mathcal{O}$) and even ($\mathcal{E}V$) subsystems, the former subject only to dephasing and the latter only to non-RWA population leakage to the doubly excited level $|e_1 e_2\rangle$:

$$\hat{H}_{\mathcal{O}} = \hbar V(t)\hat{\sigma}_x^{\mathcal{O}} + \hbar\hat{\sigma}_z^{\mathcal{O}} \otimes \hat{B},$$
$$\hat{H}_{\mathcal{E}V} = \hbar\omega_0\hat{\sigma}_z^{\mathcal{E}V} + \hbar V(t)\hat{\sigma}_x^{\mathcal{E}V}. \tag{14.16}$$

The optimization aims at finding a control $V(t)$ that maximizes the transfer fidelity in subsystem $\mathcal{O}$ [as per (14.12)–(14.14)] while concurrently minimizing the double excitation in subsystem $\mathcal{E}V$. The optimal modulation for Markovian

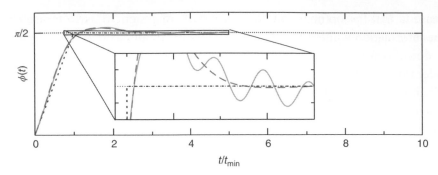

Figure 14.2 The fastest (dotted line) versus the optimal modulation for transfer that complies with the RWA (dashed line) or violates the RWA (solid line) with a final time of $t_f = 5t_{\min}$ for a non-Markovian bath. The level separation ω_0 satisfies $\omega_0 t_{\min} = \pi$, giving $\sim 2\%$ population leakage for the fastest modulation and $\sim 10\%$ loss of fidelity from decoherence. The larger the sinusoidal "wiggling" of the non-RWA transfer is, the shorter is the final time. (Reproduced with permission from Escher et al., 2011. © IOP Publishing. All rights reserved.)

and non-Markovian baths alike in Figure 14.2 shows that the optimal modulation resembles the solution for dephasing in Figure 14.1 but with wiggling throughout the time allowed for the transfer. The longer the transfer time, the smaller the wiggle amplitude. The reason is that one should complete the transfer assuming the RWA so as to minimize the information lost to the bath. Once the transfer is complete and decoherence is minimized, we can use whatever energy is left to bring the "leaked" excitation back from $|e_1 e_2\rangle$ to $|g_1 g_2\rangle$. This can be done by a weak sinusoidal modulation of frequency $2\omega_0$, inducing Rabi coupling between the doubly excited and zero-excited levels of $\mathcal{E}V$. In practice, the weak oscillation is added on top of the population-transfer modulation.

The energy needed to "undo" the non-RWA effect is inversely proportional to the transfer time $E \approx \frac{\hbar |\psi_{ee}|^2}{2t_f}$ ($|\psi_{ee}|^2$ being the population of the doubly excited state $|e_1 e_2\rangle$). Given enough time, the non-RWA effect correction requires negligible energy. If, however, the transfer time is limited, a larger fraction of the energy must be reserved to perform the non-RWA effect correction, resulting in smaller reduction of the bath-induced noise.

14.3 Optimized Control of State Transfer through Noisy Quantum Channels

QI processing and communication require reliable transfer of QI between distant nodes of a network, despite the vulnerability of the process to noise and perturbations. A simple control involves only the source and target (boundary) qubits that are weakly coupled to a chain of spins with identical couplings. Quantum states can then be transmitted with arbitrarily high fidelity at the expense of increasing the transfer time. Yet, even if identical couplings could be ensured, the required slowdown of the transfer would be detrimental, because of omnipresent decoherence.

To overcome this problem, we may optimize the trade-off between fidelity and transfer speed through noisy spin channels. This approach employs temporal modulation of the couplings between the boundary qubits and the channel spins. This modulation is treated as dynamical control of the boundary qubits that are coupled to a fermionic bath that is the source of noise. This modulation aims at realizing an optimal spectral filter that blocks transfer via the channel eigenmodes that are responsible for noise-induced leakage of the QI. We show that under optimal modulation, the fidelity and the speed of transfer can be improved by several orders of magnitude, and the fastest possible transfer is achievable (for a given fidelity).

Let us specifically consider a chain of $N + 2$ spin-1/2 particles with XX interactions between nearest neighbors. The chain and boundary-coupling Hamiltonians, H_0 and H_{bc}, are given by

$$H_0 = \hbar \sum_{i=1}^{N-1} \frac{J_i}{2} (\sigma_{xi}\sigma_{x,i+1} + \sigma_{y,i}\sigma_{y,i+1}),$$

$$H_{bc}(t) = \hbar g(t) \sum_{i \in \{0, N\}} \frac{J_i}{2} (\sigma_{xi}\sigma_{x,i+1} + \sigma_{y,i}\sigma_{y,i+1}). \tag{14.17}$$

Here $\sigma_{x(y)i}$ are the appropriate Pauli matrices, J_i are the corresponding exchange–interaction couplings, and $g(t)$ is the temporally modulated coupling strength. The magnetization-conserving H_0 can be mapped onto a noninteracting fermionic Hamiltonian that has the diagonal, particle-conserving form $H_0 = \sum_{k=1}^{N} \hbar\omega_k b_k^\dagger b_k$, where $b_k^\dagger$ populates a fermionic single-particle eigenstate $|\omega_k\rangle$ of energy ω_k.

The assumption of mirror symmetry of the couplings $J_i = J_{N-i}$ for odd N implies that there is a single nondegenerate, zero-energy fermionic mode in the quantum channel, labeled by $k = z \equiv (N+1)/2$. The two boundary qubits are resonantly coupled to this mode with temporally modulated coupling strength $\tilde{J}_z g(t)$, $\tilde{J}_z$ being the coupling $\tilde{J}_{k=z}$ renormalized by the weight of the zero-mode eigenstate $|\omega_{k=z}\rangle$ at the boundary qubit sites. This resonant coupling (tunneling) is described by the effective Hamiltonian

$$H_S(t) = \hbar \tilde{J}_z s_+(t) b_z + \text{H.c.}, \tag{14.18}$$

where

$$s_\pm(t) = g(t)(c_0^\dagger \pm c_{N+1}^\dagger), \tag{14.19}$$

$c_0^\dagger$ and $c_{N+1}^\dagger$ being the fermionic creation operators at sites 0 and $N+1$, respectively.

The treatment considers these three fermionic modes as a "system" S that interacts with a "bath" B. To this end, we rewrite the total Hamiltonian as

$$H = H_S(t) + H_B + H_{SB}(t), \qquad (14.20)$$

where

$$H_B = \sum_{k \neq z, k=1}^{N} \hbar \omega_k b_k^\dagger b_k \qquad (14.21)$$

and

$$H_{SB}(t) \equiv H_I(t) = \hbar s_+(t) \sum_{k \in k_{odd}} \tilde{J}_k b_k + \hbar s_-(t) \sum_{k \in k_{even}} \tilde{J}_k b_k + \text{H.c.} \qquad (14.22)$$

Here the system–bath interaction Hamiltonian H_I is rewritten in terms of the collective-mode operators associated with sums over the odd (even) bath modes, where $\tilde{J}_k = J_1 \langle 1|\omega_k \rangle$ and the states $|j\rangle = |0 \ldots 01_j 0 \ldots 0\rangle$ span the single-excitation subspace. The first term extends over $k_{odd} = \{1, 3, \ldots, N\}$, provided $k_{odd} \neq z$. The second term extends over $k_{even} = \{2, 4, \ldots, N - 1\}$. This form is amenable to optimal dynamical control of the multipartite system that generalizes single-qubit dynamical control by modulation of the qubit levels.

We next rewrite (14.22) in the interaction picture as

$$H_{SB}^I(t) = \hbar \sum_j S_j(t) \otimes B_j^\dagger(t). \qquad (14.23)$$

We then decompose $H_{SB}^I(t)$ into symmetric and antisymmetric system operators that are coupled to odd- and even-bath modes. The system operators $S_j(t) = \sum_{i=1}^{6} \Omega_{j,i}(t) \hat{v}_i$ are written in terms of the coefficients $\Omega_{j,i}(t)$ in a chosen basis of operators $\hat{v}_i$. We can then write the time-convolutionless second-order solution for the system density matrix $\rho_S(t)$ in the interaction picture (Ch. 11). This solution can then be used to optimize the transfer fidelity.

We wish to transfer $|\psi_0\rangle$, an arbitrary superposition of the lower $|0_0\rangle$ and upper $|1_0\rangle$ states that is initially stored on the 0 qubit, to the $N + 1$ qubit. For an isolated three-level system, perfect state transfer occurs at the time t_f, when the accumulated phase due to the modulation control

$$\phi(t) = \tilde{J}_z \int_0^t g(t') dt' \qquad (14.24)$$

satisfies

$$\phi(t_f) = \pi / \sqrt{2}. \qquad (14.25)$$

In the presence of the system–bath interaction, the state transfer is generally not perfect. The state transfer over time t_f has the average fidelity

$$\bar{f}(t_{\mathrm{f}}) = f_{0,N+1}^2(t_{\mathrm{f}})/6 + f_{0,N+1}(t_{\mathrm{f}})/3 + 1/2, \tag{14.26}$$

which is the state-transfer fidelity averaged over all possible input states $|\psi_0\rangle$. In the interaction picture, the state-transfer probability

$$f_{0,N+1}^2(t_{\mathrm{f}}) = {}_{\mathrm{s}}\langle\psi|\rho_{\mathrm{S}}(t_{\mathrm{f}})|\psi\rangle_{\mathrm{s}}, \tag{14.27}$$

where $|\psi\rangle_{\mathrm{s}} = |1_0\rangle \otimes |0_z 0_{N+1}\rangle_{\mathrm{s}}$ and $|\psi\rangle_{\mathrm{s}} \otimes |\psi\rangle_{\mathrm{B}}$ is the initial state of S+B.

14.3.1 Optimal Filter Design

We resort to modulation as a tool to minimize the infidelity $\zeta(t_{\mathrm{f}}) = 1 - f_{0,N+1}(t_{\mathrm{f}})$ by rendering the overlap between the bath and system spectra as small as possible,

$$\min \zeta(t_{\mathrm{f}}) = \min \operatorname{Re} \int_0^{t_{\mathrm{f}}} dt \int_0^t dt' \sum_{\pm} \Omega_{\pm}(t)\Omega_{\pm}(t')\Phi_{\pm}(t - t'). \tag{14.28}$$

Here $\Phi_{\pm}(t) = \sum_{k\in k\mathrm{odd(even)}} |\eta_k|^2 e^{-i\omega_k t}$ are the correlation functions associated with the odd (even) bath modes, respectively. The corresponding dynamical-control functions are $\Omega_+(t) = g(t)\cos[\sqrt{2}\phi(t)]$ and $\Omega_-(t) = g(t)$. In the energy domain, (14.28) has the form

$$\zeta(t_{\mathrm{f}}) = \int_{-\infty}^{\infty} d\omega \sum_{\pm} F_{t_{\mathrm{f}},\pm}(\omega)G_{\pm}(\omega). \tag{14.29}$$

Here,

$$G_{\pm}(\omega) = \frac{1}{2\pi} \int_{-\infty}^{\infty} dt\, \Phi_{\pm}(t)e^{i\omega t} \tag{14.30}$$

are the bath-spectrum functions, corresponding to odd (even) parity modes, and

$$F_{t_{\mathrm{f}},\pm}(\omega) = \frac{1}{2\pi} \left| \int_0^{t_{\mathrm{f}}} dt\, \Omega_{\pm}(t)e^{i\omega t} \right|^2 \tag{14.31}$$

are the filter-spectrum functions, which can be designed by the modulation control. The optimal modulation minimizes the overlap integrals of $G_{\pm}(\omega)$ and $F_{t_{\mathrm{f}},\pm}(\omega)$ for a given t_{f} by the variational Euler–Lagrange (EL) method.

We assume the channel to be symmetric with respect to the source and target qubits and the number of eigenvalues to be odd (Fig. 14.3 – top inset). Under these assumptions, the central eigenvalue is invariant under noise on the couplings, provided a gap exists between this eigenvalue and the adjacent ones, that is, they are not strongly mixed by the noise, so as not to overlap. We also assume that the discreteness of the bath spectrum of the quantum channel is smoothed out by the noise. Then, considering the central eigenvalue as part of the system spectrum, a common characteristic of $G_{\pm}(\omega)$ is to have a central gap [Fig. 14.3(a)].

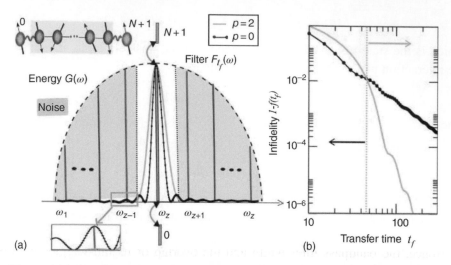

Figure 14.3 Top inset: Spin-channel chain with boundary-controlled couplings. The two boundary spins 0 and $N + 1$ are resonantly coupled to the chain by the mode z. (a) Spectrum of the effective system (rectangular bars) that interacts with the bath modes k. Dashed contour: noise spectrum described by the Wigner-semicircle (maximal disorder) line shape with a central gap around ω_z. In this gap, the optimal spectral filters $F_{t_f,-}(\omega)$ generated by boundary control with $g_p(t)$ [$p = 0$ (dotted), $p = 2$ (thin)] are shown. Bottom inset: a zoom of the tails of the filter spectrum that protect the state transfer against a general noisy bath with a central gap. (b) Infidelity as a function of transfer time t_f under optimal control (filter) with $p = 0$ (dotted) and $p = 2$ (thin). (Adapted from Zwick et al., 2014.)

In order to minimize the overlap between $G_\pm(\omega)$ and $F_{t_f,\pm}(\omega)$ for general gapped baths, and thereby the transfer infidelity (4.198), we design a bandpass filter centered on the gap. Since $G_-(\omega)$ has a narrower gap than $G_+(\omega)$, we optimize the filter $F_{t_f,-}(\omega)$ by the variational EL method (Ch. 12). We seek a narrow bandpass filter that blocks the higher frequencies. This amounts to maximizing $F_{t_f,-}(\tau) = \int_{-\infty}^{\infty} F_{t_f,-}(\omega)e^{-i\omega\tau}d\omega$ subject to the variational EL equation under the constraint of accumulated phase $\phi(t_f)$ and energy

$$E(t_f) = \hbar\tilde{J}_z^2 \int_0^{t_f} |g(t)|^2 dt \geq \hbar\phi^2(t_f)/t_f. \qquad (14.32)$$

The optimal solutions are found to be

$$g_p(t) = g_M \sin^p(\pi t/t_f). \qquad (14.33)$$

Here $p = 0, 1, 2$, where $p = 0$ corresponds to static control and $p = 1, 2$ to dynamical control. Then, (14.24) and (14.25) yield the transfer time in the form

$$t_f = \frac{c_p}{\tilde{J}_z g_M}, \qquad (14.34)$$

where

$$c_0 = \frac{\pi}{\sqrt{2}}, \quad c_1 = \frac{\pi^2}{2\sqrt{2}}, \quad c_2 = \sqrt{2}\pi. \tag{14.35}$$

Note that t_f is inversely proportional to $g_M = \max\{g_p(t)\}$, as one could expect.

The solutions in (14.33) result in sinc-like bandpass filter functions $F_{t_f,-}(\omega)$ around 0 that become narrower as t_f increases. For $p = 0$, which satisfies the minimal-energy condition $E_{\min}(t_0) = \hbar\pi^2/(2t_0)$, the corresponding filter is the narrowest around 0, but it has many wiggles on the filter tails [Fig. 14.3(a)]. These tails overlap with bath-mode frequencies that hamper the transfer. By contrast, the $p = 1, 2$ bandpass filters are wider (for the same t_f) and require more energy, $E_1 = \frac{\pi^2}{8}E_{\min}$ and $E_2 = \frac{3}{2}E_{\min}$, respectively. On the other hand, these filters are flatter and have smaller values throughout the bath-energy domain.

Hence, the bandpass filter width and the overlap of its tail-wiggles with bath energies determine which modulations $g_p(t)$ are optimal (Fig. 14.3). The shorter the t_f, the lower is the p value that yields the highest fidelity because the central peak of the filter that produces the dominant overlap with the bath spectrum is then the narrowest. However, as t_f increases, larger p values give rise to higher fidelity because the tails of the filter then make the dominant contribution to the overlap. As shown in Figure 14.3(b), the filter with $p = 2$ can reduce the transfer infidelity *by orders of magnitude* in a noisy gapped bath bounded by the Wigner semicircle, which represents fully randomized channels.

In the weak-coupling regime ($g_M \ll 1$), the transfer time is found to be

$$t_p \approx \frac{c_p\sqrt{N}}{\sqrt{2}g_M J}. \tag{14.36}$$

The infidelity then decreases with g_M according to a power law, aside from oscillations due to the discrete nature of the bath spectrum. The best trade-off between speed and fidelity in this regime is given by the optimal modulation with $p = 2$ [Fig. 14.4(a)].

The validity of this approach can be extended *to strong couplings* g_M, under the optimal filtering process that minimizes the transfer infidelity: Since the bandpass filter width increases inversely with t_f, in the strong coupling regime ($g_M \sim 1$), the filter may overlap the bath energies closest to the central (zero) eigenvalue but still block the higher bath energies, which are the most detrimental for the state transfer. Then, the participation of the closest bath energies yields a transfer time

$$t_p \approx \frac{c_p N}{\sqrt{2\pi} J}. \tag{14.37}$$

The minimal infidelity value that we denote as $g_{M_p}^{\text{opt}}$ depends on p [Fig. 14.4(a)]. The infidelity dip corresponds to a stronger filtering-out (suppression) of the higher

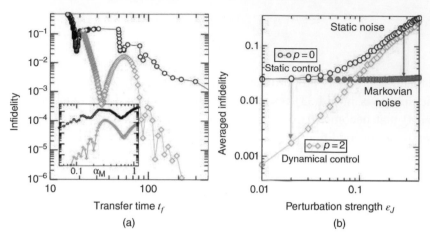

Figure 14.4 Transfer infidelity $1 - f(t_f)$ for a uniform (homogeneous) spin-chain channel [i.e., $J_i = J$ in (14.17)], whose energy eigenvalues are $\omega_k = 2J\cos(\frac{k\pi}{N+1})$, under a modulated boundary-controlled coupling $g_p(t) = g_M \sin^p(\pi t/t_f)$ as a function of (a) the transfer time t_f or g_M, (b) the perturbation strength ϵ_J of the noisy channel, averaged over 103 noise realizations for $g_{M_0}^{opt} = 0.6$ and $g_{M_2}^{opt} = 0.7$. In static noisy channels, the infidelity is strongly reduced under dynamical $p = 2$ control (empty squares). A fluctuating Markovian noisy channel is less damaging ($p = 0$, solid circles): the infidelity coincides with its unperturbed value. Here $N = 29$ and $J = 1$. (Adapted from Zwick et al., 2014.)

frequencies, retaining only those that correspond to an almost equidistant spectrum of ω_k around ω_z, and allow for coherent transfer.

The inset in Figure 14.4(a) shows that (for a fixed g_M), the dynamical control ($p = 1, 2$) of the boundary couplings suppresses the transfer infidelity *by orders of magnitude* at the expense of slowing down the transfer time at most by a factor of 2,

$$t_p/t_0 \approx c_p/c_0 \leq 2. \tag{14.38}$$

If the constraint on g_M is relaxed, that is, more energy is used, the advantages of dynamical control become even more prominent for both infidelity decrease and transfer-time reduction by orders of magnitude [Fig. 14.4(a)]. Hence, the speed–fidelity trade-off can be *drastically improved under optimal dynamical control.*

14.3.2 Robustness of Noisy Channels

We now consider the effects of noise on the coupling strengths, known as off-diagonal noise, causing

$$J_i \rightarrow J_i + J_i \tilde{\delta}_i(t), \quad i = 1, \ldots, N. \tag{14.39}$$

Here $\tilde{\delta}_i(t)$ is a uniformly distributed random variable in the interval $[-\epsilon_J, \epsilon_J]$, $\epsilon_J > 0$ characterizing the noise or disorder strength. When $\tilde{\delta}_i$ is time-dependent, the noise is *fluctuating*, whereas fixed $\tilde{\delta}_i$ corresponds to *static* noise. Such noises affect the bath-energy levels, while the central eigenvalue ω_z remains invariant. We may compare the performance of the boundary-couplings control solutions $g_p(t)$ under these types of noise:

(i) *Static noise.* Static control can suppress such noise, but *dynamical boundary control* with $p = 2$ *can render the channel even more robust* because it filters out the bath energies that damage the transfer. This is shown in Figure 14.4(b) in the strong-coupling regime for $g_{M_p} = g_{M_p}^{opt}$ where the advantage of the dynamical control with $p = 2$ compared to the static control case $p = 0$ is evident, at the expense of increasing the transfer time by only a factor of 2, $t_2/t_0 \approx 2$. In the weak-coupling regime, if we choose g_{M_p} such that the transfer fidelity is similar for control with $p = 0$ and with $p = 2$, then the modulated control with $p = 2$ yields transfer that is an order of magnitude faster. Remarkably, due to disorder-induced localization, the fidelity under static noise cannot be improved beyond the bound

$$1 - \bar{f} \propto N\epsilon_J^2 \quad (\epsilon_J \ll 1), \tag{14.40}$$

regardless of how small g_0 is.

(ii) *Markovian noise.* The worst case for quantum state transfer is the absence of an energy gap around ω_z. This case corresponds to Markovian noise characterized by $\langle \tilde{\delta}_i(t) \tilde{\delta}_i(t + \tau) \rangle = \delta(\tau)$, where the angle brackets denote the noise ensemble average. In this case, there is an analytical solution for the optimal modulation that can be approximated by

$$g(t) \approx a g_M + b \sin^q (t\pi/t_f), \tag{14.41}$$

where

$$q \sim 3.5, \; b/a \sim 1/3, \; a \sim 0.84. \tag{14.42}$$

The infidelity for this optimal modulation almost coincides with the one obtained without modulation, so that dynamical (modulation) control is not helpful in the Markovian limit. Arbitrarily high fidelities can be achieved in this limit by slowing down the transfer time, that is, by decreasing g_M. This comes about because in a Markovian bath, fast coupling fluctuations suppress the disorder-localization effects that hamper the transfer fidelity.

(iii) *Non-Markovian noise.* We consider fluctuating noise $J_i + J_i \tilde{\delta}_i(t)$, where $\tilde{\delta}_i(t) = \tilde{\delta}_i([t/t_c])$ and the integer part $[t/t_c] = n$ defines $\tilde{\delta}_i(n)$ that randomly varies within the interval $[-\epsilon_J, \epsilon_J]$. The transfer fidelity converges to its value without noise as the noise correlation time t_c decreases [Fig. 14.4(b)]. Consequently, the fidelity can be substantially improved by reducing g_M. The effective noise strength scales down as $t_c^{1/2}$. By contrast to the Markovian limit $t_c \to 0$, dynamical control can strongly suppress the infidelity in the non-Markovian regime that lies between the static and Markovian limits, provided the bath spectrum is gapped.

14.4 Discussion

Our analysis of state-transfer optimization within hybrid open systems, from a noisy qubit to its quiescent counterpart, has revealed an intriguing interplay between the ability to avoid dephasing errors by acting within the bath-memory time and the limitations imposed by leakage out of the operational subspace. Counterintuitively, under no circumstances is the fastest transfer optimal for a given transfer energy.

We discussed a general, optimal, and robust dynamical control of the trade-off between the speed and fidelity of qubit-state transfer through the central (zero) energy global mode of a quantum channel in the presence of either static or fluctuating noise. Counterintuitively, static noise is found to be more detrimental than fluctuating noise on the spin–spin couplings. Dynamical boundary control has been used to design an optimal spectral filter. The resulting infidelity can be suppressed and/or the transfer speed can be enhanced by orders of magnitude, while their robustness against noise on the spin–spin couplings is maintained or even improved. This general approach is applicable to quantum channels that can be mapped onto Hamiltonians quadratic in bosonic or fermionic operators.

Part II

Control of Thermodynamic Processes in Quantum Systems

15

Entropy, Work, and Heat Exchange Bounds for Driven Quantum Systems

The first and second laws of thermodynamics must apply even far from equilibrium. Therefore, in accordance with the second law, closed "super-systems" (composed of a system and a bath) must approach thermodynamic equilibrium regardless of how far from it they start out. Here, we consider the entropy change of a driven (dynamically controlled) quantum system that is expected to equilibrate because of its contact with a bath (Chs. 2, 5, 6). We derive a bound on the entropy change of a driven system in contact with a (thermal or nonthermal, e.g., squeezed) bath. This bound may be much tighter than that obtained from the second law in scenarios where states capable of delivering work via a unitary (isentropic) process, dubbed nonpassive states, are involved. This bound applies even outside the realm of H-theorems, which are concerned with bounds satisfied by *stationary, both nonequilibrium* and equilibrium, thermodynamic functions and distributions. For quantum open systems, the theorems of Lindblad and Spohn play the role of H-theorems: Lindblad's theorem for any completely positive (CP) map and Spohn's theorem for the more restricted case of Markovian evolution. The transgression of the validity of these theorems is discussed.

15.1 Entropy Change in Markovian and Non-Markovian Processes

In nonequilibrium thermodynamics, a dissipative process is deemed *irreversible* if the total entropy of the system and the bath combined increases. Here we discuss the characteristics of this entropy increase, known as entropy production.

15.1.1 The Lindblad Theorem and Markovian Entropy Change

The evolution of a quantum system interacting with another quantum system [here a bath (B)] can be shown to conform to a completely positive (CP), trace-preserving map from the initial state of the system, $\rho(0)$, to the evolving state $\rho(t)$, provided that the system and the bath are initially uncorrelated, that is, they are described by a product state $\rho(0)\rho_B$, ρ_B being the bath state. If, however, the evolution involves measurements (Chs. 9, 10), the map may decrease the trace. In what follows, we consider non-trace-increasing (NTI) maps $\mathcal{M}$, satisfying $\mathrm{Tr}(\mathcal{M}\rho) \leq 1$, which include both above types of evolution as special cases.

The general form of a CP-NTI map $\mathcal{M}$ is

$$\mathcal{M}\rho = \sum_j K_j^\dagger \rho K_j, \tag{15.1}$$

where K_j are known as the *Kraus operators*. The CP map $\mathcal{M}$ is trace preserving provided that $\sum_j K_j K_j^\dagger = I$, but more generally, for a CP-NTI map, $\sum_j K_j K_j^\dagger \leq I$.

For any CP-NTI map $\mathcal{M}$, Lindblad's *H-theorem* is the inequality

$$S(\mathcal{M}\rho||\mathcal{M}\tilde{\rho}) \leq S(\rho||\tilde{\rho}) \tag{15.2}$$

for the *relative entropy* of arbitrary states ρ and $\tilde{\rho}$, defined as

$$S(\rho||\tilde{\rho}) = k_B \mathrm{Tr}\big(\rho \ln \rho - \rho \ln \tilde{\rho}\big), \tag{15.3}$$

where k_B is the Boltzmann constant.

Lindblad's theorem (15.2) is the most general statement of the second law for the evolution of quantum open systems if we assume that any proper quantum evolution (including measurements) must conform to a CP-NTI map. It is known as the statement of the monotonicity of the quantum relative entropy. This theorem implies that, if a CP-NTI map $\mathcal{M}$ acting on a state ρ has a steady (invariant) state ρ_{ss}, such that

$$\mathcal{M}\rho_{ss} = \rho_{ss}, \tag{15.4}$$

then the *relative entropy* with respect to the steady state ρ_{ss} does not increase under $\mathcal{M}$:

$$S(\mathcal{M}\rho||\rho_{ss}) \leq S(\rho||\rho_{ss}). \tag{15.5}$$

Consequently, $S(\rho(t)||\rho_{ss}) \leq S(\rho(0)||\rho_{ss})$.

Surprisingly, (15.4) and (15.5) hold only in certain cases. In particular, (15.4) is valid for Markovian evolution described by the Lindblad equation, which implies sufficiently weak coupling. More generally, (15.4) holds when the steady state of the system and the bath is in a product form, $\rho_{ss}\rho_B$. The latter assumption is valid even for appreciable coupling of the system to a thermal bath if the temperature T is sufficiently high, so that the system–bath coupling $|\langle H_{SB}\rangle|$ is small compared to the thermal energy $k_B T$ (Ch. 6).

A CP trace-preserving map is generated by evolution governed by the Linblad Markovian master equation (MME) (11.61), which assumes that the system is weakly coupled to a stationary (but not exclusively thermal) bath. Under such evolution, the entropy production σ by the system in the state $\rho(t)$ obeys Spohn's inequality

$$\sigma \equiv -\frac{d}{dt}\mathcal{S}(\rho(t)||\rho_{\mathrm{ss}}) \geq 0, \tag{15.6}$$

where the steady state of the system ρ_{ss} is attained by combined control (driving) and bath effects.

In view of (15.3), inequality (15.6) for the Lindblad MME can be recast in another useful form, known as Spohn's H-theorem,

$$\sigma = k_{\mathrm{B}}\mathrm{Tr}\big[(\mathcal{L}\rho)(\ln \rho_{\mathrm{ss}} - \ln \rho)\big] \geq 0. \tag{15.7}$$

The time-integrated inequality (15.6) with $t \to \infty$ then yields for the relaxation

$$\rho(0) \mapsto \rho_{\mathrm{ss}} \tag{15.8}$$

the total (time-integrated) entropy production that is likewise nonnegative,

$$\Sigma \equiv \int_0^\infty dt\, \sigma(t) = \mathcal{S}(\rho(0)||\rho_{\mathrm{ss}}) \geq 0. \tag{15.9}$$

Yet, the Lindblad MME that underlies this result was shown in Chapter 11 to be the Markovian limit of a more general non-Markovian ME. The question we pose in what follows is whether Spohn's inequality (15.6) and its integrated counterpart (15.9) hold beyond the Markovian regime.

15.1.2 Entropy Change under Non-Markovian Dynamics

As shown below, *Spohn's inequality* (15.6) *may not hold in the non-Markovian and/or strong-coupling regimes*, in which correlations or entanglement between the system and the bath may be appreciable. By contrast, since the relative entropy is *nonnegative* by definition, (15.9) is valid for *any system-bath coupling*, not only in the Born–Markov regime. Yet, the relative entropy $\mathcal{S}(\rho(t)||\rho_{\mathrm{ss}})$ is *non-monotonic under non-Markovian evolution* of $\rho(t)$, which is not a CP map of $\rho(t')$ $(0 < t' < t)$, being dependent on the joint system–bath state.

For a control Hamiltonian $H(t)$ that varies slowly compared to the thermalization time, the corresponding ME is

$$\dot{\rho} = \mathcal{L}(t)\rho, \tag{15.10}$$

where $\mathcal{L}(t)$ is a superoperator of the same form as in the Lindblad equation [cf. (11.61)], but with time-dependent coefficients and an invariant state $\rho_{\mathrm{ss}}(t)$ that satisfies

$$\mathcal{L}(t)\rho_{\mathrm{ss}}(t) = 0. \tag{15.11}$$

The generalization of inequality (15.6) to the case of (15.10) then yields, following integration, the entropy change bound,

$$\Delta\mathcal{S} = \mathcal{S}(\rho_{\mathrm{ss}}(\infty)) - \mathcal{S}(\rho(0)) \geq -k_B \int_0^\infty \mathrm{Tr}\left\{\left[\mathcal{L}(t)\rho(t)\right]\ln\rho_{\mathrm{ss}}(t)\right\}dt. \tag{15.12}$$

If $H(t)$ acts on a system that interacts with a thermal bath at temperature T, then the steady state is a thermal state for $H(t)$,

$$\rho_{\mathrm{ss}}(t) = \rho_{\mathrm{th}}(t) = \frac{1}{Z(t)}\exp\left[-\frac{H(t)}{k_B T}\right]. \tag{15.13}$$

Inequality (15.12) then corresponds to the following entropy-change bound,

$$\Delta\mathcal{S} \geq \frac{1}{T}\int_0^\infty \mathrm{Tr}\left[\dot{\rho}(t)H(t)\right]dt = \frac{\mathcal{E}_{\mathrm{d}}}{T}, \tag{15.14}$$

where $\mathcal{E}_{\mathrm{d}}$ is the long-time limit ($t \to \infty$) of the energy dissipated by the thermal bath.

For a TLS, the entropy production $\sigma(t)$, as defined in (15.6), has the following form, with $(\rho_{\mathrm{ss}})_{ee}$ as the long-time (steady-state) excitation probability,

$$\sigma = -\dot{\rho}_{ee}\ln\frac{\rho_{ee}[1 - (\rho_{\mathrm{ss}})_{ee}]}{(1 - \rho_{ee})(\rho_{\mathrm{ss}})_{ee}}. \tag{15.15}$$

The condition to have $\sigma < 0$, in violation of Spohn's inequality, corresponds to

$$\dot{\rho}_{ee} > 0 \quad \text{for} \quad \rho_{ee} > (\rho_{\mathrm{ss}})_{ee} \tag{15.16}$$

or

$$\dot{\rho}_{ee} < 0 \quad \text{for} \quad \rho_{ee} < (\rho_{\mathrm{ss}})_{ee}. \tag{15.17}$$

Correspondingly, $\sigma(t)$ is negative if and only if

$$\frac{d}{dt}|\rho_{ee}(t) - (\rho_{\mathrm{ss}})_{ee}| \geq 0. \tag{15.18}$$

Since $|\rho_{ee}(t) - (\rho_{\mathrm{ss}})_{ee}|$ vanishes at $t \to \infty$, condition (15.18) means that the function $\rho_{ee}(t)$ is non-monotonic (i.e., it has, at least, one maximum or minimum). Conversely, when $\rho_{ee}(t)$ is monotonic, we have $\sigma(t) > 0$. The foregoing conditions are consistent with the interpretation of the relative entropy $\mathcal{S}(\rho||\rho_{\mathrm{ss}})$ as a measure of distance from equilibrium. Thus, whenever the oscillatory $\rho_{ee}(t)$ drifts away from its initial or final steady state, σ takes negative values (Fig. 15.1, inset).

Spohn's inequality (15.6) may be violated for non-Markovian evolution of $\rho(t)$ under ME (11.79a) or its numerical counterpart (Fig. 15.1). The oscillations of the TLS excitation probability $\rho_{ee}(t)$ in this regime give rise to oscillations of the entropy production $\sigma(t)$ that have alternating sign (Fig. 15.1, inset). The TLS excitation oscillations reflect a reversible energy transfer between the system and the

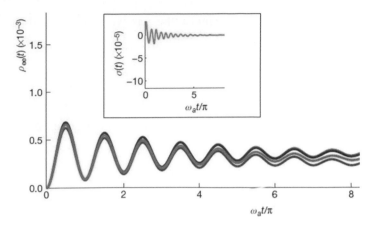

Figure 15.1 System and bath evolution as a function of time. The main panel shows the excited-level population as a function of time for the initial zero-temperature product state of the system and bath. The population relaxes to its equilibrium value after a few oscillations. We observe agreement between the results of the second-order master equation (lower line), a simplified numerical solution (two-quanta exchange with a discrete bath) (upper line), and the exact numerical solution for a discrete bath of 40 modes (middle line). Inset: negative of the rate of change of relative entropy, $\sigma(t)$. The discrete bath approximates the Lorentzian peak (11.123) with $T = 0$, $\kappa = \omega_a/5$, and $\gamma_a = 0.07\omega_a$. (Adapted from Erez et al., 2008.)

bath. As a result, the total (system+bath) state alternates between a product state and an entangled state, yielding oscillations of the system entropy, which increases for a product state and decreases for an entangled state. This explains the reason for the alternating sign of entropy production $\sigma(t)$.

15.2 Passivity and Nonpassivity

A *passive state* of a system governed by Hamiltonian H is the state Π that satisfies

$$\mathrm{Tr}(\Pi H) \leq \mathrm{Tr}(U \Pi U^\dagger H) \qquad (15.19)$$

under any unitary operator U that represents reversible external driving of the system. Inequality (15.19) means the *impossibility of extracting work* from state Π by a unitary operation. For a given Hamiltonian H, each state ρ has a *unique* passive counterpart provided both H and ρ are nondegenerate,

$$\Pi = U \rho U^\dagger, \qquad (15.20)$$

attainable by a unitary transformation U, which maps the eigenstates of ρ onto the eigenstates of H, such that the eigenvalues of Π decrease as the corresponding eigenvalues of H increase. Accordingly, the passive state Π is diagonal in the Hamiltonian (energy) eigenbasis,

$$\Pi = \sum_{j=1}^{n} \lambda_j |j\rangle\langle j|, \tag{15.21}$$

where the real, positive coefficients λ_j are the monotonically descending eigenvalues of Π:

$$\lambda_j \geq \lambda_{j+1} \quad \text{if } E_j \leq E_{j+1}, \tag{15.22}$$

$E_j|j\rangle$ being the jth energy eigenvalue. For degenerate H or ρ, two or more states satisfy (15.19), (15.21), and (15.22), so that any of those may be chosen as a passive state.

Continuous phase-space distributions, such as the Wigner function or the Glauber–Sudarshan P-function, are deemed passive if they are peaked at the origin (the vacuum) and fall off *monotonically and isotropically*. If, on the contrary, the distribution peak is displaced from the origin or if the falloff is non-monotonic or anisotropic, then it is nonpassive (Ch. 22).

Gibbs states $\rho_{\text{th}} = e^{-\beta H}/Z$ are passive, as their mean energy is the lowest attainable by any unitary operation for a fixed H. There are many other passive states, for example, the microcanonical distribution

$$\Pi[E] = \mathcal{N}^{-1} \sum_{\{j; E_j \leq E\}} |j\rangle\langle j|, \tag{15.23}$$

$\mathcal{N}^{-1}$ being the normalizing constant.

Generally, a product of passive states pertaining to distinct subsystems can be a nonpassive state, which allows for work extraction from the composite system. Only Gibbs states, however, possess *complete passivity*, namely, also their n-fold product $\rho_{\text{th}}^{\otimes n}$, describing the collective state of n identical systems, is passive with respect to an n-fold sum of the system Hamiltonians, for any $n = 1, 2, 3, \ldots$. Possible consequences of this distinction between passive and totally passive states are discussed in Section 15.5 and Chapters 18 and 22.

As shown here, dissipative changes in a passive state correspond to only heat exchange, whereas in a nonpassive state dissipation may be associated with either work or heat exchange (or both).

15.3 Work and Heat Exchange between a Driven System and a Bath

Let us consider an initially prepared state $\rho(0)$ of a quantum system that evolves into a state $\rho(t)$ under the action of a Hamiltonian $H(t)$ and a stationary bath that may be in either a thermal or a nonthermal state. The change in the system energy

$$\Delta E(t) = E(t) - E(0), \tag{15.24}$$

where

$$E(t) = \text{Tr}[\rho(t)H(t)], \tag{15.25}$$

consists of two thermodynamically different contributions:

$$\Delta E(t) = W(t) + \mathcal{E}_{\text{d}}(t). \tag{15.26}$$

The contribution

$$W(t) = \int_0^t \text{Tr}[\rho(t')\dot{H}(t')]dt' \tag{15.27a}$$

is identified as the work that is either extracted or invested by an external drive. The other contribution

$$\mathcal{E}_{\text{d}}(t) := \int_0^t \text{Tr}[\dot{\rho}(t')H(t')]dt' \tag{15.27b}$$

is the dissipative energy change of the system induced by its interaction with the bath. As shown below, $\mathcal{E}_{\text{d}}$ may consist of both dissipation (heat transfer) and transfer of ordered energy (work). In order to separate the two processes, we decompose $\mathcal{E}_{\text{d}}(t)$ into passive and nonpassive contributions, as discussed below.

15.3.1 Heat, Work, and Passive Energy

The energy E of a state ρ can be decomposed into *passive energy* E_{pas} and non-passive energy, alias *ergotropy*, $\mathcal{W}$. Ergotropy is the maximum amount of work extractable from ρ by unitary transformations. It reads

$$\mathcal{W}(\rho, H) \equiv \text{Tr}(\rho H) - \min_U \text{Tr}(U\rho U^\dagger H) \geq 0, \tag{15.28}$$

the minimization extending over all possible unitary transformations U. Passive energy, which cannot be extracted as useful work in a cyclic, unitary, fashion, is given by

$$E_{\text{pas}} = \text{Tr}(U_{\text{p}}\rho U_{\text{p}}^\dagger H) = \text{Tr}(\Pi H), \tag{15.29}$$

where U_{p} is the unitary that transforms ρ into its passive counterpart Π and thereby minimizes the second term on the right-hand side of (15.28). The energy of a state ρ can thus be divided as follows:

$$E = E_{\text{pas}} + \mathcal{W}. \tag{15.30}$$

Since ρ and Π are unitarily related, their von Neumann entropies coincide, $S(\rho) = S(\Pi)$. We therefore decompose the dissipative energy change (15.27b) as

$$\mathcal{E}_{\mathrm{d}}(t) = \mathcal{Q}(t) + \Delta\mathcal{W}_{\mathrm{d}}(t), \tag{15.31}$$

where

$$\mathcal{Q}(t) = \int_0^t \mathrm{Tr}[\dot{\Pi}(t')H(t')]dt' \tag{15.32a}$$

corresponds to a change in the passive state and thus a change in entropy. Because of its entropy-changing character, we refer to (15.32a) as *heat* exchange: $\mathcal{Q}(t)$ is the *non-unitary* change in *passive energy*. The other contribution in (15.31),

$$\Delta\mathcal{W}_{\mathrm{d}}(t) = \int_0^t \mathrm{Tr}\left[\left(\dot{\rho}(t') - \dot{\Pi}(t')\right)H(t')\right]dt', \tag{15.32b}$$

is the dissipative (non-unitary) change in the *ergotropy* due to the interaction of the system with the bath. If the system state remains always passive, so that $\Delta\mathcal{W}_{\mathrm{d}}(t) = 0$, the dissipative energy change is entirely heat, $\mathcal{E}_{\mathrm{d}}(t) = \mathcal{Q}(t)$. Then (15.32a) and (15.27b) coincide.

The system ergotropy may increase in a non-unitary fashion due to its interaction with a bath and be subsequently extracted from the system as work via a suitable unitary process. Any *unitary* change in the passive energy of a system driven by a time-dependent Hamiltonian changes the work (15.27a). The decomposition of the exchanged energy (15.31) into dissipative changes in passive energy (15.32a) and ergotropy (15.32b) is an unraveling of the first law of thermodynamics for quantum systems out of equilibrium.

The energies (15.32) are, in general, process variables and thus depend on the evolution path. Only under the action of a constant Hamiltonian, they become *path-independent* and amount to the following changes in the passive energy (which we have identified as heat exchange),

$$\mathcal{Q}(t) = \Delta E_{\mathrm{pas}}(t) = \mathrm{Tr}[\Pi(t)H] - \mathrm{Tr}[\Pi_0 H], \tag{15.33a}$$

and in the ergotropy,

$$\Delta\mathcal{W}_{\mathrm{d}}(t) = \Delta\mathcal{W}(t) = \mathcal{W}(\rho(t)) - \mathcal{W}(\rho(0)), \tag{15.33b}$$

respectively.

15.4 Heat Currents and Entropy Change

Consider a system interacting (simultaneously or alternately) with two or more baths labeled by k and driven (controlled) by fields (Ch. 11). Then $\rho(t)$ satisfies (15.10), where

$$\mathcal{L}(t)\rho = -\frac{i}{\hbar}[H(t), \rho] + \sum_k \mathcal{L}_k(t)\rho. \tag{15.34}$$

Here $H(t)$ is the system Hamiltonian containing the driving field, and

$$\mathcal{L}_k(t)\rho = \frac{1}{2}\sum_j \left\{[\hat{V}_{kj}(t)\rho, \hat{V}_{kj}^\dagger(t)] + [\hat{V}_{kj}(t), \rho\hat{V}_{kj}^\dagger(t)]\right\}, \tag{15.35}$$

where $\hat{V}_{kj}(t)$ are the time-dependent driven Lindblad jump operators (Ch. 11).

We introduce the notion of a *heat current*, defined as

$$\mathcal{J}(t) \equiv \frac{d}{dt}\mathcal{Q}(t) = \text{Tr}\left[H(t)\frac{d}{dt}\rho(t)\right] = \sum_k \mathcal{J}_k(t). \tag{15.36}$$

This total heat current is the sum over the heat currents contributed by all heat baths interacting with the system, each contribution being

$$\mathcal{J}_k(t) = \text{Tr}\left[H(t)\mathcal{L}_k(t)\rho(t)\right]. \tag{15.37}$$

By analogy with Spohn's H-theorem (15.7), we obtain the following inequalities for each k in the Markovian approximation,

$$\sigma_k(t) \equiv -k_B \text{Tr}\left\{[\mathcal{L}_k(t)\rho(t)]\left[\ln\rho(t) - \ln\rho_{ss}^{(k)}(t)\right]\right\} \geq 0, \tag{15.38}$$

where $\rho_{ss}^{(k)}(t)$ satisfies

$$\mathcal{L}_k(t)\rho_{ss}^{(k)}(t) = 0 \tag{15.39}$$

and σ_k may be dubbed the kth partial entropy production. The sum of the inequalities (15.38) over k yields the following inequality for the entropy production $\sigma = \sum_k \sigma_k$,

$$\sigma(t) = \frac{d}{dt}\mathcal{S}(t) - k_B \sum_k \mathcal{L}_k(t)\rho(t)\ln\rho_{ss}^{(k)}(t \geq 0. \tag{15.40}$$

In the case of thermal baths with temperatures T_k, $\rho_{ss}^{(k)}(t) = \rho_{th}^{(k)}(t)$ are given by (15.13) with $T \to T_k$, and the inequalities (15.38) become

$$\sigma_k(t) = -k_B \text{Tr}[\mathcal{L}_k(t)\rho(t)]\ln\rho(t) - \frac{\mathcal{J}_k(t)}{T_k} \geq 0. \tag{15.41}$$

Summing these inequalities over k yields the Markovian version of the second law for the system, whereby

$$\sigma(t) = \frac{d}{dt}\mathcal{S}(t) - \sum_k \frac{1}{T_k}\mathcal{J}_k(t) \geq 0. \tag{15.42}$$

Equation (15.40) extends (15.42) to the case where one or more of the baths are nonthermal. However, the inequalities for the partial entropy production (15.38) [or (15.41)] provide generally stricter limitations on the dissipation process than (15.40) [(15.42)].

When the steady states associated with all $\mathcal{L}_k(t)$ are the same, $\rho_{\text{ss}}^{(k)}(t) = \rho_{\text{ss}}(t)$, (15.40) reduces to (15.6), where ρ_{ss} generally depends on t. When the baths are thermal and have the same temperature T or there is one thermal bath, (15.42) reduces to the generalization of the Spohn relation (15.6) under driving,

$$\sigma(t) = \frac{d}{dt}\mathcal{S}(t) - \frac{\mathcal{J}(t)}{T} \geq 0. \tag{15.43}$$

We stress that the foregoing inequalities (15.38), (15.41), (15.42), and (15.43) may fail under non-Markovian evolution, as shown by (15.15)–(15.17).

15.4.1 Entropy Change of Passive and Nonpassive States in a Thermal Bath

Spohn's inequality, which requires nonnegative entropy production, may overestimate the *actual* system entropy change. This can be seen for the relaxation of an initially nonpassive state $\rho(0)$ to a (passive) thermal state ρ_{th} through its interaction with a thermal bath at temperature T when the Hamiltonian is constant. According to the decomposition (15.31) of $\mathcal{E}_d$, (15.14) can then be written as

$$\Delta\mathcal{S} \geq \frac{\mathcal{E}_d}{T} = \frac{\mathcal{Q} + \Delta\mathcal{W}_d}{T}. \tag{15.44}$$

Since entropy is a state variable, $\Delta\mathcal{S} = \mathcal{S}(\rho_{\text{th}}) - \mathcal{S}(\rho(0))$ is only determined by the initial state $\rho(0)$ and the thermal steady state ρ_{th}. Yet, the time-integrated Spohn's inequality (15.9) *may give rise to different inequalities for the same $\Delta\mathcal{S}$ depending on the evolution paths* from $\rho(0)$ to ρ_{th}. For example, we can choose a path that *does not involve any dissipation of ergotropy* to the bath. This path corresponds to a unitary transformation to the passive state, $\rho(0) \mapsto \Pi_0$, followed by the interaction of the system in this state with a thermal bath, where it relaxes to the thermal steady-state ρ_{th}. Inequality (15.9) yields for this path,

$$\Delta\mathcal{S} \geq \frac{\mathcal{Q}}{T}, \tag{15.45}$$

where $\mathcal{Q}$ is the same heat exchange as in (15.44).

As a result of the relaxation caused by the thermal bath, the ergotropy of the initial state $\rho(0)$ decreases from $\mathcal{W}_0 \geq 0$ down to vanishing ergotropy in the thermal steady state, hence

$$\Delta\mathcal{W}_d = \Delta\mathcal{W} = -\mathcal{W}_0 \leq 0, \tag{15.46}$$

where we used (15.33b). Then, inequality (15.45) yields a tighter bound than inequality (15.44).

15.4.2 Entropy Change of a Driven System in a Thermal Bath

Both $\mathcal{Q}$ and $\Delta\mathcal{W}_{\mathrm{d}}$ on the right-hand side of (15.44) are *path-dependent* if the driving Hamiltonian $H(t)$ slowly varies. In particular, we may choose a path such that $H(t)$ generates a *nonpassive state* at some point, although the initial state is passive and the bath is thermal. We may also choose a path void of *initial ergotropy* by extracting the ergotropy of the initial state by a unitary process *prior* to placing the system in contact with the bath, thereby generating the passive state Π_0. Subsequently, the system in this passive state interacts with a thermal bath, resulting in the steady state $\rho_{\mathrm{th}}(\infty)$. The inequality put forward here yields for the latter step,

$$\Delta\mathcal{S} \geq \frac{\mathcal{Q}'}{T}, \tag{15.47}$$

where

$$\mathcal{Q}' = \int_0^\infty \mathrm{Tr}[\dot{\varrho}(t)H(t)]dt \tag{15.48}$$

is the heat exchanged with the bath along the chosen path. Here $\varrho(t)$ is the solution of the same ME that governs $\rho(t)$ but with the passive initial condition $\varrho_0 = \Pi_0$. If the Hamiltonian is constant, (15.47) coincides with (15.45).

15.4.3 Entropy Change of a System in a Squeezed Bath

If a system interacts with a nonthermal (e.g., squeezed) bath while its Hamiltonian is kept constant, the system may relax to a *nonpassive* steady state

$$\rho_{\mathrm{ss}} = U_{\mathrm{s}}\Pi_{\mathrm{ss}}U_{\mathrm{s}}^\dagger, \tag{15.49}$$

where U_{s} is the appropriate unitary operator. When the system is an oscillator linearly coupled to a squeezed bosonic bath, then U_{s} is the squeezing operator. The total (time-integrated) entropy production (15.9) generated during the relaxation process $\rho(0) \mapsto \rho_{\mathrm{ss}}$ then evaluates to

$$\Sigma = \mathcal{S}(\rho(0)||\rho_{\mathrm{ss}}) = \mathcal{S}(\tilde{\rho}(0)||\Pi_{\mathrm{ss}}) \geq 0, \tag{15.50}$$

which is the total entropy production during the *fictitious* relaxation process

$$\tilde{\rho}(0) \equiv U_{\mathrm{s}}^\dagger \rho(0) U_{\mathrm{s}} \mapsto \Pi_{\mathrm{ss}}. \tag{15.51}$$

In order to obtain the tightest bound for the entropy change $\Delta\mathcal{S}$, the following relative-entropy inequality can be adopted, instead of (15.9),

$$\mathcal{S}(\Pi_0||\Pi_{\mathrm{ss}}) \geq 0. \tag{15.52}$$

Let us specifically consider a single-mode harmonic oscillator in a Gaussian state, the passive steady state being thermal, $\Pi_{\mathrm{ss}} = \rho_{\mathrm{th}}$. The system evolution is

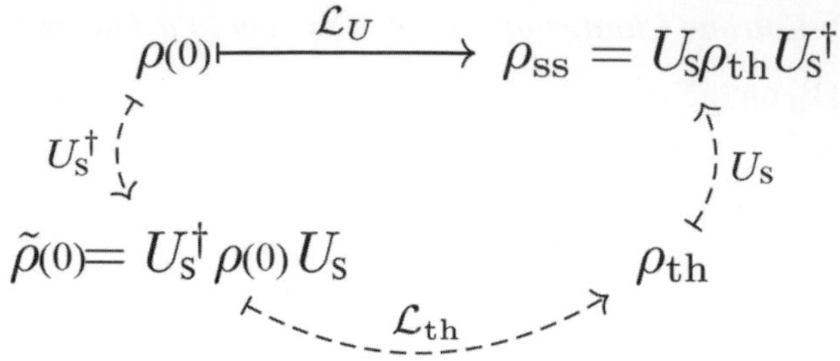

$$\rho(0) \xrightarrow{\quad \mathcal{L}_U \quad} \rho_{ss} = U_s \rho_{th} U_s^\dagger$$

$$U_s^\dagger \quad\quad\quad\quad\quad\quad\quad\quad\quad U_s$$

$$\tilde{\rho}(0) = U_s^\dagger \rho(0) U_s \quad\quad\quad\quad \rho_{th}$$

$$\mathcal{L}_{th}$$

Figure 15.2 Two alternative evolution paths between $\rho(0)$ and ρ_{ss}. The solid (physical) path corresponds to the interaction of the system with a nonthermal (e.g., squeezed) bath, whereas the dashed path involves two unitary transformations and the interaction with a thermal bath. (Adapted from Niedenzu et al., 2018.)

determined by a Markovian ME driven by Liouvillian $\mathcal{L}_U$. This evolution is unitarily equivalent (Fig. 15.2) to that of a transformed state $\tilde{\rho}(t)$ with a *thermal* bath, which is described by the following Liouville superoperator $\mathcal{L}_{th}$,

$$\mathcal{L}_{th}\tilde{\rho} = U_s^\dagger [\mathcal{L}_U (U_s \tilde{\rho} U_s^\dagger)] U_s. \tag{15.53}$$

Inequality (15.50) then yields

$$\Delta \mathcal{S} \geq \frac{\tilde{\mathcal{E}}_d}{T}, \tag{15.54}$$

where $\tilde{\mathcal{E}}_d$ is the (dissipative) change in the energy of the unitarily transformed state $\tilde{\rho}$, as it relaxes from $\tilde{\rho}(0)$ to ρ_{th}, corresponding to the second step in the dashed path in Figure 15.2. By contrast, inequality (15.52) conforms to the bound (15.45).

Thus, the bounds (15.45) and (15.47) also apply to cases where the nonthermal (squeezed) bath relaxes the system to a state of the form (15.49) (which is a squeezed thermal state in the case of an oscillator in a squeezed bath).

15.5 Discussion

The fundamental aim of this chapter has been to understand the role played by the first and second laws of thermodynamics in quantum dissipative processes. The second law as applied to quantum relaxation processes is commonly considered to be synonymous with Spohn's inequality that holds in the Markovian approximation, whereby the entropy change of a system that interacts with a thermal bath is bounded from below by the exchanged energy divided by the bath temperature. We have shown that this inequality may fail under non-Markovian evolution, as further elaborated in Chapter 16. However, even in the Markovian approximation, a much tighter bound than Spohn's on entropy change in dissipative processes is obtainable when the state of the system is *nonpassive*. The energy of a nonpassive state can be unitarily reduced until the state becomes passive, thereby extracting work.

Nonpassive states, such as harmonic-oscillator Fock, squeezed, or coherent states, have "work capacity" or "ergotropy" and may be thought of as *"quantum batteries."* The significance of nonpassivity as a work resource in bath-powered engines will be discussed in Chapters 18 and 22.

16

Thermodynamics and Its Control on Non-Markovian Timescales

Thermodynamics in the quantum domain is conventionally based on long-time Markovian (Lindblad) master equations (Ch. 11), which describe convergence to equilibrium at a constant rate. The consideration of non-Markovian thermodynamic effects in quantum settings is scanty at best. Here we explore such non-Markovian processes: Impulsive, frequent QND measurements of the TLS energy or phase shifts of the TLS state disturb its equilibrium with the bath, thereby abruptly changing the TLS temperature. The subsequent evolution of the TLS that is coupled to the bath alternates between heating and cooling at times comparable to the inverse TLS resonance frequency. Such effects are at odds with the standard notions of thermodynamics, whereby temperature and entropy monotonically converge to their equilibrium values (Ch. 15).

16.1 QND Impulsive Disturbances of the Equilibrium State

The initial, equilibrium, state of the total (TLS plus bath) ensemble is the density matrix (6.9), $\rho_{\mathrm{Eq}} \propto e^{-\beta H_{\mathrm{tot}}}$. Its off-diagonal elements in the energy basis of the TLS express quantum *correlations* (coherences) between the system and the bath, as $\langle e|\rho_{\mathrm{Eq}}|g\rangle \neq 0$. Although concealed when observing only the TLS state, $\rho = \mathrm{Tr}_{\mathrm{B}}\rho_{\mathrm{Eq}}$, these coherences (correlations) are manifest through the temperature (mixedness) increase of ρ as the system–bath coupling grows (Ch. 6).

In the equilibrium state, ρ_{Eq}, the system–bath mean interaction energy is *negative*, as shown in Chapter 6,

$$\langle H_{\mathrm{I}}\rangle \equiv \langle H_{\mathrm{SB}}\rangle = \langle H_{\mathrm{tot}}\rangle - \langle H_{\mathrm{S}} + H_{\mathrm{B}}\rangle < 0. \tag{16.1}$$

This comes about since the correction to the ground-state energy of H_{tot} due to a weakly perturbing system–bath interaction is negative (to the second order in system–bath coupling).

To change the TLS temperature, one needs first to disturb the equilibrium state or the system–bath mean interaction energy (16.1). If the disturbance is to be a QND effect so that ρ retains the σ_z diagonal form it has in equilibrium, one can choose between a σ_z-rotation (phase shift) and a projective measurement in the σ_z (TLS energy) basis. For both types of QND disturbances, it is essential that the meter–system coupling Hamiltonian $H_{\mathrm{SM}} \propto \sigma_z$ does not commute with the system–bath coupling Hamiltonian $H_{\mathrm{SB}} \propto \sigma_x$, thus leading to *nonclassical* effects. The alternative is $H_{\mathrm{SM}} \propto \sigma_x$, which exerts a classical-like force on ρ, changing it in a non-QND fashion.

This scenario is governed by the total Hamiltonian of the system that interacts with the bath, with intermittent perturbations by the coupling of the system to the meter (the measuring apparatus):

$$H(t) = H_{\mathrm{tot}} + H_{\mathrm{SM}}(t), \tag{16.2}$$

$H_{\mathrm{SM}}(t)$ being the time-dependent measurement Hamiltonian that couples the system to a meter. We assume that the meter is composed of energy-degenerate ancillae (Sec. 9.3). We *do not* invoke the RWA in the coupling Hamiltonians (H_{SB}, H_{SM}), namely, we do not impose energy conservation between the system and the bath or the meter on the timescales considered.

We show in this chapter that an impulsive σ_z-disturbance *always* produces *heating* of the equilibrium state immediately thereafter, or even a disturbance of finite, albeit brief, duration τ. An impulsive QND disturbance triggers departure from equilibrium toward heating, but the subsequent (free or driven) evolution may cause cooling through the system–bath energy exchange.

16.1.1 Measurement-Induced Interruption of System–Bath Correlation

A nearly impulsive projective quantum measurement of S correlates the TLS energy eigenstates $|g\rangle$, $|e\rangle$ with mutually orthogonal states of a meter, and the latter is then averaged (traced) over. As in Section 9.3, we describe the measurement process as a CNOT operation that retains the energy degeneracy of the meter $\langle H_{\mathrm{M}} \rangle = 0$, although its von Neumann entropy increases.

Such a measurement can be effected by a strong, pulsed probe that discriminates between $|e\rangle$ and $|g\rangle$ by their *distinct symmetries* (say, their different magnetic numbers or angular momenta) but not by their energies, which are ill-defined during the brief disturbance [Fig. 16.1(b)].

The measurement (described in Sec. 9.3) consists in letting the TLS interact with the meter (another degenerate TLS) via $H_{\mathrm{SM}}(t)$. The measurement outcomes are averaged over (for nonselective measurements) by tracing out the meter. The total effect on the system density operator is then as per (9.76); the diagonal elements

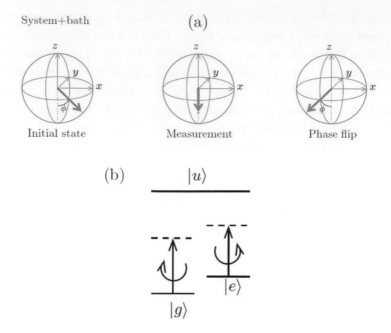

Figure 16.1 (a) Schematic diagram of the total (TLS + bath) state: the z-axis represents the TLS state populations, and the x- and y-axes represent the TLS-bath coherences. The initial state (left) has TLS-bath coherence. A projective σ_z measurement (center) eliminates the coherence without changing the TLS state populations. A σ_z phase shift (right) preserves the system population, but flips the TLS-bath coherences. (b) QND measurement of level populations by off-resonant EM pulses whose polarization is rotated in opposite senses (clockwise or counterclockwise) for $|e\rangle$ or $|g\rangle$. (Reprinted from Gordon et al., 2009.)

of ρ are unchanged, and the off-diagonals are erased. Since the TLS is entangled with the bath, the effect of the measurement is

$$\rho_{\text{tot}} \mapsto \rho_{\text{tot,M}} = \text{Tr}_{\text{M}}\left[U_{\text{C}}(\rho_{\text{tot}} \otimes |0\rangle_{\text{MM}} \langle 0|)U_{\text{C}}^{\dagger}\right] \tag{16.3}$$

$$= |e\rangle \langle e| \rho_{\text{tot}} |e\rangle \langle e| + |g\rangle \langle g| \rho_{\text{tot}} |g\rangle \langle g| \equiv \rho_e^{\text{B}} |e\rangle \langle e| + \rho_g^{\text{B}} |g\rangle \langle g| \, .$$

Here ρ_{tot} is the total state (of the system and bath combined), $\rho_{\text{tot,M}}$ is the state of the system and bath right after the measurement of duration τ,

$$U_{\text{C}} = e^{-(i/\hbar) \int_0^{\tau} dt \, H_{\text{SM}}(t)} \tag{16.4}$$

denotes the controlled-not (CNOT) operation performed by a TLS (meter) as the control (target) qubit, and $\rho_{e(g)}^{\text{B}}$ are the bath states correlated to the system projectors $|e\rangle \langle e|$ and $|g\rangle \langle g|$.

Had $H_{SM}(t)$ commuted with H_{SB} and H_S (and hence with H_{tot}), we should have had at $t = \tau$

$$\langle H_{tot}\rangle (\tau) = \text{Tr}\left[\rho_{tot,M} H_{tot}\right] \tag{16.5}$$

$$= \text{Tr}\left\{U(\tau)\left[\rho_{tot}\otimes|0\rangle_{MM}\langle 0|\right]U^\dagger(\tau)H_{tot}\right\} \tag{16.6}$$

$$= \text{Tr}\left\{\left[\rho_{tot}\otimes|0\rangle_{MM}\langle 0|\right]U^\dagger(\tau)H_{tot}U(\tau)\right\} \tag{16.7}$$

$$= \text{Tr}\left[\rho_{tot}H_{tot}\right] \equiv \langle H_{tot}\rangle (0), \tag{16.8}$$

where

$$U(\tau) = e^{-(i/\hbar)\int_0^\tau dt[H_{tot}+H_{SM}(t)]}. \tag{16.9}$$

The cyclic property of the trace was used in (16.7), and the commutativity of H_{SM} and H_{tot} in (16.8). In this model, the mean total energy is conserved.

By contrast, in our *noncommutative* model, wherein

$$H_{SM}(t) \propto \sigma_z, \qquad H_{SB} = \hbar\sigma_x\hat{B}, \tag{16.10}$$

$\hat{B}$ being a bath operator, the mean energy is not conserved, as shown in what follows. Since $H_{SM}(t) \propto \sigma_z$ commutes with H_S, it suffices to consider the measurement-induced evolution of $\langle H_{SB}\rangle (\tau)$ rather than $\langle H_{tot}\rangle (\tau)$. In the impulsive-measurement limit ($\tau \to 0$), we can drop H_{tot} in the exponent of (16.9) and then use (16.3) to obtain

$$\langle H_{SB}\rangle (\tau \to 0) = \text{Tr}[\rho_{tot,M} H_{SB}] = b_e \langle e|\sigma_x|e\rangle + b_g \langle g|\sigma_x|g\rangle = 0, \tag{16.11}$$

where $b_i = \text{Tr}(\hat{B}\rho_i^B)$. This expresses the vanishing of $\text{Tr}[\rho_{tot,M} H_{SB}]$ due to the diagonality of $\rho_{tot,M}$ with respect to S. Hence, the measurement-induced interruption of the mean interaction energy, $\langle H_{SB}\rangle (\tau) = 0$, stems from the noncommutativity of H_{SB} and $H_{SM}(t)$.

The same result is obtainable by explicitly considering a projective measurement of σ_z, taking into account that the total state after the measurement is changed from ρ_{Eq} according to

$$\rho_{tot} = \rho_{Eq} \quad \to \quad \rho_{tot,M} = \frac{1}{2}(\rho_{Eq} + \sigma_z\rho_{Eq}\sigma_z). \tag{16.12}$$

Thus,

$$\langle H_{SB}\rangle (\tau) = \frac{1}{2}\langle H_{SB}\rangle_{Eq} + \frac{\hbar}{2}\text{Tr}(\hat{B}\sigma_z\sigma_x\sigma_z\rho_{Eq})$$

$$= \frac{1}{2}\langle H_{SB}\rangle_{Eq} - \frac{1}{2}\langle H_{SB}\rangle_{Eq} = 0, \tag{16.13}$$

where we have used the identity $\sigma_z\sigma_x\sigma_z = -\sigma_x$.

This measurement uses the energy supplied by $H_{SM}(t)$ (the system–meter coupling) during the measurement at $0 \leq t \leq \tau$ to eliminate the mean system–bath interaction energy, whose pre-measurement value was negative, as obtained in Chapter 6 by perturbation theory:

$$\langle H_{\text{SB}} \rangle (0) < 0 \; \rightarrow \; \langle H_{\text{SB}} \rangle (\tau) = 0. \tag{16.14}$$

A nearly impulsive QND measurement in the basis of the system (energy) eigenstates, chosen to be the "pointer basis" of the meter, thus transfers energy via the meter–system coupling so as to momentarily interrupt (override) the system–bath interaction. This energy transfer results in a change in system–bath entanglement and thereby triggers the distinctly quantum dynamics of both the system and the bath, which subsequently redistributes their mean energy and entropy in anomalous ways.

16.1.2 Post-Measurement Short-Time Heating

Any nonselective measurement must increase the entropy of ρ_{tot}. Since prior to the measurement ρ_{tot} was assumed to be in a Gibbs state, which is the maximal entropy state among all the states with the same mean energy $\langle H_{\text{tot}} \rangle$, the post-measurement entropy increase implies an increase of $\langle H_{\text{tot}} \rangle$, that is, heating.

A QND measurement changes ρ_{tot}, yet leaves the system and bath states, ρ and ρ_{B}, unchanged; hence neither $\langle H_{\text{S}} \rangle$ nor $\langle H_{\text{B}} \rangle$ is changed. Therefore, the increase in $\langle H_{\text{tot}} \rangle$ must correspond to an equal increase in $\langle H_{\text{SB}}(t) \rangle$: The projective measurement pumps into the system + bath an energy of

$$\delta \langle H_{\text{tot}} \rangle_{\text{M}} = - \langle H_{\text{SB}} \rangle_{\text{Eq}} . \tag{16.15}$$

This energy is of the order of the temperature-dependent bath-induced Lamb-shift (Ch. 11). Owing to the nonunitary nature of the projection, the mixedness of the total state is increased. In the weak-coupling limit, the increase in mixedness due to a measurement is $\simeq O(\epsilon^4)$, namely, it is fourth-order in the system–bath coupling strength ϵ (Ch. 2) and hence negligible.

Since the off-diagonal elements of ρ_{tot} vanish as a result of a measurement, the time derivative of the system state ρ immediately after the completion of a measurement has the form

$$\dot{\rho}(\tau) = i \left(e^{-i\hbar\omega_{\text{a}}\tau} |e\rangle\langle g| - e^{i\hbar\omega_{\text{a}}\tau} |g\rangle\langle e| \right) \text{Tr} \left[\hat{B}(\tau) \left(\rho_e^{\text{B}} - \rho_g^{\text{B}} \right) \right], \tag{16.16}$$

where

$$\hat{B}(\tau) = e^{(i/\hbar)H_{\text{B}}\tau} \hat{B} e^{-(i/\hbar)H_{\text{B}}\tau} . \tag{16.17}$$

Equation (16.16) results from the partial trace over the bath of the Schrödinger equation for ρ_{tot} in the interaction picture, where (16.3) is used on the right-hand side.

In what follows we consider an oscillator bath

$$H_B = \hbar \sum_k \omega_k b_k^\dagger b_k \qquad (16.18)$$

with

$$\hat{B} = \sum_k \eta_k (b_k + b_k^\dagger) \rightarrow \hat{B}(\tau) = \sum_k \eta_k (b_k e^{-i\hbar\omega_k\tau} + b_k^\dagger e^{i\hbar\omega_k\tau}). \qquad (16.19)$$

For $\rho_{tot}(t) = |\Psi_{n,g}(t)\rangle\langle\Psi_{n,g}(t)|$ (Ch. 6), we then have

$$\rho_{eg}(t) = \langle e| \rho(t) |g\rangle = \text{Tr}_B \langle e|\rho_{tot}(t)|g\rangle = \langle B_{n,e}^{\text{even}}(t)|B_{n,e}^{\text{odd}}(t)\rangle = 0, \qquad (16.20)$$

where $|B_{n,e}^{\text{even}}(t)\rangle$, $|B_{n,e}^{\text{odd}}(t)\rangle$ are, respectively, the even- and odd-excitation parts of the bath state (Ch. 6). Hence, the system state ρ is diagonal at any time t. The same argument applies upon permuting $e \leftrightarrow g$ everywhere for $\rho_{tot} = |\Psi_{n,e}(t)\rangle\langle\Psi_{n,e}(t)|$. We then have, by virtue of (16.16) and (16.19),

$$\dot{\rho}(\tau) \propto \text{Tr}_B \left[\hat{B}(\tau)\langle e(g)|\rho_{tot}|e(g)\rangle \right] = \langle B_{n,e}^{\text{even(odd)}}(\tau)|\hat{B}|B_{n,e}^{\text{even(odd)}}(\tau)\rangle = 0. \qquad (16.21)$$

By linearity, (16.20) and (16.21) imply that the diagonality of $\rho(t)$ and the vanishing of $\dot{\rho}$ immediately after the measurement are satisfied for the total S+B state.

For the *factorizable* thermal state,

$$\rho_{Eq,0} = Z^{-1} e^{-\beta H_0} = Z_B^{-1} e^{-\beta H_B} Z_S^{-1} e^{-\beta H_S}, \qquad (16.22)$$

we have

$$\rho_e^B \equiv \langle e|\rho_{Eq,0}|e\rangle = \langle e|Z_S^{-1} e^{-\beta H_S}|e\rangle Z_B^{-1} e^{-\beta H_B} = \rho_{ee}\rho_B,$$
$$\rho_g^B = \rho_{gg}\rho_B. \qquad (16.23)$$

For $\rho_{Eq,0}$, the second derivative of ρ immediately after the measurement is found from (16.3) and (16.23) to be

$$\ddot{\rho}(\tau) = 2\sigma_z \text{Tr}\left[\hat{B}^2(\tau)(\rho_g^B - \rho_e^B) \right], \qquad (16.24)$$

where

$$\text{Tr}\left[\hat{B}^2(\tau) \left(\rho_g^B - \rho_e^B \right) \right] = \text{Tr}\left[\hat{B}^2(\tau)\rho_B \right] (\rho_{gg} - \rho_{ee}) > 0. \qquad (16.25)$$

The first factor in (16.25) is positive by virtue of the positivity of the Hermitian operator $\hat{B}^2$, and the second is positive if and only if there is no population inversion in the TLS.

The total (system-and-bath) equilibrium state satisfies

$$\rho_{Eq} = Z^{-1} e^{-\beta H_{tot}} = Z^{-1} e^{-\beta[H_0 + O(H_{SB}^2)]}. \qquad (16.26)$$

Thus, for weak system–bath coupling, $\rho_{Eq} \approx \rho_{Eq,0}$ becomes factorizable [see (16.22)].

After the measurement [as $H_{SM}(t \geq \tau) = 0$], time-energy uncertainty during a time interval $\Delta t \lesssim 1/\omega_a$ results in the breakdown of the RWA (i.e., $\langle H_S + H_B \rangle$ is then not conserved). Only $\langle H_{tot} \rangle$ is conserved, by unitarity. The post-measurement $\langle H_{SB} \rangle$ decreases with Δt, signifying the restoration of the mean system–bath interaction as they tend to equilibrate,

$$\langle H_{SB} \rangle (\tau) = 0 \rightarrow \langle H_{SB} \rangle (\tau + \Delta t) < 0. \tag{16.27}$$

This decrease of $\langle H_{SB} \rangle$ is at the expense of an *increase* of $\langle H_S + H_B \rangle = \langle H_{tot} \rangle - \langle H_{SB} \rangle$, that is, *heating* of the system and the bath, combined:

$$\frac{d}{dt} (\langle H_S \rangle + \langle H_B \rangle) \Big|_{\tau + \Delta t} > 0, \quad \frac{d}{dt} \langle H_{SB}(t) \rangle \Big|_{\tau + \Delta t} < 0. \tag{16.28}$$

The post-measurement evolution of the system alone, described by $\rho = \mathrm{Tr}_B \rho_{Eq}$, obeys the Taylor expansion at short evolution times, $\Delta t \ll 1/\omega_a$:

$$\rho(\tau + \Delta t) \simeq \rho(\tau) + \Delta t \dot{\rho}(\tau) + \frac{\Delta t^2}{2} \ddot{\rho}(\tau) + \dots. \tag{16.29}$$

The lowest- (0th) order term is *unchanged* by the measurement, $\rho(\tau) = \rho(t \leq 0)$. The first derivative $\dot{\rho}$ *vanishes* at $t = \tau$ due to the definite parity of the bath density operator, which is correlated to $|g\rangle \langle g|$ or $|e\rangle \langle e|$. This vanishing of $\dot{\rho}(\tau)$ is the QZE condition (Ch. 10). The time evolution of ρ is then governed by its second time derivative $\ddot{\rho}(\tau)$, which can be shown to have the same sign as the population difference operator of the TLS, $\sigma_z = |g\rangle\langle g| - |e\rangle\langle e|$. Hence, the second derivative in (16.29) is *positive* shortly after the measurement, consistently with (16.28), provided there is no initial population inversion of the system (i.e., for nonnegative temperature).

16.1.3 Phase-Shift Induced Equilibrium Interruption

The total (system plus bath) state following a rotation of the qubit by a phase ϕ about the z-axis [Fig. 16.1(a)] is given by

$$\rho_{Eq} \rightarrow \rho_{tot}^{\phi} = e^{-i\phi\sigma_z/2} \rho_{Eq} e^{i\phi\sigma_z/2}$$
$$= |e\rangle \langle e| B_{ee} + |g\rangle \langle g| B_{gg} + (e^{-i\phi} |e\rangle \langle g| B_{eg} + \mathrm{H.c.}). \tag{16.30}$$

Following this rotation, the mean value of the system–bath interaction energy changes as follows:

$$\langle H_{SB} \rangle_{Eq}^{\phi} \equiv \mathrm{Tr}(\sigma_x \hat{B} e^{-i\phi\sigma_z} \rho_{Eq} e^{i\phi\sigma_z}) = \mathrm{Tr}(\hat{B} e^{i\phi\sigma_z} \sigma_x e^{-i\phi\sigma_z} \rho_{Eq})$$
$$= \cos\phi \langle H_{SB}(t) \rangle_{Eq} + \sin\phi \, \mathrm{Tr}(\hat{B}\sigma_y \rho_{Eq}). \tag{16.31}$$

Let us rewrite the total Hamiltonian as

$$H_{\text{tot}} = \hat{A}\sigma_z + \hat{B}_1 + \hat{B}_2\sigma_x, \tag{16.32}$$

where $[\hat{A}, \hat{B}_i] = 0$ and $[\hat{B}_1, \hat{B}_2] \neq 0$. One can show that to second order in H_{SB},

$$\text{Tr}(\hat{B}\sigma_y\rho_{\text{Eq}}) = 0. \tag{16.33}$$

Hence, the mean system–bath interaction energy after a σ_z rotation is given by

$$\langle H_{\text{SB}}(t)\rangle^{\phi}_{\text{Eq}} \approx \cos\phi \, \langle H_{\text{SB}}(t)\rangle_{\text{Eq}}. \tag{16.34}$$

The change in the mean energy of the system by the rotation is then

$$\delta\langle H\rangle^{\phi}_{\text{tot}} = -(1 - \cos\phi)\langle H_{\text{SB}}(t)\rangle_{\text{Eq}}. \tag{16.35}$$

Thus, while $\langle H_{\text{S}}\rangle$ and $\langle H_{\text{B}}\rangle$ are the same before and after the rotation, the change in $\langle H_{\text{SB}}\rangle$ changes the total mean energy. The state mixedness remains unchanged, that is,

$$\text{Tr}(\rho^{\phi})^2 = \text{Tr}\rho^2_{\text{Eq}}. \tag{16.36}$$

Physically, such a rotation corresponds to an AC Stark shift of the TLS level separation induced by an external field.

The energy pumped into the system plus bath is maximal when the rotation angle is $\phi = \pi$, which corresponds to the phase-flip operation. Since (16.30) is not a Gibbs state, which is the minimal-energy state for that entropy, it corresponds to higher mean energy; hence

$$\delta\langle H_{\text{tot}}\rangle^{\phi} > 0, \tag{16.37}$$

thus completing the general proof of heating following a QND disturbance.

16.1.4 Non-Markovian Alternating Heating and Cooling

The evolution of ρ at longer times (in the regime of weak system–bath coupling) may be approximately described by the second-order non-Markovian master equation (ME) derived in Chapter 11. The ME for ρ, on account of its diagonality, can be cast into the TLS-population rate equations. Upon dropping the subscript S, setting the measurement time to be $t = 0$, and following the notation of Chapters 10 and 11, we then find the ME (11.79a), where the relaxation rates of the excited (ground) states $\gamma_{e(g)}(t)$, given by the real parts in (11.72), can be recast in the form,

$$\gamma_{e(g)}(t) = 2t \int_{-\infty}^{\infty} d\omega \, G_T(\omega) \, \text{sinc}[(\omega \mp \omega_{\text{a}})t]. \tag{16.38}$$

Thus, the dynamics is determined by the relaxation rates $\gamma_{e(g)}(t)$, whose non-Markovian time dependence yields three distinct regimes:

(i) At ultrashort times $t \ll 1/\omega_\mathrm{a} \ll t_c$ the sinc function in (16.38) is much broader than $G_T(\omega)$. The $|e\rangle \to |g\rangle$ and $|g\rangle \to |e\rangle$ transition rates, $\gamma_e(t)$ and $\gamma_g(t)$, are then *equal at any temperature*, indicating the complete breakdown of the RWA, since these transition rates then do not depend on the thermal quantum occupancy of the bath, but reflect the time-energy uncertainty. The rates $\gamma_{e(g)}(t)$ then become *linear in time*, which is the QZE signature (Ch. 10):

$$\gamma_{e(g)}(t \ll t_c) \approx 2\dot{\gamma}_0 t, \tag{16.39}$$

where

$$\dot{\gamma}_0 \equiv \int_{-\infty}^{\infty} d\omega \, G_T(\omega) = \langle \hat{B}^2 \rangle. \tag{16.40}$$

This ultrashort time regime then gives rise to the *universal Zeno heating rate*:

$$\frac{d}{dt}\left(\rho_{ee} - \rho_{gg}\right) \approx 4\dot{\gamma}_0 t(\rho_{gg} - \rho_{ee}). \tag{16.41}$$

In the ultrashort time limit that coincides with the QZE regime, where $\mathrm{sinc}[(\omega \mp \omega_\mathrm{a})t] \approx 1$, drastically different transition rates are obtained within the RWA, namely,

$$\gamma_e^{\mathrm{RWA}} = 2t \int_0^{\infty} d\omega \, [\bar{n}_T(\omega) + 1] G_0(\omega), \tag{16.42}$$

$$\gamma_g^{\mathrm{RWA}} = 2t \int_0^{\infty} d\omega \, \bar{n}_T(\omega) G_0(\omega), \tag{16.43}$$

and without the RWA,

$$\gamma_e = \gamma_g = \gamma_e^{\mathrm{RWA}} + \gamma_g^{\mathrm{RWA}}. \tag{16.44}$$

Thus, at ultrashort times after a measurement, the energy uncertainty is so large that the TLS couples to all modes of the bath, at both positive and negative energies. Then, there is no distinction between the ground and excited states, $|g\rangle$ and $|e\rangle$, as regards their coupling to the bath. By contrast, the RWA increasingly distinguishes between $|g\rangle$ and $|e\rangle$ as t decreases.

(ii) At intermediate non-Markovian times, $t \gtrsim 1/\omega_\mathrm{a}$, when the sinc function and $G_T(\omega)$ in (16.38) have comparable spectral widths, the change in their spectral overlap with time results in damped aperiodic oscillations of $\gamma_e(t)$ and $\gamma_g(t)$, near the frequencies $\omega_0 - \omega_\mathrm{a}$ and $\omega_0 + \omega_\mathrm{a}$, respectively. The rotating-wave (RW) and counterrotating (CR) terms induce different oscillations of $\gamma_{e(g)}(t)$. These changes are confirmed when the rate equations are integrated. In the notation of Chapter 11, we find

$$\rho_{ee}(t) = e^{-J(t)}\left[\int_0^t dt' e^{J(t')} \gamma_g(t') + \rho_{ee}(0)\right]. \tag{16.45}$$

Here the relaxation (decay) factor is

$$J(t) = J_g(t) + J_e(t) \tag{16.46}$$

with

$$J_{e(g)}(t) \equiv t\bar{\gamma}_{e(g)}(t) = \int_0^t dt' \gamma_{e(g)}(t'), \tag{16.47}$$

where $\bar{\gamma}_{e(g)}(t)$ are the time-averaged rates. This oscillatory time dependence stems from time-energy uncertainty, which determines the temporal width and oscillations of the sinc function, and conforms neither to QZE nor to the converse AZE of relaxation speedup. It will henceforth be dubbed the *oscillatory Zeno effect* (OZE). The negativity of the sinc function in (16.38) between its consecutive maxima allows for a *negative instantaneous relaxation rate*, which is completely forbidden by the RWA. Since $\mathrm{sinc}[(\omega + \omega_a)t]$ is much further shifted from the peak of $G_T(\omega)$ than $\mathrm{sinc}[(\omega - \omega_a)t]$, the $|g\rangle \to |e\rangle$ rate $\gamma_g(t)$ is more likely to be *negative* than its inverse, $\gamma_e(t)$. Hence, $\rho_{gg}(t)$ may grow at the expense of $\rho_{ee}(t)$ more than allowed by the detailed balance at equilibrium. This growth of $\rho_{gg}(t)$ corresponds to *transient cooling*, following the heating (Figs. 16.2, 16.3). Although $\gamma_{e(g)}(t)$ can be negative, their integrals (16.47) are nonnegative. As a result, the TLS density matrix ρ does not have negative eigenvalues, which would be unphysical.

(iii) At long times compared to the correlation (memory) time of the bath, $t \gg t_c$, the relaxation rates attain their Golden-Rule (Markovian) values (Chs. 5, 11):

$$\gamma_{e(g)}(t \gg t_c) \simeq 2\pi G_T(\pm\omega_a). \tag{16.48}$$

The TLS state populations then approach those of an equilibrium Gibbs state whose temperature is equal to that of the thermal bath (Fig. 16.4).

In the Markovian long-time limit, where $\mathrm{sinc}((\omega \pm \omega_a)t) \approx \delta(\omega \pm \omega_a)$, the transition rates obtained with and without the RWA have the same constant values,

$$\gamma_e = \gamma_e^{\mathrm{RWA}} = 2\pi[\bar{n}_T(\omega_a) + 1]G_0(\omega_a), \tag{16.49}$$

$$\gamma_g = \gamma_g^{\mathrm{RWA}} = 2\pi \bar{n}_T(\omega_a)G_0(\omega_a), \tag{16.50}$$

the ratio of these rates conforming to the thermal equilibrium (Gibbs) state of the system,

$$\frac{\gamma_e}{\gamma_g} = e^{\beta\hbar\omega_a}. \tag{16.51}$$

The coupling strength of N TLS to a cavity field at the antinode is $N^{1/2}\eta_{\max}/(2\pi)$. An optical beam will rotate its polarization [Fig. 16.1(b)], thus performing an ultrafast QND measurement that resolves $|g\rangle$ and $|e\rangle$, depending on their contrasting magnetic numbers, whereas their energies remain unresolved. The optical Rabi frequency that can cause the ensemble heating or cooling via QND

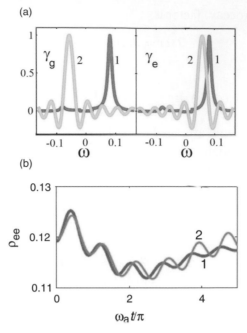

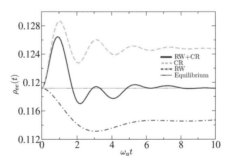

Figure 16.2 (a) $\gamma_g(t)$ and $\gamma_e(t)$ depicted as spectral overlaps of the relevant $G_T(\omega)$ (line 1) and $\text{sinc}(\omega \pm \omega_a)$ (line 2) functions. (b) A TLS undergoes first heating and then oscillatory cooling, according to the master equation (line 1) and the exact numerical solution for a bath of 40 modes (line 2). The discrete bath approximates the Lorentzian spectrum (11.123) at $T = 0$, $\kappa = \omega_a/5$, and $\gamma_a = 0.07\omega_a$. (Adapted from Erez et al., 2008.)

Figure 16.3 Evolution of a TLS coupled to a bath consisting of 10^3 oscillators under the Hamiltonian (16.1) with $H_{SB} \propto \sigma_x$. Excited-state population free evolution (as a function of $\omega_a t$) induced by only rotating (RW) terms (dash-dot), only CR terms (dashed), and the entire Hamiltonian (solid). Here we assume a Lorentzian coupling spectrum (11.52) with $G_0(\omega) = \eta\Gamma_B^2(\omega/\omega_a)/[\Gamma_B^2 + (\omega - \omega_0)^2]$ with $\beta \equiv 1/(k_B T) = 2/(\hbar\omega_a)$, $\eta = \omega_a/25$, $\Gamma_B \equiv 1/t_c = \omega_a/10$, and $\omega_0 = 2\omega_a$. (Reprinted from Gordon et al., 2009.)

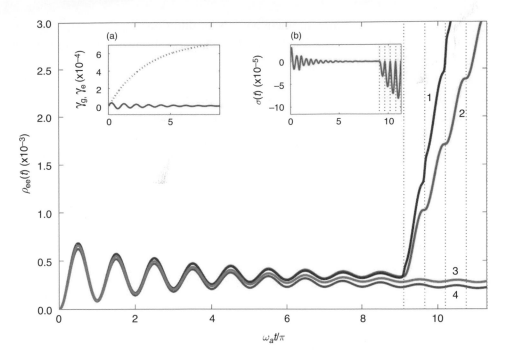

Figure 16.4 TLS evolution in a thermal bath as a function of time. Main panel: Excited-level population relaxation to equilibrium as a function of time (in units of π/ω_a) for initially zero-temperature product state of the TLS and the bath. Upon reaching equilibrium, the TLS is subjected to a series of measurements (at times marked by vertical dashed lines). Measurements of finite duration ($\Delta t_m = 0.11/\omega_a$) (line 1) result in higher heat-up than impulsive measurements (line 2), but the qualitative effect is the same for both. There is complete agreement between the second-order master equation (line 4), simplified numerical solution (two-quanta exchange with a discrete bath) (lines 1, 2), and the numerical solution for a discrete bath of 40 modes (line 3). (a) Relaxation rates, γ_g (solid line), γ_e (dotted line), as a function of time. (b) Time dependence of entropy production $\sigma(t)$. The bath parameters are the same as in Figure 16.2. (Adapted from Erez et al., 2008.)

disturbances (measurements or phase shifts) should satisfy $\Omega > \omega_a > N^{1/2}\eta_{max}$. A train of optical pulses at $\sim$ ns intervals, each with a Rabi frequency $\Omega \lesssim 1$ GHz, can effect the required QND polarization measurements or phase shifts for $\omega_a \lesssim 1$ MHz that would cause alternating heating and cooling.

16.2 Non-Markovian TLS Heating or Cooling by Repeated QND Disturbances

16.2.1 Repeated Measurements

After equilibrium has been reached, we may perform a series of quantum non-demolition (QND) measurements of the TLS energy states at times separated by $\tau_m = t_{m+1} - t_m$. Each measurement has a brief duration $\Delta t_m \ll 1/\omega_a$. Our aim is to explore the evolution as a function of the time separation between consecutive measurements in the non-Markovian domain, $\tau_m \ll t_c$.

In order to realistically model the repeated measurements, we assume a smooth temporal profile of finite duration for the coupling to the meter (detector). The mth measurement then occurs at time t_m and has a duration of Δt_m. The population evolution via projective (impulsive) measurements differs from its counterpart under finite-duration measurements. Finite-duration measurements increase the Zeno heating as compared to impulsive ones because of the extra energy supplied by the TLS coupling to the meter (detector). However, the basic effect is seen (Fig. 16.4 – main panel) to be the same and is governed by the abrupt changes of $\langle H_I \rangle = \langle H_{SB} \rangle$. Counterintuitively, finite-duration measurements are also able to *increase the cooling*, despite the extra energy pumped in by the meter.

If we repeat this procedure often enough, the TLS can either increasingly heat up or cool down, upon choosing the time intervals τ_m to coincide with either the peaks or the troughs of the ρ_{ee} oscillations in time, respectively. The oscillatory entropy production rate σ can also be chosen to progressively lower or raise the TLS entropy with each measurement [Fig. 16.4(b)]. Since consecutive measurements affect the bath and the system differently, the two may acquire different temperatures. These temperatures then become the initial conditions for subsequent QZE heating or OZE cooling. Remarkably, the system may heat up solely on account of the QZE, although the *bath is colder*, or cool down solely on account of the OZE or AZE, although the *bath is hotter* (Fig. 16.5). The bath undergoes changes in temperature and entropy too.

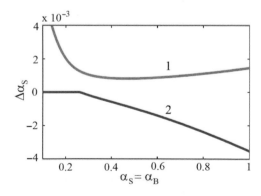

Figure 16.5 Maximal QZE temperature rise (heating) $\Delta\alpha_S > 0$ (line 1) and subsequent maximal QZE temperature drop (cooling) $\Delta\alpha_S < 0$ (line 2) as a function of the common initial temperature of the TLS and the bath $\alpha_{S(B)} = \hbar\omega_a/\beta_{S(B)}$. The bath parameters are the same as in Figure 16.2. (Adapted from Erez et al., 2008.)

During the $m + 1$ interval after m measurements separated by τ, we have in (16.38) the transition rates

$$\gamma_{e(g)}(t) = 2(t - m\tau) \int_{-\infty}^{\infty} d\omega \, G_T(\omega) \, \text{sinc}[(\omega \mp \omega_a)(t - m\tau)], \qquad (16.52)$$

the upper (lower) sign standing for γ_e (γ_g). Here we have assumed that each measurement decorrelates the qubit and the bath anew, and thus *resets* the time to $t - m\tau$.

As discussed in Section 16.1, these rates remain time dependent at longer but still non-Markovian times $1/\omega_a \ll t \lesssim t_c$, where t_c, the bath correlation (memory) time, is the inverse width of $G_0(\omega)$.

The relaxation integrals (16.47) may be written as [see (11.94a) and (11.97a)]

$$J_{e(g)}(t) = 2\pi t \int_{-\infty}^{\infty} d\omega \, G_T(\omega) F_t(\pm\omega - \omega_a), \qquad (16.53)$$

where the upper (lower) sign stands for $e(g)$ and the spectral function in the integrals is [cf. (11.102) and (11.101)]

$$F_{t=m\tau}(\omega) = \frac{\tau}{2\pi} \text{sinc}^2 \frac{\omega\tau}{2}. \qquad (16.54)$$

The function $F_t(-\omega - \omega_a)$ in (16.53) gives rise to the faster, counter-resonant (CR) (alias antiresonant) term into oscillation of $\rho_{ee}(t)$. This function is peaked near the frequency $\omega_0 + \omega_a$, where ω_0 is the frequency at which $G_T(\omega)$ is maximal. The CR term accounts for *heating* (Fig. 16.3). It is dominant for ultrashort time intervals, $\tau \ll 1/\omega_a$, consistently with the conclusion (discussed in Sec. 16.1) that heating prevails immediately after a disturbance. On this timescale, the energy spread $1/\tau$ exceeds ω_a; hence $|e\rangle$ and $|g\rangle$ cannot be discriminated, so that the TLS may then absorb or emit quanta at the expense of the system–bath interaction energy, not of the bath.

The faster CR term in the oscillation averages out at lower measurement rates, $1/\tau \lesssim \omega_a$. For such measurement rates, the CR term is surpassed by the RW terms in the oscillation of $\rho_{ee}(t)$, which is peaked near the frequency $\omega_0 - \omega_a$. This RW oscillation term is seen in Figures 16.2 and 16.3 to be *responsible for cooling*. RW processes are resonant: Any change in the TLS energy is compensated for by the bath. Hence, we may conclude that *non-Markovian cooling of the TLS comes at the expense of the bath heating*. Although RW and CR oscillations occur on different timescales, cooling can be properly analyzed *only by allowing for both* oscillations: The CR oscillation restricts the amount of cooling that might have occurred had only RW terms been in force.

16.2.2 Repeated Phase Shifts (Rotations)

Free evolution intermittent with m impulsive phase shifts (ϕ rotations of σ_z) at intervals τ still conforms at the $m + 1$ interval to (16.45), but the transition rates vary *continuously*, without being reset to zero every τ, as they do in the case of measurements. The spectral filter function, to be used in (16.53) as the counterpart of (16.54) in the case of m repeated phase shifts ϕ, can be described as in Chapter 11: The phase-modulation function $\epsilon(t)$ given by (11.67) takes the form

$$\epsilon(t) = e^{i[t/\tau]\phi}, \tag{16.55}$$

where [...] denotes the integer part of a number. For this $\epsilon(t)$, (11.92) and (11.95) yield the spectral filter function

$$F_{t=m\tau}^{\phi}(\omega) = \frac{\tau}{2\pi}\text{sinc}^2\left(\frac{\omega\tau}{2}\right)\frac{\sin^2[m(\omega\tau - \phi)/2]}{m\sin^2[(\omega\tau - \phi)/2]}. \tag{16.56}$$

At ultrashort times, $t \lesssim 1/\omega_a$, $F_t^{\phi}(\omega)$ causes sinc^2 oscillations of the relaxation integrals $J_{e(g)}(t)$ (Fig. 16.6), so that the detailed-balance modifications are then rather similar for phase shifts and measurements [cf. (16.54)]. At longer times, $t \gg 1/\omega_a$, $F_t^{\phi}(\omega)$ in (16.56) becomes a weighted sum of two peaks, corresponding

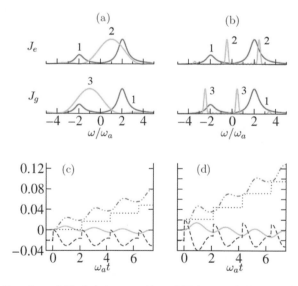

Figure 16.6 Overlap of $G_T(\omega)$ (curves 1) and $F_t(\omega - \omega_a)$ (curves 2) or $F_t(-\omega - \omega_a)$ (curves 3) in $J_e(t)$ and $J_g(t)$, for (a) measurements; (b) π phase flips. (c) Evolution of $\langle H_S(t)\rangle$ (solid), $\langle H_B(t)\rangle$ (dash-dot), $\langle H_{SB}(t)\rangle$ (dashed), and $\langle H_{\text{tot}}(t)\rangle$ (dotted) under repeated measurements at intervals $\omega_a\tau = 2$. Here $\langle H_S(0)\rangle = 0$ and $\langle H_B(0)\rangle = 0$. (d) Likewise, under repeated π-phase flips at intervals $\omega_a\tau = 2$. Parameters are the same as in Figure 16.3. (Adapted from Gordon et al., 2009.)

to $\delta(\omega - \phi/\tau)$ and $\delta(\omega - (\phi - 2\pi)/\tau)$. They correspond to the splitting of the excited energy level of the TLS into quasienergy levels shifted from ω_a by ϕ/τ and $(\phi - 2\pi)/\tau$. For phase-flips $\phi = \pi$, (16.56) becomes the sum of two terms with equal magnitude and *opposite frequency shifts*, $\pm\pi/\tau$, whereas for small phase increments, $|\phi| \ll 1$, the term that yields a unidirectional frequency shift, ϕ/τ, dominates (Ch. 11).

If one resorts to impulsive π-phase flips, the resulting redistribution is determined by the modulation of the TLS level-distance by spectral shifts π/τ: $\delta \langle H_S \rangle$ then undergoes *parametric modulation*, during its exchange with $\delta \langle H_B(t) \rangle + \delta \langle H_{SB}(t) \rangle$.

Phase flips allow for more pronounced cooling than measurements at a similar rate π/τ (Fig. 16.6): If the interval τ is kept fixed, the amplitude of $\delta \langle H_S(t) \rangle$ oscillations is maximized for phase flips compared to measurements or small phase shifts, since the energy transfer from the system to the bath is the largest for $\phi = \pi$. Yet, does this imply that the cumulative cooling following many disturbances is also maximized for phase flips? As shown below, this is not necessarily the case.

16.2.3 How Does Non-Markovian Cooling Occur?

We are now in a position to interpret and set the rules for non-Markovian cooling by various QND disturbances. The cooling process may be viewed as the time-dependent redistribution of the excess total energy imparted to the system plus bath by the impulsive disturbance. This redistribution causes the cooling of $\langle H_S \rangle$ to be the *deficit* between the heating of $\langle H_B \rangle$ and the cooldown of $\langle H_{SB} \rangle$, subject to the following conservation law for their respective changes (δ):

$$\delta \langle H_S(t) \rangle = -[\delta \langle H_B(t) \rangle + \delta \langle H_{SB}(t) \rangle]. \tag{16.57}$$

RW-induced oscillations amount to $\delta \langle H_S \rangle = -\delta \langle H_B \rangle$, that is, energy exchange between the *system and bath only*. By contrast, CR-induced oscillations give rise to rapid variation of $-[\delta \langle H_{SB} \rangle + \delta \langle H_B \rangle]$, causing heating, $\delta \langle H_S \rangle > 0$.

During the free evolution between measurements, the relaxation integrals in (16.47) oscillate as $\mathrm{sinc}^2[(\omega_a \pm \omega_0)\tau/2]$ at $\tau \le 1/\omega_a$, ω_0 being the peak frequency of the bath response. These underdamped *frequency beats* that govern the alternating heating and cooling attest to the remarkably simple *quasi-reversibility* of the system–bath energy exchange, on the timescale of ω_a^{-1}, irrespective of the bath complexity. Namely, all the bath oscillators *oscillate in unison* on this timescale that is much shorter than the non-Markovian bath correlation (memory) time t_c [Fig. 16.6(c)]. Hence, their energy exchange with the system is then much more effective than at longer (Markovian) times when the bath oscillators are out of tune. Although maximal cooling occurs at time intervals $\tau \simeq \pi/(\omega_a - \omega_0)$ between consecutive QND disturbances, poorer time resolution does not invalidate this cooling procedure, attesting to its robustness.

16.3 Control of Steady States by QND Disturbances

16.3.1 Thermal Equilibration of TLS in the Markovian Limit

In the Markovian limit, $t \gg t_c$, the sinc function is much narrower than the coupling spectrum $G_T(\omega)$ and can be approximated by a delta function in frequency, resulting in

$$\rho_{ee}(t \gg t_c) = \rho_{ee}(0)e^{-2\gamma t} + (\rho_{ee})_{\text{Eq}}(1 - e^{-2\gamma t}). \tag{16.58}$$

Here the Markovian (Golden-Rule) decay rate is given by

$$\gamma \equiv \frac{1}{2}(\gamma_e + \gamma_g) = \pi[G_T(\omega_{\text{a}}) + G_T(-\omega_{\text{a}})]. \tag{16.59}$$

Correspondingly, (16.58) shows the convergence of ρ_{ee} at the Golden-Rule rate γ to the thermal equilibrium state

$$(\rho_{ee})_{\text{Eq}} = \frac{e^{-\beta\hbar\omega_{\text{a}}/2}}{2\cosh(\beta\hbar\omega_{\text{a}}/2)}. \tag{16.60}$$

16.3.2 Convergence to a Steady State by Frequent QND Monitoring

Here we study the dynamics of a monitored (measured) TLS coupled to a bath, so as to examine whether the steady state under repeated measurements is still the equilibrium state (16.60). To this end, we define the monitoring superoperator $\hat{M}_\tau$,

$$\hat{M}_\tau \rho_{\text{tot}} = \hat{P}\hat{L}_\tau \rho_{\text{tot}}, \tag{16.61}$$

composed of a unitary propagator of the total Hamiltonian over time τ, $U(\tau)$,

$$\hat{L}_\tau \rho_{\text{tot}} = U(\tau)\rho_{\text{tot}}U^\dagger(\tau), \tag{16.62}$$

followed by a projection superoperator $\hat{P}$ corresponding to a *nonselective*, impulsive measurement in the basis of the free-system energy states $|j\rangle$,

$$\hat{P}\rho_{\text{tot}} = \sum_j |j\rangle\langle j| \rho_{\text{tot}} |j\rangle\langle j|. \tag{16.63}$$

Here ρ_{tot} is the total (generally entangled system + bath) density matrix. The evolution following m consecutive applications of (16.61) is given by

$$\rho_{\text{tot}}(t = m\tau) = \hat{M}_\tau^m \rho_{\text{tot}}(0). \tag{16.64}$$

The asymptotics of (16.64) as $m \to \infty$ yields the steady state of the evolution,

$$\chi(\tau) = \lim_{m \to \infty} \text{Tr}_\text{B}[\hat{M}_\tau^m \rho_\text{tot}(0)], \tag{16.65}$$

where Tr_B is the trace over the bath.

In the case of a TLS coupled to a bath and repeated projections of the total state onto the energy states of the TLS, the convergence to the steady state assumes the exponential form:

$$\rho_{ee}(t = m\tau) = e^{-2\bar{\gamma}(\tau)t} \rho_{ee}(0) + \chi(\tau)[1 - e^{-2\bar{\gamma}(\tau)t}]. \tag{16.66}$$

Here $\bar{\gamma}(\tau)$ is the time-dependent relaxation rate $\gamma(t)$, (11.71), averaged over the interval $[0, \tau]$,

$$\bar{\gamma}(\tau) = \frac{1}{2\tau} \int_0^\tau dt' [\gamma_g(t') + \gamma_e(t')], \tag{16.67}$$

and

$$\chi(\tau) = \frac{\int_0^\tau dt' e^{2\bar{\gamma}(t')t'} \gamma_g(t')}{\int_0^\tau dt' e^{2\bar{\gamma}(t')t'} [\gamma_g(t') + \gamma_e(t')]}. \tag{16.68}$$

As discussed above, measurement-induced steady states depend on whether the interval between measurements is short enough to give rise to non-Markovian effects and whether the counterrotating (CR) terms affect the system–bath interaction. As τ varies in the non-Markovian regime, $\bar{\gamma}(\tau)$ alternates between AZE speedup and QZE slowdown. Likewise, the steady state of the evolution has a non-monotonic dependence on τ in this regime. One can then reach a steady state that has higher or lower entropy than the thermal equilibrium [i.e., one can either mix (heat up) or purify (cool down) the state upon choosing the appropriate measurement interval].

The convergence to steady state according to (16.66) rests on the crucial fact that quantum nonselective, impulsive measurements obliterate any TLS coherence: They "reset the clock" so that consecutive steps of the evolution depend only on the interval τ and not on t. Since the first measurement erases the off-diagonal terms, they are irrelevant for all subsequent evolution. Hence, initial states that are diagonal in the free-Hamiltonian eigenbasis are the pointer states of the evolution (Ch. 9).

For the Hamiltonian considered here, an off-diagonal element ρ_{eg} can be shown to decay to zero at a rate $\bar{\gamma}(\tau)$. The relaxation rate $2\bar{\gamma}(\tau)$ and the steady-state excitation $\chi(\tau)$ are determined by the free evolution during τ, which is unitary for a closed TLS and non-unitary for an open TLS due to the tracing-out of the bath.

We may contrast (16.67) with the result of monitoring a closed TLS governed by the Hamiltonian

$$H_\text{closed} = -\frac{\hbar\Omega}{2}\sigma_x, \tag{16.69}$$

Ω being the Rabi flipping frequency between $|e\rangle$ and $|g\rangle$. The convergence rate and the steady state are then

$$\gamma_{\text{closed}}(\tau) = -\frac{\ln|\cos(\Omega\tau)|}{\tau},$$

$$\chi_{\text{closed}} = \frac{1}{2} \quad \forall\tau. \tag{16.70}$$

Namely, the steady state is fully mixed.

16.3.3 The Quantum Zeno (QZE) Regime

The timescale of the closed-TLS dynamics is the Rabi oscillation period $2\pi/\Omega$ that defines the QZE condition (Ch. 10): For $\Omega\tau \to 0$, the evolution converges at the linear rate $\gamma_{\text{closed}} \simeq \Omega^2\tau \to 0$. Thus, for a fixed time and increasing number of measurements, the evolution tends to "freeze." However, there is a physical limit to the smallness of τ. The finiteness of τ results in the TLS convergence, albeit slow, to the only steady state of the evolution, which is the fully mixed state.

The shortest relevant timescale is the natural oscillation period of the system, $2\pi/\omega_{\text{a}}$. More-frequent measurements, $\tau \ll 1/\omega_{\text{a}} \ll t_c$, yield:

$$\gamma_{\text{open}}(\tau \ll 1/\omega_{\text{a}}) = 4A\tau, \quad \chi_{\text{open}}(\tau \ll 1/\omega_{\text{a}}) \approx \frac{1}{2}, \tag{16.71}$$

where $A = \int_{-\infty}^{\infty} d\omega\, G_T(\omega)$ is the spectrally integrated (total) system–bath coupling strength. This timescale reveals the closest analog to the closed-system QZE, limit of (16.70), in that the convergence rate is linear in τ and the steady state is the fully mixed state. Thus, the QZE in open quantum systems paradoxically results in a fully mixed steady state and not in "freezing" the evolution.

The non-Markovian dynamics by itself or, equivalently, the condition $\tau \ll t_c$ does not suffice to ensure the QZE in open systems. According to (16.71), it requires such ultrashort measurement intervals that it must account for the CR terms. Disregarding this fact and taking the RWA in the QZE regime, $\omega_{\text{a}}\tau \ll 1$, yields a convergence $\propto \tau$, at only *half* the convergence rate obtained from the complete Hamiltonian. More importantly, the steady state obtained under the RWA strongly depends on the bath-coupling spectrum, $G_T(\omega)$, contrary to the true steady state obtained without the RWA that is the fully mixed state.

We can understand the convergence to the steady state by considering the two competing aspects of open-system monitoring. The projective measurements of σ_z have quantum non-demolition (QND) effects on the system and hence preserve its entropy. The bath-induced evolution, on the other hand, can either increase or decrease its entropy, depending on the bath-coupling spectrum, the evolution interval τ, and the state of the system at that time. The RW terms preserve the combined system–bath state within the same energy subspaces, thus acting as if these were multiple closed systems. The coupling strength and the measurement interval determine the relative weights of these subspaces that are invariant under measurements within the RWA. By contrast, the CR terms mix these energy subspaces, thereby turning the state of system plus bath into that of a *single closed system*, which should therefore conform to (16.70). The shorter the measurement time interval, the larger the contribution of the CR terms, which results in an increasing entropy (heat-up) of the steady state.

16.3.4 Cooling by Monitoring

By applying m repeated QM disturbances at intervals τ and taking the limit $m \to \infty$, we asymptotically approach a *fixed point* of the evolution, given by

$$\rho_{ee}(t = m\tau) \xrightarrow{m \to \infty} \frac{\int_0^\tau dt\, e^{J(t)} \gamma_g(t)}{\int_0^\tau dt\, e^{J(t)} [\gamma_g(t) + \gamma_e(t)]} \equiv \rho_{ee}^{(\infty)}(\tau). \tag{16.72}$$

Here $J(t) = J_g(t) + J_e(t)$, whereas $\gamma_{e(g)}(t)$ and $J_{e(g)}(t)$ are defined by (16.52) and (16.53) with $F_t(\omega)$ given by (16.54) for measurements and by (16.56) for phase shifts.

We thus obtain the following universal condition for the TLS-state purification (cooling) that increases with m:

$$\rho_{ee}(m\tau) > \chi \equiv \min_\tau \rho_{ee}^{(\infty)}(\tau), \tag{16.73}$$

where the *minimal achievable excitation*, χ (that corresponds to the coldest achievable temperature of the TLS) is found upon varying τ [Fig. (16.7)(a)].

Provided that the TLS-bath coupling is weak enough, $\epsilon = (\eta_k)_{\max}/\omega_a \ll 1$, (16.73) is simplified to

$$\rho_{ee}(m\tau) > \chi \simeq \min_\tau \frac{J_g(\tau)}{J(\tau)}. \tag{16.74}$$

The *universal cooling bound* expressed by (16.74) is the central result of this section. It is noteworthy that the *coupling strength* ϵ^2 *does not affect the cooling condition or the bound* χ, as both the numerator and denominator of (16.74) are proportional to it.

The excited-state population of the TLS after time t is given by

$$\rho_{ee}(t) = e^{-J(t)} \rho_{ee}(0) + e^{-J(t)} \int_0^t dt'\, e^{J(t')} \gamma_g(t'), \tag{16.75}$$

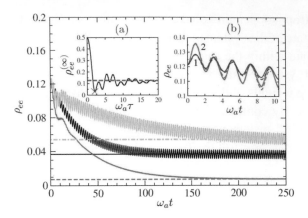

Figure 16.7 Main panel: excitation falloff to the cooling bound χ^M (solid) for measurements, χ^π (dash-dot) for phase flips, and $\chi^{\pi/10}$ (dashed) for small phase flips. Parameters are the same as in Figure 16.3. (a) Asymptotic fixed point, $\rho_{ee}^{(\infty)}(\tau)$, as a function of measurement intervals, τ, with χ marked. (b) Excitation as a function of time for repeated measurements (curve 1) and phase flips (curve 2), the same parameters except for $\eta = \omega_a/100$, $\Gamma_B = \omega_a/2$. Numerical simulations (dashed) for a 28-mode bath fully agree with the ME solutions (solid). (Adapted from Gordon et al., 2009.)

where $\gamma_g(t)$ and $J(t)$ depend on the specific disturbances and the interval τ between them. In the limit $t \to \infty$, the first term in (16.75) vanishes, and the universal cooling bound is obtained:

$$\rho_{ee}(t) \xrightarrow{\;t \to \infty\;} \frac{\int_0^t dt' e^{J(t')} \gamma_g(t')}{\int_0^t dt' e^{J(t')} \gamma(t')} \equiv \rho_{ee}^{(\infty)}(\tau). \qquad (16.76)$$

In this limit, the cooling condition is simplified to

$$\chi < \rho_{ee}(0) \;\Rightarrow\; \frac{J_e(t)}{J_g(t)} > \left(\frac{J_e}{J_g}\right)_{\mathrm{Eq}} = \left(\frac{\gamma_e}{\gamma_g}\right)_{\mathrm{Eq}} = \frac{\bar{n}_T(\omega_a) + 1}{\bar{n}_T(\omega_a)}, \qquad (16.77)$$

where the relaxation integrals, $J_{e(g)}(t = m\tau) = \int_0^\infty dt' \gamma_{e(g)}(t')$, are explicitly given by

$$J_g(t = m\tau) = m\tau^2 \left[\int_0^\infty d\omega\, G_0(\omega)\bar{n}_T(\omega)\mathrm{sinc}^2((\omega - \omega_a)\tau/2) \right.$$
$$\left. + \int_0^\infty d\omega\, G_0(\omega)[\bar{n}_T(\omega) + 1]\mathrm{sinc}^2((\omega + \omega_a)\tau/2) \right],$$

$$J_e(t = m\tau) = m\tau^2 \left[\int_0^\infty d\omega\, G_0(\omega)\bar{n}_T(\omega)\mathrm{sinc}^2((\omega + \omega_a)\tau/2) \right.$$
$$\left. + \int_0^\infty d\omega\, G_0(\omega)[\bar{n}_T(\omega) + 1]\mathrm{sinc}^2((\omega - \omega_a)\tau/2) \right]. \quad (16.78)$$

Equivalently, (16.77) and (16.78) yield the cooling condition

$$\int_0^\infty d\omega\, G_0(\omega)\, \mathrm{sinc}^2\left(\frac{\omega - \omega_a}{2}\tau\right)[\bar{n}_T(\omega_a) - \bar{n}_T(\omega)]$$
$$> \int_0^\infty d\omega\, G_0(\omega)\, \mathrm{sinc}^2\left(\frac{\omega + \omega_a}{2}\tau\right)[\bar{n}_T(\omega_a) + \bar{n}_T(\omega) + 1], \quad (16.79)$$

where the left-hand side stems from RW terms of the interaction Hamiltonian H_{SB} and the right-hand side stems from CR terms therein. This inequality underscores the need to account for both RW- and CR-induced effects when considering cooling.

In the high-temperature limit, $\bar{n}_T(\omega) = 1/(\beta\hbar\omega) \gg 1$, a necessary condition to satisfy the inequality (16.79) is that the spectral bath-response $G_0(\omega)$ be concentrated in the frequency range

$$\omega_a < \omega < \omega_{\max}, \quad (16.80)$$

where

$$\omega_{\max} = \frac{1}{\beta\hbar}\left(1 + \frac{\beta\hbar\omega_a + \sqrt{4 + 12\beta\hbar\omega_a + (\beta\hbar\omega)^2}}{2}\right). \quad (16.81)$$

Condition (16.80) implies that the bath spectrum should not be detuned too far above ω_a, if we wish to achieve cooling. Conversely, there are frequency ranges where there is no cooling, regardless of the shape of $G_0(\omega)$.

Upon using the appropriate expressions for the relaxation integrals $J_{e(g)}(t)$, we obtain the following cooling condition:

$$\frac{\int_0^\infty d\omega\, G_0(\omega) F_t(\omega_a - \omega)[\bar{n}_T(\omega_a) - \bar{n}_T(\omega)]}{\int_0^\infty d\omega\, G_0(\omega) F_t(\omega_a + \omega)[\bar{n}_T(\omega_a) + \bar{n}_T(\omega) + 1]} > 1. \quad (16.82)$$

For small phase shifts $|\phi| \ll 1$, this condition reduces to

$$\phi > 0. \quad (16.83)$$

This cooling condition is completely insensitive to the coupling spectrum $G_0(\omega)$, in sharp contrast to the same condition for measurements and π-phase flips.

For $m \gg 1$ repeated measurements, the minimal achievable excitation is found to be

$$\chi_M = \min_\tau \frac{\int_{-\infty}^\infty d\omega\, G_T(\omega)\, \mathrm{sinc}^2((\omega + \omega_a)\tau)}{\int_{-\infty}^\infty d\omega\, G_T(\omega)[\mathrm{sinc}^2((\omega - \omega_a)\tau) + \mathrm{sinc}^2((\omega + \omega_a)\tau)]}. \quad (16.84)$$

After m π-phase flips, such that $m\tau \gg 1/\omega_a$, but $\pi/\tau > \omega_a$, the minimal excitation in the low-temperature limit for the bath is given by

$$\chi_\pi = \min_\tau \frac{G_-(n_- + 1) + G_+ n_+}{G_+(2n_+ + 1) + G_-(2n_- + 1)} \qquad (16.85)$$

$$\xrightarrow{\beta \to \infty} \min_\tau \frac{G_- + G_+ n_+}{G_+ + G_-}, \qquad (16.86)$$

where $G_\pm \equiv G_0(\pi/\tau \pm \omega_a)$ and $n_\pm \equiv \bar{n}_T(\pi/\tau \pm \omega_a)$. This result is an average between $n_- + 1 \simeq 1$ and n_+, weighted by G_- and G_+, respectively.

The counterpart of (16.86), under the same conditions but for repeated small phase shifts $\phi \ll 1$, is found to be

$$\chi_{\phi \ll 1} = \frac{\bar{n}_T(\omega_a + \phi/\tau)}{2\bar{n}_T(\omega_a + \phi/\tau) + 1} \xrightarrow{\beta \to \infty} \left[1 + e^{\beta\hbar(\omega_a + \phi/\tau)}\right]^{-1}. \qquad (16.87)$$

This minimal excitation χ sets the maximal achievable purity, $1 - \chi$, by QND means, for a given bath temperature and the chosen bath-coupling spectrum. This cooling bound should be contrasted with the Markovian cooling bound, set by the $|e\rangle \leftrightarrow |g\rangle$ transition rates $\gamma_{e(g)}(t \to \infty)$:

$$\chi_{\text{Markov}} = \frac{\bar{n}_T(\omega_a)}{2\bar{n}_T(\omega_a) + 1} \xrightarrow{\beta \to \infty} \left(1 + e^{\beta\hbar\omega_a}\right)^{-1}. \qquad (16.88)$$

Whereas both χ_M and χ_π can yield ultrafast cooling, it is $\chi_{\phi \ll 1}$ that is able to reduce the temperature to *the lowest limit*. The small-ϕ limit (16.87) may allow cooling down to *any* temperature, provided $\beta\hbar\phi/\tau \gg 1$. The convergence to χ_M and χ_π is through *oscillatory* falloff of the excitation, whereas $\chi_{\phi \ll 1}$ is approached through *stepwise* falloff by ϕ/τ (Fig. 16.7).

The requirements for the cooling of a TLS coupled to a bath via frequent QND disturbances of the TLS state can be summarized as follows:

(i) Cooling the TLS below the bath temperature by frequent measurements requires the bath-coupling spectrum $G_0(\omega)$ to have higher-frequency components than the level separation, namely, $G_0(\omega > \omega_a)$ must be non-negligible to allow cooling. For a bell-shaped $G_0(\omega)$ this means that the peak of G_0 at $\omega = \omega_0$ must be detuned above ω_a [Fig. 16.8(a)], so as to maximize $\gamma_e(t)$ while minimizing $\gamma_g(t)$.

(ii) For phase flips, (16.85) yields cooling only if $G_+/G_- \gtrsim \exp(\beta\omega_a)$.

(iii) For cooling by small phase shifts, (16.87) implies that we must wait longer than $t \gg 1/|\omega_0 - \omega_a|$ so that $F_t(\omega)$ in (16.56) is narrow enough to be approximated by $\delta(\omega - \omega_a - \phi/\tau)$. Otherwise, $\chi_{\phi \ll 1}$ is *insensitive* to the coupling spectrum $G_0(\omega)$, as opposed to the cooling bounds (i), (ii), achievable by phase flips or measurements, respectively. The lowest temperature is achievable for any bath by small phase shifts according to (16.87): by

(a)   **(b)**

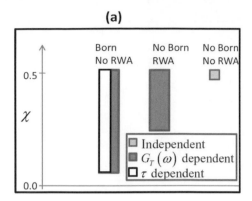
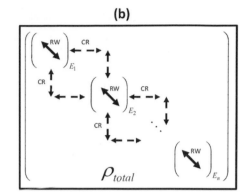

Figure 16.8 (a) Diagram of steady states in different scenarios. The range of steady states for an open system within the Born approximation but without the RWA is determined by both the coupling spectrum $G_T(\omega)$ and the measurement interval τ. Without the Born approximation but with the RWA, this range depends only on the coupling spectrum. Without either approximation, the only steady state is the fully mixed ($T = \infty$) state. (b) The combined system–bath density matrix without the Born approximation. RW terms only mix states within the same energy subspace, whereas CR terms mix states in different energy subspaces, thereby causing ρ_{tot} to converge to the fully mixed state. (Adapted from Gordon et al., 2010.)

choosing $\beta\hbar\phi/\tau \geq 2$, we may reduce the equilibrium excitation $e^{-\beta\hbar\omega_a}$ by more than an order of magnitude.

Since the TLS temperature at equilibrium is higher than that of the bath for an appreciable coupling strength (Ch. 6), we should not only undo the increase in the TLS temperature caused by its coupling to the bath, but also cool it below the bath temperature $T = 1/\beta$. For a sufficiently high T, non-Markovian cooling (by phase flips more than by measurements) can cool the TLS below the bath temperature (Fig. 16.9). On the other hand, when the temperature approaches zero ($\beta \to \infty$), only the extra temperature of the TLS due to its coupling to the bath may be reduced by phase flips or measurements, the TLS may not be cooled below the bath temperature, since the net cooling that one then obtains is a *non-monotonic* function of the bath temperature and the coupling spectrum (Fig. 16.9, inset). By contrast, cooling by small phase shifts is a monotonic function of the bath temperature, so that by taking $\beta\phi/\tau$ to be large enough, an arbitrarily low temperature is attainable by the TLS (within the Born approximation).

16.3.5 Steady States Attainable by Monitoring beyond the Born Approximation

The major approximation that underlies the analysis above is the Born approximation, under which the bath does not change its temperature from measurement to

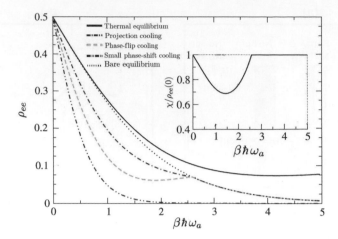

Figure 16.9 Initial excitation probability $\rho_{ee}(0)$ (solid) and asymptotic excitation probability $\rho_{ee}^{(\infty)}$ for the optimal rate $1/\tau$ (as in Fig. 16.7) of measurements (dash-dot), phase flips (dashed), and small phase shifts (dash-double-dot) as a function of $\beta\hbar\omega_a$ (the inverse temperature in units of the inverse TLS energy difference $\hbar\omega_a$). The excitation probability falls off non-monotonically with $\beta\hbar\omega_a$. Inset: fractional cooling bound, $\chi/\rho_{ee}(0)$, attainable by measurements for a coupling spectrum (11.52) with a Lorentzian $G_0(\omega)$ centered at ω_0 with width Γ_B plotted as a function of $\beta\hbar\omega_a$ (solid line). (Adapted from Gordon et al., 2009.)

measurement (or from one phase shift to another). A wide range of steady states is then allowed, as discussed above. Yet, these are merely quasi-steady states, since they are contingent on the Born approximation. Namely, the bath changes are typically slower than the system changes, but there exists a timescale such that changes in the bath state and consequently in the quasi-steady states of the system are non-negligible. This timescale is the Born time t_B, which depends on the bath size: The Born approximation holds and the convergence to (16.67) applies for $t \ll t_B$, whereas for $t \geq t_B$ the accumulated bath changes start to influence the system dynamics.

The monitored (system + bath) dynamic equations (16.61)–(16.64) still hold even without the Born approximation, but it is unclear whether they yield convergence to a steady state that depends on τ. On the one hand, repeated QND measurements of the system erase the system–bath coherences, thereby increasing the total entropy of the total (system + bath) state. On the other hand, the free evolution between measurements is unitary and thus preserves the entropy in this total Hilbert space. Where does this interplay lead to?

To answer this question, let us examine the total density matrix, ρ_{tot}, of the system plus the N-mode bath in two cases:

(i) If the RWA applies [i.e., $\tau \gg 1/\omega_a$—cf. (16.71)], then ρ_{tot} is block-diagonal in $N \times N$ subspaces, with total energy E_n, $n = 0, 1, \ldots$. These energy subspaces remain closed and accumulate the entropy increase following the measurements [Fig. 16.8(b)]. The steady state of each such closed subspace is that of the fully mixed state, regardless of τ. The overall steady state in this scenario depends on the relative weights of the E_n-subspaces. These weights, in turn, depend solely on the bath-coupling spectrum $G_T(\omega)$, which determines the steady state.

(ii) If the RWA does not apply, the CR terms mix the foregoing energy subspaces and effectively render the total state that of a *single* closed system [Fig. 16.8(b)]. The steady state of such an effective closed system, subject to entropy-increasing measurements and the entropy-preserving evolution, can only be the fully mixed ($T = \infty$) state. Hence, repeated measurements of a system coupled to a bath will inevitably force them to converge to a fully mixed state.

16.3.6 Freezing Initial States by Monitoring

An important corollary of the non-Markovian analysis of open-TLS monitoring is the intriguing possibility to "freeze" an initial state by choosing τ in (16.66) such that

$$\chi(\tau) = \rho_{ee}(0). \tag{16.89}$$

The possibility of such freezing arises from the presence of both the CR and the RW terms in the system–bath interaction. The CR terms necessarily heat up (i.e., increase the entropy of the system), while the RW terms make it alternate between cooling and heating, that is, below or above the equilibrium temperature. The combined effect of the RW and CR terms allows the system to return to its initial entropy at a later time τ. Then, within the Born approximation, the system and the bath at time τ are exactly in the initial state, $\rho_{\text{tot}}(0) = \rho_{\text{tot}}(\tau)$. The total state of the system plus bath proceeds in the time interval $\tau \leq t \leq 2\tau$ along the same dynamical path as in the time interval $0 \leq t \leq \tau$, so that the state of the system obeys

$$\rho(2\tau) = \rho(\tau) = \rho(0). \tag{16.90}$$

Hence, upon repeating measurements at intervals τ, the initial state of the system remains "frozen." The "freezing" of a given initial state depends on the bath spectrum $G_T(\omega)$, but there is always a broad range of initial states other than the thermal equilibrium state that can be "frozen."

Similar "freezing by observation" holds for frequently measured open multilevel systems (MLSs), which exhibit convergence to a steady state determined by the non-Markovian τ-dependent measurement-induced evolution. A remarkable trend of such open MLSs is their *nonthermal equilibration by observation*. The steady state of an n-level MLS has the vector form,

$$\boldsymbol{\chi}(\tau) = [\boldsymbol{I} - \boldsymbol{F}_\mathrm{M}(\tau)]^{-1} \boldsymbol{F}_\mathrm{S}(\tau), \tag{16.91}$$

where $\boldsymbol{\chi}(\tau) = \{\chi_n(\tau)\}$ is the n-level population vector, $\boldsymbol{F}_\mathrm{M}(\tau)$ is the single-measurement relaxation factor,

$$\boldsymbol{F}_\mathrm{M}(\tau) = \mathrm{T}_+ \, e^{-\boldsymbol{\gamma}(\tau)\tau}, \tag{16.92}$$

T_+, as before, denoting chronological ordering, and

$$\boldsymbol{F}_\mathrm{S}(\tau) = \int_0^\tau dt' \boldsymbol{F}_\mathrm{M}(\tau) \boldsymbol{F}_\mathrm{M}^{-1}(t') \boldsymbol{\gamma}_g(t'). \tag{16.93}$$

In (16.92), $\boldsymbol{\gamma}(\tau)$ is the matrix analog of the TLS convergence rate $\gamma(\tau)$, and $\boldsymbol{\gamma}_g = \{\gamma_g^{(n)}\}$ is the n-vector of downward $|n\rangle \to |g\rangle$ transition rates.

As can be seen from (16.91)–(16.93) and Figure 16.8, the open-MLS steady-state populations strongly depend on τ and not only on the bath temperature (through the relaxation rates). Hence, multilevel equilibration by observation has a distinctly *nonthermal* character in that it can be (non-monotonically) steered by the measurement interval. This is in contrast to the TLS steady state, which can always be cast in thermal (Gibbs) form with a measurement-dependent temperature.

Another situation that contrasts the steady-state behavior in an MLS to that of a TLS occurs when the MLS is partly degenerate. For example, in a three-level system, with two degenerate levels, measurements in the basis comprised of different orthogonal superpositions of these degenerate states may lead to different steady-state level populations. Thus, the choice of the basis for the measurements plays a crucial role in determining the steady states of a partly degenerate MLS.

One cannot attain arbitrary ratios of excited-level populations in an MLS if the observation period, τ, remains fixed. Yet, the repertoire of steady states can be extended by multiple nonselective measurements, each addressing another subset of levels at different time intervals (Fig. 16.10).

Hence, choosing observation times such that $\boldsymbol{\chi}(\tau) = \rho(0)$ results in "freezing" the initial state for any number of measurements, even for nonthermal states. This "freezing" is unrelated to the QZE (Ch. 10), since it does not require extremely short measurement intervals. On the contrary, this non-Markovian "freezing" occurs only for a specific value of τ and for *any* number of measurements, unlike the QZE.

Let us write the $N+1$-level density matrix in the following form:

$$\rho = \rho_{gg} |g\rangle \langle g| + \sum_{n,n'=1}^N \rho_{nn'} |n\rangle \langle n'| + \sum_{n=1}^N (\rho_{gn} |g\rangle \langle n| + \rho_{ng} |n\rangle \langle g|), \tag{16.94}$$

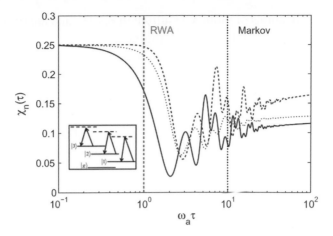

Figure 16.10 Steady-state level populations of a monitored open four-level system as a function of $\omega_a\tau$, where ω_a is the lowest transition frequency and τ is the interval between consecutive measurements. The level populations are shown to have different, non-monotonic, dependence on τ, thus manifesting their non-thermal characteristics: $\chi_1(\tau)$ (dashed), $\chi_2(\tau)$ (dotted), and $\chi_3(\tau)$ (solid). Here the same bath spectrum as assumed in Figure 16.3 with the parameters: coupling strengths $\eta = \omega_a\{0.01, 0.01, 0.02\}$ corresponding to $\omega_{n=1,2,3} = \omega_a\{0.7, 0.8, 1.0\}$, inverse temperature $\beta = 2/(\hbar\omega_a)$, correlation time of the bath $t_c = 20/\omega_a$, and maximal bath-coupling frequency $\omega_0 = 1.5\omega_a$. Inset: two-photon (Raman-based) measurements addressing the different levels. (Reprinted from Gordon et al., 2010.)

where we have separated the ρ_{gg} and ρ_{gn} terms from the $\rho_{nn'}$ terms, where n, n' denote the MLS excited states. The equations for ρ_{gn} are independent of those involving $\rho_{nn'}$ and ρ_{gg}. Hence, once the ρ_{gn} are set to zero by a measurement, they remain so after the measurement and are henceforth disregarded.

For the level-populations vector that evolves under measurements, $\boldsymbol{\rho}_M = \{\rho_{11}, \ldots, \rho_{NN}\}^T$, the evolution between consecutive measurements of the energy separated by τ can therefore be solved to second order in the system-bath coupling, in the form

$$\boldsymbol{\rho}_M(\tau) = \boldsymbol{F}_M(\tau)\boldsymbol{\rho}_M(0) + \boldsymbol{F}_S(\tau), \qquad (16.95)$$

where $\boldsymbol{F}_M(\tau)$ and $\boldsymbol{F}_S(\tau)$ are given by (16.92) and (16.93) respectively. The convergence-rate matrix elements in (16.92) are then

$$\gamma_{nn'}(t) = \frac{1}{\tau}\int_0^t dt'[\delta_{nn'}\gamma_n(t') + \gamma_g^{(n)}(t')], \qquad (16.96)$$

where $\delta_{nn'}$ is Kronecker's delta, and the transition rates are $\gamma_n(t)$ from $|n\rangle$ to $|g\rangle$ and $\gamma_g^{(n)}(t)$ from $|g\rangle$ to $|n\rangle$.

A QND measurement of the MLS energy levels at $t = 0$ eliminates all off-diagonal terms in the density matrix. Namely, $\rho_{nn'}(0) = 0$ for $n \neq n'$ so that we can integrate the equations of motion, resulting in $\rho_{nn'}(t)$ that are of fourth order in the system–bath coupling and are thus negligible.

Consecutive impulsive measurements of the energy levels n, separated by τ, yield, from (16.91)–(16.96):

$$\boldsymbol{\rho}_{\mathrm{M}}(t = m\tau) = \boldsymbol{F}_{\mathrm{M}}^m(\tau)\boldsymbol{\rho}_{\mathrm{M}}(0) + [\boldsymbol{I} - \boldsymbol{F}_{\mathrm{M}}(\tau)]^{-1}[\boldsymbol{I} - \boldsymbol{F}_{\mathrm{M}}^m(\tau)]\boldsymbol{F}_{\mathrm{S}}(\tau). \quad (16.97)$$

By virtue of (16.92) and $\Gamma(\tau)$ being a real and positive definite matrix, $\boldsymbol{F}_{\mathrm{M}}^m(\tau) \to 0$ as $m \to \infty$. The state $\boldsymbol{\rho}_{\mathrm{M}}$ therefore converges to a steady state:

$$\boldsymbol{\chi}(\tau) = [\boldsymbol{I} - \boldsymbol{F}_{\mathrm{M}}(\tau)]^{-1}\boldsymbol{F}_{\mathrm{S}}(\tau). \quad (16.98)$$

Hence, (16.97) and (16.98) show that an MLS weakly coupled to a bath always *exponentially converges* to a steady state, $\boldsymbol{\chi}(\tau)$, as the number of energy measurements increases.

16.4 TLS Cooling Control in a Bath

16.4.1 Optimal Coding Control

Here our goal is to maximize the system–bath coupling and thereby the system cooling (purification). We apply the approach of constrained optimization (Ch. 12) to the linear entropy (the state impurity),

$$\mathcal{S}_{\mathrm{L}} = 2(1 - \mathrm{Tr}\varrho^2), \quad (16.99)$$

of a qubit, where $\mathcal{S}_{\mathrm{L}}$ has been normalized to the range from 0 to 1. Let us assume the TLS is in an initial mixed state,

$$\varrho(0) = p|1\rangle\langle 1| + (1 - p)|0\rangle\langle 0|, \quad (16.100)$$

of the ground and excited states $|0\rangle$ and $|1\rangle$). Here $0 \leq p \leq 0.5$ is related to $\mathcal{S}_{\mathrm{L}}$ by $p = (1 - \sqrt{1 - \mathcal{S}_{\mathrm{L}}})/2$. With σ_j ($j = 1, 2, 3$) denoting the Pauli matrices, (16.100) can be written in terms of $H_0 = \frac{\hbar\omega_0}{2}\sigma_3$ as

$$\varrho(0) = \frac{e^{-\beta H_0}}{\mathrm{Tr}\, e^{-\beta H_0}} = \frac{|1\rangle\langle 1|}{1 + e^{\beta\hbar\omega_0}} + \frac{|0\rangle\langle 0|}{1 + e^{-\beta\hbar\omega_0}}, \quad (16.101)$$

where the inverse temperature $\beta = \ln(1/p - 1)/(\hbar\omega_0)$. Our goal is a constrained optimization of $\Delta\mathcal{S}_{\mathrm{L}}$, that is, $\hat{P} = -4\varrho(0)$ in (12.11). Unlike the gate error (13.22), $\Delta\mathcal{S}_{\mathrm{L}}$ can be negative or positive, indicating cooling or heating, respectively.

The minimization of $\Delta\mathcal{S}_{\mathrm{L}}$ for the initial state (16.100) is achieved by the effective evolution control

$$U(\tau) = e^{-\frac{i}{2}f_3(\tau)\sigma_3}e^{-\frac{i}{2}f_2(\tau)\sigma_2}e^{-\frac{i}{2}f_1(\tau)\sigma_3}. \quad (16.102)$$

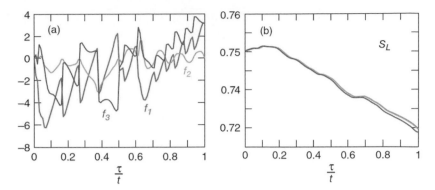

Figure 16.11 Evolution of the TLS initial mixed state (16.100) with $p = 0.25$ of a function of τ for optimized cooling of (a) effective control (f_1, f_2, f_3 of the respective Pauli matrices σ_x, σ_y, σ_z) and (b) linear entropy $\mathcal{S}_L$: upper and lower curves show numerical integration of the time-convolutionless ME (Ch. 11) and the second-order approximation of the solution. The two curves coincide. Here the system–bath coupling strength $\kappa = 10^{-2}$, $\omega_0 = 2\pi/t$, $t = 10t_c$, and the energy constraint $E = 100$. (Adapted from Clausen et al., 2012.)

The complicated control functions $f_j(\tau)$ ($j = 1, 2, 3$) for minimized $|\Delta \mathcal{S}_L|$, leading to $\mathcal{S}_L$ reduction, are shown in Figure 16.11.

For uncorrelated bath components, that is, $G_{jk}(\omega) = 0$ for $j \neq k$, $G_j \equiv G_{jj}(\omega)$. The components $F_j \equiv (\mathbf{F}_t)_{jj}(\omega)$ of the system modulation spectrum contributing to the spectral overlap (12.36) depend on the initial state $\varrho(0)$ via the matrix $\boldsymbol{\Xi}$ (12.31). Cooling or heating is achieved by maximizing the negative or positive spectral overlap, respectively, of F_j and G_j (Fig. 16.12). Such maximized overlap is the opposite of system–bath dynamical decoupling (Ch. 12), where the goal is to minimize the spectral overlap.

The plots in Figure 16.12 illustrate the role of the energy constraint E. Upon increasing E, we allow for spectral overlap with higher-frequency components of the bath spectrum. This implies that for a bath spectrum with a finite frequency cutoff, the increase of E above a certain saturation value does not lead to further improvement of the optimization (Fig. 16.13).

The achievable change $\Delta \mathcal{S}_L$ for a given bath spectrum is $\varrho(0)$-dependent. Thus, for a maximally mixed state ($p = 0.5$), the matrix $\boldsymbol{\Xi}$ in (12.31) vanishes and with it $\Delta \mathcal{S}_L$ (Fig. 16.13).

Constraint-independent upper and lower cooling bounds can be given for the maximum change $P = \mathrm{Tr}(\hat{P}\Delta\varrho)$ that is achieved also for given bath spectra and $\hat{P}$ under condition (12.22) for the quadratic ($d^2 - 1$)-dimensional matrices $\boldsymbol{\epsilon}$, $\boldsymbol{\Xi}$, and $\boldsymbol{G}(\omega)$. Assuming that the bath spectral response has a finite norm, $\max \mathrm{Tr}\, \boldsymbol{G}(\omega) < \infty$, we can estimate P by using the fact that $\mathrm{Tr}(\boldsymbol{AB}) \leq \mathrm{Tr}\boldsymbol{A}\,\mathrm{Tr}\boldsymbol{B}$ for positive

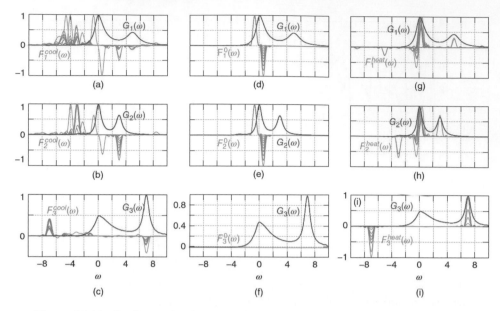

Figure 16.12 Cooling or heating of a TLS corresponding to a maximized change of linear entropy $\mathcal{S}_L$: maximal negative change [left: (a), (b), (c)] or positive change [right: (g), (h), (i)], respectively, for given bath spectra (G_j) with $j = 1, 2, 3$ denoting the x, y, and z G_j components (upper, middle, or lower rows, respectively). The optimized modulation spectra (F_j) are shown for an energy constraint $E = 10^2$ (left column) and $E = 10^3$ (right column). These are contrasted to the $\omega = 0$ dip of the unmodulated Hamiltonian $\hat{H}_0$ ($E = 0$), [middle column: (d), (e), (f)]. All graphs are individually scaled to 1 with respect to their maximum value. The scaling of ω is the same as in Figure 12.5. (Adapted from Kurizki, 2013.)

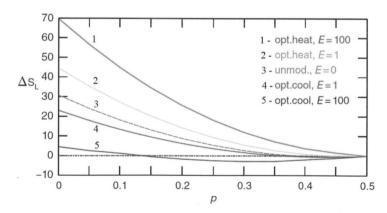

Figure 16.13 Change of linear entropy $\Delta\mathcal{S}_L$ in units of the system–bath coupling strength: minimization (cooling) or maximization (heating) under different constraints ($E = 0, 1, 100$ in units of $\hbar/t_c$) as a function of the initial p for the bath spectra in Figure 16.12. (Adapted from Kurizki, 2013.)

semidefinite matrices A, B, and applying Holder's inequality in the form of

$$\int_{-\infty}^{\infty} d\omega |f(\omega)g(\omega)| \leq \sup\{|g(\omega)|\} \int_{-\infty}^{\infty} d\omega |f(\omega)|. \tag{16.103}$$

Upon decomposing $\Xi = \Xi_1 - \Xi_2$ into positive semidefinite matrices Ξ_i and making use of the identity $\frac{1}{t}\int_{-\infty}^{\infty} d\omega \epsilon_t^{\dagger}(\omega)\epsilon_t(\omega) = I$, we then obtain the constraint-independent cooling bounds

$$-P_2 \leq P \leq P_1, \quad P_i = t \sup[\mathrm{Tr}\, G(\omega)] \,\mathrm{Tr}\, \Xi_i. \tag{16.104}$$

This result reveals that for given t and $G(\omega)$, the cooling bounds P_i are determined by $\varrho(0)$ and Ξ.

16.5 Discussion

We have addressed a distinctly quantum mechanical scenario of repeated disturbances of the thermal equilibrium between a TLS and a bath by frequent and brief QND measurements or phase changes of the TLS states in the energy basis. If the measurements are frequent enough, they can induce either the quantum Zeno or the anti-Zeno effect (the QZE or the AZE), corresponding to the slowdown or speedup of the TLS relaxation (Ch. 10), respectively. The resulting entropy and temperature of both the system and the bath are then found to swing up and down with the interval between disturbances, that is, in a fashion completely unrelated to what is expected according to standard thermodynamical rules that hold in the Markovian domain. The practical advantage of these anomalies is the possibility of *ultrafast control* of heat and entropy, allowing for cooling and state-purification of quantum systems much sooner than their thermal equilibration time.

Although it is possible to use the information gained by measurements to sort out system subensembles according to their measured energy in order to extract work or entropy change, in the spirit of Maxwell's demon (Ch. 17), here we let the entire TLS ensemble evolve regardless of the measured result (i.e., we trace out the meter states). The quantum-mechanical noncommutativity of the system–meter and system–bath interactions causes the heat-up of the system at the expense of the meter–system coupling, but not at the expense of the coupling to the bath, only at very short time intervals τ compatible with the QZE. A *transition from heating to cooling* of the TLS ensembles as we vary the interval between consecutive measurements from $\tau \ll 1/\omega_a$ to $\tau \sim 1/\omega_a \ll t_c$, where t_c is the bath-correlation (memory) time, marks the transition from the QZE to the AZE.

Our analysis has shown that TLS entanglement with the bath at equilibrium is destroyed by short-time quantum dynamics induced upon disturbing this equilibrium. The bath dynamics induced by either projective measurements or phase shifts at a rate comparable to the TLS transition frequency has been shown to be nearly coherent, or quasireversible, and hence amenable to temperature control. The universal bound on non-Markovian cooling derived here demonstrates the importance of the system–bath coupling spectrum and the limitations of the RWA when designing efficient cooling/purification.

We have shown that open quantum systems that relax via coupling to a bath while being monitored by an energy meter inevitably reach a quasi-steady state controllable by the rate of monitoring. In the non-Markovian regime, this approach allows for the "freezing" of states by choosing monitoring rates that set a nonthermal steady state to be the desired one. For measurement rates high enough to cause the QZE, the only steady state is the fully mixed state, due to the RWA breakdown. Regardless of the monitoring rate, the quasi-steady states of an observed open quantum system live only as long as the Born approximation holds, namely, the bath entropy does not change. Beyond the Born time, the steady state no longer depends on the measurement interval. Rather, under the RWA, it depends solely on the bath-coupling spectrum, whereas without the RWA, both the system and the bath converge to their fully mixed states.

Since, under monitoring, any off-diagonal elements present in the initial state decay to zero, one can identify the energy eigenstates of the system with the pointer basis (Ch. 9). However, in the non-Markovian regime, one can choose specific pointer states that are maximally stable, for a given observation rate. Whereas one can engineer the quantum jump operators through which the system couples to the bath so as to achieve the desired end state via Markovian dynamics, our treatment draws on the richness of non-Markovian dynamics to achieve diverse steady states. Furthermore, manipulating the jump operators is equivalent to an external control over the system–bath interaction. In the present approach, such control is not required: instead, manipulations are effected by the meter, which couples to the system alone.

The interplay between measurement-induced and bath-induced rotating wave (RW) and counterrotating (CR) terms in non-Markovian dynamics essentially revises the notion of evolution freezing that is typically associated with the QZE.

A possible experimental scenario, where these effects may play out, involves an atom or molecule with transition frequency ω_a (in the MHz range) that is weakly coupled to a near-resonant microwave cavity whose finite-temperature radiation constitutes a bath with the finite-temperature coupling spectrum $G_T(\omega)$. Measurements or phase flips can be exerted on such a TLS ensemble with resonance frequency ω_a at time intervals τ chosen according to (16.74)–(16.87) by optical pulses that undergo different Raman-induced phase shifts $\Delta\phi_e$ or $\Delta\phi_g$ depending on the different symmetries of $|e\rangle$ and $|g\rangle$. The relative abundance of $\Delta\phi_e$ and $\Delta\phi_g$ after m disturbances would then reflect the ratio $\rho_{ee}(t_m)/\rho_{gg}(t_m)$. QND probing or phase flips may be performed with time duration *much shorter than* ω_a^{-1}, *without resolving the energies* of $|e\rangle$ and $|g\rangle$. Appreciable cooling is then obtainable well within the thermalization time, on the ps timescale. By comparison, ordinary (optical) cooling takes longer than the lifetime of the populated state decaying to $|g\rangle$ (i.e., μs at the shortest). We may use probing pulses of 100 fs duration at ps intervals. One can then achieve considerable cooling on a ps timescale.

In vapor or solid-state ensembles, the relaxation timescale $T_1 \gtrsim 1$ μs may be much longer than the decoherence (dephasing) time $T_2 \sim$ 10–100 ns. Yet, if the manipulations considered here are on a ps timescale, T_2 processes do not affect the cooling.

17

Work–Information Relation and System–Bath Correlations

Here we discuss the rapport of information and its control with the concepts of thermodynamics, focusing on the ability of an observer to extract work from a system upon *modifying* it according to the results of its measurement. Such an omniscient observer, dubbed "Maxwell's demon," allegedly demonstrates that information can alleviate the restrictions imposed by the second law of thermodynamics on the energy exchanged between a system and its environment. The second law can, however, be extended by incorporating information and clarifying its physical manifestations. Namely, information acquisition by measurement and its erasure are endowed with thermodynamic cost. Szilárd's version of Maxwell's demon exploits a bit of information (the outcome of a yes/no measurement) to perform a cyclic process that extracts $k_B T \ln 2$ of work from a single bath at temperature T. Here we extend these concepts to massive extraction of work by homodyne measurements of a thermal (totally passive) state of a field mode. It is thereby converted to a known nonpassive state. Upon feeding back this information to a unitary transformer, the work stored in this state can be extracted.

According to the Landauer principle, work obtainable from a measurement cannot exceed the energy cost of erasing its record from the observer's memory. Here we present a possibility to extract work even if the information obtained from a measurement is *not* read out [i.e., for nonselective measurements (NSM)]. A larger amount of work than that allowed by the Landauer principle is shown to be extractable by exploiting the energy obtained from the change of the system–bath correlation through an impulsive, strictly quantum measurement, provided we act within the memory (non-Markovian) time of the bath.

17.1 Information and the Second Law of Thermodynamics

Information acquired by a measurement on a system characterized by a set of microstates x corresponds to the update of the statistical state, $\rho(x)$, to $\rho(x|m)$, where m represents the outcome of that measurement. In the Szilárd engine, the statistical state $\rho(x|m)$ of an atom in a box after a measurement is confined to either the left or the right half of the box. This updated state $\rho(x|m)$ is in general out of equilibrium, even if the pre-measurement state $\rho(x)$ was in equilibrium. Hence, information moves the system away from equilibrium at no apparent energy cost. The consideration of such processes calls for an extension of the definitions of entropy and free energy to post-measurement nonequilibrium states.

17.1.1 Bound on Free Energy, Work, and Entropy

The Shannon entropy of a random variable X characterized by probability density $\rho(x)$ is defined as

$$\mathcal{H}(X) = -\sum_x \rho(x) \ln \rho(x). \tag{17.1}$$

If X represents the state of an open quantum system, then the Shannon entropy, multiplied by Boltzmann's constant k_B, is identical with the non-equilibrium thermodynamic von Neumann entropy

$$\mathcal{S}(\rho) = k_B \mathcal{H}(X) = -k_B \sum_x \rho(x) \ln \rho(x). \tag{17.2}$$

Another key notion is the *nonequilibrium* free energy of a system in contact with a thermal bath described by a statistical state ρ and Hamiltonian H_0

$$\mathcal{F}(\rho; H_0) \equiv \langle H_0 \rangle_\rho - T\mathcal{S}(\rho). \tag{17.3}$$

This free energy is the analog of its equilibrium counterpart in a non-equilibrium isothermal process, wherein the system is in contact with a thermal bath at temperature T. The system, however, may not have a well-defined temperature when it is under control (Ch. 12). The minimal (average) work W necessary to isothermally drive the system from one state to another is the difference, $\Delta\mathcal{F}$, between the nonequilibrium free energies in (17.3). The excess work with respect to this minimum is the dissipated or irreversible work, W_{diss}. It can be associated with the total isothermal entropy production of the system and the bath. The second law as applied to isothermal processes connecting nonequilibrium states yields the inequality

$$T\Delta S_{\text{tot}} = W_{\text{diss}} \equiv W - \Delta\mathcal{F} \geq 0. \tag{17.4}$$

17.1.2 Mutual Information and the Second Law

The mutual information $I(\mathcal{U}; \mathcal{V})$ between random variables $\mathcal{U}$ and $\mathcal{V}$, defined as

$$I(\mathcal{U}; \mathcal{V}) = \sum_{u,v} \rho(u, v) \ln \frac{\rho(u, v)}{\rho(u)\rho(v)} = \mathcal{H}(\mathcal{U}) + \mathcal{H}(\mathcal{V}) - \mathcal{H}(\mathcal{U}, \mathcal{V}), \qquad (17.5)$$

is positive and symmetric. It vanishes if and only if $\mathcal{U}$ and $\mathcal{V}$ are statistically uncorrelated (independent). It can be rewritten as

$$I(\mathcal{U}; \mathcal{V}) = \mathcal{H}(\mathcal{U}) - \mathcal{H}(\mathcal{U}|\mathcal{V}) = \mathcal{H}(\mathcal{V}) - \mathcal{H}(\mathcal{V}|\mathcal{U}), \qquad (17.6)$$

where

$$\mathcal{H}(\mathcal{U}|\mathcal{V}) = -\sum_{u,v} \rho(v)\rho(u|v) \ln \rho(u|v) \qquad (17.7)$$

is the conditional entropy. Since $\mathcal{H}(\mathcal{U})$ quantifies the uncertainty in $\mathcal{U}$, and $\mathcal{H}(\mathcal{U}|\mathcal{V})$ is the uncertainty in $\mathcal{U}$ for a given $\mathcal{V}$, the mutual information expresses the reduction in the uncertainty of $\mathcal{U}$ when we know $\mathcal{V}$, and vice versa. The mutual information is therefore a measure of the correlations. For error-free measurements of a system variable $\mathcal{V}$ by an appropriate meter variable $\mathcal{U}$, their correlation is perfect, namely, $\mathcal{U}$ uniquely determines $\mathcal{V}$, so that

$$\mathcal{H}(\mathcal{V}|\mathcal{U}) = 0, \quad I(\mathcal{U}; \mathcal{V}) = \mathcal{H}(\mathcal{V}). \qquad (17.8)$$

We next evaluate the change in nonequilibrium free energy due to QND measurements, by which neither the Hamiltonian of the system nor its microstate are affected. Such measurements do not change the average energy, but increase the nonequilibrium free energy

$$\Delta \mathcal{F}_{\mathrm{M}} = -T \Delta \mathcal{S}_{\mathrm{M}} = k_{\mathrm{B}} T I(X; M). \qquad (17.9)$$

Because the mutual information is positive, a measurement that yields information necessarily increases the free energy and thereby the work that can be isothermally extracted. This work increase, as suggested by (17.4), allows us to identify (17.9) with *entropy change* (Ch. 14).

We can now explicitly incorporate information in the second law. To this end, we consider a system initially in equilibrium state with a bath at temperature T. The "demon" performs a measurement at time t_{M} and makes use of the acquired information to change the system parameters $\lambda_m(t)$, according to the measurement outcome m. The increase in free energy at $t \geq t_{\mathrm{M}}$ due to the measurement (17.9) then yields the second law formulation for such a feedforward process,

$$W - \Delta \mathcal{F} \geq -k_{\mathrm{B}} T I(X(t_{\mathrm{M}}); M). \qquad (17.10)$$

In (17.10), the work W done on the system corresponds to the average over measurement outcomes m. In a cyclic process, where $\lambda_m(\tau) = \lambda_m(0)$ for all m and the system relaxes back to thermal equilibrium, (17.10) reduces to

$$\Delta \mathcal{F} = 0, \quad W \geq -k_B T I(X(t_M); M). \tag{17.11}$$

Hence, we can extract work proportional to the information obtained by the measurement. It follows from (17.8) for error-free measurements that the right-hand side of (17.11) is proportional to the Shannon entropy of the measurement, which is the upper bound on work extractable from a measurement,

$$I(X(t_M); M) = \mathcal{H}(M). \tag{17.12}$$

This bound is reached by the Szilárd engine, since M is then the left- or right-hand location of the particle in the vessel, with probability 1/2, so that $\mathcal{H}(M) = \ln 2$.

17.2 The Landauer Principle

17.2.1 Memory and Landauer's Principle

The outcome of a measurement is recorded in a memory, with two or more *distinguishable* states in which the information is stored, classically (in bits) or quantum mechanically (in qubits). Reliable information storage necessitates these states to be metastable under the system Hamiltonian $H(x)$ and have long lifetimes in the presence of environmental noise (i.e. bath effects). Robustness against bath effects means that under this Hamiltonian, ergodicity breaks down on the timescale of information storage. Equivalently, such robustness requires then that the phase space of the system be split into separate ergodic regions for the different informational states m. The non-ergodicity needed for encoding information can be induced, for example, by a phase transition from single-particle to collective excitations, such as the transition from single-spin to magnon (ferromagnetic) excitations in magnetic memories, or their dipolar atomic analogs (Chs. 3–4).

We consider cyclic memories that after any manipulation end up having the same initial Hamiltonian $H(x)$. The average work that must be invested in order to change the state of a memory from M to M', corresponding to a change in the distribution of informational states from p_m to p'_m, satisfies the following inequality bound according to the second law (17.4):

$$W \geq \mathcal{F}(M') - \mathcal{F}(M). \tag{17.13}$$

In symmetric memories, all informational states possess the same equilibrium free energy. Then, (17.13) corresponds to the change in information content:

$$W \geq k_B T [\mathcal{H}(M) - \mathcal{H}(M')]. \tag{17.14}$$

We may apply (17.14) to the resetting or erasure of memory, which means that, regardless of the initial distribution $\{p_m\}$, the memory is driven to occupy the chosen state $m = 0$, such that $p'_0 = 1$ and all other $p'_m = 0$. For a symmetric memory, the minimal work required to implement this operation is, according to (17.14),

$$W_{\text{reset}} \geq k_{\text{B}}T\mathcal{H}(M), \tag{17.15}$$

since $\mathcal{H}(M') = 0$. For a completely random bit, with $p_0 = p_1 = 1/2$, (17.15) yields the Landauer bound: The work required for resetting a random qubit satisfies

$$W_{\text{reset}} \geq W_{\text{L}} = k_{\text{B}}T \ln 2. \tag{17.16}$$

The converse process is an increase of the memory randomness in order to allow work extraction. Thus, an N-bit symmetric memory can increase its Shannon entropy, thereby allowing to extract an amount of work

$$W = k_{\text{B}}T[\mathcal{H}(M') - \mathcal{H}(M)]. \tag{17.17}$$

Hence, a low-entropy (ordered) memory can serve as the "fuel" of an information engine. The work extractable from such an engine by completely randomizing a pure-state qubit is then bounded by the same Landauer limit on work extraction,

$$W^{\text{ext}} \leq W_{\text{L}} = k_{\text{B}}T \ln 2. \tag{17.18}$$

If a d-level quantum system is in a pure state, one can use it to draw $k_{\text{B}}T \ln d$ work out of the heat bath. If the system is in a mixed state ρ with entropy $\mathcal{S}(\rho)$, then the amount of work that can be extracted is

$$W^{\text{ext}} = k_{\text{B}}T[\ln d - \mathcal{S}(\rho)/k_{\text{B}}], \tag{17.19}$$

where the expression in square brackets is the information content of the state. This bound is commonly construed to follow from the second law.

17.2.2 Cost of Measurement

According to (17.6)–(17.9), the nonequilibrium free energy of the compound XM consisting of a system X and a memory (or meter) M that only interact impulsively during the measurement is given by

$$\mathcal{F}(XM) = \mathcal{F}(X) + \mathcal{F}(M) + k_{\text{B}}TI(X; M). \tag{17.20}$$

Thus mutual information accounts for the difference between the nonequilibrium free energy of the compound and its constituents X, M.

Assuming that the measurement does not change $\mathcal{F}(X)$, the change in the nonequilibrium free energy of the compound following the measurement is

$$\Delta\mathcal{F}_{\text{M}}^{(\text{tot})} = \Delta\mathcal{F}_{\text{M}} + k_{\text{B}}TI(X; M), \tag{17.21}$$

where $\Delta\mathcal{F}_M$ is the change in the free energy of the memory. The second law (17.4) associates the free-energy increase given by (17.21) with work that satisfies the inequality

$$W_M \geq \Delta\mathcal{F}_M + k_B TI(X, M). \tag{17.22}$$

For a symmetric memory with error-free measurement

$$-\Delta\mathcal{F}_M = k_B T \Delta\mathcal{H}(M) = k_B TI(X; M), \tag{17.23}$$

which means that a measurement can be performed with zero work, $W_M = 0$.

Following the measurement that has correlated X and M, feedforward can be used to extract the free energy associated with the information in the form of work from the system. To this end, we can drive the system in a manner that depends on the memory state, while keeping the memory fixed. If the feedforward obliterates all correlations, then, according to (17.20), the change in the nonequilibrium free energy of the compound is given by

$$\Delta\mathcal{F}^{(tot)} = \Delta\mathcal{F} - k_B TI(X; M), \tag{17.24}$$

where, as in (17.10), $\Delta\mathcal{F}$ (of the system) is averaged over the measurement outcomes m. Upon applying the second law (17.4) to (17.24), (17.10) is recovered for the average work extracted via feedforward W.

Since the entropy production is given by the difference between the work and the (nonequilibrium) free-energy change of the compound, inequalities (17.22) and (17.10) are equivalent to the entropy production being non-negative (Ch. 15). The measurement and feedforward processes are thermodynamically reversible when (17.22) and (17.10) become equalities.

We must keep in mind that the information that allows the feedforward has a cost: The work and free energy over the complete measurement-feedforward cycle, obtained by summing the inequalities (17.10) and (17.22), yields

$$W_M + W \geq \Delta\mathcal{F}_M + \Delta\mathcal{F}. \tag{17.25}$$

This means that the mutual information cancels out and the work has been invested only to increase the free energy of the memory, which is subsequently extracted by the system.

If the memory is reset to its initial state M' after the feedforward so that it can be used again for another feedforward loop, we have $\Delta\mathcal{F}_{reset} = -\Delta\mathcal{F}_M$. The *work required to store the measurement result and then reset the memory* is then found to satisfy

$$W_M + W_{reset} \geq k_B TI(X; M). \tag{17.26}$$

Thus, the work required for memory operation is determined by the mutual information of the system and the meter.

17.3 Work Extraction from Passive States by Information Feedforward

Passive states cannot be used for extracting useful work if only unitary transformations are allowed (Ch. 15). Here we study the possibility of work extraction of a thermal state of a harmonic-oscillator (field) mode assuming that one can perform quantum measurements of its state and feedforward of the acquired information. To elucidate the idea, consider a macroscopic pendulum whose phase is completely random and the amplitude has a Gaussian distribution. This state is equivalent to a thermal state with a high temperature, that is, a completely passive state (Ch. 15). However, if we observe both the position and the momentum of the pendulum, we project it onto a nearly coherent state, from which work can be extracted by bringing the pendulum to rest. In the quantum domain, the position and the momentum measurements have a cost: They cannot be done perfectly and are limited by the quantum uncertainty. Nevertheless, we show below that a passive (thermal) signal can be used to extract work via homodyne measurements followed by information feedforward.

17.3.1 Model Description

We consider a homodyne measurement of two quadratures of a split fraction of the incoming field. The remaining part of the field is projected to a state from which work can be extracted by a unitary transformation (displacement). To find the maximum extractable work, one has to take into account the energy cost of the measurements and the influence of the quantum noise entering the scheme.

A thermal state of a harmonic oscillator can be represented as a mixture of coherent states $|\alpha\rangle$, with α having Gaussian distribution with $\langle|\alpha|^2\rangle = \bar{n}$. Assume first that a coherent state $|\alpha\rangle$ with the complex coherent amplitude α,

$$\alpha = \frac{1}{\sqrt{2}}(x + ip), \tag{17.27}$$

enters the scheme as in Figure 17.1. After the first beam splitter with splitting ratio $\eta^2/(1 - \eta^2)$, the state $|\eta\alpha\rangle$ is transmitted and the state $|\sqrt{1 - \eta^2}\alpha\rangle$ is reflected toward a homodyne detector for estimating the quadratures $\hat{x}$ and $\hat{p}$ of the input state.

Let us assume that the local oscillator is in a coherent state with real quadrature amplitude β and its imaginary-quadrature counterpart $i\beta$. The modes 0, 1, 2, 3, 4 behind the beam splitters are then in a multimode (product) coherent state of the form

$$|\psi\rangle = |\gamma_0\rangle|\gamma_1\rangle|\gamma_2\rangle|\gamma_3\rangle|\gamma_4\rangle, \tag{17.28}$$

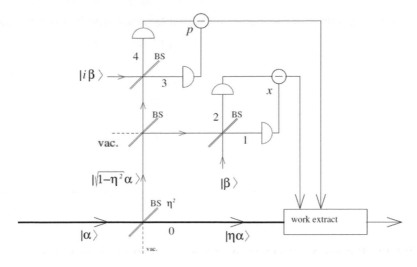

Figure 17.1 Work extraction via homodyne measurement and feedforward: a coherent-state component $|\alpha\rangle$ of a thermal state enters a beam splitter, which transmits a fraction η^2 of the input energy and reflects the remaining fraction, $\sqrt{1 - \eta^2}$. A homodyne measurement is performed on the reflected fraction to estimate the quadratures x and p of the input amplitude α.

where

$$\gamma_0 = \eta\alpha, \quad \gamma_{1,2} = \frac{1}{\sqrt{2}}\left(\sqrt{\frac{1 - \eta^2}{2}}\alpha \pm \beta\right), \quad \gamma_{3,4} = \frac{1}{\sqrt{2}}\left(\sqrt{\frac{1 - \eta^2}{2}}\alpha \pm i\beta\right).$$

$$(17.29)$$

Photodetection of the states in modes 1–4 yields Poissonian statistics with mean values $\bar{n}_k = |\gamma_k|^2$,

$$\bar{n}_{1,2} = \frac{1 - \eta^2}{8}\left[\left(x \pm \frac{2\beta}{\sqrt{1 - \eta^2}}\right)^2 + p^2\right], \qquad (17.30)$$

$$\bar{n}_{3,4} = \frac{1 - \eta^2}{8}\left[x^2 + \left(p \pm \frac{2\beta}{\sqrt{1 - \eta^2}}\right)^2\right]. \qquad (17.31)$$

The photocount differences $\Delta n_x \equiv n_1 - n_2$ and $\Delta n_p \equiv n_3 - n_4$ carry information on the quadratures of the input field. In particular, their mean values are

$$\langle\Delta n_x\rangle = \sqrt{1 - \eta^2}\beta x, \quad \langle\Delta n_p\rangle = \sqrt{1 - \eta^2}\beta p, \qquad (17.32)$$

and their variances are

$$\langle\Delta n_x^2\rangle - \langle\Delta n_x\rangle^2 = \langle\Delta n_p^2\rangle - \langle\Delta n_p\rangle^2 = \frac{1 - \eta^2}{4}(x^2 + p^2) + \beta^2. \quad (17.33)$$

The probability distribution for the photocount differences can be expressed as the Skellam distribution of the difference of two statistically independent variables, each with a Poissonian distribution,

$$P(\Delta n_x | \alpha) = e^{-\bar{n}_1 - \bar{n}_2} \left(\frac{\bar{n}_1}{\bar{n}_2} \right)^{\Delta n_x / 2} I_{\Delta n_x} (2\sqrt{\bar{n}_1 \bar{n}_2}), \qquad (17.34)$$

$$P(\Delta n_p | \alpha) = e^{-\bar{n}_3 - \bar{n}_4} \left(\frac{\bar{n}_3}{\bar{n}_4} \right)^{\Delta n_p / 2} I_{\Delta n_p} (2\sqrt{\bar{n}_3 \bar{n}_4}), \qquad (17.35)$$

where $I_k(z)$ is the modified Bessel function of the first kind. The coherent-state amplitude distribution enters these equations through the dependence of $\bar{n}_k$ on x and p. Upon multiplying these functions we get the conditional probability distribution for Δn_x and Δn_p on the condition that the input state is $|\alpha\rangle$,

$$P(\Delta n_x, \Delta n_p | \alpha) = P(\Delta n_x | \alpha) P(\Delta n_p | \alpha). \qquad (17.36)$$

Let us now take the input state to be a mixture of coherent states,

$$\hat{\varrho} = \int \int P(\alpha) |\alpha\rangle \langle \alpha| d^2\alpha. \qquad (17.37)$$

For such a statistical state, the probability distribution of photodetections is

$$P(\Delta n_x, \Delta n_p) = \int \int P(\Delta n_x, \Delta n_p | \alpha) P(\alpha) d^2\alpha, \qquad (17.38)$$

where $P(\alpha)$ is the Glauber–Sudarshan distribution function of the input state. Specifically, for a thermal input state with the mean number of photons $\bar{n}$ this function is Gaussian

$$P(\alpha) = \frac{1}{\pi \bar{n}} \exp\left(-\frac{|\alpha|^2}{\bar{n}} \right). \qquad (17.39)$$

Upon inverting relation (17.38) and using the Bayes rule, we find the conditional distribution of α, on condition that the photon number differences Δn_x and Δn_p are detected, as

$$P(\alpha | \Delta n_x, \Delta n_p) = \frac{P(\Delta n_x, \Delta n_p | \alpha) P(\alpha)}{P(\Delta n_x, \Delta n_p)}. \qquad (17.40)$$

The state of the unmeasured (remaining) field mode (conditional on the detected Δn_x and Δn_p) can be expressed as

$$\hat{\varrho}(\Delta n_x, \Delta n_p) = \int \int P(\alpha | \Delta n_x, \Delta n_p) |\eta\alpha\rangle \langle \eta\alpha| d^2\alpha$$

$$= \frac{1}{\eta^2} \int \int P\left(\frac{\alpha}{\eta} \Big| \Delta n_x, \Delta n_p \right) |\alpha\rangle \langle \alpha| d^2\alpha. \qquad (17.41)$$

This state has in general nonvanishing mean values of quadratures $\hat{x}$ and $\hat{p}$,

$$\langle x \rangle = \text{Tr}\left[\hat{x}\hat{\varrho}(\Delta n_x, \Delta n_p)\right] = \eta \int\int x P(\alpha | \Delta n_x, \Delta n_p) d^2\alpha,$$

$$\langle p \rangle = \text{Tr}\left[\hat{p}\hat{\varrho}(\Delta n_x, \Delta n_p)\right] = \eta \int\int p P(\alpha | \Delta n_x, \Delta n_p) d^2\alpha. \quad (17.42)$$

This resulting state may well be nonpassive, that is, store work (ergotropy) as described in Chapter 15.

17.3.2 Gaussian Approximation

Assuming that the mean photon numbers are large enough so that their probability distributions can be approximated by Gaussians, we can evaluate (17.40) to be

$$P(\Delta n_x, \Delta n_p | \alpha) \approx \frac{1}{2\pi \left[\frac{(1-\eta^2)|\alpha|^2}{2} + \beta^2\right]} \exp\left[-\frac{\Delta_x^2 + \Delta_y^2}{(1-\eta^2)|\alpha|^2 + 2\beta^2}\right], \quad (17.43)$$

where

$$\Delta_x = \Delta n_x - \sqrt{2(1-\eta^2)}\beta \, \text{Re}\,\alpha, \quad \Delta_y = \Delta n_p - \sqrt{2(1-\eta^2)}\beta \, \text{Im}\,\alpha. \quad (17.44)$$

Even though we may have started from a state with Gaussian $P(\alpha)$, the resulting state is not Gaussian as α enters (17.43) not only as an exponential of a quadratic function. One can, however, resort to the Gaussian approximation in (17.43) by replacing $|\alpha|^2$ with $\bar{n}$ for $\bar{n} \gg 1$. This allows the analytical integration of (17.38). One then obtains the approximate Gaussian

$$P(\Delta n_x, \Delta n_p) \approx \frac{1}{2\pi\sigma_{\Delta n}^2} \exp\left(-\frac{\Delta n_x^2 + \Delta n_p^2}{2\sigma_{\Delta n}^2}\right), \quad (17.45)$$

where the variance is

$$\sigma_{\Delta n}^2 = \beta^2 + \bar{n}(1-\eta^2)\left(\beta^2 + \frac{1}{2}\right). \quad (17.46)$$

17.3.3 Work Extraction Bound

By unitarily displacing (downshifting) the state (17.41) such that the mean quadratures of the final state are zero, one can extract the work

$$W^{\text{ext}}(\Delta n_x, \Delta n_p) = \frac{\hbar\omega}{2}\left(\langle x \rangle^2 + \langle p \rangle^2\right), \quad (17.47)$$

where ω is the input mode frequency. The mean net work obtained in this process can be found by averaging this expression over all values of Δn_x, Δn_p and

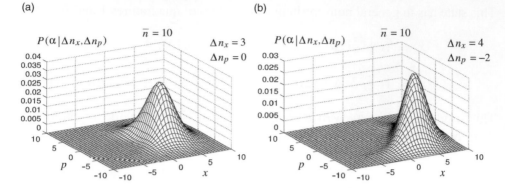

Figure 17.2 Conditional probability of the field quadratures x, p on condition of detected photon number differences Δn_x and Δn_p, with $\bar{n} = 10$ and optimized values of $\beta = 0.780$ and $\eta = 0.902$.

subtracting the invested energy of the two local oscillators. The net extractable work is then

$$W_{\text{net}}^{\text{ext}} = \sum_{\Delta n_x} \sum_{\Delta n_x} W^{\text{ext}}(\Delta n_x, \Delta n_p) P(\Delta n_x, \Delta n_p) - 2\hbar\omega\beta^2. \qquad (17.48)$$

Consistently with (17.45), one finds (Fig. 17.2)

$$P(x, p|\Delta n_x, \Delta n_p) \approx \frac{1}{2\pi\sigma_x^2} \exp\left[-\frac{(x - \bar{x}_{\Delta nx})^2 + (p - \bar{p}_{\Delta np})^2}{2\sigma_x^2}\right], \qquad (17.49)$$

with

$$\bar{x}_{\Delta nx} = \frac{\Delta n_x}{\beta\sqrt{1 - \eta^2}\left[1 + \frac{1}{\bar{n}(1-\eta^2)} + \frac{1}{2\beta^2}\right]}, \qquad (17.50)$$

$$\bar{p}_{\Delta np} = \frac{\Delta n_p}{\beta\sqrt{1 - \eta^2}\left[1 + \frac{1}{\bar{n}(1-\eta^2)} + \frac{1}{2\beta^2}\right]}, \qquad (17.51)$$

$$\sigma_x^2 = \frac{\bar{n}}{1 + \frac{2\beta^2\bar{n}(1-\eta^2)}{2\beta^2 + \bar{n}(1-\eta^2)}}. \qquad (17.52)$$

Then, upon approximating $\langle x \rangle$ and $\langle p \rangle$ in (17.42),

$$\langle x \rangle \approx \eta\bar{x}_{\Delta nx}, \qquad \langle p \rangle \approx \eta\bar{p}_{\Delta np}, \qquad (17.53)$$

and replacing the summation in (17.48) by an integral, one finds for the net extractable work (Figs. 17.3, 17.4)

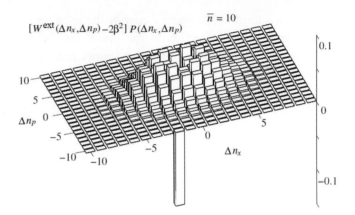

Figure 17.3 Product of the conditional probability and work, $[W^{\text{ext}}(\Delta n_x, \Delta n_p) - 2\beta^2] P(\Delta n_x, \Delta n_p)$, to be used in the expression for the net extractable work. Here $W_{\text{net}} = 2.31\hbar\omega$ for $\bar{n} = 10$. Only zero photon-difference detection (at the origin) yields negative extractable work.

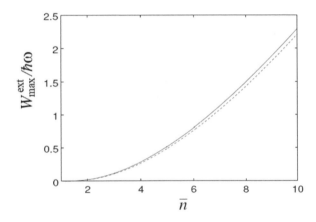

Figure 17.4 Optimized net work (mean value) in dependence on $\bar{n}$. Solid line: numerical optimization of the exact expression (17.48). Dashed line: same in the Gaussian approximation.

$$W_{\text{net}}^{\text{ext}} \approx \frac{\hbar\omega}{2} \int\int \left(\langle x \rangle^2 + \langle p \rangle^2 \right) P(\Delta n_x, \Delta n_p) d\Delta n_x \, d\Delta n_p - 2\hbar\omega\beta^2$$

$$\approx 2\hbar\omega\beta^2 \left[\frac{\eta^2(1-\eta^2)\bar{n}^2}{2\beta^2 + (1-\eta^2)(1+2\beta^2)\bar{n}} - 1 \right]. \tag{17.54}$$

This expression can be optimized with respect to η and β. The net maximal work extractable from a thermal state with $\bar{n} \gg 1$ is then found to be

$$W_{\text{max}}^{\text{ext}} \approx \hbar\omega(\bar{n} - 4\sqrt{\bar{n}}). \tag{17.55}$$

One can identify the $4\sqrt{\bar{n}}$ term as the energy cost of the measurement: $\sqrt{\bar{n}}$ is the cost of splitting off the measured fraction at the first beam splitter, $2\sqrt{\bar{n}}$ is the vacuum fluctuation noise entering through the two empty input parts of the beam splitters, and $2\beta^2 = \sqrt{\bar{n}}$ is the energy cost of the two local oscillators corresponding to the two measured quadratures. Unlike standard homodyne probing, here the local-oscillator energy is much weaker than $\hbar\omega\bar{n}$. It is thus possible to extract nearly all of the energy stored in the thermal input as net work, by averaging over an ensemble of homodyne measurements, each measurement followed by appropriate feedforward. Since each such step is very fast, it yields high useful power from thermal noise. The alternative conversion of the thermal energy into work is via heat-machine cycles, which are typically much slower and thus yield low power (Ch. 11).

17.4 The Landauer Principle Revisited for Non-Markovian System–Bath Correlations

Let us consider a brief QND measurement that decorrelates the system from the bath, thus altering their correlation energy $\langle H_{SB}\rangle$. As shown here, subsequent periodic modulation of the system energy levels (e.g. the TLS transition frequency) allows for work extraction even if the measurement result is *unread*, that is, for a nonselective measurement (NSM), provided that the cycle is completed *within the bath memory time*. This work extraction comes about because of temporal changes in the system–bath correlation energy that only take place on non-Markovian timescales.

17.4.1 No Work Can Be Extracted from a Single Bath in a Markovian Cycle

We consider a system that is in contact with a bath and is subject to driving that changes its level populations and energies. Under Markovian conditions, the bath-induced transition rates are positive, obeying detailed balance. In a TLS, they satisfy

$$\gamma_e(t)\rho_{ee}^{Eq}(t) = \gamma_g(t)\rho_{gg}^{Eq}(t) \tag{17.56}$$

for the (Gibbs) state $\rho_{jj}^{Eq}(t) = Z^{-1}(t)e^{-\beta E_j(t)}$. One can then prove, under Markovian conditions, the following inequality for the rate of the (von Neumann) entropy change due to driving the system away from equilibrium,

$$\dot{S} \equiv -k_B \sum_j \dot{\rho}_{jj} \ln \rho_{jj} \geq -k_B \sum_j \dot{\rho}_{jj} \ln \rho_{jj}^{Eq} = \frac{1}{T}\mathcal{J}, \tag{17.57}$$

where the heat current is $\mathcal{J} = \dot{Q} = \sum_j \dot{\rho}_{jj} E_j$. In a cycle, the initial and final states have the same entropy and energy. Then, the work extractable in a cycle satisfies

$$W_{\text{cycle}}^{\text{ext}} = Q \leq 0. \tag{17.58}$$

This means that the work extracted in a Markovian cycle must be supplied to the system from the outside. Here, as opposed to Section 17.2, work-information trade-off has not been invoked. In what follows we consider such trade-off under non-Markovian conditions.

17.4.2 Work–Information Trade-off in a Non-Markovian Cycle

The joint, *entangled* system–bath state at equilibrium, ρ_{Eq}, is changed to a product state $\rho_S \otimes \rho_B$ by the measurement performed by a meter M. The total Hamiltonian describing this setup is

$$H_{\text{tot}} = H_{\text{S+B}} + H_{\text{BM}} + H_{\text{SM}}, \tag{17.59}$$

where H_{SM} is the system–meter interaction and H_{BM} is the bath–meter interaction. This total Hamiltonian is τ-periodic, $H_{\text{tot}}(\tau) = H_{\text{tot}}(0)$. Hence, the work extracted in a cycle (which has the opposite sign with respect to the work invested in the cycle by the bath and the meter) amounts to the difference between the initial and the final mean energies,

$$W_{\text{tot}}^{\text{ext}}(\tau) = -W_{\text{cycle}} = -\text{Tr}[U(\tau)\rho_{\text{tot}}(0)U^{\dagger}(\tau)H_{\text{tot}}(0) - \rho_{\text{tot}}(0)H_{\text{tot}}(0)]. \tag{17.60}$$

The time evolution imposed by H_{tot} is unitary, so that the entropy of the joint (system–bath–meter) density operator $\rho_{\text{tot,M}}$ is constant. Since the mean energy at constant entropy is minimized for a thermal equilibrium state, (17.60) yields

$$W_{\text{tot}}^{\text{ext}} = -W_{\text{cycle}} = -\oint \text{Tr}(\rho_{\text{tot}} \dot{H}_{\text{tot}})dt = -\Delta E_{\text{M}} + W_{\text{cycle}}^{\text{ext}} < 0, \tag{17.61}$$

where ΔE_{M} is as in (17.63) and $W_{\text{cycle}}^{\text{ext}}$ is as in (17.66). Thus, no work can be *extracted* from a single bath by the entangled evolution of the (system-bath-detector) "supersystem," consistently with the second law.

However, when the meter is traced out, as in the case of an NSM, the entropy and energy of the system plus bath may change in a manner that violates the negativity of (17.61), as detailed below.

The extractable work in the cycle satisfies, consistently with Alicki's formula,

$$W_{\text{cycle}}^{\text{ext}} = -W_{\text{cycle}} = -\oint \text{Tr}(\rho \dot{H}_{\text{S}})dt = -\oint w(t)\dot{\omega}(t)dt. \tag{17.62}$$

Here (Fig. 17.5) $\omega(t)$ is the time-dependent TLS transition frequency and $w(t)$ is the level population difference, whereas (17.58) shows that no work can be extracted in a Markovian cycle. This may be the case if the modulation occurs on a non-Markovian timescale (i.e., for a modulation rate Δ that satisfies $\Delta \gg 1/t_c$).

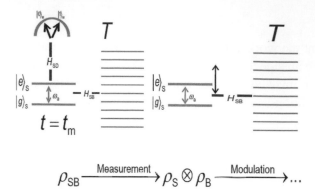

Figure 17.5 Work-extraction cycle induced by a QND nonselective measurement (NSM) of the TLS energy, followed by its cyclic energy modulation effected by a classical piston. Throughout the cycle, the TLS is in contact with a bath at temperature T. (Reprinted from Gelbwaser-Klimovsky et al., 2015.)

As discussed in Chapter 16, an impulsive projective measurement decorrelates the system and the bath, thereby increasing their correlation energy by

$$\Delta E_M = - \langle H_{SB} \rangle_{Eq} > 0. \tag{17.63}$$

This scenario stands in contrast to Landauer's, where system–bath correlations are not accounted for. If the cycle duration is shorter than the bath memory time, $t_{cycle} < t_c$, but longer than the time needed to perform the measurement that triggers the cycle, then the maximal amount of extractable work without measurement readout (for an NSM) is given by

$$(W_{NSM}^{ext})_{max} = \Delta E_M - T \Delta \mathcal{S}_M, \tag{17.64}$$

where $\Delta \mathcal{S}_M$ is the entropy increase due to the NSM.

A cycle allowing to extract work following the NSM ($W_{cycle}^{ext} > 0$) can be designed such that the piston corresponds to a periodic modulation of the TLS transition frequency that is chosen to have the form

$$\omega(t) = \omega_0 + \lambda \sin \Delta t. \tag{17.65}$$

In the weak-modulation regime, $\lambda \ll \Delta$, the work extracted within a cycle evaluates to

$$W_{cycle}^{ext} \approx -\hbar \lambda \int_0^{2\pi/\Delta} J_g(t) \Delta \cos(\Delta t) dt, \tag{17.66}$$

with $J_g(t) \equiv \int_0^t \gamma_g(t') dt'$. Upon modulating the system on non-Markovian timescales (typically for $\Delta \gg \lambda \sim \omega_0$), (17.66) allows for either positive or negative work extraction, owing to its sign oscillation with Δ. For $W_{cycle}^{ext} > 0$

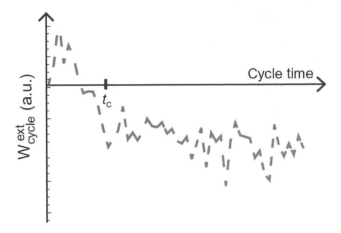

Figure 17.6 Extractable work in the NSM-based cycle of Figure 17.5 as a function of the cycle duration (in units of the bath correlation time t_c). For cycles longer than t_c, work may not be extracted. (Reprinted from Gelbwaser-Klimovsky et al., 2015.)

(i.e. work production by the system), $\dot{w}(t)$ has to oscillate out of phase with $w(t)$. This means that the energy invested by the meter that performs the NSM in (17.63) can be partly extracted as work, but only in such a non-Markovian cycle for an appropriate choice of Δ and λ (Fig. 17.6). This input energy can be in the form of *noisy pulses*, described by *passive states*, so that the input cannot store work (Ch. 15). The cycle described above converts such passive input into *active* output capable of delivering work. Such a cycle exemplifies the conclusion of Section 15.2 that, upon entangling the initially uncorrelated passive (but nonthermal) states of distinct subsystems, here the system (S), the bath (B), and the meter (M), the state of one subsystem (here S) may become nonpassive and thus deliver work.

17.4.3 Work–Information Relation for a Non-Markovian Cycle

The possibility of work extraction following a NSM in a non-Markovian cycle allows for a modified work-information trade-off in a non-Markovian cycle. The relation between $(W_{\text{sel}}^{\text{ext}})_{\text{max}}$ (for a selective measurement) and $(W_{\text{NSM}}^{\text{ext}})_{\text{max}}$ (for a nonselective measurement) shows that the maximum work (per cycle) extractable from a selective measurement is *higher* than the Landauer bound W_{L} in Section 17.2,

$$(W_{\text{sel}}^{\text{ext}})_{\text{max}} = (W_{\text{NSM}}^{\text{ext}})_{\text{max}} + W_{\text{L}}. \tag{17.67}$$

The extra work $(W_{\text{NSM}}^{\text{ext}})_{\text{max}}$ stems from non-Markovian system–bath correlations or entanglement unaccounted for by the Landauer principle. Such entanglement is

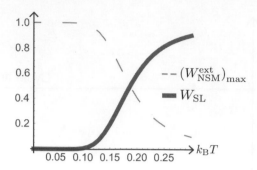

Figure 17.7 Comparison of the Landauer bound on extractable work with work extractable in the NSM-based scheme of Figure 17.5, as a function of the bath temperature $k_B T$. At low T, the NSM-based work can greatly exceed the Landauer bound. (Reprinted from Gelbwaser-Klimovsky et al., 2015.)

excluded in the Markovian treatment, wherein the maximum extractable work (by a selective measurement) adheres to the Landauer principle.

Remarkably, the NSM allows for work extraction from a bath at $T = 0$ *without information gain*: The system–bath correlation energy is always negative, even at $T = 0$. Hence, decorrelation of the system and the bath through a measurement increases the total energy, allowing the cycle to be triggered, yielding the extractable work.

$$(W_{\text{sel}}^{\text{ext}})_{\text{max}} = (W_{\text{NSM}}^{\text{ext}})_{\text{max}} > 0. \tag{17.68}$$

In this case, all extractable work stems from this correlation energy change (Fig. 17.7).

17.4.4 Experimental Implementation

We may envisage TLS (atoms or molecules) with resonance frequency ω_0 in the microwave domain in a high-Q cavity with controllable finite temperature. QND measurements (monitoring) of the TLS-level population can be performed by optical pulses at time intervals shorter than ω_0^{-1} and much shorter than the cavity bath memory (correlation) times $t_c \sim 10^{-4}$ s. The piston may be realized by an off-resonant coherent microwave signal that modulates the TLS transition frequency at a rate $\Delta \gg 1/t_c$. An amplification of the coherent piston mode then signifies work extraction (Fig. 17.8).

This scenario may transform very noisy (thermal) probe pulses that induce the QND measurements into work, manifest by coherent amplification of the piston field. It may be useful when the coupling of the TLS to two different baths [as required by heat machines (Chs. 18–20)] is hard to implement experimentally.

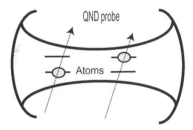

Figure 17.8 A setup for the realization of the NSM-based scheme in Figure 17.5, using two-level atoms in a microwave cavity (that serves as a thermal bath) subject to periodic QND measurements and off-resonant energy modulation by optical probes. (Reprinted from Gelbwaser-Klimovsky et al., 2015.)

17.5 Discussion

We have discussed the notions of entropy, free energy, work extraction, and their second-law bounds for post-measurement nonequilibrium states:

(i) The minimal work required to isothermally drive a system from one state to another is the difference between free energies. The dissipated or irreversible work associated with the isothermal entropy production [cf. (17.4)] is related by the second law to this minimal work.

(ii) QND measurements increase the free energy and thereby the work that can be extracted isothermically. The upper bound on extractable work following a measurement is proportional to the Shannon-entropy information of the measurement (17.12).

(iii) The minimal work required to reset a symmetric memory is the Shannon entropy of the measurement generated by the meter [cf. (17.15)], which yields the Landauer bound for work required to reset a completely random bit. The same work is extractable by completely randomizing a pure-state qubit.

(iv) The minimal work required to store the measurement result and subsequently reset the memory is proportional to the mutual information of the system and the meter.

These results imply that the thermal equilibrium of a system and a bath cannot be changed by mere observation of the system, unless the result is recorded (stored) and some action is taken pursuant to the observation (feedforward). QND measurements, however, affect neither the state of the system nor that of the bath. Hence, any change in the thermal equilibrium without measurement-result readout storage and feedforward may be suspected to be a violation of either the first or the second law. Namely, the first law implies that QND observations affect neither the energy nor the heat exchange between the system and the bath. Kelvin's formulation of the second law precludes the extraction of work in a cycle by a system that is coupled to a single bath.

In a Szilárd engine, by contrast, work extraction is obtained through feedforward, which represents specific actions depending on the result of the observation. If information gathered by observations is commuted into work, this work must, in accordance with the Landauer principle, not exceed the cost incurred by the erasure (resetting) of the demon's memory.

Maxwell's demon can be viewed as a system that biases interstate transitions seemingly without energy cost. Smoluchowski and later Feynman considered such an autonomous Maxwell demon based on a ratchet mechanism that can rectify thermal fluctuations. Such a demon, however, requires the system to be out of equilibrium by alternate coupling of the system to two thermal baths at different temperatures. An *extended version* of Maxwell's demon considered here is effected by phase and amplitude estimation of a thermal state of a field mode, followed by appropriate feedforward. In this version, the detectors with their unused modes constitute the cold bath, and the thermal input mode is drawn from a hot bath.

Yet, as shown here, the foregoing tenets do not hold if the quantum system is subject to dynamical control by frequent observations or phase shifts, even if they do not directly affect the system. Such anomalies require these acts of control to be shorter than all the relevant timescales and their intervals to be within the memory (correlation) time of the bath. Thus, the violation of the tenets of information thermodynamics is intimately connected with the breakdown of the Markovian approximation. Then, frequent unread measurements of a quantum system coupled to a thermal bath may draw work in a cycle from the system–bath interaction (correlation) energy, unaccounted for by the Landauer bound. Namely, work and information are related by (17.19) only under Markovian assumptions, but their relation is changed by system–bath correlations within non-Markovian time intervals. Remarkably, the convolution changes caused by measurements can change the passive (yet nonthermal) state of the system into an active state that partly converts the invested passive energy into work.

Equivalently, measurements can control the temperature and entropy of a quantum system that is entangled with a bath if these measurements are performed at intervals shorter than the bath correlation time. Such measurements may enable the system to do more work in a cycle than permitted by the Landauer principle.

Such work extraction above the Landauer bound is at the expense of the change in the system–bath correlation energy (consistently with the first law). The second law is adhered to when including the measuring device in the calculation of the energy and entropy changes. This is possible even if no information is gathered or if the bath is at $T = 0$, provided the cycle is non-Markovian. The first and second laws still remain intact and may be colloquially summarized by a single tenet: "There are no free lunches."

18

Cyclic Quantum Engines Energized by Thermal or Nonthermal Baths

The efficiency of cyclic heat engines is limited by the Carnot bound that follows from the second law of thermodynamics. This bound is attained by engines that operate between two heat baths under the reversibility condition whereby the total entropy (of the system and baths combined) is conserved. By contrast, the efficiency of engines powered by nonthermal baths may surpass the Carnot bound. The key to understanding the performance of such engines is a proper division of the energy supplied by the nonthermal bath to the working medium (WM), that is, the system, into heat and work. This division depends on the change in the system entropy and ergotropy. The crucial point is that the efficiency bound for cyclic engines powered by a nonthermal bath does not solely follow from the laws of thermodynamics, so that the thermodynamic Carnot bound is inapplicable to such engines.

18.1 Universal Efficiency Bound

Let us consider a broad class of cyclic quantum engines (Fig. 18.1) that operate between a cold thermal bath (at temperature T_c) and an energizing (thermal or nonthermal) bath while being subject to the action of a work-extracting "piston." Here and in Chapters 19–21 we treat the piston classically, representing its action as a time-dependent drive, whereas in Chapters 22 and 23 the piston is quantized.

If the energizing bath is nonthermal, it may produce a nonpassive state of the working medium (WM); see Chapter 15. The passive counterpart of this state is a thermal state with temperature $T_h > T_c$. For a harmonic-oscillator (HO) WM that interacts with a squeezed-thermal bath, T_h is the temperature that the bath would have in the absence of squeezing.

Let the driven-WM Hamiltonian $H(t)$ slowly change during the interaction with the baths so that the WM attains its steady state by the end of the interaction.

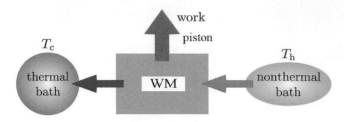

Figure 18.1 In a broad class of cyclic engines, a fraction of the energy obtained from a nonthermal (e.g., squeezed-thermal) bath is converted by the working medium (WM) into useful work that is extracted by a piston. The rest of the energy is dumped into the cold thermal bath. (Adapted from Ghosh et al., 2018.)

In a reciprocal cycle, that is, a cycle that consists of strokes (Ch. 19), the interaction with a nonthermal (e.g., squeezed-thermal) bath is the energizing stroke that unitarily evolves the WM from a thermal to a nonpassive state that carries ergotropy,

$$\rho_{ss}(t \to \infty) = U \rho_{th} U^\dagger. \tag{18.1}$$

To obtain work from the engine, the ergotropy from the WM must be extracted by the piston via a suitable unitary transformation prior to the WM interaction with the cold bath. In continuous cycles, where both baths are simultaneously coupled to the WM, part of the ergotropy is then dissipated into the cold bath, thus hampering the cycle efficiency. On the other hand, in a continuous cycle, $H(t)$ may be taken to *commute with itself at all times* (Ch. 19), whereas in stroke (reciprocal) cycles, Hamiltonians do not commute with themselves at different times, thereby reducing the efficiency via "quantum friction." Thus, it is a priori unclear whether stroke cycles outperform continuous cycles or vice versa. Here we assume a strokes cycle that is not hampered by quantum friction (i.e. take $H(t)$ to be self-commuting as in Chapter 19).

In the two strokes of the WM–bath interactions, the entropy changes obey, according to the second law,

$$\Delta S_c \geq \frac{\mathcal{E}_c}{T_c}, \quad \Delta S_h \geq \frac{\mathcal{Q}_h}{T_h}. \tag{18.2}$$

Here $\mathcal{E}_c \leq 0$ is the dissipative change in the WM energy due to its interaction with the cold (thermal) bath and $\mathcal{Q}_h \geq 0$ is the heat that the WM would receive if the initial state were passive and the energizing bath were thermal. The WM cyclically returns to its initial state, hence

$$\Delta S = \Delta S_c + \Delta S_h = 0 \tag{18.3}$$

over a cycle. The conditions (18.2) and (18.3) yield the inequality

$$\frac{\mathcal{E}_c}{T_c} + \frac{\mathcal{Q}_h}{T_h} \leq 0. \tag{18.4}$$

Correspondingly, the first law of thermodynamics (energy conservation) over a cycle yields

$$\mathcal{E}_c + \mathcal{E}_h + W_{\text{cycle}} = 0, \tag{18.5}$$

where $\mathcal{E}_h$ is the dissipative energy change of the WM due to its interaction with the energizing bath (Fig. 18.1) and W_{cycle} is the work performed on the WM in the cycle, whereas $W^{\text{ext}} = -W_{\text{cycle}}$ is the net work extracted in the cycle. An engine functions properly when $W_{\text{cycle}} < 0$ ($W^{\text{ext}} > 0$).

The engine efficiency is the ratio of the extracted work to the *total energy* (heat and ergotropy) $\mathcal{E}_h$ imparted to the WM by the energizing bath. If the energy portion imparted to the WM that remains after the work extraction is dumped into the cold bath and is denoted by $\mathcal{E}_c$, then

$$\mathcal{E}_c \leq 0, \quad \mathcal{E}_h \geq 0. \tag{18.6}$$

The engine efficiency is then given by

$$\eta := \frac{-W}{\mathcal{E}_h} = 1 + \frac{\mathcal{E}_c}{\mathcal{E}_h}. \tag{18.7}$$

Since, from condition (18.4),

$$\mathcal{E}_c \leq -\frac{T_c}{T_h} \mathcal{Q}_h, \tag{18.8}$$

the efficiency is restricted by the general bound

$$\eta \leq 1 - \frac{T_c}{T_h} \frac{\mathcal{Q}_h}{\mathcal{E}_h} \equiv \eta_{\text{max}}. \tag{18.9}$$

The efficiency bound (18.9) is universal in the sense that it applies to cyclic engines energized by any thermal or nonthermal bath. This bound depends not only on the two temperatures, T_c and T_h, but also on the ratio $\mathcal{Q}_h/\mathcal{E}_h$ of the heat input to the total energy input from the energizing bath. Assuming that the energizing bath can only increase the WM entropy, we have $\mathcal{Q}_h \geq 0$ and, according to (18.6), $\mathcal{E}_h > 0$. The maximal efficiency (18.9) is then limited by unity,

$$\eta_{\text{max}} \leq 1. \tag{18.10}$$

Unity efficiency is reached in the "mechanical"-engine limit, $\mathcal{Q}_h \to 0$, where the energizing nonthermal bath only provides ergotropy, as in the case of the squeezed-vacuum bath state (Sec. 18.2). In the heat-engine limit, on the other hand,

$\mathcal{Q}_h \rightarrow \mathcal{E}_h$, only heat, but no ergotropy, is supplied by the energizing bath, whence (18.9) reduces to the Carnot bound

$$\eta_{\text{Carnot}} = 1 - T_c/T_h. \tag{18.11}$$

As opposed to the regime that conforms to (18.6), there exists a regime wherein $\mathcal{E}_c > 0$, $\mathcal{Q}_h < 0$. Such a machine acts simultaneously as an engine and a refrigerator that cools the cold bath at the expense of energy supplied by the cold bath to the WM. The efficiency then attains the limit

$$\eta = \frac{-W_{\text{cycle}}}{\mathcal{E}_h + \mathcal{E}_c} = \frac{\mathcal{E}_h + \mathcal{E}_c}{\mathcal{E}_h + \mathcal{E}_c} = \eta_{\text{max}} = 1. \tag{18.12}$$

This efficiency does not depend on $\Delta\mathcal{S}$, so that the second law is then inapplicable.

18.2 Quantum Machines Powered by Nonthermal Bath with Ergotropy

We here consider examples of stroke-based ("reciprocal") bath-powered engine cycles that are not restricted by the second law, but by other constraints on their entropy.

18.2.1 Modified Otto Cycle Powered by a Squeezed-Thermal Bath

We first consider machines powered by a squeezed-thermal bath that obey a modified Otto cycle [Fig. 18.2(a)]. The two isentropic strokes (adiabatic compression and decompression of the WM) and the two isochoric strokes (interaction with the baths under a fixed Hamiltonian) of the standard Otto cycle are supplemented by an *ergotropy-extraction stroke*. For a harmonic-oscillator (HO) WM, ergotropy extraction may be implemented by abruptly ramping up the HO frequency and subsequently ramping this frequency down.

As the interaction of the WM with the energizing bath is isochoric, we have

$$\mathcal{E}_h = \mathcal{Q}_h + \Delta\mathcal{W}_h, \tag{18.13}$$

where $\mathcal{Q}_h$ is the heat (passive energy) change and $\Delta\mathcal{W}_h$ is the ergotropy change of the WM, respectively. The efficiency bound (18.9) of this Otto-like cycle then reads

$$\eta_{\text{max}} = 1 - \frac{T_c}{T_h} \frac{\mathcal{Q}_h}{\mathcal{Q}_h + \Delta\mathcal{W}_h}. \tag{18.14}$$

To evaluate the ergotropy transfer $\Delta\mathcal{W}_h$, we consider the interaction with the energizing bath to be sufficiently long for the WM to reach steady state. The bound (18.14) only requires the interaction of the WM HO with the squeezed-thermal bath, which is governed by the master equation (Ch. 11)

$$\dot{\rho} = \kappa[(N+1)\mathcal{D}(a, a^\dagger) + N\mathcal{D}(a^\dagger, a) - M\mathcal{D}(a, a) - M\mathcal{D}(a^\dagger, a^\dagger)]\rho. \tag{18.15}$$

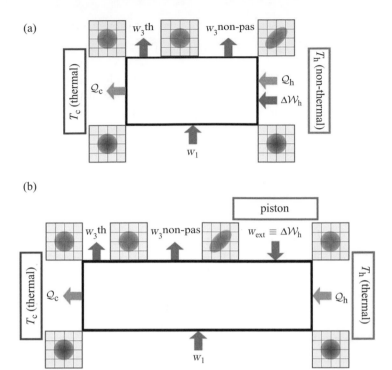

Figure 18.2 (a) Modified Otto cycle for an oscillator WM powered by a squeezed-thermal bath: In the first stroke, the work W_1 is invested by the piston to compress the WM, thermalized to temperature T_c. In the second stroke, the WM interacts with a squeezed-thermal bath, from which passive energy (heat) $\mathcal{Q}_h$ and ergotropy $\Delta \mathcal{W}_h$ are imparted to the WM. The standard Otto cycle is modified in that a unitary operation is performed on the WM to extract its ergotropy as work $W_3^{\text{non-pas}}$. The WM then becomes thermal again and work W_3^{th} is extracted by the piston by expanding the WM, as in the standard Otto cycle. The cycle is closed by the interaction of the WM with the cold thermal bath at temperature T_c into which the heat $\mathcal{Q}_c$ is dumped. (b) Equivalent hybrid engine: Instead of being energized by a squeezed-thermal bath, the WM receives the heat $\mathcal{Q}_c$ from a hot thermal bath at temperature T_h. Subsequently, an external work reservoir (e.g., a battery) performs the work W_{ext} to squeeze the WM, thereby increasing the WM ergotropy by the same amount $\Delta \mathcal{W}_h$ as that supplied in (a) by the squeezed-thermal bath. (Adapted from Ghosh et al., 2018.)

Here the dissipater is defined for operators A and B as $\mathcal{D}(A, B)\rho := 2A\rho B - BA\rho - \rho BA$, a and $a^\dagger$ are the WM creation and annihilation operators, and κ denotes the WM–bath exchange rate. Upon taking the (arbitrary) squeezing phase to be zero, the effects of squeezing are quantified as follows,

$$N := \bar{n}_h(\cosh^2 r + \sinh^2 r) + \sinh^2 r \qquad (18.16a)$$

$$M := -\cosh r \sinh r(2\bar{n}_h + 1), \qquad (18.16b)$$

where

$$\bar{n}_h = [\exp(\hbar\omega/[k_B T_h]) - 1]^{-1} \qquad (18.17)$$

is the thermal excitation occupancy of the bath and r is the squeezing parameter.

The steady state of (18.15) is a squeezed-thermal state of the WM with a mean number of quanta that deviates from the thermal $\bar{n}_h$ by

$$\Delta\bar{n}_h = (2\bar{n}_h + 1)\sinh^2(r) > 0. \qquad (18.18)$$

The efficiency bound in (18.14) then evaluates to

$$\eta_{max} = 1 - \frac{T_c}{T_h}\frac{\bar{n}_h - \bar{n}_c}{\bar{n}_h + \Delta\bar{n}_h - \bar{n}_c}, \qquad (18.19)$$

where $\bar{n}_h$ and $\bar{n}_c$, the hot- and cold-bath thermal occupancies, are evaluated at the respective frequencies ω_h and ω_c. The efficiency of the modified Otto cycle then evaluates to

$$\eta = 1 - \frac{(\bar{n}_h - \bar{n}_c)\omega_c}{(\bar{n}_h + \Delta\bar{n}_h - \bar{n}_c)\omega_h}. \qquad (18.20)$$

For $\mathcal{E}_h > 0$, the machine acts as an engine that delivers work for $\bar{n}_h + \Delta\bar{n}_h \geq \bar{n}_c$, corresponding to the regime (18.6), whereas for $\mathcal{E}_c > 0$ the machine operates as an engine combined with a refrigerator for the cold bath for which $\eta = 1$ as per (18.12) (Fig. 18.3).

In the limit $T_c = T_h = 0$ ($\bar{n}_c = \bar{n}_h = 0$), the machine obeying (18.19) and (18.20) can still deliver work. Namely, the work invested in the cycle is then

$$W_{cycle} = -\hbar(\omega_h - \omega_c)\Delta\bar{n}_h < 0. \qquad (18.21)$$

The machine then acts as an *effectively mechanical* engine, energized by $\mathcal{E}_h = \hbar\omega_h\Delta\bar{n}_h > 0$, which is pure ergotropy transfer from the squeezed-vacuum bath. The WM energy increase in the second stroke is then *isentropic*, as it does not involve any heat exchange with the bath (Fig. 18.4).

The modified Otto cycle in Figure 18.2(a) is equivalent to the cycle in Figure 18.2(b) involving a hot thermal bath at temperature T_h and an auxiliary work-source reservoir (which is *not the piston*). After the second stroke this auxiliary work source acts unitarily on the thermal WM state, transforming it into the same nonpassive state that it would become via contact with a squeezed-thermal bath, provided that the work invested by the work source/reservoir is equivalent to the ergotropy supplied by the squeezed-thermal bath. Because of the equivalence of the two cycles, the heat provided by the squeezed-thermal bath is the *same* as the heat that a thermal bath at temperature T_h would provide [cf. (18.4)], whereas the energy surplus imparted by the squeezed-thermal bath can be identified as work extractable from ergotropy. The work imparted by the auxiliary work reservoir

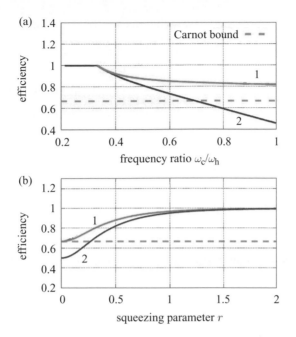

Figure 18.3 Modified Otto-cycle efficiency (18.20) (curves 2) as a function of (a) the frequency ratio ω_c/ω_h and (b) the squeezing parameter r. The bound (18.19) (curves 1) interpolates between the Carnot bound (in the heat-engine limit) and 1 (in the mechanical-engine limit). Parameters: $T_h = 3T_c$, (a) $r = 0.5$, (b) $\omega_c = \omega_h/2$. (Adapted from Ghosh et al., 2018.)

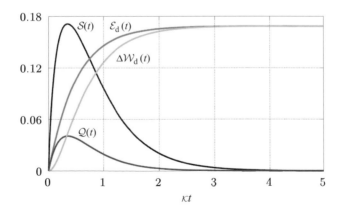

Figure 18.4 Isentropic WM–bath interaction: Entropy $\Delta \mathcal{S}(t)$, energy $\mathcal{E}_d(t)$, heat $\mathcal{Q}(t)$, and ergotropy $\Delta \mathcal{W}_d(t)$ changes [cf. (15.31) and (15.32)] for a harmonic oscillator initially in the vacuum state that interacts with a squeezed-vacuum bath according to the master equation (18.15). The energies are given in units of $\hbar\omega_h$ and the entropy in units of k_B. (Adapted from Ghosh et al., 2018.)

is not bounded by the second law, whereas the heat exchanges are. Hence, the maximal efficiency of this cycle is not bounded by thermodynamics.

18.2.2 Modified Carnot Cycle Powered by a Squeezed-Thermal Bath

We next consider a cycle powered by a squeezed-thermal bath comprised of the four strokes of the ordinary Carnot cycle and an additional ergotropy-extraction stroke (stroke 3 in Fig. 18.5). In stroke 1, the mode undergoes an adiabatic compression from frequency ω_c to frequency $\omega_2 = \omega_c T_h / T_c$. In the energizing stroke 2, the frequency is slowly reduced to $\omega_h \leq \omega_2$ while the mode interacts with the squeezed-thermal bath, attaining a squeezed-thermal steady state. Its ergotropy is extracted in the unitary stroke 3 by an "unsqueezing" operation, bringing the mode to a thermal state at temperature T_h. In stroke 4, the frequency is adiabatically reduced to $\omega_1 = \omega_h T_c / T_h$. Stroke 5 is an isothermal compression induced by a cold bath back to the initial thermal-state temperature T_c.

Stroke 2 corresponds to isothermal expansion wherein the state $\varrho(t)$ is in thermal equilibrium, so that $\mathcal{Q}_h = T_h \Delta \mathcal{S}_h$. Likewise, stroke 5 represents isothermal

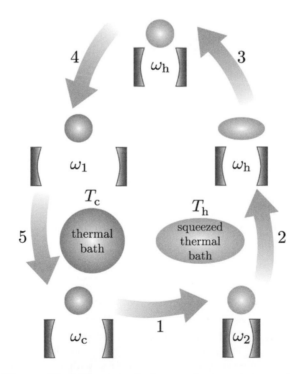

Figure 18.5 The modified Carnot cycle for a cavity-mode WM. The cavity mode starts the cycle in a thermal state at frequency ω_c and temperature T_c in the lower left corner (see text). (Reprinted from Ghosh et al., 2018.)

compression, hence $\mathcal{E}_c = T_c \Delta \mathcal{S}_c$. Since the entropy change over a cycle must vanish, the equality sign in (18.4) is obtained. Hence, the efficiency of this cycle is bounded by η_{max} in (18.9). This bound is adequate for all possible engine cycles that contain a similar energizing stroke, characterized by a slowly-changing Hamiltonian and an initial thermal state at temperature T_h. Such a Carnot-like cycle always attains the maximum efficiency bound (18.9), irrespective of the division between passive (thermal) energy and ergotropy imparted by the energizing bath.

18.3 Quantum Machines Energized by Heat from Nonthermal Baths

In a quantum cyclic machine, the WM may not draw both ergotropy and heat from a bath, but instead be thermalized by the energizing bath, in spite of the bath being nonthermal. The WM then relaxes to a thermal state at a *real* temperature T_h if left in contact with this energizing bath. If the cold bath is at temperature T_c, the first and second laws imply, respectively, that

$$\mathcal{Q}_c + \mathcal{Q}_h + W_{cycle} = 0 \qquad (18.22a)$$

and

$$\frac{\mathcal{Q}_h}{T_h} + \frac{\mathcal{Q}_c}{T_c} \leq 0. \qquad (18.22b)$$

Such a machine may operate as a *genuine heat engine* whose efficiency is restricted by the Carnot bound

$$\eta = \frac{-W_{cycle}}{\mathcal{Q}_h} \leq 1 - \frac{T_c}{T_h} \equiv \eta_{Carnot}. \qquad (18.23)$$

It should be stressed that the temperature of the energizing bath prior to its transformation into a nonthermal state plays no role; the only temperatures that matter for the efficiency bound are T_c and T_h.

Here we study the temperature control of a simple dissipative quantum system: a micromaser setup wherein a leaky cavity field mode interacts with a bath composed of three-level atoms with coherently superposed energy states that cross the cavity sequentially: the "phaseonium" bath introduced by Scully. Alternatively, the cavity mode interacts with a bath comprised of entangled pairs of atoms (dimers) that traverse the cavity one by one. The inter-level phaseonium coherence or the two-atom coherence (entanglement) affect the heat exchange of the bath and cavity mode. Such heat exchange modification can boost the efficiency bound of a Carnot cycle by replacing the cavity-wall temperature with T_h that can be made considerably higher by choosing the appropriate phase of the coherence.

18.3.1 Cavity Mode Energized by Phaseonium

Let us consider a micromaser setup wherein a beam of three-level atoms with two almost-degenerate ground states traverses the cavity. The atoms are initially in a thermal state at temperature T_a. Subsequently a small amount of coherence is injected to the ground-state manifold (Fig. 18.6). The resulting phase-coherent three-level atoms have been nicknamed "phaseonium" by Scully. The cavity field state ρ then evolves according to the master equation (ME)

$$\dot{\rho} = \left\{ \frac{\mu}{4} \left[p_{g_1} + p_{g_2} + C(\varphi) \right] + \kappa(\bar{n}+1) \right\} \mathcal{L}_c \rho + \left(\frac{\mu}{2} p_e + \kappa \bar{n} \right) \mathcal{L}_h \rho. \quad (18.24)$$

In this ME (Ch. 11), $\mathcal{L}_c$ is the Lindblad superoperator for the de-excitation (cooling) of the cavity mode via photon emission by the atoms, and $\mathcal{L}_h$ is its counterpart for the excitation (heating) of the cavity mode through photon absorption by the atoms,

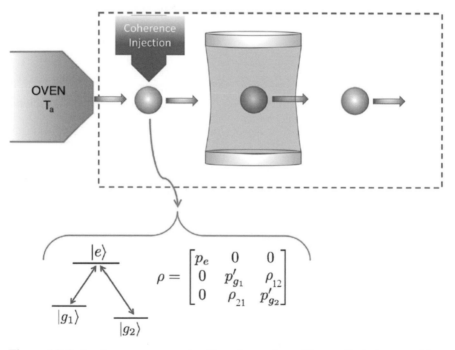

Figure 18.6 A micromaser energized by phaseonium. The cavity is traversed by a beam of thermal three-level atoms at temperature T_a (with populations p_e, p_{g_1}, and p_{g_2}) to which some coherence $C(\varphi) = 2\,\mathrm{Re}\,\rho_{12}$ is injected before they interact with the cavity mode. The ground-state populations are slightly changed by the coherence injection, to p'_{g_1} and p'_{g_2}, respectively. (Adapted with permission from Dağ et al., 2019. © 2019 American Chemical Society)

$$\mathcal{L}_c \rho = 2a\rho a^\dagger - a^\dagger a\rho - \rho a^\dagger a,$$
$$\mathcal{L}_h \rho = 2a^\dagger \rho a - aa^\dagger \rho - \rho aa^\dagger, \tag{18.25}$$

$\mu = r(g\tau)^2$ is an effective atom-field coupling rate determined by the injection rate r of the atoms, their interaction time τ with the cavity field, and their coupling strength g to the cavity mode. In (18.24), κ is the cavity loss rate and $\bar{n}$ is the thermal occupancy at temperature T in the bare-cavity mode. The standard contributions to the heating and cooling rates are determined, respectively, by p_e, p_{g_1}, and p_{g_2}, the thermally excited and ground state populations of the three-level atoms. The two ground-state populations are related by $p_{g_1} = \exp[-\hbar\Delta_g/(k_B T_a)]p_{g_2}$, where $\hbar\Delta_g = \hbar\omega_{g_1} - \hbar\omega_{g_2} > 0$ is their small energy splitting. The nonstandard contribution to the cooling rate is induced by

$$C(\varphi) = 2\,\mathrm{Re}\,\rho_{12} = 2|\rho_{12}|\cos\varphi, \tag{18.26}$$

the coherence between the two ground states, where ρ_{12} is the off-diagonal matrix element in the $g1$-$g2$ subspace (Fig. 18.6) and $\varphi = \arg\rho_{12}$.

The rate of change of $\langle n \rangle$, the mean number of photons in the cavity mode, according to the ME (18.24), then satisfies

$$\frac{d}{dt}\langle n \rangle = \mu p_e + 2\kappa\bar{n} - \left\{2\kappa + \frac{\mu}{2}\left[p_{g_1} + p_{g_2} - 2p_e + C(\varphi)\right]\right\}\langle n \rangle. \tag{18.27}$$

Only the negative term on the right-hand side of (18.27) has a contribution from the quantum coherence $C(\varphi)$. As a result, the micromaser threshold depends on the coherence via the condition

$$\kappa + \frac{\mu}{4}\left[p_{g_1} + p_{g_2} - 2p_e + C(\varphi)\right] > 0, \tag{18.28}$$

and so does the steady state

$$\langle n \rangle_{ss} = \frac{\kappa\bar{n} + \mu p_e/2}{\kappa + \frac{\mu}{4}\left[p_{g_1} + p_{g_2} - 2p_e + C(\varphi)\right]}. \tag{18.29}$$

18.3.2 Cavity Mode Energized by Entangled Dimers

Let us consider a beam of two-level atom pairs (dimers) injected into a cavity with resonant mode-frequency ω (Fig. 18.7). Each dimer is prepared in the initial state

$$\rho = \begin{pmatrix} \rho_{11} & 0 & 0 & 0 \\ 0 & \rho_{22} & \rho_{23} & 0 \\ 0 & \rho_{32} & \rho_{33} & 0 \\ 0 & 0 & 0 & \rho_{44} \end{pmatrix}. \tag{18.30}$$

Here the basis is comprised of the vectors $\{|1\rangle, |2\rangle, |3\rangle, |4\rangle\} := \{|ee\rangle, |eg\rangle, |ge\rangle, |gg\rangle\}$, where $|g\rangle$ and $|e\rangle$ are the ground and excited states of the identical atoms.

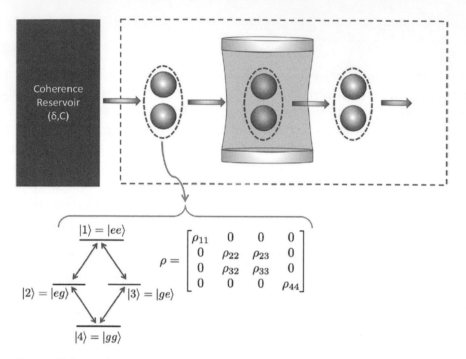

Figure 18.7 A micromaser powered by a beam of two-level atom pairs (dimers). The beam drives the cavity mode into a thermal state whose temperature is determined by the atom-pair (double-excitation) population inversion $w := 2(\rho_{11} - \rho_{44})$ and the coherence $C := 2\,\mathrm{Re}\,\rho_{23}$. (Reprinted with permission from Dağ et al., 2019. © 2019 American Chemical Society)

The state (18.30) is deemed entangled (inseparable) if at least one eigenvalue of the partially transposed ρ is negative, which holds if

$$|\rho_{23}|^2 > \rho_{11}\rho_{44}. \tag{18.31}$$

Conversely, for $|\rho_{23}|^2 \leq \rho_{11}\rho_{44}$ the state ρ is separable.

Analogously to the case of phaseonium, the ME for the cavity field mode that resonantly interacts with the dimers is given by

$$\dot{\rho} = \left[\frac{\mu r_-}{2} + \kappa(\bar{n}+1)\right]\mathcal{L}_c\rho + \left(\frac{\mu r_+}{2} + \kappa\bar{n}\right)\mathcal{L}_h\rho. \tag{18.32}$$

Here the notation is as for the phaseonium case, except that the field-atom coupling rate μ is multiplied by the two-atom excitation parameters

$$r_\pm := 1 + C(\varphi) \pm \frac{w}{2}, \tag{18.33}$$

where the two-atom coherence is denoted by

$$C(\varphi) = 2\,\mathrm{Re}\,\rho_{23}(\varphi), \tag{18.34}$$

φ being the phase of ρ_{23}, and

$$w = 2(\rho_{11} - \rho_{44}).\tag{18.35}$$

Here the double-excitation population inversion w, which ranges from $w = -2$ (for the doubly ground state $\rho = |gg\rangle \langle gg|$) to $w = 2$ (for the doubly excited state $\rho = |ee\rangle \langle ee|$) determines the energy of the dimer state (18.30) via $E = \hbar\omega(1 + w/2)$.

Below the micromaser threshold (i.e. for $r_+ < r_- + 2\kappa/\mu$ or $w < 2\kappa/\mu$), the ME (18.32) yields a thermal steady state of the cavity field, at the temperature T_h determined by

$$\frac{r_+ + 2\kappa\bar{n}/\mu}{r_- + 2\kappa(\bar{n}+1)/\mu} = \exp\left(-\frac{\hbar\omega_c}{k_B T_h}\right),\tag{18.36}$$

where the left-hand side is the ratio of the absorption and emission coefficients in (18.32). Thus, the dimer beam, which is prepared in the non-equilibrium state (18.30), acts as a heat bath at temperature T_h for the cavity field below the micromaser threshold. Namely, T_h is the real temperature of the cavity field, controlled by the two-atom coherence $C(\varphi)$.

It follows from (18.36) that $T_h > T$, the bare-cavity temperature, provided that

$$C(\varphi) > -1 - \left(\bar{n} + \frac{1}{2}\right) w.\tag{18.37}$$

This inequality implies that the benefit of heat-exchange coherence is maximal in low-temperature cavities and shows that $C(\varphi)$ and w complement each other as heating resources. The maximally allowed magnitude of coherence C is bounded by the condition

$$|C| \leq 2\sqrt{\rho_{22}\rho_{33}} \leq 1 - \frac{|w|}{2},\tag{18.38}$$

in order for ρ in (18.30) to be nonnegative.

To illustrate the coherence effects, we consider the singly excited two-atom Bell states

$$|\Psi^{\pm}\rangle := \frac{|ge\rangle \pm |eg\rangle}{\sqrt{2}}\tag{18.39a}$$

and compare them to their phase-averaged counterpart,

$$\rho_{\text{mix}} := \frac{1}{2} |ge\rangle \langle ge| + \frac{1}{2} |eg\rangle \langle eg|.\tag{18.39b}$$

All these states have the same energy, and yet they give rise to very different cavity-mode temperatures, $T_h = T_{\pm}$ for $|\Psi^{\pm}\rangle$, respectively, and $T_h = T_{\text{mix}}$ for ρ_{mix}. These temperatures satisfy the relation

$$T_+ > T_{\text{mix}} > T_- \equiv T.\tag{18.40}$$

Relation (18.40), which shows the ability of two-atom coherence to enhance heating, is obtained from (18.36) upon substituting $w = 0$ and $C = \pm 1$ [for the states (18.39a)] and $C = 0$ [for the state (18.39b)], respectively.

The fact that the Bell state $|\Psi^+\rangle$ yields the highest possible temperature among all states with $w = 0$, whereas $|\Psi^-\rangle$ cannot raise the cavity-mode temperature above T results from quantum interference that gives rise to dark and bright Dicke states (cf. Ch. 8). The bright state $|\Psi^+\rangle$ exhibits constructive interference of the exchange between the atoms and the cavity field and thus maximizes energy transfer from the atom pair to the cavity mode (for states with $w = 0$). The dark state $|\Psi^-\rangle$ exhibits destructive interference that prevents any transfer, so that for this state $r_\pm = 0$ and hence $T_- \equiv T$.

The entanglement of two identical atoms sharing a single excitation are determined by the symmetry and angular momentum of the parent (molecular) dimer. Depending on these properties, the dissociated dimer may emerge in a bright or dark Dicke state. This simple description mainly applies to dimers such as Ca_2 that dissociate into a symmetric or antisymmetric combination of the two-atom states $|^1P_m\rangle\,|^1S\rangle$, where m is the magnetic number.

The "parent" (molecular-dimer) state determines the "nascent" two-atom entanglement as follows:

- Gerade, spin-singlet dimer state (such as $^1\Sigma_g$ or $^1\Pi_g$) or ungerade, spin-triplet dimer state (such as $^3\Sigma_u$ or $^3\Pi_u$) yields a dark (Dicke-singlet) state $|\Psi^-\rangle$.
- Ungerade, spin-singlet dimer state (such as $^1\Sigma_u$ or $^1\Pi_u$) or gerade, spin-triplet dimer state (such as $^3\Sigma_g$ or $^3\Pi_g$) yields a bright atom-pair (Dicke-triplet) state $|\Psi^+\rangle$.

Thus, $|\Psi^-\rangle$ or $|\Psi^+\rangle$ may be prepared by the appropriate dimer-state dissociation, as demonstrated experimentally.

The preparation protocol of these entangled atom-pair states is simple: The dark Dicke states require a two-photon (Raman) excitation of the dimer above dissociation threshold, and the bright Dicke states require one-photon excitation of the dimer above dissociation threshold [Fig. 18.8(a)].

Under resonant atom–field interaction in a cavity, the excited dimer should dissociate into two atoms sharing an optical excitation, for example,

$$|Ca_2^*(^1\Sigma_u)\rangle \rightarrow (|Ca(^1P)\rangle + |Ca(^1S)\rangle)/\sqrt{2} \qquad (18.41)$$

or

$$|Ca_2^*(^1\Pi_g)\rangle \rightarrow (|Ca(^1P)\rangle - |Ca(^1S)\rangle)/\sqrt{2} \qquad (18.42)$$

[Fig. 18.8(a)].

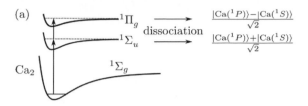

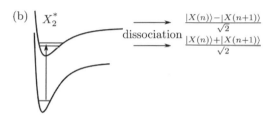

Figure 18.8 (a) Superradiant ($^1\Sigma_u$) or subradiant ($^1\Pi_g$) Dicke state prepared by one- or two-photon excitation of Ca$_2$ above the dissociation threshold. (b) Rydberg states ($n \gg 1$) of any dimer X_2 give rise to super- or subradiant Dicke states. (Reprinted with permission from Dağ et al., 2019. © 2019 American Chemical Society)

In microwave cavities, dimers that share a microwave excitation may dissociate as, for example, $|X_2^*\rangle \rightarrow (|X(n)\rangle \pm |X(n+1)\rangle)/\sqrt{2}$ [Fig. 18.8(b)]. The bright and dark Dicke states are split in energy by the two-atom resonant dipole–dipole interaction (Ch. 8). At large separations (exceeding 1 micron distance in the case of optical excitations) the two emerging atoms do not interact and the energies of the different Dicke states become equal. The dissociated fragments can traverse a 1 micron cavity waist within a time of $\tau \sim 1$ ns. There can be only one pair of fragments, belonging to the same parent molecule, in the optical cavity if the repetition rate r of the photodissociating (laser) pulses satisfies $r \leq 1/\tau \sim 1$ GHz. Then, a cavity–dipole coupling strength of $g \sim 30$ MHz corresponds to $\mu = r(g\tau)^2 \sim 1$ MHz as the effective coupling parameter. For $r \sim 1$ GHz, we find a mean free time of 1 ns between consecutive fragment pairs in the cavity. If the dissociation is effected by a 1 ns pulse just before the cavity, the fragments will be separated by a distance $R \sim 1,000$ nm when they arrive at the cavity after the dissociation, hence our model is applicable and the cooperative effects only stem from the initial state of the fragments.

As in a two-level micromaser, the cavity-mode cooling and heating rates are determined by the atomic populations. Here, however, the cooling rate by the coherence terms may be increased or decreased by adjusting the phase φ. By contrast, the heating rate is not altered by the coherence, only the ground-state manifold is affected by the coherence. The cavity temperature is thus maximized

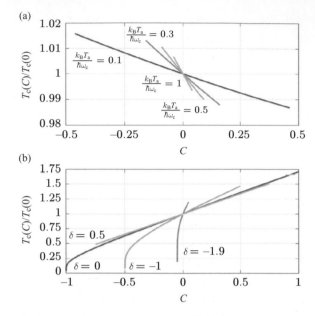

Figure 18.9 Cavity steady-state temperature for (a) phaseonium and (b) atom pairs as a function of the coherence C. The temperature is scaled by the temperature without coherence. For phaseonium the values of $C(\varphi)$ are mainly limited by the unitary injection process, whereas the HEC of the dimer is limited by the non-negativity of ρ. Parameters: $k_B T = 0.05 \hbar \omega_c$ (such that $\bar{n} \approx 0$), $\Delta_g = 0.1 \omega$, and $\kappa = \mu/2$. (Reprinted with permission from Dağ et al., 2019. © 2019 American Chemical Society)

through the reduction of the cooling rate (i.e., by choosing the phase $\varphi = \pi$). Then, the cavity-mode temperature is raised above the atomic temperature T_a, compared to that of incoherent thermal atoms. This coherence effect is, however, restricted by the maximal value of the coherence that must satisfy

$$|C| < p_{g_2} - p_{g_1} = p_{g_2} \left[1 - \exp\left(-\frac{\hbar \Delta_g}{k_B T_a} \right) \right]. \qquad (18.43)$$

While the heating ($T_h > T$) regime (18.37) does not necessarily require the atoms to be entangled, for a given w $T_h(C)/T_h(0)$ is maximized for the state with the maximally possible coherence $C = 1 - |w|/2$ (Fig. 18.9), which is always entangled.

18.4 Discussion

A major insight discussed in this chapter is that reversibility does not yield a proper efficiency bound of cyclic engines energized by nonthermal baths whenever such baths impart dissipative energy that consists of both heat and ergotropy to the working medium (WM). This inadequacy of the reversibility condition reflects the inability of the second law to discern between changes in passive energy (heat) and ergotropy in a dissipative cycle: The efficiency bound crucially depends on whether the WM is in a passive or nonpassive state. There are classical analogs to dissipative processes that involve the transfer of both disordered, passive energy (alias heat) and ordered, nonpassive (e.g., mechanical) energy (alias work). For example, frictional forces can impart velocity to a viscous liquid in a channel where one of he walls starts moving, thereby allowing liquid to do work.

Since nonpassivity and passivity also exist in a classical context, the question arises: To what extent is the cyclic-machine performance truly affected by its quantum features? To answer this question, we must resolve the following issues: (i) What are the criteria for the bath or the WM quantumness? (ii) Is there a compelling relation between such quantumness and the machine performance? (iii) Conversely, if a machine has quantum ingredients and is energized by a heat bath, does this machine conform to the rules obeyed by heat engines?

Since the dynamics of the harmonic-oscillator (HO) WM is commonly described by linear or quadratic operators, the Gaussian character of their state is preserved. Such a state is nonclassical if its P-function has negative values. According to this criterion, a squeezed bath with thermal photon number $\bar{n}$ and squeezing parameter $r > 0$ is nonclassical only if its fluctuations are below the minimum uncertainty limit, which requires

$$\bar{n} < (e^{2r} - 1)/2. \tag{18.44}$$

Whether or not the WM state is nonclassical, however, does not directly impact the machine operation – *only the energy and ergotropy of their state play a role*. The highest (near-unity) efficiency is attained by nearly *mechanical operation* of the machine. It can be effected by the ergotropy imparted by a squeezed bath to the WM. Yet, nonclassicality and nonpassivity do not generally go hand in hand: Coherent-thermal or squeezed-thermal states may be either classical or nonclassical, but both are nonpassive. In general, the extracted work and the efficiency are optimized by maximizing the ergotropy and minimizing the passive (thermal) energy of the WM, but are not necessarily related to the nonclassicality of their state.

Yet, as discussed in Chapters 15 and 17, there is another consideration to bear in mind: Quantumness proves to be indispensable for work extraction if the WM and the bath (Ch. 16) or the WM and the piston (Ch. 22) are initially in a product of passive (but not thermal) states. Then, an entangling operation may cause one of these subsystems to acquire nonpassivity and thus become amenable to work extraction.

Quantum cyclic engines may be energized by a nonthermal quantum resource (e.g. a phaseonium bath), that heats the WM to a *phase-dependent real temperature* $T_h = T(\varphi)$. The nonthermal character of the bath is advantageous for the engine operation only if T_h is elevated compared to its incoherent counterpart, which corresponds to $\Delta \bar{n}_h > 0$. Contrary to machines fuelled by ergotropy, in machines heated by nonthermal baths, $\Delta \bar{n}_h$ may be negative (e.g. when a phaseonium bath has the "wrong" phase φ).

We have compared a micromaser energized by entangled dimers with its counterpart energized by phaseonium. The achievable cavity temperature range may be larger for the dimer-based micromaser, because two-atom coherence (entanglement) is free of the constraints that phaseonium coherence is subject to [compare (18.37) and (18.38) for the entangled dimer with (18.43) and (18.28) for phaseonium].

Entangled dimer states are also more controllable than phaseonium. Increasing the coherence within the ground-state manifold of phaseonium requires weakly excited atoms, which makes the atom incapable of carrying much energy into the cavity.

Both entangled and mixed-state dimers may find applications in quantum heat machines. If the cavity mode is coupled to one environment at temperature T, it cannot operate as the WM of a heat engine (HE), but a beam of mixed-state dimers may form a local bath for the cavity mode at a temperature $T_{mix} > T$ as per (18.40), so that this mode may alternately interact with baths at T and T_{mix}. The corresponding Carnot bound is then

$$\eta_{Carnot}^{mix} = 1 - \frac{T}{T_{mix}}. \tag{18.45}$$

By contrast, a beam of entangled dimers in the Bell state $|\Psi^+\rangle$ as per (18.39a) may raise the Carnot bound to

$$\eta_{Carnot}^{+} = 1 - \frac{T}{T_+}. \tag{18.46}$$

If, for example, $\bar{n} = 0.6$ such that $\eta_{Carnot}^{mix} \approx 1/2$, it can be boosted by 33 percent to $\eta_{Carnot}^{+} \approx 2/3$. The only requirement (Fig. 18.8) is that the frequency of the photon that causes the dimer dissociation must exceed the cavity-mode frequency ω.

19

Steady-State Cycles for Quantum Heat Machines

Most quantum heat machines (QHM) studied to date are the quantum counterparts of classical heat machines that operate in *reciprocating cycles* consisting of "strokes" in which the working medium (WM) alternates between coupling to the "hot" and "cold" heat baths. Typically, such a cycle has four strokes, as in the Carnot and Otto cycles. Both these cycles are composed of two adiabatic strokes in which the WM is decoupled from any heat bath while being driven by the piston, and two heat transfer strokes (isotherms for the Carnot cycle and isochores for the Otto cycle) in which the WM is alternately coupled to one of the heat baths.

Such reciprocating cycles may pose a principal problem: *On–off switching of system–bath interactions and their nonadiabaticity (associated with finite duration) may strongly affect energy and entropy exchange*, which requires us to account for quantum friction. Furthermore, for WM totally embedded in thermal baths, decoupling from the bath may not be possible.

As an alternative, it is expedient to consider QHM that operate in a continuous cycle, without strokes. Instead, the WM is *permanently* coupled to both heat baths while being modulated by the piston. To this end, we introduce a rigorous approach to the steady-state dynamics of *periodically driven* QHM that are permanently coupled to heat baths.

Here we consider heat engines (HE) based on a quantum-mechanical WM driven by periodic Hamiltonians that commute with themselves at all times and do not induce coherence in the WM. Continuous and stroke cycles constitute the opposite limits of this theory, which encompasses also intermediate, hybrid cycles. The theory yields the speed, power, and efficiency limits attainable by incoherently operating multilevel HE depending on the cycle form and the dynamical regimes.

19.1 Reciprocating Heat Engines in Quantum Settings

Reciprocating heat engines (HE) operate in a sequence of stroke operations whose order determines the cycle. In the Markovian approximation, a reciprocating HE is defined by a product of completely positive (CP) maps (Ch. 15), which act on the WM,

$$\mathcal{M}_{\text{cyc}} = \prod_j \mathcal{M}_j, \tag{19.1}$$

where $\mathcal{M}_{\text{cyc}}$ is the cycle map and $\mathcal{M}_j$ are the stroke maps. The steady-state operation is an invariant of the cycle map (Ch. 15)

$$\mathcal{M}_{\text{cyc}}\rho_{\text{ss}} = \rho_{\text{ss}}. \tag{19.2}$$

If ρ_{ss} is single-valued and nondegenerate, the CP character of $\mathcal{M}_{\text{cyc}}$ ensures monotonic convergence to the steady-state cycle, termed the limit cycle.

The four-stroke quantum Otto cycle commonly serves for elucidating the QHM principles. It consists of two unitary strokes and two thermalization strokes: The Hamiltonian is parametrically controlled by an external parameter, which changes the level distance of the WM system (Chs. 11 and 12). For a TLS WM, this Hamiltonian reads

$$H = \frac{\hbar}{2}\omega(t)\sigma_z + \frac{\hbar}{2}\Omega\sigma_x, \tag{19.3}$$

where the level distance $\omega(t)$ in the first (energy-diagonal) term alternates between ω_{h} and ω_{c} and Ω is the driving Rabi frequency of the second term that is off-diagonal in energy (Chs. 11–13).

This cycle consists of the following strokes [Fig. 19.1(a)]:

1. In the hot *isochore* stroke: Heat is transferred from the hot bath to the WM described by the Lindblad map $\mathcal{M}_{\text{h}}$, without change in the level distance, $\omega(t) = \omega_{\text{h}}$.
2. In the expansion *adiabat* stroke: The WM reduces its level distance from ω_{h} to $\omega_{\text{c}} < \omega_{\text{h}}$ and produces work, while being isolated from both baths, as described by the unitary map $\mathcal{U}_{\text{hc}}$.
3. In the cold *isochore* stroke, described by the Lindblad map $\mathcal{M}_{\text{c}}$: Heat is transferred from the WM to the cold bath without change in the level distance, $\omega(t) = \omega_{\text{c}}$.
4. In the compression *adiabat* stroke, described by the unitary map $\mathcal{U}_{\text{ch}}$: The WM increases its level distance from ω_{c} to ω_{h}, thereby consuming power while being isolated from both baths.

The cycle map is assumed to be the product of the stroke maps,

$$\mathcal{M}_{\text{cyc}} = \mathcal{U}_{\text{ch}}\mathcal{M}_{\text{c}}\mathcal{U}_{\text{hc}}\mathcal{M}_{\text{h}}, \tag{19.4}$$

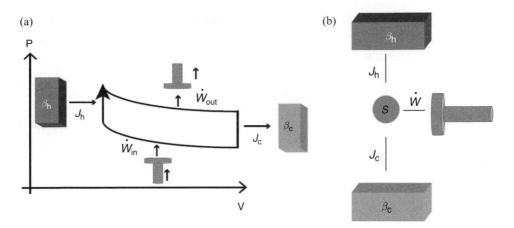

Figure 19.1 (a) A heat machine (HE) that operates in a 4-stroke reciprocating (Otto) cycle: Alternating coupling to the hot and the cold baths, interrupted by alternating compression and decompression by the piston. (b) A continuous-cycle quantum heat machine: The working medium (WM) is continuously coupled to a cold heat bath at inverse temperature β_c and to a hot heat bath at β_h. The piston periodically modulates the WM level distance, allowing for power ($\dot{W}$) extraction or supply. (Reprinted from Gelbwaser-Klimovsky et al., 2015.)

where the CP map of each stroke is the exponential of the appropriate Lindbladian $\mathcal{M}_j = e^{\int \mathcal{L}_j dt}$. The stroke maps *do not commute*, for example,

$$[\mathcal{U}_{hc}, \mathcal{M}_h] \neq 0. \tag{19.5}$$

In the adiabatic limit of the Otto cycle, the WM state is diagonal in energy throughout the cycle. The efficiency is then

$$\eta = 1 - \frac{\omega_c}{\omega_h} \leq \eta_{Carnot}, \tag{19.6}$$

whose upper bound is the Carnot efficiency

$$\eta_{Carnot} = 1 - \frac{T_c}{T_h}. \tag{19.7}$$

Here T_j is the temperature of the jth bath ($j = $ c,h).

Nonzero power requires the finite time allocated to the unitary maps $\mathcal{U}_{hc}$ and $\mathcal{U}_{ch}$, resulting in deviations from the adiabatic limit. Since the driving Hamiltonian (19.3) does not self-commute,

$$[H(t), H(t')] \neq 0, \tag{19.8}$$

coherence is generated and the WM state ρ is then not diagonal in the energy basis. Such coherence generation costs additional external work, a phenomenon termed *quantum friction* that can be understood from the notion of passivity (Ch. 15): In the adiabatic limit, the eigenvalues of the WM density matrix remain passive in the energy basis, and minimal work corresponds to the change in the energy level distance. Any deviation from adiabaticity increases the input work.

In order to have nonzero power, the time for thermalization via the CP maps $\mathcal{M}_c$ and $\mathcal{M}_h$ should be restricted, thereby avoiding infinite-time full thermalization. Optimizing the thermalization time in the adiabatic limit yields the efficiency at maximum power. The efficiency at *maximum power* is of greater practical interest than the Carnot efficiency, which corresponds to the limit of vanishing power. Under high-temperature and linear heat-transfer conditions, the efficiency at maximum power obeys the Curzon–Ahlborn bound

$$\eta_{CA} = 1 - \sqrt{\frac{T_c}{T_h}}. \tag{19.9}$$

The Curzon–Ahlborn efficiency (19.9), contrary to the Carnot bound, is not universal and may be surpassed by suitable designs, consistently with thermodynamic laws.

In the opposite, impulsive (sudden) limit of the Otto cycle, only limited action takes place in each stroke. The overall work per cycle is then reduced as the cycle time diminishes, but the power may saturate to a constant value. In this limit each stroke can be described as $\mathcal{M}_j = \exp(\mathcal{L}_j \tau_j)$, where τ_j is the stroke duration. A four-stroke cycle becomes equivalent to a continuous HE that produces finite, nonzero power as the longest stroke becomes impulsive, $\tau \to 0$:

$$\mathcal{M}_{cyc} = \mathcal{U}_{ch}\mathcal{M}_c\mathcal{U}_{hc}\mathcal{M}_h = e^{\mathcal{L}_{ch}\frac{1}{2}\tau}e^{\mathcal{L}_c\tau}e^{\mathcal{L}_{hc}\tau}e^{\mathcal{L}_h\tau}e^{\mathcal{L}_{ch}\frac{1}{2}\tau} \approx e^{(\mathcal{L}_{ch}+\mathcal{L}_c+\mathcal{L}_{hc}+\mathcal{L}_h)\tau}. \tag{19.10}$$

This limit, which holds up to the accuracy of $O(\tau^3)$, is based on the cyclic property of the HE and the Trotter formula. Work extraction in this limit requires WM coherence, so that excessive decoherence (dephasing) nulls the power.

19.2 Continuous Cycles under Periodic Modulation

A continuous QHM consists of a system (WM) with periodically modulated level distance, as per (19.3), that is *continuously* coupled to cold and hot thermal baths [Fig. 19.1(b)]. The interaction with the baths alongside the WM level-distance modulation eventually brings the WM to a periodic steady state or limit cycle. The periodicity allows the machine to operate indefinitely. Our analysis of such machines rests on the Floquet method for solving the dynamic equations for periodically driven system (WM) and thereby obtaining the heat currents and the power at steady state, upon imposing the constraints of the first and second laws of thermodynamics on these variables.

19.2.1 Floquet Theory of Periodically Driven Continuous Cycles

We consider a quantum system S (the WM) that is simultaneously coupled to cold and hot thermal baths and is periodically driven or perturbed with time period $\tau = 2\pi/\Delta$ by the time-dependent Hamiltonian $H_S(t)$:

$$H_S(t + \tau) = H_S(t). \tag{19.11}$$

We choose to have frictionless dynamics at all times, by taking $H_S(t)$ to be diagonal in the energy basis of S [unlike (19.3)], so that

$$\left[H_S(t), H_S(t')\right] = 0 \quad \forall\, t, t'. \tag{19.12}$$

The simultaneous interaction of the WM with the cold (c) and hot (h) baths is described by

$$H_I = \sum_{j=c,h} \hat{S} \otimes \hat{B}_j, \tag{19.13}$$

where the bath operators $\hat{B}_c$ and $\hat{B}_h$ commute: $\left[\hat{B}_c, \hat{B}_h\right] = 0$; and $\hat{S}$ is a system operator: $\hat{S} = \hat{\sigma}_x$ for a TLS, $\hat{S} = \hat{L}_x$ for angular-momentum models, or $\hat{S} = \hat{X}$ for a harmonic oscillator (HO). The rotating wave approximation (RWA) is not invoked in the system–bath interaction Hamiltonian (19.13). We require the two baths to have *nonoverlapping spectra*, for example, super-Ohmic spectra with distinct upper cutoff frequencies (Fig. 19.2), which are experimentally commonplace. More generally, one can always engineer nearly nonoverlapping bath spectra by a mechanism dubbed bath spectral filtering (Ch. 20). This requirement allows the WM to effectively couple (intermittently) to one or the other bath during the modulation period τ, without changing the interaction Hamiltonian to either bath. The importance of this requirement will be shown in Section 19.3. The periodicity of $H_S(t)$ is accounted for by resorting to a Floquet expansion of the Liouville operator in the harmonics of $\Delta = 2\pi/\tau$.

19.2.2 Floquet Expansion of System Operators

The system (WM) time-evolution operator reads

$$U(t, 0) \equiv T_+ \exp\left[-\frac{i}{\hbar} \int_0^t H_S(t')dt'\right], \tag{19.14}$$

where T_+ is, as before, the chronological (time-ordering) operator. According to the Floquet theorem, this time-evolution operator can be decomposed as

$$U(t, 0) = \hat{u}(t)e^{-i\bar{H}_S t}, \tag{19.15}$$

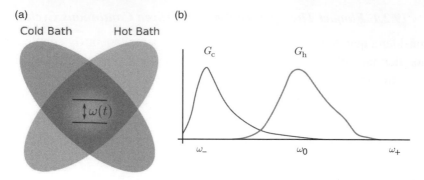

Figure 19.2 (a) A TLS with periodically modulated level distance $\omega(t)$ as the working medium (WM) in a quantum heat machine (QHM). The WM is simultaneously coupled to hot and cold baths with (b) nonoverlapping spectra. Possible realizations include a driven-atom WM coupled to spectrally filtered heat baths in an electromagnetic cavity or a driven impurity coupled to spectrally distinct phonon baths via electron–phonon coupling at the interface of two different solids. [(a) reprinted from Mukherjee et al., 2020. (b) adapted from Gelbwaser-Klimovsky and Kurizki, 2014. © 2014 American Physical Society]

where $\hat{u}(t)$ is a τ-periodic unitary operator such that $\hat{u}(0) = \hat{I}$ is the identity operator, and $\bar{H}_S$ is the average of $H_S(t)$ over its period. The part of the time-evolution operator (19.14) associated with $\bar{H}_S$ can be decomposed as

$$e^{-i\bar{H}_S t/\hbar} = \sum_n e^{-i\omega_n t} |n\rangle \langle n|. \qquad (19.16)$$

The τ-periodic operator $\hat{u}(t)$ is expanded in the Fourier series,

$$\hat{u}(t) = \sum_q \hat{\xi}(q) e^{-iq\Delta t}, \qquad (19.17)$$

where $\Delta = 2\pi/\tau$ is the fundamental modulation rate, q (integer values) label the Fourier harmonics, and

$$\hat{\xi}(q) = \frac{1}{\tau} \int_0^\tau \hat{u}(t) e^{iq\Delta t} dt \qquad (19.18)$$

is the qth harmonic contribution to the evolution operator. Upon resorting to (19.15)–(19.17), the Floquet components of the system operator can be written in the interaction picture as

$$\hat{S}(t) = U^\dagger(t,0)\hat{S}U(t,0) = \sum_q \sum_{\{\omega\}} \hat{S}_{q\omega} e^{-i(\omega+q\Delta)t}, \qquad (19.19)$$

where $\{\omega\}$ is the set of all transition frequencies between the levels of $\bar{H}_S$ and the operators $\hat{S}_{q\omega}$ are the corresponding transition operators. Expression (19.19) is an extension of (11.59) to the case of a periodically driven system.

19.2.3 Floquet Approach to Markovian Master Equation

For a periodically varying Hamiltonian (19.11), the foregoing Floquet expansions can be used to generalize the Markovian (Lindblad) (ME) (Ch. 11). Upon casting the Lindblad generator that acts on the system state ρ in the Floquet-expanded form in the interaction picture, we then have

$$\dot{\rho} = \mathcal{L}\rho, \tag{19.20}$$

where

$$\mathcal{L} = \sum_q \sum_{\{\omega\}} \mathcal{L}_{\omega q}. \tag{19.21}$$

Here the expansion is in terms of the qth-harmonic Lindblad generators. For each harmonic q, the spectral response $G_T(\omega)$ is shifted by $q\Delta$, yielding [cf. (19.19)]

$$\mathcal{L}_{\omega q}\rho = \pi G_T(\omega + q\Delta)(2\hat{S}_{\omega q}\rho\hat{S}^\dagger_{\omega q} - \hat{S}^\dagger_{\omega q}\hat{S}_{\omega q}\rho - \rho\hat{S}^\dagger_{\omega q}\hat{S}_{\omega q}). \tag{19.22}$$

The secular simplification is assumed in (19.21), which implies that all the frequencies $\omega + q\Delta$ differ from each other.

In the case of a system coupled to two baths, the Markovian ME above remains valid under the substitution

$$G_T(\omega) = G_h(\omega) + G_c(\omega). \tag{19.23}$$

Then, (19.21) becomes

$$\mathcal{L} = \mathcal{L}_h + \mathcal{L}_c = \sum_q \sum_{\{\omega\}} \sum_{j=c,h} \mathcal{L}_{j,\omega q}, \tag{19.24}$$

where

$$\mathcal{L}_{j,\omega q}\rho = \pi G_j(\omega + q\Delta)\left(2\hat{S}_{\omega q}\rho\hat{S}^\dagger_{\omega q} - \hat{S}^\dagger_{\omega q}\hat{S}_{\omega q}\rho - \rho\hat{S}^\dagger_{\omega q}\hat{S}_{\omega q}\right) \tag{19.25}$$

is associated with the jth bath.

By virtue of the detailed-balance KMS relation (Ch. 4),

$$G_j(-\omega) = e^{-\beta_j\hbar\omega}G_j(\omega), \tag{19.26}$$

where $\beta_j = (k_BT_j)^{-1}$. The expansion (19.24) decomposes the effects of the two baths into those of multiple "sub-baths" (labeled by q), which interact with the system. Each sub-bath Lindblad generator (19.25) possesses a Gibbs-like stationary state

$$(\tilde{\rho})_{jq\omega} = Z^{-1}\exp\left(-\frac{\omega + q\Delta}{\omega}\beta_j\bar{H}_S\right). \tag{19.27}$$

For a QHM operating in the steady state, the time derivative

$$\dot{\rho}(t) = (\mathcal{L}_h + \mathcal{L}_c)\rho_{ss} \tag{19.28}$$

vanishes as the steady state ρ_{ss} persists throughout the cycle.

19.2.4 From Non-Markovian to Markovian Dynamics

We assume weak system–bath coupling, consistently with the Born (but not necessarily the Markov) approximation (Chs. 11 and 16) and examine the dynamics as we transit from Markovian to non-Markovian timescales, and the ensuing change of the QHM performance, upon decreasing the modulation period. We focus on system–bath coupling durations $\tau_C = n\tau$, where $n > 1$ denotes the number of modulation periods. The relevant timescales compared to τ and τ_C are the bath correlation time t_c and the thermalization time $\tau_{th} \sim \gamma_0^{-1}$, where γ_0 is the system–bath coupling strength. We take $n \gg 1$ such that $\tau_C \gg \tau$. This allows us to adopt the approximation wherein the fast-rotating terms in the dynamics are averaged out. When τ_C is sufficiently long, $\tau_C \gg t_c$, we may perform the Born, Markov, and secular approximations and obtain the Markovian ME (19.20). If, in addition, the modulation is slow (i.e, $\tau \gg t_c$), we arrive at the time-independent Markovian (Lindblad) ME (Ch. 11).

In the opposite limit of fast modulation, $\tau \ll t_c$, the Markov approximation is inapplicable for $\tau_C = n\tau \lesssim t_c$. In this limit, we need a non-Markovian fast-modulation form of the ME. For the case of a TLS, such an ME has been derived in Chapter 11. In the present case, $H_S(t)$ is given by (11.64), and $\hat{S}$ is given by (11.65) with $\tilde{\epsilon}(t) = 1$. Correspondingly, the ME has the form (11.68) or, equivalently, (11.79). For $\rho(t)$ that is diagonal in the energy basis, which we consider below, it is sufficient to consider only the equation for the populations (11.79a),

$$\dot{\rho}_{ee} = -\dot{\rho}_{gg} = -\gamma_e(t)\rho_{ee} + \gamma_g(t)\rho_{gg}, \tag{19.29}$$

where the population relaxation rates $\gamma_{e(g)}(t)$ are given by (11.72). In the case of two baths, the use of (19.23) yields

$$\gamma_e(t) = \gamma_{h,e}(t) + \gamma_{c,e}(t),$$
$$\gamma_g(t) = \gamma_{h,g}(t) + \gamma_{c,g}(t). \tag{19.30}$$

For a periodic $\delta_a(t)$ in (11.64), the Floquet theorem is equivalent to (11.76). Using (11.39) and (11.76) in (11.72), we obtain

$$\gamma_{e(g)}(t) = 2\,\mathrm{Re}\sum_{q,q'} \epsilon_q \epsilon_{q'}^* e^{\pm i(q-q')\Delta t} \int_{-\infty}^{\infty} d\omega\, G_T(\omega) \int_0^t dt'\, e^{i(\pm\omega_{q'}-\omega)(t-t')}. \tag{19.31}$$

Here the upper (lower) sign corresponds to $\gamma_e(t)$ [$\gamma_g(t)$] and

$$\omega_q = \omega_a + \delta + q\Delta, \tag{19.32}$$

where δ is given by (11.75), and (in the present notation) Δ, τ correspond, respectively, to ω_m, T_m in Chapter 11.

Under fast modulation, the terms with $q \neq q'$ in the sum in (19.31) undergo fast oscillations, and, upon integrating (19.29), they are averaged out almost to zero after several oscillation periods. Therefore, for

$$t \gg 1/\Delta \qquad (19.33)$$

these terms can be omitted, yielding

$$\gamma_{e(g)}(t) = 2t \sum_q P_q \int_{-\infty}^{\infty} d\omega \, G_T(\omega) \, \mathrm{sinc}[(\omega \mp \omega_q)t], \qquad (19.34)$$

where $P_q = |\epsilon_q|^2$.

In the case of two baths, (19.23), (19.30), and (19.34) then yield

$$\gamma_{j,e(g)}(t) = 2t \sum_q P_q \int_{-\infty}^{\infty} d\omega \, G_j(\omega) \, \mathrm{sinc}[(\omega \mp \omega_q)t] \quad (j = \mathrm{h,c}). \qquad (19.35)$$

Formula (19.35) expresses the convolution of the jth-bath spectral-response function $G_j(\omega)$, whose spectral width is $\sim \Gamma_{\mathrm{B}} \sim 1/t_{\mathrm{c}}$, with the sinc functions centered at ω_q and broadened according to the time-energy uncertainty relation.

Let us adopt, for definiteness, modulation of the form

$$\delta_{\mathrm{a}}(t) = \lambda \Delta \sin(\Delta t), \qquad (19.36)$$

where the modulation depth λ satisfies $0 < \lambda \ll 1$. This implies that in (19.32) $\delta = 0$,

$$\epsilon_q = i^{-q} e^{i\lambda} J_q(\lambda), \qquad (19.37)$$

where $J_q()$ is the Bessel function of the order q, so that in (19.35),

$$P_q = J_q^2(\lambda) = J_{|q|}^2(\lambda). \qquad (19.38)$$

For $\lambda \ll 1$, we obtain

$$P_0 \approx 1 - \lambda^2/2, \quad P_q \approx (\lambda/2)^{2|q|} \ (q \neq 0). \qquad (19.39)$$

Thus, for modulation of the form (19.36), $q = \pm 1$ are the strongest shifted harmonics.

Owing to the finite widths ($\sim 1/\tau_{\mathrm{C}}$) of the sinc functions in the frequency domain for short coupling times ($\tau_{\mathrm{C}} \lesssim t_{\mathrm{c}}$), the WM is driven away from the steady state ρ_{ss} as follows from (19.29), causing $\rho(t)$ to evolve within the memory-time interval $0 < t \leq t_{\mathrm{c}}$. However, a cyclic QHM operating in the steady state requires cycles consisting of n modulation periods that satisfy

$$\tau_{\mathrm{C}}^{-1} \ll k_{\mathrm{B}} T_{\mathrm{c,h}}/\hbar \equiv (\hbar \beta_{\mathrm{c,h}})^{-1}, \qquad (19.40)$$

so that the Boltzmann factors satisfy

$$e^{-\frac{\hbar(\omega_a\pm\Delta+1/\tau_C)}{k_B T_{c,h}}} \approx e^{-\frac{\hbar(\omega_a\pm\Delta)}{k_B T_{c,h}}}.$$
(19.41)

19.2.5 Markovian Heat Currents

The dynamics obtained from the Floquet-expanded Lindblad superoperator (19.24) can also be derived upon averaging (coarse-graining) over a modulation period the time-dependent evolution generated by the periodically driven non-Markovian ME in Chapter 11.

The time derivative of the von Neumann entropy satisfies

$$\frac{d}{dt}S(\rho(t)) = -\text{Tr}[\dot\rho(t)\ln\rho(t)] - \sum_q \sum_{\{\omega\}} \sum_{j=c,h} \text{Tr}[\mathcal{L}_{j,\omega q}\rho(t)\ln\rho(t)],$$
(19.42)

where we have used (19.24) and $\text{Tr}[\dot\rho(t)] = 0$. We now resort to the Spohn inequality (15.7), which is the expression of the second law for any Lindblad superoperator and its stationary state ρ_{ss}. When applied to every term of the sum in (19.42), Spohn's inequality yields

$$\frac{d}{dt}S(\rho(t)) + \text{Tr}[\dot\rho(t)\ln\rho(t)] - \sum_q \sum_{\{\omega\}} \sum_{j=c,h} \text{Tr}[\mathcal{L}_{j,\omega q}\rho(t)\ln(\tilde\rho)_{jq\omega}] \geq 0.$$ (19.43)

This inequality is then compared to the dynamical version of the second law, whereby

$$\frac{d}{dt}S(\rho(t)) - \sum_j \frac{1}{k_B T_j}\mathcal{J}_j(t) \geq 0.$$
(19.44)

This comparison identifies the *heat current* $\mathcal{J}_j(t)$ (energy-flow rate) between the WM and the jth bath as

$$\mathcal{J}_j(t) = -k_B T_j \sum_q \sum_{\{\omega\}} \text{Tr}[\mathcal{L}_{j,\omega q}\rho(t)\ln(\tilde\rho)_{jq\omega}].$$
(19.45)

At steady state, these heat currents assume, upon inserting (19.27) into (19.45), the form

$$\mathcal{J}_j = \sum_q \sum_{\{\omega\}} \frac{\omega+q\Delta}{\omega} \text{Tr}[(\mathcal{L}_{j,\omega q}\rho_{ss})\bar H_S],$$
(19.46)

where $\mathcal{L}_{j,\omega q}\rho_{ss}$ is the energy-flow rate exchanged with the sub-bath (j,q) at the transition frequency ω and $\frac{\omega+q\Delta}{\omega}\bar H_S$ is the qth harmonic energy.

In this steady-state regime, the second law (19.44) yields

$$\sum_j \frac{\mathcal{J}_j}{T_j} \leq 0.$$
(19.47)

According to the first law (of energy conservation), the stationary power of an external periodic force acting on the system is

$$\dot{W} = -\sum_j \mathcal{J}_j. \tag{19.48}$$

The possible operational regimes of the QHM (i.e., its operation as an HE or a refrigerator) are determined by the signs of the cycle-averaged heat currents $\overline{\mathcal{J}_h}$, $\overline{\mathcal{J}_c}$, and power $\overline{W}$,

$$\overline{W} = \frac{1}{\tau_C} \oint_{\tau_C} \dot{W}(t)dt; \quad \overline{\mathcal{J}_j} = \frac{1}{\tau_C} \oint_{\tau_C} \mathcal{J}_j(t)dt. \tag{19.49a}$$

Negative power means that work is extracted from the QHM, that is, that it operates as an HE (Chs. 20–22), whereas positive power means that work is invested in the WM by the piston, as required for refrigerator (or heat-pump) operation (Ch. 23). These quantities and the steady-state efficiency

$$\eta = -\frac{\oint_{\tau_C} \dot{W}(t)dt}{\oint_{\tau_C} \mathcal{J}_h(t)dt} \tag{19.49b}$$

are controllable by the modulation rate Δ.

19.3 Bridging Self-Commuting Continuous and Reciprocal Cycles

The considerations outlined above underscore the need for elucidating the principal issues: (1) What is the best possible dependence of HE power or efficiency on the cycle deviation within the Markovian or the non-Markovian regimes? (2) What is the optimal cycle form for attaining the best performance: reciprocating, continuous, or intermediate (hybrid) cycles? Is there a fundamental speed limit on HE operation?

We address these issues by means of a unified theory that applies to any cycle in multilevel HEs. Our basic assumption is that in order to avoid friction, the driving Hamiltonian commutes with itself at all times and thus does not generate any quantum coherence in the WM: The driving Hamiltonian is diagonal in the WM energy basis.

In the four-stroke Otto cycle the WM intermittently couples to the hot bath at frequency ω_2 and to the cold bath at frequency ω_1. Here we aim to reproduce any cycle, including a (smoothed-out) approximation to the Otto cycle, by spectral separation of the couplings to the two baths, for an appropriate choice of the modulation harmonics $q\Delta$ and the coupling-amplitude smoothness in time.

The generic setup [Fig. 19.3(a)] is described by the parametrically modulated Hamiltonian

$$H(k, t) = H_S(k, t) + H_M(k, t) + H_B. \tag{19.50}$$

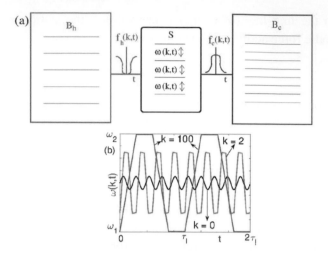

Figure 19.3 (a) A generic QHM based on a multilevel system S (WM) that under-goes periodic frequency modulation $\omega(k, t)$ and amplitude modulation of the couplings $f_h(k, t)$, $f_c(k, t)$ to thermal baths B_h, B_c with distinct (nonoverlap-ping) spectra. (b) Frequency modulation $\omega(k, t)$ for different k from continuous to Otto cycles (see text). (Reprinted from Mukherjee et al., 2016. © 2016 American Physical Society)

Here $H_S(k, t)$ is the controlled-system (WM) Hamiltonian, with modulation period, satisfying the periodicity condition

$$H_S(k, t + \tau) = H_S(k, t) = \sum_{n \geq 0} \omega_n(k, t) |n\rangle \langle n|, \qquad (19.51)$$

where we have introduced the "smoothness" parameter $0 \leq k < \infty$. This param-eter determines the cycle form, ranging from continuous through intermediate (*hybrid*) to reciprocal (stroke) cycle.

For simplicity, we assume $\omega_n(k, t) = n\omega(k, t)$, that is, take the levels to be equidistant and synchronously modulated, with time average $\overline{\omega_n(k, t)} = n\omega_0$. Such synchronous modulation of equidistant levels is applicable to HO or angular momentum WM models.

The controlled, amplitude-modulated interaction with the independent cold (c) and hot (h) baths is generally given by

$$H_M(k, t) = \sum_{j=c,h} f_j(k, t)\hat{S} \otimes \hat{B}_j. \qquad (19.52)$$

As in (19.13), the WM operator $\hat{S}$ for an n-level system is coupled to bath operators $\hat{B}_c$ and $\hat{B}_h$, satisfying $\left[\hat{B}_c, \hat{B}_h\right] = 0$. The generalization of (19.13) is here the ampli-tude modulation of the system–bath couplings $f_c(k, t)$ and $f_h(k, t)$ parameterized by the smoothness parameter k.

In order to achieve frictionless dynamics (cf. Sec. 19.2), the system Hamiltonian and the system–bath interaction Hamiltonian are taken to commute with themselves at all times:

$$\left[H_S(t), H_S(t')\right] = \left[H_M(t), H_M(t')\right] = 0. \tag{19.53}$$

We assume that the coupling amplitudes $f_j(k, t)$ are modulated with frequency $\Delta_I = 2\pi/\tau_I$, which is slow compared to the arbitrary-fast modulation rate $\Delta = 2\pi/\tau$ of $H_S(t)$. The periods τ and τ_I are commensurate. The ME for the WM state $\rho(t)$ has then the form

$$\dot{\rho} = \sum_{j=c,h} f_j^2(k, t)\mathcal{L}_j\rho. \tag{19.54}$$

Under the Markovian approximation and the KMS condition, we get for the HO WM

$$S_{q\omega} = a(\omega = \omega_0) \equiv a, \tag{19.55}$$

where a is the annihilation operator, so that

$$\mathcal{L}_{j,\pm q}(t)\rho = \frac{P_q}{2}[G_j(\omega_0 \pm q\Delta)([a\rho, a^\dagger] + [a, \rho a^\dagger])$$
$$+ G_j(-\omega_0 \mp q\Delta)([a^\dagger\rho, a] + [a^\dagger, \rho a])], \tag{19.56}$$

where

$$\omega_0 = \frac{1}{\tau}\int_0^\tau \omega(t)dt, \tag{19.57}$$

$$P_q = P_{-q} = \left|\frac{1}{\tau}\int_0^\tau e^{i\int_0^{t'}[\omega(t')-\omega_0]dt'}e^{iq\Delta t}dt\right|^2,$$

$$G_j(\omega_0 \pm q\Delta) \equiv G_j(\omega, \pm q) = \int_{-\infty}^\infty e^{i(\omega_0 \pm q\Delta)t}\left\langle\hat{B}_j(t)\hat{B}_j(0)\right\rangle dt$$
$$= e^{\beta_j\hbar(\omega_0 \pm q\Delta)}G_j(-\omega_0 \mp q\Delta). \tag{19.58}$$

As discussed below, HE operation requires spectral separation of the baths by setting $G_h(\omega_0 - q\Delta < \omega_0) \approx 0$ and $G_c(\omega_0 + q\Delta > \omega_0) \approx 0$.

In what follows we investigate the HE performance in terms of speed limits, efficiency, and power, as a function of the modulation rate Δ and the cycle form determined by $f_j(k, t)$ and $\omega(k, t)$.

19.3.1 Modeling of Cycle Forms

We choose a periodic modulation of $\omega(k, t)$ [Fig. 19.3(b)] that can reproduce both the continuous and Otto-cycle limits:

$$\omega(k, t) = \omega_{\text{Cont}}(t) \exp(-k) + \omega_0 + \omega_{\text{Otto}}(t) \exp(-1/k), \qquad (19.59)$$

where the smoothness parameter k that ranges from 0 to ∞ parameterizes both $\omega(k, t)$ and $f_j(k, t)$, as discussed below:

(a) *The chosen continuous-modulation function is*

$$\omega_{\text{Cont}}(t) = \lambda \Delta \sin(\Delta t), \qquad (19.60)$$

λ *being the modulation depth.*

(b) *The function $\omega_{\text{Otto}}(t)$ is taken to be trapezoidal:* in the Otto-cycle limit, it increases linearly from ω_1 to ω_2 for $l\tau_{\text{I}} < t \leq (l + 1/4)\tau_{\text{I}}$ with a chosen nonnegative integer l (in an isentropic compression stroke), then remains fixed at ω_2 until $t = (l + 1/2)\tau_{\text{I}}$ in an isochoric stroke where the WM is in contact with the hot bath, then decreases to ω_1 with the opposite slope until $t = (l + 3/4)\tau_{\text{I}}$ in an isentropic expansion stroke and then stays fixed until $t = (l + 1)\tau_{\text{I}}$ (in an isochoric stroke in contact with the cold bath), thus completing the cycle.

For any k, the equidistant levels of the WM oscillate synchronously and thus yield the same sideband spacings $\pm q\Delta$ [Fig. 19.4(a)].

We search for the dependence of the maximal efficiency and the efficiency at maximal power depend on the operation-cycle form and on the rate Δ. To this end, the system–bath coupling strengths $f_j(k, t)$ parameterization complies with the following requirements:

i) *periodicity:* $f_j(k, t + \tau_{\text{I}}) = f_j(k, t)$;
ii) *normalization:* $0 \leq |f_j(k, t)| \leq 1$;
iii) *the $f_j(k, t)$ smoothness varies with k,* $0 \leq k < \infty$.

The parameterization must render a constant coupling in the continuous-cycle $(k = 0)$ limit,

$$f_{\text{c}}(0, t) = f_{\text{h}}(0, t) = 1 \ \forall \, t, \qquad (19.61)$$

and stepwise variation in the $k \to \infty$ Otto cycle limit,

$$f_{\text{c}}(k, t) = \theta(t - \tau_{\text{I}}/4)\theta(\tau_{\text{I}}/2 - t),$$
$$f_{\text{h}}(k, t) = \theta(t - 3\tau_{\text{I}}/4)\theta(\tau_{\text{I}} - t) \ \text{for} \ 0 \leq t < \tau_{\text{I}}, \qquad (19.62)$$

$\theta()$ being the Heaviside step function. Interpolation between the continuous and reciprocal (stroke) cycles corresponds to intermediate values of k for which the coupling to both the cold and the hot baths is never completely turned off or on,

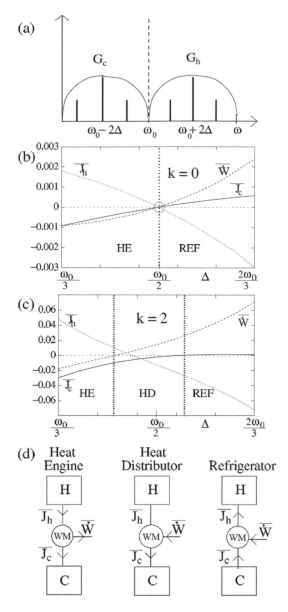

Figure 19.4 (a) Spectral response of the hot and cold baths and the modulation sidebands in the Markovian limit. (b) Heat currents and power for the continuous cycle ($k = 0$) in two regimes: HE ($\overline{\dot{W}} < 0$, $\overline{\mathcal{J}}_c < 0$, $\overline{\mathcal{J}}_h > 0$) for $\Delta < \Delta_{SL}$ (SL denoting the HE speed limit) and refrigerator (REF) ($\overline{\dot{W}} > 0$, $\overline{\mathcal{J}}_c > 0$, $\overline{\mathcal{J}}_h < 0$) for $\Delta > \Delta_{SL}$. Here $\Delta_{SL} = \omega_0/2$. (c) Same for a hybrid cycle ($k = 2$), which allows for three possible regimes. (d) HE, heat distributor (HD) ($\overline{\dot{W}} > 0$, $\overline{\mathcal{J}}_c < 0$, $\overline{\mathcal{J}}_h > 0$) and refrigerator, where $T_c = 10$, $T_h = 30$, $\omega_0 = 3$, $\lambda = 0.1$, $G_c(0 < \omega_0 - q\Delta < \omega_0) = G_h(\omega_0 + q\Delta > \omega_0) = 1$. (Adapted from Mukherjee et al., 2016. © 2016 American Physical Society)

Table 19.1 *Parameters and functions*

	Definitions
k	Smoothness parameter
$\omega(k,t)$	Level spacing
$f_{h(c)}(k,t)$	Coupling amplitude of system to hot (h) or cold (c) bath
$\Delta_I = 2\pi/\tau_I$	Modulation rate of the system–bath coupling amplitudes
$\Delta = 2\pi/\tau$	Modulation rate of the system level spacing $\omega(k,t)$
Δ_{SL}	HE modulation speed limit

so that the strokes are not fully separated in time. Such a cycle will be dubbed a hybrid cycle.

The explicit functional form of the period τ_I of $f_j(k,t)$ (Table 19.1) is chosen to satisfy

$$\tau_I = \frac{2\pi}{\Delta_I} = \frac{2\pi}{\Delta}\Phi(k); \quad \Phi(k) = \left[\frac{(k+N_1)}{(k+N_2)}\right], \tag{19.63}$$

where $\Phi(k)$ is the largest integer closest to the fraction in brackets, with $N_1 \gg N_2$. In the continuous-cycle limit $k = 0$, $\Phi(k) \gg 1$, whereas in the Otto-cycle limit $k \to \infty$, where $\Delta = \Delta_I$, $\Phi(k) = 1$. The choice of parameterization is arbitrary, but the behavior it predicts for any physical cycle form is generic.

19.3.2 Heat Flow, Work and Efficiency

Since inter-level coherences decay exponentially with time, the long-time evolution is governed by the Pauli ME for the WM level populations $\mathcal{P}_n(k,t) = \langle n|\rho(k,t)|n\rangle$. For an HO with energy-level distance ω_0, this equation reads

$$\frac{d}{dt}\mathcal{P}_n = G_0 \sum_{j,q\geq 0} \mathcal{P}_q(k)f_j^2(k,t)\Big\{\big[(n+1)\mathcal{P}_{n+1} - n\mathcal{P}_n\big]$$

$$+e^{-\beta_j\hbar(\omega_0\pm q\Delta)}\big[n\mathcal{P}_{n-1} - (n+1)\mathcal{P}_n\big]\Big\}$$

$$= \sum_{j,q\geq 0} \mathcal{P}_q(k)f_j^2(k,t)\big[\gamma_{ne}(k,t) + e^{-\beta_j\hbar\omega_j}\gamma_{na}(k,t)\big]. \tag{19.64}$$

Here we have involved the KMS condition (19.26) and taken into account that only the frequencies $\omega_j = \omega_0 - q\Delta$ ($\omega_j = \omega_0 + q\Delta$) contribute for the cold (hot) bath, by assuming that

$$G_c(0 < \omega_0 - q\Delta < \omega_0) = G_0,$$

$$G_h(\omega_0 + q\Delta > \omega_0) = G_0. \tag{19.65}$$

The $n + 1 \rightarrow n$ (emission) and $n - 1 \rightarrow n$ (absorption) rates are then, respectively,

$$\gamma_{ne}(k, t) = G_0 \left[(n + 1)\mathcal{P}_{n+1}(k, t) - n\mathcal{P}_n(k, t) \right],$$
$$\gamma_{na}(k, t) = G_0 \left[n\mathcal{P}_{n-1}(k, t) - (n + 1)\mathcal{P}_n(k, t) \right]. \tag{19.66}$$

Upon employing the above "smoothness" k-parameterization, the heat currents $\mathcal{J}_c$ and $\mathcal{J}_h$, flowing out of the cold and hot baths respectively, are obtained from (19.45) and (19.64), consistently with the second law, in the form

$$\mathcal{J}_h(t) = f_h^2(k, t) \sum_{q \geq 0} (\omega_0 + q\Delta) P_q(k) G_h(\omega_0 + q\Delta) \mathcal{F}_h(q, k, t),$$

$$\mathcal{J}_c(t) = f_c^2(k, t) \sum_{q \geq 0} (\omega_0 - q\Delta) P_q(k) G_c(\omega_0 - q\Delta) \mathcal{F}_c(q, k, t). \tag{19.67}$$

Here the harmonic (sideband) weights are $P_q(k)$, $G_{h(c)}(\omega_0 \pm q\Delta)$ is the spectral bath response [cf. (19.58)],

$$\mathcal{F}_{j=h,c}(q \geq 0, k, t) = \sum_n n \left[\gamma_{ne}(k, t) + e^{-\beta_j \hbar(\omega_0 + q\Delta)} \gamma_{na}(k, t) \right], \tag{19.68}$$

while $\mathcal{F}_{h(c)}(k, t)$ express the h (c) contributions to the detailed balance between heat emission and absorption.

A crucial condition of the present treatment of diverse cycles is the spectral separation of the hot and cold baths, so that the sidebands $\pm(\omega + q\Delta)$ only couple to the hot bath and the $\pm(\omega - q\Delta)$ sidebands only couple to the cold bath [Fig. 19.4(a)]. This spectral separation is compatible with the Markovian limit and is required to selectively control the heat currents. Such separation for all contributing harmonics [Fig. 19.4(a)], so as to allow HE operation, amounts to the requirement

$$\mathcal{F}_h > 0, \quad \mathcal{F}_c < 0,$$
$$G_h(\omega_0 - q\Delta \leq \omega_0) \approx 0; \quad G_c(\omega_0 + q\Delta \geq \omega_0) \approx 0. \tag{19.69}$$

The possible operational regimes of the QHM are (i) HE, which corresponds to $\overline{\dot{W}} < 0$, $\overline{\mathcal{J}_c} < 0$, $\overline{\mathcal{J}_h} > 0$, (ii) heat distributor, in which $\overline{\dot{W}} > 0$, $\overline{\mathcal{J}_c} < 0$, $\overline{\mathcal{J}_h} > 0$, and (iii) refrigerator ($\overline{\dot{W}} > 0$, $\overline{\mathcal{J}_c} > 0$, $\overline{\mathcal{J}_h} < 0$) (Figs. 19.4 and 19.5).

We may use (19.67) to calculate the cycle-averaged power output

$$\overline{\dot{W}} = \frac{1}{\tau_I} \oint_{\tau_I} \dot{W}(t) dt \tag{19.70a}$$

and the steady-state efficiency

$$\eta = -\frac{\oint_{\tau_I} \dot{W}(t) dt}{\oint_{\tau_I} \mathcal{J}_h(t) dt} \tag{19.70b}$$

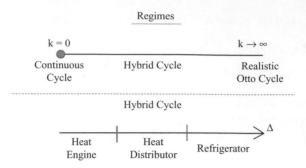

Figure 19.5 A continuous cycle (dot) corresponds to $k = 0$, whereas $k \to \infty$ describes a realistic Otto cycle. For a hybrid cycle, the QHM operates in the HE, heat distributor, or refrigerator regimes depending on the modulation rate Δ. The heat distributor regime does not exist for the continuous cycle $k = 0$. (Adapted from Mukherjee et al., 2016. © 2016 American Physical Society)

as functions of the modulation rate Δ, the cycle duration τ_I, and the smoothness parameter k. The dependence of η and $\bar{\dot{W}}$ (Figs. 19.6 and 19.7) does not originate from quantum coherence-related features: Although they reflect the quantized WM level structure, they have classical counterparts.

19.4 Speed Limits from Continuous to Otto Cycles

The abrupt turn-on and turnoff of the strokes in a traditional Otto cycle (which we dub TOC) is obviously idealized. The abrupt strokes also entail friction. By contrast, a frictionless realistic Otto cycle (here dubbed ROC) is reproduced by allowing a large number of harmonics to contribute as $\Delta_m \to 0$, so that the hot bath effectively couples to the WM only at

$$\omega_2 = \omega_0 + q_{\text{Otto}} \Delta \tag{19.71}$$

and the cold bath at

$$\omega_1 = \omega_0 - q_{\text{Otto}} \Delta. \tag{19.72}$$

There is no speed limit of HE operation in a "perfect" TOC. By contrast, there is a speed limit (SL) for any realistic cycle, including ROC. This speed limit comes about because a modulation rate Δ above Δ_{SL} results in the machine acting as a heat distributor, or a refrigerator of the cold bath, and thus consuming, rather than generating, power: $\bar{\dot{W}} > 0$. Yet, Δ_{SL} decreases with increasing k. The highest speed limit is compatible with a continuous cycle. A frictionless, finite-duration ROC demands an increasingly slower modulation in order to produce power ($\bar{\dot{W}} < 0$) [Fig. 19.6(b)].

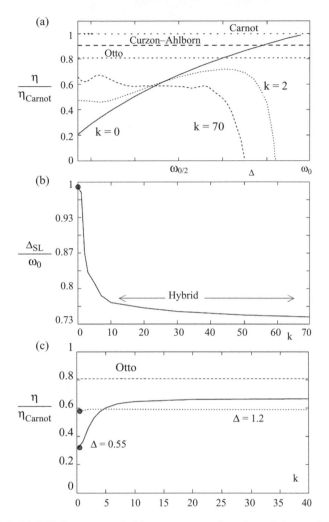

Figure 19.6 (a) Efficiency η scaled by η_{Carnot} as a function of the modulation rate Δ for different k values. In the $k = 0$ limit, $\eta(k = 0)$ may exceed the Curzon–Ahlborn bound and the Otto-cycle efficiency η_{Otto}, whereas cycles with larger k fail to do so. Here $\Delta_{\text{SL}} \approx 0.98\omega_0$, so that for the continuous $k = 0$ cycle, the refrigerator regime is only attainable for very large Δ, beyond the range plotted here. (b) Speed limit: Δ_{SL} as a function of k. (c) Efficiency η scaled by η_{Carnot} as a function of k: η tends to η_{Otto} as Δ decreases. Dots mark the efficiency in a continuous cycle. Here $\omega_1 = 1$, $\omega_2 = 5$, $\omega_0 = 3$, $T_c = 1$, $T_h = 100$, $G_c(0 < \omega_0 - q\Delta < \omega_0) = G_h(\omega_0 + q\Delta > \omega_0) = 1$. (Adapted from Mukherjee et al., 2016. © 2016 American Physical Society)

In the continuous limit $k \to 0$, where only the first harmonic (sideband) significantly contributes, the efficiency attains the value

$$\eta_{\text{Cont}} = \frac{2\Delta}{\omega_0 + \Delta}. \tag{19.73}$$

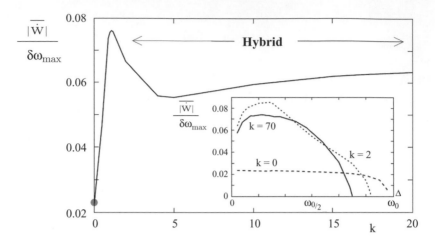

Figure 19.7 Absolute average power $|\overline{\dot{W}}|$ scaled by $\delta\omega_{\max}(k)$, the maximum modulation amplitude of $\omega(k, t)$, as a function of k. A hybrid HE cycle ($k \approx 1.1$) yields higher power than either a continuous cycle, $k = 0$ (shown by the dot), or a realistic Otto cycle with $k \to \infty$. Here $\Delta = 1.2$. Inset: Same as a function of Δ. In the limit $\Delta \to 0$ (and $\tau_{\mathrm{I}} \to \infty$), the average power vanishes for large k, as indicated in the plot. (Same parameters as in Fig. 19.6.) (Adapted from Mukherjee et al., 2016. © 2016 American Physical Society)

This expression is bounded by the Carnot efficiency, whereas the Curzon–Ahlborn limit may be surpassed. For large k values, $\eta(k)$ saturates to the lower Otto-cycle efficiency

$$\eta_{\mathrm{Otto}}(k \to \infty) = 1 - \frac{\omega_1}{\omega_2}. \tag{19.74}$$

The frictionless ROC then satisfies

$$q_{\max} \to \infty, \quad \Delta \to 0, \quad q_{\max}\Delta \lesssim \Delta_{\mathrm{SL}}, \tag{19.75}$$

in order to keep the power $\overline{\dot{W}} < 0$.

The maximal power is attained near the k value where $\omega(k, t)$ in (19.59) changes from $\omega(k, t) \approx \omega_{\mathrm{Cont}}(t) + \omega_0$ to $\omega(k, t) \approx \omega_{\mathrm{Otto}}(t) + \omega_0$. It is thus possible to engineer a hybrid-cycle HE that outperforms both the continuous and the Otto limits (Fig. 19.7) because of its optimal cost of coupling to and decoupling from the bath.

We are now able to answer the principal questions raised in Section 19.3: The optimal cycle form is a hybrid cycle. The scaled power $|\overline{\dot{W}}|/\delta\omega_{\max}(k)$ peaks at $k \approx 1.1$ for the chosen parameterization. Thus, smooth strokes are advantageous compared to abrupt ones in terms of both efficiency and power.

19.5 Discussion

In this chapter, we have derived the master equation (ME) for a periodically driven (modulated) system that interacts with spectrally separated hot and cold heat baths. This HE has been used to evaluate the time-dependent heat currents and power. The system Hamiltonian has been taken to be self-commuting at all times (i.e. diagonal in the same energy basis at any time). This choice avoids quantum friction that inevitably accompanies coherence generation (off-diagonal terms of the density matrix of the system in the energy basis). Such self-commuting Hamiltonians, strictly speaking, correspond to continuous cycles, although they may well resemble reciprocating (strokes) cycles. In a reciprocating cycle consisting of abrupt strokes, coherence is generated at the price of friction. However, this price of generating coherence can be eliminated if at the end of a stroke a passive state is restored (in the energy basis). Such frictionless protocols are termed shortcuts to adiabaticity. These protocols allow to achieve adiabatic-like solutions in finite time for the stroke maps.

Continuous-cycle models are often more realistic than reciprocating cycles. Consider a gas of atoms driven by a laser, which acts as the piston. The driven atoms represent the system (WM). They interact through collisions with another atomic species (a "buffer gas"), which represents the hot bath. Concurrently, the WM atoms are coupled to the electromagnetic vacuum (acting as the cold bath) through spontaneous emission. Since spontaneous emission and collisions with the buffer gas cannot be avoided, the coupling to or decoupling from the bath is impossible, so that a reciprocating-cycle model is inapplicable to such scenarios, as opposed to continuous-cycle models.

In the present continuous-cycle description, the ME terms that stand for the system-bath interactions can be decomposed into "sub-bath" contributions according to Floquet's theorem. Each "sub-bath" contribution stands for the interaction of the periodically modulated system with a heat bath at a shifted Bohr frequency corresponding to a harmonic Floquet sideband. The heat currents can likewise be decomposed into Floquet harmonics and are (by construction) compatible with the second law of thermodynamics.

This framework has been the basis of a unified treatment of quantum heat machines (QHM), namely, heat engines (HE) or refrigerators with a periodically modulated multilevel quantum-mechanical WM. The theory allows us to interpolate between opposite limits of cycle scheduling (cycle forms): continuous and smoothed multi-stroke cycles.

Several generic features emerge from this unified treatment: (a) The QHM may operate as a HE, a heat distributor, or a refrigerator, depending on the modulation rate Δ and the cycle smoothness parameter k. (b) The efficiency increases regardless of the cycle form with the WM modulation-rate Δ, attaining a maximum that is bounded by the Carnot bound, before becoming ill-defined at the speed limit set by the modulation rate Δ_{SL}, above which the machine stops acting as an HE. (c) Remarkably, hybrid cycles that interpolate between continuous and stroke cycles may exhibit modulation-induced HE power boost: Such cycles possess an optimal trade-off between speed and the cost of coupling to the baths, which are never turned off or on completely.

The expressions obtained here for the efficiency and power bounds are independent of the WM quantized level structure and have classical counterparts, since quantum coherence effects (and thus friction) are absent as a consequence of the system-driving Hamiltonian $H_S(t)$ commuting with itself at all times.

These results suggest that traditional thermodynamics is adhered to, whereas quantum coherence is neither essential nor advantageous for HE performance. Incoherently operating HE may thus be optimal designs in various experimental scenarios: (i) optomechanical HE machines or (ii) HE based on multilevel WM, for example, a molecular rotator, modulated by electromagnetic fields and interacting with intracavity heat baths, (iii) HE based on a Rydberg-atom WM.

20

Two-Level Minimal Model of a Heat Engine

Here we apply the principles discussed in Chapters 15–19 to the analysis of a minimal model of a heat machine. It consists of a two-level system with periodically modulated energy splitting that is continuously, weakly, coupled to two spectrally separated heat baths at different temperatures. This machine can act as either a heat engine (HE) or a refrigerator (heat pump) depending on a single parameter – the modulation rate. The HE efficiency grows with this rate up to the Carnot bound, but at a higher rate it acts as a refrigerator. The power scales non-monotonically with this rate. Remarkably, for modulation rates that fall within the non-Markovian regime, power boosts are induced by the anti-Zeno effect (AZE) (Chs. 10, 16). Such boosts signify quantum advantage over heat-machines that commonly operate in the Markovian regime, where the quantumness of the system–bath interaction plays no role. The AZE-induced power boost stems from the time-energy uncertainty relation in quantum mechanics, which may result in enhanced system–bath energy exchange for modulation periods comparable to the bath correlation time.

20.1 Model and Treatment Principles

The Hamiltonian of the QHM in question can be written as

$$H_{QHM} = H_S(t) + H_B + H_{SB} \tag{20.1}$$

with

$$H_{SB} = \sigma_x (B_h + B_c). \tag{20.2}$$

Here the two-level system (TLS) acts as the working medium (WM). It is weakly, simultaneously coupled to two baths via $H_I \equiv H_{SB}$, where B_h and B_c are respectively the operators of a hot bath (h) and a cold bath (c) and σ_x is the x-component

of the spin-1/2 operator (Pauli matrix), as in Chapter 11. This S-B coupling is
responsible for bath-induced transitions between the TLS levels (Ch. 3) and thus
allows for the control of heat flow between the two baths and heat-to-work conversion via periodic modulation of the TLS frequency $\omega(t)$ about its resonance value
ω_a. This control is effected by the system Hamiltonian that is diagonal in the TLS
energy basis (Ch. 11)

$$H_S(t) = \frac{\hbar}{2}\omega(t)\sigma_z. \tag{20.3}$$

This control Hamiltonian can be realized by adiabatically eliminating a highly
detuned level of a three-level system and periodically applying an AC Stark shift by
a time-dependent control field (piston). In the fully quantized version of this model,
the classical time-dependent control field is replaced by a quantum harmonic-
oscillator field that is dispersively coupled to the TLS (as discussed in Ch. 22). In
the heat engine (HE) regime, the modulation converts part of the heat-flow energy
from the h-bath to the c-bath into work extractable by the control field. The entire
process then consists in energy transfer from the h-bath to the control field (piston).

The following scenario captures the essence of the model (Fig. 20.1): A quantum
oscillator in a double-well potential that acts as the WM is "sandwiched" between
two media at different temperatures serving as the c- and h-baths. The oscillation
is periodically modulated by an off-resonant source of radiation that controls the
heat current between the baths via the WM. Part of the energy that is released by
the WM coherently amplifies the radiation source, thus producing work.

Whether this minimal QHM model yields work or refrigeration is determined
by the direction of heat flow from the h-bath (the heat current) to the c-bath in
the HE regime or vice versa in the refrigerator regime. The heat current is given
by the polarization rate of the TLS, obtained from the steady-state solution of the
master equation (ME) for the TLS density operator. This ME (Chs. 11, 15) allows
for non-Markovian (bath-memory) effects and is accurate to second order in the
S-B coupling (20.2).

To evaluate work extraction or consumption by the piston P, we may invoke
energy conservation (the first law of thermodynamics)

$$\mathcal{J}_h + \mathcal{J}_c - \frac{d\langle H_P\rangle}{dt} = 0, \tag{20.4}$$

where $\mathcal{J}_{h(c)} = dQ_{h(c)}/dt$ are the heat-flow rates or currents from h (c) to S. Power
generation (the rate of work) is commonly identified with $d\langle H_P\rangle/dt = \dot{W}$, the
mean piston-energy change rate, when the piston is classical rather than quantum-
mechanical and therefore can be treated as a driven work reservoir. In the quantum
domain, the relation between $d\langle H_P\rangle/dt$ and $\dot{W}$ is more subtle (Ch. 22).

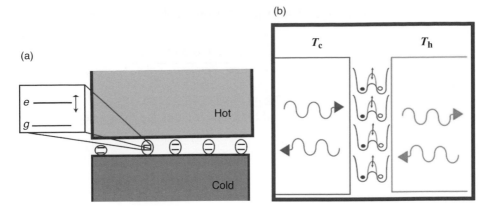

Figure 20.1 (a) A QHM with WM comprised of many TLS (e.g., quantum dots) sandwiched between layers of different materials with different phonon spectra and temperatures acting as hot and cold baths. The inset shows the periodic AC Stark shift modulation of the TLS-level distance. (b) WM consisting of many double-well qubits with periodically modulated tunneling barrier embedded between hot and cold baths. [(a) Reprinted from Gelbwaser-Klimovsky et al., 2015. (b) Reprinted from Gelbwaser-Klimovsky et al., 2013. © 2013 American Physical Society]

20.1.1 Steady-State Continuous-Cycle in the Markovian Limit

The Floquet decomposition of the Markovian master equation presented in Chapter 19 provides a general framework for the analysis of continuously operated HE based on a periodically modulated TLS WM.

In the Markovian limit under periodic driving, detailed balance of transition rates between the WM levels, as well as the periodic modulation rate, determine, according to the first and second laws, the heat currents between the (hot and cold) baths and thereby the power that is either produced (in the HE regime) or consumed (in the refrigerator regime). The Markovian regime arises under sufficiently slow modulation, such that the WM-baths coupling duration within a cycle, τ_C, exceeds the bath correlation time: $\tau_C \gg t_c$.

The reduced density operator for the TLS WM then evolves according to the Floquet-expanded Markovian ME

$$\dot{\rho} = \mathcal{L}\rho = \sum_q \sum_{j=c,h} \mathcal{L}_{jq}\rho, \qquad (20.5)$$

where the q-harmonic Lindblad operators are given in Chapter 19. The steady-state solution of the ME (20.5) is diagonal in the energy basis of the TLS WM:

$$\rho_{ss} = \begin{pmatrix} (\rho_{ee})_{ss} & 0 \\ 0 & (\rho_{gg})_{ss} \end{pmatrix}. \qquad (20.6)$$

The ratio of excited- to ground-state populations of the TLS in the steady state satisfies

$$w = \frac{(\rho_{ee})_{ss}}{(\rho_{gg})_{ss}} = \frac{\sum_{q,j} P_q\, G_j(\omega_a + q\Delta)\, e^{-\beta_j \hbar(\omega_a + q\Delta)}}{\sum_{q,j} P_q\, G_j(\omega_a + q\Delta)}, \tag{20.7}$$

with β_j, P_q, and $G_j(\omega_a + q\Delta)$ being as in Chapter 19. According to Chapter 19, the heat currents between the TLS and the c and h baths are then given by

$$\mathcal{J}_{c(h)} = \sum_q \hbar(\omega_a + q\Delta) P_q G_{c(h)}(\omega_a + q\Delta) \frac{e^{-\beta_{c(h)}\hbar(\omega_a + q\Delta)} - w}{w + 1}. \tag{20.8}$$

The power invested in the QHM is given, according to the first law, by

$$\dot{W} = \mathcal{J}_c + \mathcal{J}_h = \sum_{q \in \mathbb{Z}} \sum_{j \in \{c,h\}} \hbar(\omega_a + q\Delta) P_q G_j(\omega_a + q\Delta) \frac{w - e^{-\beta_j \hbar(\omega_a + q\Delta)}}{w + 1}. \tag{20.9}$$

A negative sign of $\dot{W}$ corresponds to power extraction (at expense of the hot bath), for example, to the QHM acting as an HE. The gain (HE) condition $\dot{W} < 0$ will be investigated for generic cases in Section 20.2.

20.2 Periodic Modulation, Filtered Bath Spectra, and the HE Condition

We here investigate the HE operation condition $\dot{W} < 0$ for sinusoidal modulation of $\omega(t)$ and for the bath response spectra $G_c(\omega)$ and $G_h(\omega)$ that are engineered by generic spectral filtering. The control parameter is the modulation rate Δ that allows us to operate the QHM on demand as either an engine or a refrigerator.

We consider a sinusoidally modulated TLS transition frequency [cf. (19.36)]

$$\omega(t) = \omega_a + \lambda\Delta \sin(\Delta t). \tag{20.10}$$

Three Fourier harmonics ($q = 0, \pm 1$) contribute to the system–bath coupling in the weak-modulation limit $0 \le \lambda \ll 1$, with respective weights [cf. (19.39)]

$$P_0 \approx 1 - \frac{\lambda^2}{2}, \qquad P_{\pm 1} \approx \frac{\lambda^2}{4}. \tag{20.11}$$

The Carnot efficiency bound can be attained if only two harmonics contribute to the sums in (20.8) and (20.9). This requires a spectral separation of the baths, so that the TLS is only coupled to the cold bath at a specific sideband $\omega_a + q_1\Delta$ and to the hot bath only at another sideband $\omega_a + q_2\Delta$ (Fig. 20.2). Such restriction of the coupling to only two sidebands can be achieved for well-separated response spectra of the c- and h-baths, satisfying

$$G_c(\omega) \simeq 0 \text{ for } \omega \approx \omega_a \pm \Delta; \qquad G_h(\omega) \simeq 0 \text{ for } \omega \le \omega_a. \tag{20.12}$$

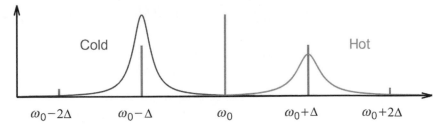

Figure 20.2 Quasi-Lorentzian nearly separated bath-response spectra required for high-efficiency continuous-cycle QHM operation. The TLS WM with resonance frequency ω_a is coupled to the cold (hot) bath at the left (right) Floquet sideband only. The vertical lines indicate the coupling weights P_q to these sidebands. (Adapted from Gelbwaser-Klimovsky et al., 2015.)

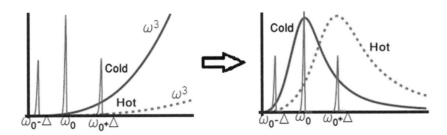

Figure 20.3 Harmonic (Floquet) peaks of the response superimposed on ω^3-dependent cold and hot bath-coupling spectra (left) and on the filtered spectra of the *same baths* reshaped by filter modes that transform these bath spectra into skewed Lorentzians (right). The filtered bath spectra are chosen to satisfy spectral separation. (Adapted from Gelbwaser-Klimovsky et al., 2013. © 2013 American Physical Society)

The coupling of the two-level system (TLS) *to spectrally separated baths* by *spectrally filtering* the bath response (as in Fig. 20.3) is achievable by coupling the TLS through a harmonic-oscillator mode of frequency ω_j ($j = $ h,c) ("the filter") to each bath. Then the TLS becomes *effectively* coupled to two baths with response spectra

$$G_{jf}(\omega) = \frac{\gamma_{jf}}{\pi} \frac{[\pi G_j(\omega)]^2}{\{\omega - [\omega_{jf} + \Delta_{jT}(\omega)]\}^2 + [\pi G_j(\omega)]^2} \quad (j = \text{h,c}). \quad (20.13)$$

Here γ_{jf} is the coupling rate of the TLS to the respective filter mode, $G_j(\omega)$ is the real part of the original (unfiltered) bath response (coupling spectrum), and $\Delta_{jT}(\omega)$ is the imaginary part of the bath response (associated with the bath-induced Lamb shift) as in (11.41) and (11.42),

$$\Delta_{jT}(\omega) = P \int_0^\infty d\omega' \, \frac{G_j(\omega')}{\omega - \omega'}. \tag{20.14}$$

Having imposed condition (20.12), filtering allows us to reduce the expressions for the heat currents and the power, (20.8) and (20.9), to

$$\mathcal{J}_{\mathrm{h}} = \mathcal{K}(\omega_{\mathrm{a}} + \Delta), \quad \mathcal{J}_{\mathrm{c}} = -\mathcal{K}\omega_{\mathrm{a}}, \quad \dot{W} = -\mathcal{K}\Delta, \tag{20.15a}$$

where, to lowest order in λ, we have defined

$$\mathcal{K} = \frac{\hbar\lambda^2}{4} \frac{e^{-\beta_{\mathrm{h}}\hbar(\omega_{\mathrm{a}}+\Delta)} - e^{-\beta_{\mathrm{c}}\hbar\omega_{\mathrm{a}}}}{1 + e^{-\beta_{\mathrm{c}}\hbar\omega_{\mathrm{a}}}} G_{\mathrm{h}}(\omega_{\mathrm{a}} + \Delta). \tag{20.15b}$$

A critical modulation rate defines the quantum speed limit (SL) for work extraction depending on the hot (T_{h}) and cold (T_{c}) bath temperatures,

$$\Delta_{\mathrm{SL}} = \omega_{\mathrm{a}} \frac{T_{\mathrm{h}} - T_{\mathrm{c}}}{T_{\mathrm{c}}}. \tag{20.16}$$

Namely, according to (20.15), the machine acts as a HE that generates power, at the expense of heat flow from the hot bath to the cold bath, provided that

$$\Delta < \Delta_{\mathrm{SL}} \implies \mathcal{J}_{\mathrm{c}} < 0, \ \mathcal{J}_{\mathrm{h}} > 0, \ \dot{W} < 0. \tag{20.17}$$

The HE efficiency is then

$$\eta = \frac{\Delta}{\omega_{\mathrm{a}} + \Delta} \quad (\Delta \le \Delta_{\mathrm{SL}}). \tag{20.18}$$

At $\Delta = \Delta_{\mathrm{SL}}$, the Carnot bound is attained according to (20.16), whereas the power and the heat currents vanish, $\dot{W} = \mathcal{J}_{\mathrm{c}} = \mathcal{J}_{\mathrm{h}} = 0$, so that reversibility is reached (Chs. 17, 18). For $\Delta > \Delta_{\mathrm{SL}}$, the machine acts as a refrigerator (Ch. 23).

Provided that the bath response spectra fulfill (20.12) and $0 < \lambda \ll 1$, only the $q = 1$ harmonic exchanges energy with the hot bath at frequencies $\pm\omega_1 = \pm(\omega_{\mathrm{a}} + \Delta)$. The $q = -1$ harmonic then exchanges energy with the cold bath at frequencies $\pm\omega_{-1} = \pm(\omega_{\mathrm{a}} - \Delta)$. The contributions of the higher-order sidebands ($|q| > 1$) are then negligible.

20.2.1 Maximum Power

The maximal power and the corresponding efficiency of the HE are attainable at the optimal modulation frequency Δ_{max} under the following simplifying assumptions:

(a) The spectral density $G_{\mathrm{h}}(\omega)$ is flat (i.e. $\frac{d}{d\omega} G_{\mathrm{h}} \simeq 0$ near $\omega = \omega_{\mathrm{a}} + \Delta_{\mathrm{max}}$).
(b) High temperature of the hot bath is assumed: $\exp(\frac{\omega}{T_{\mathrm{h}}}) \approx 1 + \frac{\omega}{T_{\mathrm{h}}}$.

Under these assumptions, the modulation Δ_{max} that yields the maximal power $\dot{W}_{\mathrm{max}}$ and the corresponding efficiency are found to be

$$\Delta_{\text{max}} = \frac{1}{2}\Delta_{\text{SL}}, \quad \eta(\dot{W}_{\text{max}}) = \frac{1 - \frac{T_c}{T_h}}{1 + \frac{T_c}{T_h}} \geq \eta_{\text{CA}}, \tag{20.19}$$

where the Curzon–Ahlborn efficiency at maximum power of a macroscopic Carnot-type engine is

$$\eta_{\text{CA}} = 1 - \left(\frac{T_c}{T_h}\right)^{1/2}. \tag{20.20}$$

20.3 Minimal QHM Model beyond Markovianity

20.3.1 Anti-Zeno Effect (AZE) in a Non-Markovian Cycle

Let us consider the following two-stroke non-Markovian cycle:

i) Stroke 1: The QHM operates by keeping the WM and the baths coupled over n modulation periods, from time $t = 0$ to $t = n\tau = \tau_C \lesssim t_c$ ($n \gg 1$, $\tau \ll t_c$). In this stroke, spectral separation of the hot and cold baths is required as in Section 20.2 for work extraction.

ii) Stroke 2: At $t = n\tau = \tau_C$, we decouple the WM from the hot and cold baths. The WM and the thermal baths are then kept uncoupled (non-interacting) for a time interval $\bar{t} \gtrsim t_c$, so as to eliminate all the transient memory effects (i.e., decorrelate the WM and the baths).

After this decoupling interval, the WM is recoupled to the hot and cold thermal baths. The WM is then further periodically modulated by the Hamiltonian (20.3) [cf. (19.11)]. The setup is initialized after time $\tau_C + \bar{t}$, provided that n is such that after n modulation periods $\rho(\tau_C + \bar{t}) = \rho(0)$, so as to close the cycle with the WM returning to its initial state. The QHM may then run indefinitely in this steady-state non-Markovian cyclic regime, as discussed below.

In the non-Markovian cycle, the QHM performance crucially depends on the relative widths of the bath-coupling spectra and the sinc-function spectral filter (Chs. 10, 15), as their overlap controls the WM evolution. The convolution (19.35) that determines the coupling strength to a bath may lead to its drastic heat exchange enhancement, owing to the time-energy uncertainty relation that manifests itself by the anti-Zeno effect (AZE) in the non-Markovian domain (Ch. 15). In turn, this AZE enhancement can boost the heat currents and modulation power.

A slow modulation (with period $\tau \gg t_c$) results in sinc functions, which are nonzero only over a narrow frequency range, wherein $G_j(\omega)$ ($j =$ h,c) is nearly constant (flat), which leads to time-independent relaxation rates $\gamma_{j,e(g)}$ and Markovian dynamics. In this limit of long coupling duration $\tau_C = n\tau \gg t_c$, where the sinc functions approach the delta-function limit, and (19.35) reduces to the standard Markovian form

$$\gamma_{j,e(g)} = 2\pi G_j(\pm\omega_a).$$ (20.21)

By contrast, fast modulation ($\tau \lesssim t_c$) corresponds to broad sinc functions, for which $G_j(\omega)$ strongly varies over the frequency range $\sim \tau^{-1}$ for which the sinc functions are nonzero. Appropriate choices of modulation may then yield a power and heat-currents boost compared to the Markovian limit, whenever the sinc functions have enhanced overlap with $G_{h,c}(\omega)$. This regime is a consequence of the time-energy uncertainty relation of quantum mechanics and is associated with the anti-Zeno effect (AZE) (Chs. 10, 15), which results in dynamically enhanced system–bath energy exchange.

For this power boost to occur, the bath-response spectral functions should be peaked at frequencies sufficiently detuned from $\omega_a \pm \Delta$, so that their overlap with the sinc functions may appreciably increase under fast modulation in the AZE regime: This requirement amounts to $\tau^{-1}, \delta \gtrsim \Gamma_B$.

In order to study the QHM performance as a function of the modulation rate, we here consider the example of two nonoverlapping quasi-Lorentzian spectral response functions of the two baths displaced by δ with respect to ω_q of width Γ_B, with the peak at $\omega_h = \omega_a + \Delta + \delta$ ($\omega_c = \omega_a - \Delta - \delta$). Another example we consider is that of two nonoverlapping super-Ohmic spectral response functions of the two baths, with their origins shifted from $\omega = 0$ by $\omega_h = \omega_a + \Delta - \delta$ and $\omega_c = \omega_a - \Delta + \delta$ respectively, under the assumptions (20.39b). The dependence of $\omega_{h,c}$ on Δ ensures that any enhancement in heat currents and power under fast modulation results from the broadening (rather than the shift) of the sinc functions, which are centered at $\omega_a \pm \Delta$.

The non-Markovianity is quantified by the spectral widths of the sinc functions compared to the width $1/t_c$ of the spectral bath response spectrum $G(\omega)$. Upon keeping the cycle duration fixed, then the non-Markovianity scales with the spectral width of $G(\omega)$. Super-Ohmic bath spectra with their clear cutoff provide salient examples of the non-Markovian effects described here.

In the examples shown here, the quantum advantage in HEs powered by baths with quasi-Lorentzian or super-Ohmic spectral functions can increase the power by a factor larger than two or seven, respectively (Figs. 20.4 and 20.5), without changing the efficiency.

At $t \sim \tau_{th}$ the WM attains the diagonal state

$$\rho(t) = p_e(t) |e\rangle \langle e| + p_g(t) |g\rangle \langle g|,$$ (20.22)

that, according to Chapter 19, satisfies the rate equations [cf. (19.29), where $\rho_{ee(gg)} = p_{e(g)}$]

$$\dot{p}_e = -\dot{p}_g = \gamma_g(t)p_0 - \gamma_e(t)p_e.$$ (20.23)

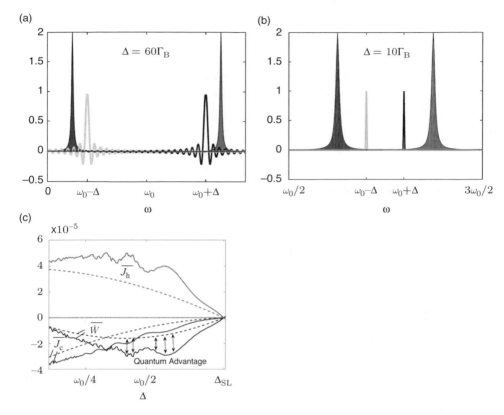

Figure 20.4 (a), (b) The quasi-Lorentzian bath-response spectra of the cold and hot baths, $G_c(\omega)$ and $G_h(\omega)$ (left and right filled curves, respectively), and the functions sinc $[(\omega - \omega_a + \Delta)t]$ and sinc $[(\omega - \omega_a - \Delta)t]$ (left- and right-peaked solid curves, respectively), for (a) fast modulation, $\Delta = 60\Gamma_B$, and (b) slow modulation, $\Delta = 10\Gamma_B$, at $t = 10\tau$. Fast modulation results in non-Markovian bath response caused by broadening of the sinc functions, thus leading to enhanced overlap with the spectral functions compared to the Markovian limit of slow modulation, where the sinc functions are δ-function-like. (c) Power $\overline{W}$ given by the sum of the heat currents $\overline{\mathcal{J}_h}$ and $\overline{\mathcal{J}_c}$ averaged over $n = 10$ modulation periods (solid lines) and the same obtained under the Markovian approximation for long cycles, that is, with $n \to \infty$ modulation periods (dashed lines), as functions of the modulation rate Δ. AZE for $\tau_C \lesssim t_c$ results in output power boost (shown by double-arrowed lines) by up to more than a factor of 2, signifying quantum advantage in the HE regime. The horizontal line corresponds to zero power and currents. Here $\lambda = 0.2$, $\omega_a = 20$, $\gamma_g = 1$, $\Gamma_B = 0.2$, $N = 1$, $\delta = 3$, $\epsilon = 0.01$, $\beta_h = 0.0005$, $\beta_c = 0.005$. (Adapted from Mukherjee et al., 2020.)

Here, upon allowing for (19.30), (19.35), and (19.39), we have

$$\gamma_g(t) = \frac{\lambda^2}{4} \left[\mathcal{I}_h(-\omega_a - \Delta, t) + \mathcal{I}_c(-\omega_a + \Delta, t) \right],$$

$$\gamma_e(t) = \frac{\lambda^2}{4} \left[\mathcal{I}_h(\omega_a + \Delta, t) + \mathcal{I}_c(\omega_a - \Delta, t) \right], \qquad (20.24)$$

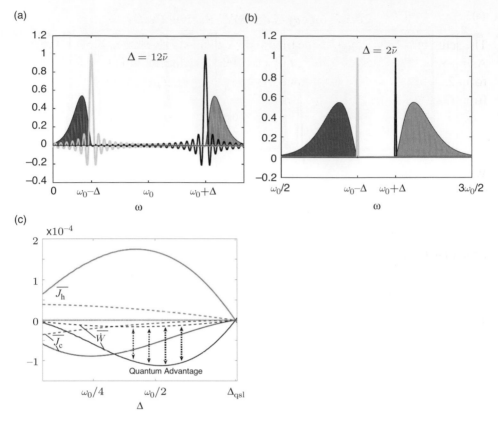

Figure 20.5 (a), (b) Overlap of super-Ohmic spectral functions $G_c(\omega)$ and $G_h(\omega)$ (left and right filled curves, respectively) with cutoff frequency $\bar{\nu}$ [see (20.38)], with the modulation response functions, sinc$[(\omega - \omega_a + \Delta)t]$ and sinc$[(\omega - \omega_a - \Delta)t]$ (left- and right-peaked solid curves, respectively), for (a) fast modulation, $\Delta = 12\bar{\nu}$, and (b) slow modulation, $\Delta = 2\bar{\nu}$, at $t = 10\tau$. Fast modulation results in broad sinc functions, and thus enhanced overlap with the spectral functions. (c) Power $\overline{W}$ and heat currents $\overline{\mathcal{J}_h}$ and $\overline{\mathcal{J}_c}$ averaged over $n = 10$ modulation periods (solid lines) as compared to their counterparts under the Markovian approximation for a long cycle, $n \to \infty$ (dashed lines), versus the modulation frequency Δ. Quantum advantage is obtained for $\tau_C \lesssim t_c$, when broadening of the sinc functions yields an output power boost (shown by double-arrowed lines) of up to a factor greater than 7, in the HE regime. The horizontal dotted line corresponds to zero power and currents. Here $\bar{\nu} = 1$, $\delta = 0.1$, $\epsilon = 0.1$, $\alpha = 1$, $\omega_a = 20$, $\gamma_g = 1$, $\beta_h = 0.0005$, $\beta_c = 0.005$. (Adapted from Mukherjee et al., 2020.)

where the convolution is given by

$$\mathcal{I}_j(\omega, t) = 2t \int_{-\infty}^{\infty} d\omega' \, G_j(\omega') \, \text{sinc}[(\omega' - \omega)t]. \qquad (20.25)$$

In what follows, the steady-state non-Markovian cycle will be investigated.

20.3.2 Steady-State AZE-Regime Cycles

The setup may be operated in the time-independent steady state in the AZE regime. As $t \to \infty$, the integral $\mathcal{I}_j\left(\omega_q, t\right)$ [see (20.25)] reduces to the time-independent form $2\pi G_j(\omega_q)$. On the other hand, for $t \sim n\tau \lesssim t_c$, $\mathcal{I}_j\left(\omega_q, t\right)$ has contributions from $G_j(\omega_q + \omega)$, where

$$|\omega| \lesssim \frac{1}{n\tau}. \tag{20.26}$$

We here assume that $1/(n\tau) \ll \omega_a \pm \Delta, T_c, T_h$. In this limit the KMS condition yields

$$G_j(-\omega_q - \omega) \approx e^{-\beta_j \hbar(\omega_q + \omega)} G_j(\omega_q + \omega) \approx e^{-\beta_j \hbar \omega_q} G(\omega_q + \omega), \tag{20.27}$$

which leads to

$$\mathcal{I}_j(-\omega_q, t) \approx e^{-\beta_j \hbar \omega_q} \mathcal{I}_j(\omega_q, t) \quad (q = 0, \pm 1). \tag{20.28}$$

Then

$$w = \frac{p_e}{p_g} \approx \frac{e^{-\beta_h \hbar(\omega_a + \Delta)} \mathcal{I}_h(\omega_a + \Delta, t) + e^{-\beta_c \hbar(\omega_a - \Delta)} \mathcal{I}_c(\omega_a - \Delta, t)}{\mathcal{I}_h(\omega_a + \Delta, t) + \mathcal{I}_c(\omega_a - \Delta, t)}, \tag{20.29}$$

where we have considered the two sidebands $q = 1, -1$ only.

The condition

$$\mathcal{I}_h\left(\omega_a + \Delta, t\right) \approx K \mathcal{I}_c\left(\omega_a - \Delta, t\right), \tag{20.30}$$

which holds for mutually symmetric bath spectral functions up to a multiplicative factor

$$G_h(\omega_a + x) \approx K G_c(\omega_a - x) \tag{20.31}$$

for any real x, then leads to the time-independent steady state ρ_{ss} (Fig. 20.6) of the WM at times much longer than the thermalization time τ_{th}, under the condition of weak WM-baths coupling. This non-equilibrium steady state $\rho \to \rho_{ss}$ has the energy-diagonal form:

$$\rho_{ss} = p_{e,ss} |e\rangle \langle e| + p_{g,ss} |g\rangle \langle g|, \tag{20.32}$$

where

$$\frac{p_{e,ss}}{p_{g,ss}} \equiv w = \frac{K e^{-\beta_h \hbar(\omega_a + \Delta)} + e^{-\beta_c \hbar(\omega_a - \Delta)}}{1 + K}. \tag{20.33}$$

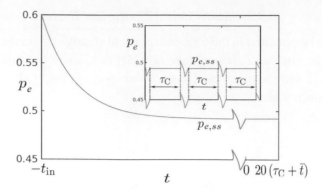

Figure 20.6 Time-evolution of the excited-state probability $p_e(t)$ of a TLS WM in a non-Markovian cycle. The WM is first connected to the hot and cold baths, with quasi-Lorentzian spectral functions, at a negative time $-t_{in}$ ($t_{in} \gg \tau_{th}$) and attains the steady-state value $p_{e,\,ss}$ at $t + t_{in} \gg \tau_{th}$. The WM is subsequently decoupled from both baths at a time $-\bar{t} \lesssim -t_c < 0$ and recoupled again to the baths at time $t = 0$, such that the WM is noninteracting with the baths over the time interval $-\bar{t} \leq t < 0$, shown by the break line. The QHM thereby completes a cycle with total duration $\bar{t} + \tau_C$, where the coupling time duration $\tau_C = n\tau$. The probability p_e remains unchanged at the steady-state value, even after multiple cycles. Inset: Same as the main plot, zoomed in for three consecutive cycles. The WM is noninteracting with the baths over time intervals $\bar{t} \gtrsim t_c$ between two consecutive cycles, shown by the break lines. Here $\lambda = 0.2$, $\omega_a = 20$, $\Delta = 10$, $n = 10$, $\beta_h = 0.0005$, $\beta_c = 0.005$. The bath spectral functions are quasi-Lorentzian with $\gamma_g = 1$, $\Gamma_B = 0.2$, $\delta_h = \delta_c = 1$. (Adapted from Mukherjee et al., 2020.)

The efficiency in the heat-engine regime is given by (Fig. 20.7)

$$
\eta = \frac{\oint_{\tau_C} [\mathcal{J}_h(t) + \mathcal{J}_c(t)] dt}{\oint_{\tau_C} \mathcal{J}_h(t) dt}
$$

$$
= \frac{(\omega_a + \Delta)\zeta_h \oint_{\tau_C} \mathcal{I}_h(\omega_a + \Delta, t) dt + (\omega_a - \Delta)\zeta_c \oint_{\tau_C} \mathcal{I}_c(\omega_a - \Delta, t) dt}{(\omega_a + \Delta)\zeta_h \oint_{\tau_C} \mathcal{I}_h(\omega_a + \Delta, t) dt},
$$

(20.34)

where we have defined

$$
\zeta_h = \frac{e^{-\beta_h \hbar(\omega_a + \Delta)} - w}{w + 1}, \quad \zeta_c = \frac{e^{-\beta_c \hbar(\omega_a - \Delta)} - w}{w + 1}.
$$

(20.35)

The integral $\mathcal{I}_h(\omega_a + \Delta, t)$ [see (20.25)] can be approximated by

$$
\mathcal{I}_h(\omega_a + \Delta, t) = 2 \int_{-\infty}^{\infty} G_h(\omega) \frac{\sin(\omega - \omega_a - \Delta)t}{\omega - \omega_a - \Delta} d\omega
$$

$$
\approx 2 \int_{-\Delta}^{\infty} G_h(\omega_a + \Delta + x) \frac{\sin(xt)}{x} dx,
$$

(20.36)

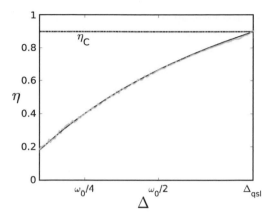

Figure 20.7 HE efficiency η versus the modulation rate Δ in the AZE regime (wiggling solid line) is the same as in the long coupling-time limit $\tau_C \rightarrow \infty$ (smooth solid line). The efficiency approaches the Carnot limit $\eta_C = 1 - \beta_h/\beta_c$ (horizontal dotted line) at $\Delta = \Delta_{SL}$. (Adapted from Mukherjee et al., 2020.)

where we have taken into account that $G_h(\omega) = 0$ for $0 < \omega \leq \omega_a$ and $\sin(xt)/x$ is small for large $|x|$. Similarly, we have

$$\mathcal{I}_c(\omega_a - \Delta, t) \approx 2 \int_{-\Delta}^{\infty} G_c(\omega_a - \Delta - x) \frac{\sin(xt)}{x} dx, \qquad (20.37)$$

where we have taken into account that $G_c(\omega) = 0$ for $\omega \geq \omega_a$. For symmetric bath spectral functions, we have $\mathcal{I}_h(-\omega_a - \Delta, t) \approx K\mathcal{I}_c(-\omega_a + \Delta, t)$, which in turn results in the AZE-regime efficiency being approximately equal to that obtained in the Markovian regime (Fig. 20.7).

20.3.3 Example: Super-Ohmic Bath Spectra

We consider super-Ohmic bath spectral functions of the form

$$G_h(\omega \geq 0) = K\gamma_g \frac{(\omega - \omega_h)^s}{\bar{\nu}^{s-1}} e^{-(\omega - \omega_h)/\bar{\nu}} \theta(\omega - \omega_h),$$

$$G_c(\omega \geq 0) = \gamma_g \frac{(\omega_c - \omega)^s}{\bar{\nu}^{s-1}} e^{-(\omega_c - \omega)/\bar{\nu}} \theta(\omega_c - \omega)\theta(\omega - \epsilon),$$

$$G_{h,c}(-\omega) = G_{h,c}(\omega) e^{-\beta_{h,c}\hbar\omega}, \qquad (20.38)$$

with the origin shifted from $\omega = 0$ by

$$\omega_h = \omega_a + \Delta - \delta,$$
$$\omega_c = \omega_a - \Delta + \delta. \qquad (20.39a)$$

Here $s > 1$, and the assumptions

$$0 < \delta \ll \Delta, \omega_a, \omega_a - \Delta \qquad (20.39b)$$

ensure that $G_{h,c}(\omega)$ is nonzero at the maxima of the two sinc functions at $\omega_a \pm \Delta$. A small $\epsilon > 0$ guarantees that $G_c(\omega = 0) = 0$. The Δ-dependent ω_h and ω_c ensure that any heat-current and power enhancement are due to the broadening of the sinc functions under fast modulation, rather than the shifting of their peaks.

Conditions (19.40), along with the KMS condition (Ch. 15), imply that

$$\mathcal{I}_h(-\omega_a - \Delta, t) \approx e^{-\frac{\omega_a + \Delta}{T_h}} \mathcal{I}_h(\omega_a + \Delta, t),$$
$$\mathcal{I}_c(-\omega_a + \Delta, t) \approx e^{-\frac{\omega_a - \Delta}{T_c}} \mathcal{I}_c(\omega_a - \Delta, t). \qquad (20.40)$$

Equations (20.40), in turn, guarantee that (20.33) yields the steady state even at short times and thus eliminates any time dependence in ρ (Fig. 20.6).

The heat currents $\mathcal{J}_c$ and $\mathcal{J}_h$, flowing out of the cold and hot baths, respectively, are obtained consistently with the second law in the form

$$\mathcal{J}_h(t) = \frac{\lambda^2}{4\Delta^2} (\omega_0 + \Delta)\mathcal{I}_h(\omega_a + \Delta, t) \frac{e^{-\beta_h \hbar(\omega_a + \Delta)} - w}{w + 1},$$
$$\mathcal{J}_c(t) = \frac{\lambda^2}{4\Delta^2} (\omega_0 - \Delta)\mathcal{I}_c(\omega_a - \Delta, t) \frac{e^{-\beta_c \hbar(\omega_a - \Delta)} - w}{w + 1}, \qquad (20.41)$$

where we have used $P_{\pm 1} = \lambda^2/(4\Delta^2)$.

20.3.4 The QZE Regime

In contrast to the advantageous AZE regime, ultrafast modulations that satisfy t_c, $\delta^{-1} \gg \tau_C$ lead to the regime governed by the quantum Zeno effect (QZE) (Chs. 10, 15), where the convolutions

$$\mathcal{I}_{c,h}(\omega_q, t) \to 0, \qquad (20.42)$$

resulting in vanishing heat currents and power (Fig. 20.8).

Hence, in the regime of ultrafast modulation the QZE leads to vanishing heat currents and power, thus implying that such a regime is incompatible with thermal machine operation. However, the QZE regime may allow for work extraction in the presence of appreciable system–bath correlations, as detailed in Chapter 15.

20.4 Discussion

A TLS that interacts with both hot and cold baths via off-diagonal (σ_x-) coupling, while its transition frequency is periodically modulated, has been shown to be a minimal model of a QHM. Its HE and refrigerator regimes can be interchanged by varying the modulation rate. When approaching the quantum speed-limit

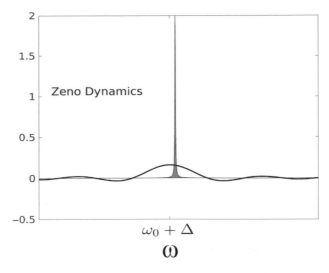

Figure 20.8 Zeno dynamics, when the sinc function is much broader than $G_h(\omega)$ and approaches zero for nearly all ω. (Adapted from Mukherjee et al., 2020.)

modulation rate Δ_{SL} from below, the HE reaches Carnot efficiency bound, whereas the heat currents and the power output vanish. Any further increase of the modulation rate turns the engine into a refrigerator. The engine's efficiency at *maximal* power output can surpass the Curzon–Ahlborn bound at an appropriate modulation rate.

This model is inherently nonadiabatic and thus circumvents the difficulty of abrupt on-off switching of the WM-bath interactions in alternating strokes, which may strongly affect their energy and entropy exchange and thereby their quantum state (Ch. 19).

In order to efficiently operate the QHM described above, a necessary requirement on the bath-coupling spectra is their separation. Hot and cold phonon baths associated with two different materials that have an upper Debye cutoff, with the desired spectral separation can be achieved if the respective Debye frequencies lie in the vicinity of the material temperatures,

$$\omega_D^c \simeq \frac{1}{\beta_c \hbar}, \quad \omega_D^h \simeq \frac{1}{\beta_h \hbar}, \tag{20.43}$$

as for a TLS WM realized by a quantum dot sandwiched between two different materials (Fig. 20.1).

A general prescription for obtaining spectrally separated response of the two baths consists of constructing "filter" modes through which the system interacts with the baths. These filter modes can be easily tuned to create the desired *effective* bath spectra without the need to directly engineer the materials' autocorrelation functions.

Quantum advantage has been shown to be achievable when the TLS energy-level distance is modulated faster than what is allowed by the Markov approximation. To this end, one can invoke methods of quantum-system control via frequent coherent (e.g. phase-flipping or level-modulating) operations as well as their incoherent counterparts (e.g. projective measurements or noise-induced dephasing). Such control has been shown, both theoretically and experimentally (Chs. 10, 15), to yield non-Markovian dynamics that conforms to either the quantum Zeno effect (QZE), whereby the bath effects on the system are drastically suppressed or slowed down, or the anti-Zeno effect (AZE) that implies the enhancement or speedup of the system–bath energy exchange.

The analysis of cyclic TLS-based heat machines has shown that the AZE can drastically enhance the HE power output, without affecting its efficiency.

The reason for this power boost is that the sinc functions in the convolutions with the bath-coupling spectra are sufficiently broad in the AZE regime so as to enhance the convolutions compared to the Markovian case, where these sinc functions are spectrally narrow enough to be approximated by delta functions. This AZE enhancement of the convolution leads to an increase in the TLS relaxation rates and the rates of heat exchange and entropy production and hence to a power boost. This boost is of quantum nature, since the broadening of the sinc factors is due to the quantum energy-time uncertainty relation.

Yet, the QHM rate of operation (as measured by the power output) speeds up in the AZE regime, which constitutes a practical quantum advantage.

This power boost relies on transient heat fluxes that change within the cycle. Hence, the combined system WM+baths is not in a steady state. Yet, we can incorporate this transient dynamics within steady-state cycles by decoupling the WM from the baths, allowing the WM-bath correlations to vanish within the bath memory time and then recoupling the WM again to the baths when they all have resumed their initial states. These cycles can be repeated without restriction, thereby allowing us to operate the QHM with quantum-enhanced power over arbitrarily many cycles. At the same time, the efficiency (i.e., the ratio of the work output to the heat input) remains the same as in the Markovian regime.

Since non-Markovianity and AZE, in particular, require non-flat bath spectral functions, suitable hot and cold baths are provided by microwave cavities and waveguides or dielectric gratings, with distinct cutoff and bandgap frequencies. The required WM modulations are then compatible with MHz periodic driving of superconducting (transmon) qubits or NV-center qubits in diamonds or nano-mechanical oscillators.

21

Quantum Cooperative Heat Machines

In this chapter, we study the impact of quantum cooperative (many-body) effects on the operation of periodically modulated heat engines (HE). Periodic arrays of N two-level systems (TLS) with appropriate spacing are shown to exchange energy with the modulating field and with the (hot and cold) thermal baths at a faster rate than disordered TLS. Such cooperative energy exchange stems from the spatial symmetry of the periodically arranged TLS that causes them to behave as a collective spin-$N/2$ particle. This cooperativity boosts the HE power compared to that of N independent TLS-based HEs.

21.1 Many-Body Heat Engine (HE) with Permutation Symmetry

We consider an ensemble of N TLS subject to a driving field that periodically modulates the TLS transition frequency, namely,

$$H_S(t) = \frac{\hbar\omega(t)}{2} \sum_{i=1}^{N} (\sigma_z)_i, \qquad (21.1a)$$

such that

$$H_S(t + \tau) = H_S(t), \qquad (21.1b)$$

with the modulation period $\tau = 2\pi/\Delta$. The atoms are all *identically* (indistinguishably) coupled to the cold and hot (j = c,h) thermal baths, via the TLS–bath coupling Hamiltonians

$$(H_I)_j \equiv (H_{SB})_j = \sum_{i=1}^{N} (\sigma_x)_i \otimes B_j, \qquad (21.1c)$$

where B_j is the jth bath coupling operator.

The setup governed by (21.1) has been studied for a single TLS in Chapter 20, where it has been shown to constitute a universal QHM that can be operated on demand as either an HE (converting heat from the hot bath into power) or a refrigerator (consuming power to refrigerate the cold bath). The classical field that induces the periodic modulation in the Hamiltonian (21.1a) acts as a piston that extracts power from the HE.

We now introduce the collective spin operator

$$\hat{J} := \sum_{i=1}^{N} \frac{1}{2}\sigma_i, \tag{21.2}$$

where σ is the vector of the Pauli matrices. In terms of these collective operators, the Hamiltonians (21.1) adopt the simple forms,

$$H_S(t) = \hbar\omega(t) J_z, \tag{21.3a}$$

$$(H_{SB})_j = 2\hbar J_x \otimes B_j. \tag{21.3b}$$

21.1.1 Irreducible Collective-Spin Subspaces

The set of unitary matrices of the form $e^{i\hat{J}\cdot a}$, where a is an arbitrary real vector, forms a representation of the SU(2) group. This representation can be decomposed into a direct sum of irreducible representations. Correspondingly, the 2^N-dimensional joint Hilbert space of N TLS is decomposed into irreducible subspaces. In a basis of vectors spanning these subspaces, the collective-spin operator (21.2) acquires a block-diagonal form,

$$\hat{J} = \bigoplus_{k=1}^{m} S_k. \tag{21.4}$$

Here S_k is a spin operator of dimension $\leq N+1$, and the number of the irreducible subspaces is

$$m = \frac{N!}{\left[\frac{N}{2}\right]! \left(N - \left[\frac{N}{2}\right]\right)!}, \tag{21.5}$$

where $\left[\frac{N}{2}\right]$ is the integer part of $\frac{N}{2}$. Each of the irreducible subspaces corresponds to an eigenvalue $j(j+1)$ of $\hat{J}^2$ and has the respective dimension $2j+1$, where $j = 0, 1, \ldots, \frac{N}{2}$ for N even and $j = \frac{1}{2}, \frac{3}{2}, \ldots, \frac{N}{2}$ for N odd. We will adopt the notation j_k to denote the j belonging to the kth subspace ($k = 1, \ldots, n$) when necessary. The corresponding basis of the Hilbert space consists of entangled collective-spin (Dicke) states (Ch. 8).

The maximum-cooperativity irreducible subspace is spanned by $N+1$ fully symmetric (under exchange or permutation) states $|j\rangle$ that carry j excitations.

In terms of the N-TLS states, these Dicke states read

$$|0\rangle = |ggg\ldots g\rangle \tag{21.6a}$$

$$|1\rangle = \hat{S}_{\text{sym}}\,|egg\ldots g\rangle \tag{21.6b}$$

$$|2\rangle = \hat{S}_{\text{sym}}\,|eeg\ldots g\rangle \tag{21.6c}$$

$$\vdots$$

$$|N\rangle = |eee\ldots e\rangle\,, \tag{21.6d}$$

where $\hat{S}_{\text{sym}}$ is the symmetrization operator.

The largest possible spin j $= N/2$ is uniquely formed by the totally symmetric N-TLS states, but, in general, different irreducible subspaces may share the same j. The block-diagonal structure of the collective operators corresponds to subspaces of eigenstates with different symmetry.

Here we show that the symmetry of the Hamiltonian (21.3) can be exploited as a quantum-thermodynamic resource that boosts the HE power of cooperative (interference) effects (Fig. 21.1).

21.2 Cooperative and Noncooperative Master Equations (ME)

A Lindblad ME for the periodically modulated TLS governed by the Hamiltonian (21.1) can be derived using Floquet theory (see Ch. 11 for a TLS and Ch. 19 for a multilevel system). The ME for $N > 1$ TLS is obtained upon replacing the single

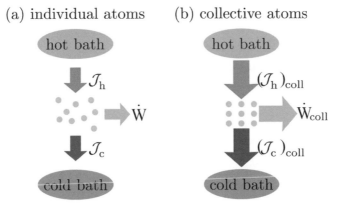

Figure 21.1 (a) An HE comprised of N irregularly spaced TLS is equivalent to N single-TLS HEs whose energy currents are additive and scale linearly in N. (b) In geometries where the TLS are ordered and indistinguishably coupled to the baths, the energy currents and the power output can be cooperatively enhanced by constructive interference and scale faster than linearly in N. (Adapted from Niedenzu and Kurizki, 2018.)

TLS operators by their collective counterparts according to (21.2). The reason is that $\boldsymbol{\sigma}$ and $\hat{\boldsymbol{J}}$ adhere to the same Lie algebra (i.e., the same commutation relations).

Hence, the multi-TLS ME in the interaction picture obeys the Floquet decomposition in harmonics

$$\dot{\rho} = \sum_{j=c,h} \sum_q \mathcal{L}_{j,q} \rho \tag{21.7}$$

with the qth harmonic Liouvillian (in this chapter, we denote $\omega_a \equiv \omega_0$)

$$\mathcal{L}_{j,q} := \frac{1}{2} P_q G_j(\omega_0 + q\Delta) \left[\mathcal{D}(\hat{J}_-) + e^{-\beta_j \hbar(\omega_0 + q\Delta)} \mathcal{D}(\hat{J}_+) \right]. \tag{21.8}$$

Here $\beta_j = (k_B T_j)^{-1}$, T_j being the temperature of the jth bath, whereas the dissipater $\mathcal{D}$ (Chs. 11–14) defined by

$$\mathcal{D}(A)\rho := 2A\rho A^\dagger - A^\dagger A\rho - \rho A^\dagger A \tag{21.9}$$

is a function of the transition operators

$$\hat{J}_\pm := \sum_{i=1}^N (\sigma_\pm)_i \tag{21.10}$$

and the qth harmonic coupling weight P_q (normalized to 1) is determined (as in Chs. 15–20) by the modulation form $\omega(t)$. The first term in the brackets in (21.8) describes the emission of quanta governed by the transition operator $\hat{J}_-$, and the second term describes their absorption governed by the transition operator $\hat{J}_+$. The rates at the Floquet sideband frequencies of quanta emitted to or absorbed by the baths are determined by the bath-coupling spectra $G_j(\omega)$ and the modulation rate Δ.

The Markovian approximation, which requires the bath autocorrelation time to be much shorter than the relaxation time of the system, breaks down for very large N (Ch. 13). Here we assume, however, that the Markovian (Lindblad) ME faithfully describes the cooperative effects.

The decay rates between the levels of the collective spin system $|j, m\rangle \leftrightarrow |j, m-1\rangle$ ($m = -j, \dots, j$) scale as

$$\gamma(m) \propto |\langle j, m-1| \hat{J}_- |j, m\rangle|^2 = (j - m + 1)(j + m). \tag{21.11}$$

These rates increase as we approach the central level: For maximum cooperativity ($j = N/2$), we obtain the superradiant N^2 scaling only for a Hamiltonian that is perfectly symmetric under TLS permutations. The irreducible subspaces are dynamically invariant, so that the ME (21.7) solely involves the collective operators $\hat{J}_\pm$.

This ME does not couple irreducible subspaces with different symmetry that correspond to the blocks in (21.4): Since $[\hat{J}_\pm, \hat{\boldsymbol{J}}^2] = 0$, these subspaces are *invariant* under the ME evolution, such that every $(2j_k + 1)$-dimensional subspace evolves individually. Hence, the steady-state solution of the ME (21.7) is a weighted direct sum of Gibbs-like states of these individual spin subspaces,

$$\rho_{ss} = \bigoplus_{k=1}^{n} \langle \Pi_k \rangle_{\rho_0} (\rho_{ss})_k. \tag{21.12}$$

Here

$$(\rho_{ss})_k = Z_k^{-1} \exp\left(-\beta_{\text{eff}} \hbar \omega_0 S_{kz}\right), \tag{21.13}$$

where

$$Z_k := \text{Tr} \exp\left(-\beta_{\text{eff}} \hbar \omega_0 S_{kz}\right). \tag{21.14}$$

Here Π_k denotes the projector onto the kth invariant subspace and ρ_0 is the initial TLS condition such that $\sum_{k=1}^{n} \langle \Pi_k \rangle_{\rho_0} = 1$. The inverse effective temperature β_{eff} is defined via the "global" (two-bath) detailed-balance condition,

$$\exp(-\beta_{\text{eff}} \omega_0) := \frac{\sum_{j=c,h} \sum_q P_q G_j(\omega_0 + q\Delta) e^{-\beta_j \hbar(\omega_0 + q\Delta)}}{\sum_{j=c,h} \sum_q P_q G_j(\omega_0 + q\Delta)}. \tag{21.15}$$

The numerator in Eq. (21.15) is the total absorption rate in the ME (21.7), whereas the denominator corresponds to the total emission rate therein.

The *non-uniqueness* of the steady state (21.12) is thus crucial for the establishment of collective effects – only then are the TLS correlated and not acting independently.

The steady state (21.12) thus reflects the fact that the initial populations $\langle \Pi_k \rangle_{\rho_0}$ of the invariant subspaces (i.e., the initial weights of the individual subspaces) cannot change dynamically under the ME (21.7). Each subspace relaxes to its own Gibbs-like steady state (21.13). The steady state (21.12) is attained in the interaction picture with respect to the Hamiltonian (21.1a). In the Schrödinger picture this state corresponds to a limit cycle with periodicity $\tau = 2\pi/\Delta$.

If the TLS permutation symmetry is broken, the ME involves "cross-decay" and "cross-absorption" terms and has the form

$$\dot{\rho} = \sum_{i,i'=1}^{N} c_{ii'} [2\sigma_{i-}\rho\sigma_{i'+} - \sigma_{i'+}\sigma_{i-}\rho - \rho\sigma_{i'+}\sigma_{i-}$$

$$+ e^{-\beta_{\text{eff}} \hbar \omega_0} (2\sigma_{i+}\rho\sigma_{i'-} - \sigma_{i'-}\sigma_{i+}\rho - \rho\sigma_{i'-}\sigma_{i+})], \tag{21.16}$$

where the matrix of multi-spin cross-decay/absorption coefficients $\{c_{ii'}\}$ is Hermitian and positive. Only under permutation symmetry, which holds when all the

$c_{ii'}$ are equal, are the eigenvalues of the multi-spin $c_{ii'}$ matrix all zero, except for a single eigenvalue that corresponds to collective decay or absorption via the collective-spin operators $\hat{J}_{\pm}$.

Imperfections may result in partial symmetry, such that some eigenvalues of the $c_{ii'}$ matrix vanish. In contrast to the perfect-symmetry case of only a single nonzero eigenvalue, only a sub-ensemble of the N TLS behaves collectively under partial symmetry and hence only the initial condition of that TLS sub-ensemble appears in the steady-state solution.

Under completely broken symmetry, such that all eigenvalues of the matrix are nonzero, the HE is run by N independent TLS, devoid of collective effects, with a *unique* steady state of the TLS [unlike (21.12)].

21.3 Collective Energy Currents

The steady state (21.12) is an out-of-equilibrium WM state that is maintained by the interplay of the heat current $\mathcal{J}_c$ from the cold bath, the heat current $\mathcal{J}_h$ from the hot bath, and the power $\dot{W} = -(\mathcal{J}_c + \mathcal{J}_h)$. The heat currents from the two baths to the WM are (cf. Ch. 19)

$$\mathcal{J}_j = \sum_q (\omega_0 + q\Delta) \operatorname{Tr}\left[\left(\mathcal{L}_{j,q}\rho_{ss}\right) J_z\right] \qquad (j = c, h). \qquad (21.17)$$

According to this expression, under Δ-periodic driving, quanta are exchanged not only at the "bare" transition frequency ω_0 but also at the Floquet sidebands $\omega_0 + q\Delta$ with steady-state probabilities corresponding to the TLS interaction with the jth bath at the qth harmonic sideband.

Upon inserting the steady state (21.12) into (21.17), the heat currents and the power evaluate to

$$\mathcal{J}_j = \sum_{k=1}^n \langle \Pi_k \rangle_{\rho_0} \mathcal{J}_j(j_k), \qquad (21.18a)$$

$$\dot{W} = \sum_{k=1}^n \langle \Pi_k \rangle_{\rho_0} \dot{W}(j_k). \qquad (21.18b)$$

Here

$$\mathcal{J}_j(j_k) = F(j_k) \sum_q (\omega_0 + q\Delta) P_q G_j(\omega_0 + q\Delta) \left[e^{-\beta_j \hbar(\omega_0+q\Delta)} - e^{-\beta_{\mathrm{eff}}\hbar\omega_0}\right] \quad (21.19)$$

is the heat current induced by the kth invariant subspace and

$$\dot{W}(j_k) = -\left[\mathcal{J}_c(j_k) + \mathcal{J}_h(j_k)\right] \qquad (21.20)$$

is the corresponding power. The prefactor in (21.19) is given by

$$F(\mathbf{j}_k) := \sum_{j=0}^{2\mathbf{j}_k-1} p_{j,k}^{ss}(j+1)(2\mathbf{j}_k - j),\qquad(21.21)$$

where $p_{j,k}^{ss}$ is the (thermal) population of the jth level of the kth subspace at inverse temperature β_{eff}.

The prefactor (21.21) evaluates to

$$F(\mathbf{j}_k) = \frac{\sum_{j=0}^{2\mathbf{j}_k-1} e^{-j\beta_{\text{eff}}\hbar\omega_0}(j+1)(2\mathbf{j}_k - j)}{\sum_{j=0}^{2\mathbf{j}_k} e^{-j\beta_{\text{eff}}\hbar\omega_0}}.\qquad(21.22)$$

As shown below, this prefactor expresses the collective enhancement of the energy currents.

The energy currents (21.18) account for the fact that the TLS in the decomposition (21.4) pertaining to different irreducible subspaces act as non-interacting WM in the HE. Their contributions are weighted by their initial populations that are conserved since the irreducible subspaces are dynamically invariant. The sign of each individual contribution (21.19) to the heat currents is determined by the difference between the Boltzmann factor of each bath [at temperature T_j and energy $\hbar(\omega_0 + q\Delta)$] and that of the TLS [at temperature $(k_B\beta_{\text{eff}})^{-1}$ and energy $\hbar\omega_0$].

The collective heat currents (21.19) can be cast into the simple form

$$\frac{\mathcal{J}_j(\mathbf{j}_k)}{\mathcal{J}_j\left(\frac{1}{2}\right)} = \frac{F(\mathbf{j}_k)}{F\left(\frac{1}{2}\right)},\qquad(21.23)$$

where the heat current $\mathcal{J}_j(\frac{1}{2})$ is induced by a single TLS. The ratio on the right-hand side of (21.23) does not depend on the bath index j. Hence, both heat currents $\mathcal{J}_h(\mathbf{j}_k)$ and $\mathcal{J}_c(\mathbf{j}_k)$ are equally amplified, and so is the power originating from the spin-$\mathbf{j}_k$ subspace with respect to the power generated by a single TLS,

$$\frac{\dot{W}(\mathbf{j}_k)}{\dot{W}\left(\frac{1}{2}\right)} = \frac{F(\mathbf{j}_k)}{F\left(\frac{1}{2}\right)}.\qquad(21.24)$$

21.4 Cooperative Power Enhancement

The initial condition ρ_0 can play a crucial role in cooperative many-body thermal machines. The maximal benefit from cooperativity is achievable for an HE initially prepared such that only the maximal cooperativity subspace associated with the largest possible spin-$N/2$ is populated, that is, $\langle \Pi_K \rangle_{\rho_0} = 1$ ($\mathbf{j}_K = \frac{N}{2}$). This condition corresponds to perfect constructive interference of the fields scattered by the TLS. This unique subspace comprises all N TLS being initially excited

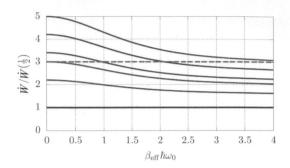

Figure 21.2 Ratio of the HE collective power for $N = 3$ TLS under different initial conditions, to the power generated by an HE based on a single TLS. The initial overlap with the spin-3/2 subspace is $\langle \Pi_1 \rangle_{\rho_0} = \{1, 0.8, 0.6, 0.5, 0.3, 0\}$ (from top to bottom). The horizontal dashed line is the power generated by three independent TLS. Power above this line indicates a cooperative boost, owing to constructive interference. Power below this line indicates suppression due to destructive interference. (Adapted from Niedenzu and Kurizki, 2018.)

or all being initially in the ground state, as per (21.6). The fully symmetric Dicke states spanning this subspace are eigenstates of the system Hamiltonian. Any initial condition within this subspace guarantees that in the steady state (21.12), the N TLS act as one collective spin-$N/2$ WM of the HE. The specific initial condition within this subspace does not affect the HE steady-state operation, which is a periodic limit cycle.

The collective power $\dot{W}_{\text{coll}} := \dot{W}\left(\frac{N}{2}\right)$ and its noncooperative counterpart $\dot{W} := N\dot{W}\left(\frac{1}{2}\right)$ established by N individual atoms (cf. Fig. 21.1) then fulfill the relation

$$\frac{\dot{W}_{\text{coll}}}{\dot{W}} = \frac{1}{N}\frac{F\left(\frac{N}{2}\right)}{F\left(\frac{1}{2}\right)}. \tag{21.25}$$

The same ratio (21.25) (Fig. 21.2) also holds for the ratio of the collective heat currents $(\mathcal{J}_j)_{\text{coll}} := \mathcal{J}_j(\frac{N}{2})$ to their individual counterpart $\mathcal{J}_j := N\mathcal{J}_j(\frac{1}{2})$ ($j = $ c, h). Hence, there is no collective effect on the engine efficiency since the ratio $\eta = -\dot{W}/\mathcal{J}_{\text{h}}$ coincides in the collective and individual cases, $-\dot{W}_{\text{coll}}/(\mathcal{J}_{\text{h}})_{\text{coll}} = -\dot{W}/\mathcal{J}_{\text{h}}$.

Equation (21.25) demonstrates that the cooperative power is a consequence of quantum coherence between the N TLS. Consequently, if this coherence is destroyed by dephasing (Chs. 11–14) all the TLS interact independently with the baths and the driving field. The power output of such an HE equals the power output of N individual HE with a single TLS WM.

The ratio (21.25) for a fixed N becomes, in the low-temperature limit,

$$\lim_{\beta_{\text{eff}}\hbar\omega_0 \to \infty} \frac{\dot{W}_{\text{coll}}}{\dot{W}} = 1. \tag{21.26}$$

For such low effective temperatures, both the individual TLS and the spin-$N/2$ (collective subspace) are mostly in their respective ground state [cf. (21.13)]. Since these states coincide, both give rise to the same heat current, $(\mathcal{J}_j)_{\text{coll}} = \mathcal{J}_j$, and thus to the same power.

By contrast, in the high effective-temperature limit $\beta_{\text{eff}}\hbar\omega_0 \to 0$ many levels of the spin-$N/2$ subspace become excited. The corresponding transition strengths between level $|j\rangle$ (carrying j excitations) and levels $|j \pm 1\rangle$ are enhanced (compared to the TLS case $\sigma_- = |g\rangle\langle e|$) by the matrix elements of the collective-spin lowering operator,

$$S_- = \sum_{j=0}^{N-1} \sqrt{(j+1)(N-j)}\,|j\rangle\langle j+1|. \tag{21.27}$$

This enhancement is also reflected by the amplification prefactor (21.21). The resulting power enhancement (21.25) is then

$$\lim_{\beta_{\text{eff}}\hbar\omega_0 \to 0} \frac{\dot{W}_{\text{coll}}}{\dot{W}} = \frac{N+2}{3}. \tag{21.28}$$

The scaling behavior $\dot{W}_{\text{coll}} \sim N^2 \dot{W}(\frac{1}{2})$ is thus established for sufficiently high effective temperatures, such that the spin-$N/2$ subspace is considerably excited. This N^2 scaling, similar to Dicke superradiance, is a direct consequence of the fact that full cooperativity among the TLS causes an inverted TLS ensemble to radiate in a short burst with peak intensity proportional to N^2.

By contrast, here we have steady-state or "persistent" superradiance: Quanta are continuously exchanged between the baths, the piston (driving field), and the TLS, which gives rise to collectively enhanced steady-state energy flows.

For a given value of β_{eff}, we find the saturation relation

$$\lim_{N \to \infty} \frac{\dot{W}_{\text{coll}}}{\dot{W}} = \coth\left(\frac{\beta_{\text{eff}}\hbar\omega_0}{2}\right). \tag{21.29}$$

As expected (Fig. 21.3), in the low effective-temperature limit $\beta_{\text{eff}}\hbar\omega_0 \gg 1$ the right-hand side of (21.29) tends to unity, so that even for large N there is no power enhancement compared to the individual-TLS case. By contrast, in the high-temperature limit, $\beta_{\text{eff}}\hbar\omega_0 \to 0$, the right-hand side of (21.29) diverges as $2(\beta_{\text{eff}}\hbar\omega_0)^{-1}$, so that collective effects significantly enhance the power (Fig. 21.4). The saturation value (21.29) is thus sensitive to the value of $\beta_{\text{eff}}\hbar\omega_0$: There is always a maximum (saturation) N value above which the power is not further amplified. Consequently, the ideal superradiant scaling (21.28) for all N is only achieved in the infinite-temperature limit $\beta_{\text{eff}}\hbar\omega_0 \to 0$.

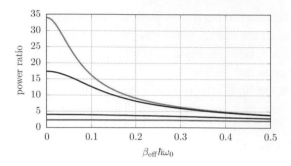

Figure 21.3 Ratio (21.25) of the HE power generated by N collective TLS to the power generated by N individual TLS as a function of the inverse effective temperature (21.15) and the TLS number. $N = \{5, 10, 50, 100\}$ (from bottom to top). The maximum enhancement factor $(N + 2)/3$ is attained for $\beta_{\text{eff}} \hbar \omega_0 \rightarrow 0$ [Eq. (21.28)]. (Adapted from Niedenzu and Kurizki, 2018.)

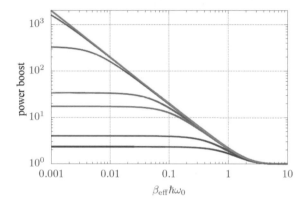

Figure 21.4 Power boost (21.25) for different TLS numbers $N=\{5, 10, 50, 10^2, 10^3, 10^4, \infty\}$ (from bottom to top). $N = \infty$ corresponds to the maximum (saturation) power boost (21.29) for a given $\beta_{\text{eff}} \hbar \omega_0$. The saturation value (21.29) cannot be surpassed even by adding more TLS: For $\beta_{\text{eff}} \hbar \omega_0 = 0.2$, for example, the power boost can only be a factor of 10 even if there were thousands of cooperative TLS. (Adapted from Niedenzu and Kurizki, 2018.)

21.4.1 Cooperative Power Boost under Sinusoidal Modulation

As an example, let us consider the sinusoidal frequency modulation

$$\omega(t) = \omega_0 + \lambda \sin(\Delta t) \tag{21.30}$$

in the Hamiltonian (21.1a) under the condition $0 \leq \lambda \ll \Delta \leq \omega_0$. The weights P_q in the ME (21.7) then evaluate to

$$P_0 \simeq 1 - \frac{1}{2}\left(\frac{\lambda}{\Delta}\right)^2, \tag{21.31a}$$

$$P_{\pm 1} \simeq \left(\frac{\lambda}{2\Delta} \right)^2. \tag{21.31b}$$

The higher Floquet sidebands $|q| > 1$ do not contribute significantly and can be neglected.

HE operation requires the quanta that are exchanged by the TLS with the hot bath to carry more energy than those exchanged with the cold bath. This requirement can be satisfied by spectrally separated baths, for example, by imposing the condition

$$G_c(\omega) \approx 0 \quad \text{for} \quad \omega \geq \omega_0, \tag{21.32a}$$

$$G_h(\omega) \approx 0 \quad \text{for} \quad \omega \leq \omega_0. \tag{21.32b}$$

The heat currents (21.19) then evaluate to

$$\mathcal{J}_c(j_k) = F(j_k)(\omega_0 - \Delta) \left(\frac{\lambda}{2\Delta} \right)^2 G_c(\omega_0 - \Delta) \left[e^{-\beta_c \hbar (\omega_0 - \Delta)} - e^{-\beta_{\text{eff}} \hbar \omega_0} \right], \tag{21.33a}$$

$$\mathcal{J}_h(j_k) = F(j_k)(\omega_0 + \Delta) \left(\frac{\lambda}{2\Delta} \right)^2 G_h(\omega_0 + \Delta) \left[e^{-\beta_h \hbar (\omega_0 + \Delta)} - e^{-\beta_{\text{eff}} \hbar \omega_0} \right], \tag{21.33b}$$

$$\dot{W}(j_k) = - [\mathcal{J}_c(j_k) + \mathcal{J}_c(j_k)], \tag{21.33c}$$

where, according to Eq. (21.15),

$$\exp(-\beta_{\text{eff}} \hbar \omega_0) = \frac{G_c(\omega_0 - \Delta) e^{-\beta_c \hbar (\omega_0 - \Delta)} + G_h(\omega_0 + \Delta) e^{-\beta_h \hbar (\omega_0 + \Delta)}}{G_c(\omega_0 - \Delta) + G_h(\omega_0 + \Delta)}. \tag{21.34}$$

The difference of the Boltzmann factors in (21.33) determines the signs of the respective heat currents, that is, whether the machine acts as an HE ($\mathcal{J}_c < 0$, $\mathcal{J}_h > 0$, $\dot{W} < 0$) or a refrigerator ($\mathcal{J}_c > 0$, $\mathcal{J}_h < 0$, $\dot{W} > 0$) (Ch. 23). The crossover from HE to refrigerator operation (Fig. 21.5) occurs when the energy currents (21.33) vanish at the same inverse critical temperatures as in the single-TLS case (Ch. 20),

$$\beta_h^{\text{crit}} = \frac{\omega_0 - \Delta}{\omega_0 + \Delta} \beta_c. \tag{21.35}$$

These results imply that the largest possible power boost (21.28) caused by a cooperative quantum effect requires the two bath temperatures to be similar and sufficiently large compared to ω_0: $\beta_c \hbar \omega_0 \ll 1$ and $\beta_h \hbar \omega_0 \ll 1$ such that $\beta_{\text{eff}} \hbar \omega_0 \to 0$. The steady state (21.13) of the spin-$N/2$ subspace is then the *maximum-entropy state with equally populated energy levels*. The energy currents may, however, be small despite the superradiant boost (Fig. 21.6).

If, however, the cold-bath temperature is fixed and the task is to maximize the power, irrespective of how large the cooperative boost is, then it is more favorable

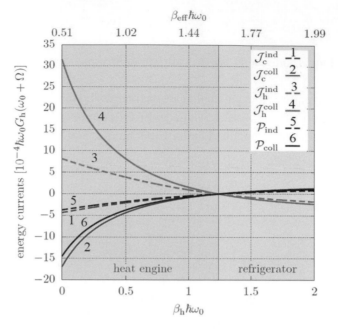

Figure 21.5 Energy currents generated by $N = 100$ individual TLS (dashed) and collective TLS (solid) as a function of $\beta_h \hbar \omega_0$ under the conditions (21.31) and (21.32). The maximum power boost of ≈ 4 is attained for a high temperature of the hot bath ($\beta_{\text{eff}} \hbar \omega_0 \approx 0.51$). This maximum boost cannot be increased by adding more TLS since the saturation value (21.29) also evaluates to 4 (cf. Fig. 21.4). Parameters: $\exp(-\beta_c \hbar \omega_0) = 0.1$ ($\beta_c \hbar \omega_0 = 2.3$), $\Delta = 0.3\omega_0$, $G_c(\omega_0 - \Delta) = G_h(\omega_0 + \Delta)$, $\lambda = 0.01\Delta$. (Adapted from Niedenzu and Kurizki, 2018.)

to simply increase the hot-bath temperature as much as possible so as to generate the largest possible temperature difference, as for any HE, even though the cooperative boost may then be reduced but still present (Fig. 21.5).

For example, the parameters in Figure 21.6 allow for very small values of $\beta_{\text{eff}} \hbar \omega_0$, but the maximum enhancement (21.29) still saturates to roughly 56, which may be much lower than the maximum enhancement $(N + 2)/3$ that is attained in the limit $\beta_{\text{eff}} \hbar \omega_0 \to 0$ of (21.28). The latter limit requires both baths to have comparable temperatures much higher than the TLS transition frequency ω_0, but then the power decreases significantly compared to the case of distinct temperatures shown in Figure 21.5.

Hence, there is a trade-off between a large cooperative boost and a large power output, since the maximum boost (21.28) is extremely sensitive to the bath temperatures (see Figs. 21.3 and 21.4, where for small $\beta_{\text{eff}} \hbar \omega_0$ and large N the boost increases very steeply).

21.5 Discussion

We have investigated the impact of quantum cooperative effects on HE operation. In suitable geometries and for a carefully chosen initial condition, N TLS may

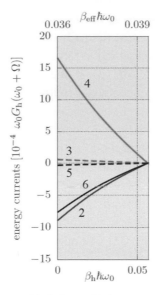

Figure 21.6 Same as Figure 21.5 for a higher temperature of the cold bath, $\exp(-\beta_c\hbar\omega_0) = 0.9$ (corresponding to $\beta_c\hbar\omega_0 = 0.11$). Here the minimal value of $\beta_{\mathrm{eff}}\hbar\omega_0$ is 0.036, which results in a power boost of ≈ 28. Contrary to Figure 21.5, there is an advantage to adding more TLS since here the saturation value (21.29) evaluates to 56 (Fig. 21.4). Owing to the smaller difference between the two bath temperatures, the power is less than in Figure 21.5, yet the cooperative power boost is larger. Note the different scaling of the axes compared to Figure 21.5. (Adapted from Niedenzu and Kurizki, 2018.)

behave as a giant collective spin-$N/2$ system. The HE based on N TLS then acts as based on a single spin $N/2$-system. Although there is coherence (entanglement) in the N-TLS state among TLS, there is no coherence in the energy basis of the spin-$N/2$ system: off-diagonal elements in this basis would only reduce the power output. This behavior leads to enhanced energy currents and to a non-extensive scaling of the power output. Namely, the power generated by an HE that involves N TLS in the collective regime can greatly surpass the power generated by N individual engines, each based on a single TLS WM.

The collective behavior that results in the enhancement of the steady-state energy currents is a consequence of the indistinguishability of the TLS with respect to the two baths, which implies an SU(2) symmetry and the existence of invariant irreducible subspaces. The coupling Hamiltonian (21.1c) thus describes the optimal (ideal) situation. However, its strict symmetry is not necessarily required to obtain a symmetry-preserving Markovian ME of the form (21.7). In order to clarify this point, let us consider the interaction of N two-level atoms with the surrounding EM bath in the dipole approximation (Ch. 8),

$$H_{\mathrm{SB}} = -\sum_{i=1}^{N}\sum_{\mathbf{k}} g_{\mathbf{k}}\sigma_x \otimes \left(b_{\mathbf{k}}e^{i\mathbf{k}\cdot\mathbf{x}_i} + b_{\mathbf{k}}^{\dagger}e^{-i\mathbf{k}\cdot\mathbf{x}_i}\right), \qquad (21.36)$$

where $\mathbf{x}_i$ is the ith atom position, $g_{\mathbf{k}}$ the atom–EM-mode coupling, and $b_{\mathbf{k}}$ annihilates a photon in mode $\mathbf{k}$ of the EM field. Owing to the position-dependent phases, the operator (21.36) cannot be cast in the form (21.3b). However, in the Markovian approximation that leads to the ME (21.7), the atoms effectively interact only with the resonant field mode at frequency ω_0 (without modulation) or at frequencies $\omega_0 + q\Delta$ with modulation.

In one-dimensional geometries, the atoms can effectively be indistinguishable with respect to the resonant field mode if they are placed at distances of integer multiples of the resonance wavelength $2\pi c/\omega_0$, such that all atoms scatter in phase. Since, however, the atoms not only exchange photons with the bath at the frequency ω_0 but also at the Floquet sidebands, we must additionally require $2\pi c/(\omega_0 + q\Delta) \approx 2\pi c/\omega_0$ for the relevant $q = \pm1$ in Section 11.4. Hence, in the Markovian approximation, the Hamiltonian (21.36) may effectively be replaced by the fully symmetric Hamiltonian (21.1c).

Superradiance under the interaction (21.36) in three dimensions for atoms confined within a volume much smaller than the cubed wavelength encounters the complication that the bath-induced dipole–dipole interaction diverges (Ch. 8). By contrast, it has been theoretically and experimentally demonstrated that in one dimension maximal collective coupling to the bath can be achieved while entirely suppressing the dipole–dipole interaction by placing the atoms in a chain at distances $d = 2\pi c/\omega_0$ or integer multiples thereof.

The general requirement that the atoms should appear identical to the bath at all Floquet sidebands can be lifted by imposing a spectral separation of the two baths. For the modulation (21.30) and the spectral separation (21.32) the two baths may be formed by two overlapping low-finesse cavities. In either scenario, the baths must have different temperatures T_c and T_h and mode frequencies resonant with $\omega_0 \pm \Delta$. Another scenario may involve two commensurate cavity modes at different temperatures where the free spectral range of the cavity is 2Δ, similar to the two-mode Tavis–Cummings model (Ch. 6). The feasibility of cooperative strong coupling in multimode cavities has been experimentally demonstrated for superconducting qubits (in microwave cavities) and for atoms in a photonic crystal waveguide.

The boosted energy currents are due to steady-state ("persistent") superradiance. Whereas the power generated by an HE that involves N individual TLS as its WM essentially does not involve quantum-interference effects, the collective power generated under constructive-interference conditions is a genuine quantum many-body effect that involves entanglement among the TLS in the WM.

22

Heat-to-Work Conversion in Fully Quantized Machines

The simple models of QHMs analyzed in Chapters 18-20 are not fully quantum mechanical but rather semiclassical, since the WM modulation or drive (the piston) is generated by an external classical field.

Here we analyze fully quantized HE and their compliance with thermodynamic laws. As shown, the nonpassivity of the piston quantum states allows to classify them according to their contribution to work and power extraction.

Under nonlinear (quadratic) pumping, the piston mode evolves into a squeezed thermal state that strongly enhances its work capacity (ergotropy) and therefore the output power, whereas its efficiency is still limited by the Carnot bound. This enhanced power output may be viewed as *catalysis*, whereby a small amount of catalyst (here a weak pump) strongly enhances the heat-to-work conversion.

Lasers and masers may be viewed as heat engines that rely on population inversion or coherence in the WM. Here we introduce a simple electromagnetic heat-powered engine that bears basic differences to existing masers or lasers in that it does not rely on population inversion or coherence in its two-level WM nor does it require any external driving of the piston (signal) mode.

22.1 Principles of Work Extraction in Fully Quantized HE

For a fully quantized heat machine with a quantum piston instead of an external modulation (Fig. 22.1), the total Hamiltonian is

$$H_{\text{tot}} = H_{\text{S}} + H_{\text{P}} + H_{\text{SP}} + \sum_{j=c,h} (H_{\text{Bj}} + H_{\text{SBj}}). \qquad (22.1)$$

The WM Hamiltonian H_{S} is now time independent. The (free) Hamiltonian of the piston (P), H_{P}, and the interaction Hamiltonian between the WM, S, and the piston, H_{SP}, have been added to the Hamiltonian in Chapters 19 and 20.

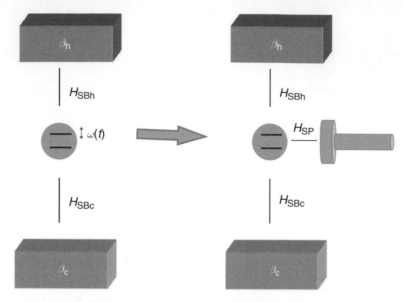

Figure 22.1 The driving field (left) in the semiclassical model of Chapter 20 is replaced in a fully quantized heat machine by a quantized piston (right). (Reprinted from Gelbwaser-Klimovsky et al., 2015.)

In semiclassical models (Chs. 17–21), the power [(19.48)], all the energy interchanged with the external field (piston), is considered to be work. If the piston is quantized, however, this may lead to contradictions with the second law. As an example, suppose that a piston is in a thermal state throughout its evolution, but with an increasing temperature. The piston gains energy, but since only one bath would suffice for that, this energy cannot be considered to be work; otherwise, we might extract work from a single bath, in violation of Lord Kelvin's version of the second law. This shows that work in a fully quantized setup must be carefully analyzed to avoid spurious violations of the second law.

The presumption that $\frac{d\langle H_P \rangle}{dt}$ represents the rate of work production is shown here to fail for a quantized P, for which this rate is a *combination of heat production rate and power (work-production rate)*. The ratio of heat- to work-production rates strongly depends on the evolving piston state $\rho_P(t)$ that, in turn, determines $\langle H_P(t) \rangle$.

Under weak system–bath coupling, $\langle H_P(t) \rangle$ undergoes quasi-cyclic, slowly-drifting evolution, treated here in the Markovian approximation.

The maximum extractable work from ρ_P is given by its ergotropy

$$\mathcal{W}(\rho_P) = \langle H_P(\rho_P) \rangle - \langle H_P(\pi_P) \rangle. \tag{22.2}$$

Here π_P is the passive state that corresponds to ρ_P (Ch. 15). The two states are related by the unitary transformation

$$\rho_P \rightarrow U\rho_P U^\dagger = \pi_P. \tag{22.3}$$

This transformation maximizes the energy change while preserving the entropy, thereby maximizing the *extractable* work (ergotropy) $\mathcal{W}$, since $\mathcal{W}(\pi_P) = 0$, by definition. Equivalently, the *passive state* π_P is the state with the least energy that is accessible from ρ_P, while keeping the Hamiltonian constant.

Let the piston be initially prepared in a nonpassive state $\rho_P(0)$ with ergotropy $\mathcal{W}(\rho_P(0))$. The QHM evolves the state of the piston so that at time t_m it is in the state $\rho_P(t_m)$ with ergotropy $\mathcal{W}(\rho_P(t_m))$. The work extracted by the piston is then the increase in the piston ergotropy,

$$W_{\text{ext}} = \mathcal{W}(\rho_P(t_m)) - \mathcal{W}(\rho_P(0)). \tag{22.4}$$

An HE functions properly if the extractable work $W_{\text{ext}} > 0$.

The passive state π_P corresponding to the Gaussian ρ_P is the thermal (Gibbs) state

$$\pi_P(t) = Z^{-1} e^{-\frac{H_P}{k_B T_P(t)}}, \tag{22.5}$$

since a Gibbs state has the minimal energy at a given entropy. For such ρ_P, we may identify π_P with a Gibbs state associated with a time-dependent temperature $T_P(t)$. One can calculate the maximum extractable power as follows:

$$\left(\frac{dW_{\text{ext}}}{dt}\right)_{\text{max}} = \frac{d}{dt}\langle H_P\rangle_{\rho_P} - k_B T_P \frac{d}{dt}\mathcal{S}_P. \tag{22.6}$$

We may conclude from (22.6) that the smaller the entropy production by a state is, the better this state is suited for work extraction. Resilience against thermalization or heat-up is thus the relevant property for work extraction.

Equation (22.6) clarifies that the energy interchanged by the piston is not entirely work – part of it involves an entropy change, heating up the piston. If the piston is in a thermal (Gibbs) state, then

$$\frac{d}{dt}\langle H_P\rangle_{\rho_P} = k_B T_P \frac{d}{dt}\mathcal{S}_P \implies \left(\frac{dW}{dt}\right)_{\text{max}} = 0. \tag{22.7}$$

Thus, a thermal state is unable to store and deliver work.

The *bound* for the total entropy-production rate of S+P may be made tighter than that involved by the Clausius version of the second law or Spohn's inequality (Ch. 15) by associating entropy change with the passive state of S+P (Ch. 18). Assuming a small ratio of the S-P coupling strength g to the piston resonant frequency Δ, so that the system and the piston states are nearly uncorrelated,

$$\rho_{S+P} = \rho_S \otimes \rho_P + O\left(\frac{g}{\Delta}\right)^2, \tag{22.8}$$

their entropy production is essentially additive:

$$\dot{S}_{S+P} = \dot{S}_S + \dot{S}_P + O\left(\frac{g}{\Delta}\right)^4. \tag{22.9}$$

After coarse-graining, $\dot{S}_S = 0$ in the steady state and the only entropy production is that of the piston, $\dot{S}_P$, which satisfies (Ch. 18)

$$\dot{S}_P \geq \frac{\mathcal{J}_h}{k_B T_h} + \frac{\mathcal{J}_c}{k_B T_c}, \tag{22.10}$$

provided the heat currents are associated with the change in the passive state that corresponds to the evolving $\rho_P(t)$. In what follows, this inequality will be used to infer efficiency bounds that allow for entropy and work production by P.

The energy stored in the WM grows exponentially in the linear-gain regime and is state-independent so that even an initial passive (e.g., thermal) state may be amplified in energy. This is not the case for ergotropy, which is highly sensitive to the initial quantum state and may grow only for nonpassive states.

If the piston is initially in a passive state, it will not extract work in the course of the HE action: as an analysis of its Markovian evolution in Section 22.2 shows, the evolving piston state does not increase its ergotropy. Yet, an initially passive piston state can be "ignited" by a displacement in the phase plane away from zero energy, which renders the state nonpassive and thereby allows for work extraction (Sec. 22.2). Another route to nonpassivity from an initially passive state is via a homodyne measurement followed by feedforward control that extracts work (Ch. 17). The most "classical" among such nonpassive states is the coherent state. Remarkably, even if its phase is averaged out, it still retains a higher capacity for work extraction (ergotropy) than other, nonclassical, states.

22.2 Two-Level Quantum Amplifier (Laser) as Heat Engine

Here we discuss a remarkably simple, heat-powered maser or laser. The common view, consistently with the Scovil and Schulz-DuBois (SSD) model (discussed below), is that at least two transitions are needed for a maser/laser, one of which must be population-inverted [Fig. 22.2(a)]. However, the setup considered here uses only one (two-photon, i.e., Raman) transition, and its WM is well approximated by uninverted TLSs. The work output of this setup is the useful (noiseless) portion of the amplified signal (piston state). The output signal cannot be fully coherent, which hampers the efficiency and the power production.

The basic ingredients of the setup are [Fig. 22.2(b, c)] (i) hot and cold heat baths, realized by ambient thermal radiation that is filtered into spectrally distinct modes

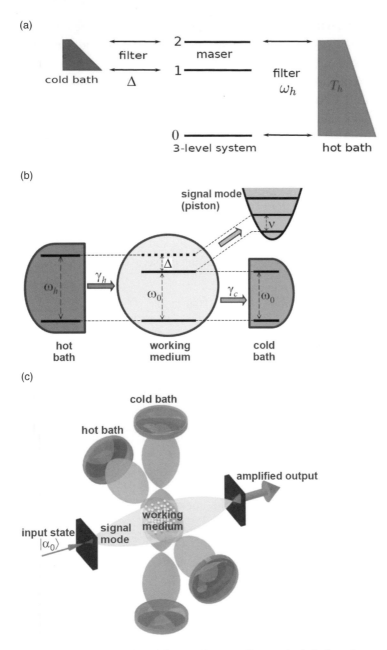

Figure 22.2 (a) The three-level SSD scheme relies on bath-induced population inversion between levels 2 and 1 of the WM to amplify a signal that is resonant with the 2 – 1 transition. This scheme is principally different from the two-level scheme in (b), where population inversion exists between the baths, not the WM levels. (b) Two-level HE acting as a laser/maser. Two narrow-band cavities spectrally filter the hot and cold baths. The TLSs with resonant frequency $\omega_a \equiv \omega_0$ interact with an off-resonant signal (piston) at frequency Δ. (c) An outline of the setup. (Adapted from Ghosh et al., 2018b.)

of narrow-band cavities; (ii) a WM consisting of TLSs that are continuously (as in Chs. 19 and 20) coupled to the hot and cold baths (cavity modes); and (iii) a signal mode that acts as a "piston" that extracts the work.

The two levels of the WM are coupled to the cold bath near a frequency ω_a and (via a two-photon transition) to the hot bath, near the frequency ω_h as well as to the signal (piston) at $\Delta = \omega_h - \omega_a$. The heat-to-work conversion consists in photon absorption from the hot bath, reemission into the cold bath, and production of a signal photon at the difference frequency.

22.2.1 The Hamiltonian Derivation

The system–bath (S-B) coupling, H_{SB}, has the spin-boson form. The TLS coupling to the harmonic-oscillator (HO) piston (P) is represented by H_{S+P}. A Hamiltonian of this kind that can yield work can be written as

$$H = H_{S+P} + H_{SB} + \sum_{j=h,c} H_{B_j}. \tag{22.11}$$

Here the S-P Hamiltonian has the form

$$H_{S+P} = \frac{\hbar}{2}\omega_a\sigma_z + \hbar\Delta a_P^\dagger a_P + H_{SP}, \tag{22.12}$$

and

$$H_{SB} = \hbar\sigma_x \otimes (B_c + B_h). \tag{22.13}$$

The σ_x-coupling in H_{SB} allows for work extraction from the bath via S. By contrast, $\sigma_z(B_c+B_h)$ coupling may only cause dephasing (Chs. 11–14) rather than contribute to work, because it commutes with the energy of S and therefore cannot transfer energy (from the bath via S) into P.

A spin-boson Hamiltonian of the interaction between the piston and a TLS, H_{SP}, may be taken to have the so-called optomechanical form

$$H_{SP} = \hbar g\sigma_z \otimes (a_P + a_P^\dagger), \tag{22.14}$$

g being the coupling strength and a_P, $a_P^\dagger$, respectively, the annihilation and creation operators of the harmonic oscillator P. This Hamiltonian holds, provided the P-S coupling is *dispersive*, so that P does not induce transitions in the TLS. P here is assumed not to have direct interactions with the heat baths, although it is affected by the baths via (22.13), as shown below.

The analysis of the model (22.14) is simplified by using a new set of canonical operators obtained from a_P, $a_P^\dagger$, σ_k by a unitary *dressing transformation* that diagonalizes the Hamiltonian.

The transformation

$$a_{\mathrm{P}} \mapsto b = U^\dagger a_{\mathrm{P}} U, \quad \sigma_k \mapsto \tilde{\sigma}_k = U^\dagger \sigma_k U, \tag{22.15}$$

where

$$U = e^{\frac{g}{2\Delta}(a_{\mathrm{P}}^\dagger - a_{\mathrm{P}})\sigma_z}, \tag{22.16}$$

diagonalizes the Hamiltonian in (22.12), (22.14) to the form

$$H = H_{\mathrm{S}} + H_{\mathrm{P}} \tag{22.17}$$

with

$$H_{\mathrm{S}} = \frac{\hbar}{2}\omega_{\mathrm{a}}\sigma_z, \quad H_{\mathrm{P}} = \hbar\Delta b^\dagger b - \hbar\left(\frac{g}{2}\right)^2 \frac{1}{\Delta}. \tag{22.18}$$

The Pauli matrix σ_x in H_{SB} (22.11) is given in terms of the new dynamical variables as

$$\sigma_x = \tilde{\sigma}_+ e^{\frac{g}{\Delta}(b^\dagger - b)} + e^{-\frac{g}{\Delta}(b^\dagger - b)}\tilde{\sigma}_-. \tag{22.19}$$

The Heisenberg-picture Fourier decomposition of σ_x to lowest order in g/Δ, can be obtained in the form

$$\sigma_+(t) = e^{iHt/\hbar}\sigma_+ e^{-iHt/\hbar} = e^{i\omega_{\mathrm{a}}t}\tilde{\sigma}_+ e^{\frac{g}{\Delta}(b^\dagger e^{i\Delta t} - b e^{-i\Delta t})} \approx$$
$$\tilde{\sigma}_+ e^{i\omega_{\mathrm{a}}t} + \frac{g}{\Delta}\left(S_{+1}^\dagger e^{i(\omega_{\mathrm{a}}+\Delta)t} - S_{-1}^\dagger e^{i(\omega_{\mathrm{a}}-\Delta)t}\right), \tag{22.20}$$

where the transformed S+P raising and lowering operators are

$$S_{+1}^\dagger = \tilde{\sigma}_+ b^\dagger, \quad S_{-1}^\dagger = \tilde{\sigma}_+ b. \tag{22.21}$$

The approximation made in (22.20) is valid for low excitations of the piston, which will be shown to correspond to its linear amplification or dissipation

$$\frac{g}{\Delta}\sqrt{\langle b^\dagger b \rangle} = g\sqrt{\frac{\langle H_{\mathrm{P}} \rangle}{\hbar\Delta^3}} \ll 1. \tag{22.22}$$

Since the σ_x operator that couples the system to the bath in (22.11), is mixed, according to (22.19) and (22.20), with $\tilde{\sigma}_+ b^\dagger$ and $\tilde{\sigma}_+ b$, H_{P} in (22.18) is affected by the system–bath coupling.

Here we focus on the case where the WM $|e\rangle \leftrightarrow |g\rangle$ transition at frequency ω_{a} is near-resonant with a narrow-band cold bath (c) near frequency ω_{c}. The two-photon (Raman) resonance condition is assumed to be satisfied by the signal (piston) mode at frequency Δ combined with the hot-bath (h) modes near frequency ω_{h} [Fig. 22.2(b)],

$$\omega_{\mathrm{a}} \simeq \omega_{\mathrm{c}}, \quad \Delta \simeq \omega_{\mathrm{h}} - \omega_{\mathrm{c}}. \tag{22.23}$$

We next consider another Hamiltonian that couples S, P, and B

$$H_{(S+P)B} \equiv \hbar V = \hbar \sum_k g_{ug,k} a_k |u\rangle \langle g| + \hbar g_{eu} a_P^\dagger |e\rangle \langle u| + \text{H.c.} \tag{22.24}$$

Here we adopt the RWA and denote a virtual level that enables the Raman transition by $|u\rangle$ and its energy by E_u, and the corresponding dipolar couplings by $g_{ug,k}$ and g_{eu}, respectively, a_k being the annihilation operator of the hot bath. The hot-bath states have the narrow-band frequencies $\omega_k \simeq \omega_h$, wave vectors k, and occupation numbers n_k; Δ-frequency piston mode has occupation number n_P; $|g, n_k, n_P\rangle = |\Psi_i\rangle$ is the initial state, $|e, n_k - 1, n_P + 1\rangle = |\Psi_f\rangle$ is the final state, and they are coupled by a Raman process. The probability amplitude for the Raman transition $|\Psi_i\rangle \rightarrow |\Psi_f\rangle$ after a time t is given, to second order in the interaction V, by

$$\mathcal{A}_{i \rightarrow f} \simeq - \int_0^t dt_1 \int_0^{t_1} dt_2 \langle \Psi_f | V(t_1) V(t_2) | \Psi_i \rangle, \tag{22.25}$$

since $\langle \Psi_f | \Psi_i \rangle = \langle \Psi_f | V | \Psi_i \rangle = 0$. In (22.25), $V(t)$ is the operator V in the interaction representation. Explicitly,

$$
\begin{aligned}
\mathcal{A}_{i \rightarrow f} &\simeq - \sum_k \int_0^t dt_1 \int_0^{t_1} dt_2 g_{eu} g_{ug,k} \big\langle n_k - 1, n_P + 1 \big| e^{it_1(E_e - E_u)} \\
&\quad \times a_P^\dagger e^{i\Delta t_1} e^{it_2(E_u - E_g)} a_k e^{-i\omega_k t_2} \big| n_k, n_P \big\rangle \\
&= - \sum_k g_{eu} g_{ug,k} \frac{-i\sqrt{n_k}\sqrt{n_P + 1}}{i(E_u - E_g - \omega_k)} \left\{ \frac{e^{it[(E_e - E_g) - (\omega_k - \Delta)]} - 1}{(E_e - E_g) - (\omega_k - \Delta)} \right. \\
&\quad \left. - \frac{e^{it(E_e - E_u + \Delta)} - 1}{E_e - E_u + \Delta} \right\},
\end{aligned}
\tag{22.26}
$$

where the level energies satisfy $E_u > E_e > E_g$, $0 < \Delta < \omega_k$, and the frequencies are nonresonant (with single-photon transitions). In the limit $t \rightarrow \infty$, this transition amplitude reduces to

$$\mathcal{A}_{i \rightarrow f} = 2\pi i \sum_k g_{eu} g_{ug,k} \frac{\delta[(E_e - E_g) - (\omega_k - \Delta)]}{(E_u - E_g - \omega_k)} \sqrt{n_k}\sqrt{n_P + 1}. \tag{22.27}$$

The same transition amplitude is obtainable from first-order perturbation theory (Fig. 22.3), using the two-photon Raman Hamiltonian

$$V_R(t) = \sum_k \left[g_{Rk} |g\rangle \langle e| a_P^\dagger a_k e^{-i(\omega_k - \omega_a - \Delta)t} + \text{H.c.} \right], \tag{22.28}$$

with the Raman coupling

$$g_{Rk} = 2\pi \hbar \frac{g_{eu} g_{ug,k}}{E_u - E_g - \hbar \omega_h}, \tag{22.29}$$

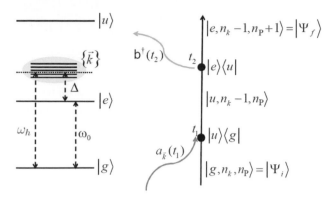

Figure 22.3 Raman-coupling Hamiltonian (Reprinted from Ghosh et al., 2018b.)

taken to be real and $\omega_k \approx \omega_h$, approximated near the two-photon Raman resonance. This Raman Hamiltonian is analogous to the spin-boson Hamiltonian (22.11)–(22.14) that couples S, P, and B, as can be seen from the analogous form of the operators in (22.28) and in (22.20), (22.21).

22.2.2 Piston Evolution via the Lindblad ME

The energy exchanged between the system and the hot (h) and cold (c) baths is compensated by the energy exchanged between the system and the piston. The underlying bath-induced dynamics is described by the Lindblad generator, which adheres to the second law.

The corresponding master equation (ME) for the state of S+P is

$$\frac{d\rho_{S+P}(t)}{dt} = \sum_{q=0,\pm 1} (\mathcal{L}_{q,h} + \mathcal{L}_{q,c})\rho_{S+P}(t). \tag{22.30}$$

Here $q = 0, \pm 1$ labels the harmonics ω_a and, respectively,

$$\omega_\pm = \omega_a \pm \Delta. \tag{22.31}$$

The generators associated with these harmonics in the two baths, $\mathcal{L}_{qj}$ (j=h,c), have the following Lindblad form [upon denoting the bath-response rates by $G_j(\omega_q)$ and setting $\rho \equiv \rho_{S+P}$],

$$\mathcal{L}_{0,j}\rho = \frac{1}{2}\Big\{ G_j(\omega_a)\big([\tilde{\sigma}_-\rho, \tilde{\sigma}_+] + [\tilde{\sigma}_-, \rho\tilde{\sigma}_+]\big) + $$
$$ G_j(-\omega_0)\big([\tilde{\sigma}_+\rho, \tilde{\sigma}_-] + [\tilde{\sigma}_+, \rho\tilde{\sigma}_-]\big)\Big\}, \tag{22.32}$$

$$\mathcal{L}_{q,j}\rho = \frac{g^2}{2\Delta^2}\Big\{G_j(\omega_q)\big([S_q\rho, S_q^\dagger] + [S_q, \rho S_q^\dagger]\big) +$$
$$G_j(-\omega_q)\big([S_q^\dagger\rho, S_q] + [S_q^\dagger, \rho S_q]\big)\Big\}, \quad q = \pm1. \qquad (22.33)$$

Under this bath-induced dynamics, the reduced density matrix $\rho_{S+P}(t)$ (obtained by tracing out the baths) allows us to compute the heat currents $\mathcal{J}_j$ from the expression

$$\mathcal{J}_j = \dot{Q}_j = \sum_{q=0\pm1} \text{Tr}\big(H_{S+P}\mathcal{L}_{q,j}\rho_{S+P}\big) \quad (j = \text{c,h}). \qquad (22.34)$$

The bath-induced dynamics of S+P has two distinct timescales for $\frac{g}{\Delta} \ll 1$: the fast one is related to the change of the TLS state populations and the slow one to change of the piston state. The ratio of the ρ_{ee}/ρ_{gg} TLS populations quickly equilibrates for each element of the piston density matrix $\rho_P = \text{Tr}_S \rho_{S+P}$.

Then upon tracing-out the TLS, one obtains the Lindblad ME for the slowly-changing harmonic oscillator (HO) that represents the piston transformed according to (22.15), described by the annihilation and creation operators b, $b^\dagger$, respectively. It has the form

$$\dot{\rho}_P = \frac{\Gamma+D}{2}\big([b, \rho_P b^\dagger] + [b\rho_P, b^\dagger]\big) + \frac{D}{2}\big([b^\dagger, \rho_P b] + [b^\dagger\rho_P, b]\big), \qquad (22.35)$$

where

$$\Gamma = \left(\frac{g}{\Delta}\right)^2 \{[G(\omega_h) - G(\omega_-)]\rho_{ee} + [G(-\omega_-) - G(-\omega_h)]\rho_{gg}\},$$
$$D = \left(\frac{g}{\Delta}\right)^2 [G(\omega_-)\rho_{ee} + G(-\omega_h)\rho_{gg}]. \qquad (22.36)$$

Here Γ and D are the drift and diffusion rates, respectively, and ω_- is defined in (22.31). They depend, according to (22.36), on the sum of the cold- and hot-baths response spectra $G(\omega) = G_c(\omega) + G_h(\omega)$, sampled at the appropriate combination (cycle) frequencies $\omega_\pm$ for the S-P coupling Hamiltonian H_{S+P}.

To investigate the dependence of work on the state of P in this model we let S reach *steady-state*. In the coherent-state basis,

$$\rho_P = \int d^2\alpha\, \mathbf{P}(\alpha, \alpha^*)|\alpha\rangle\langle\alpha|, \qquad (22.37)$$

where $\mathbf{P}(\alpha, \alpha^*)$ is the phase-space (quasiprobability) distribution, the ME (22.35) for ρ_P is a Fokker–Planck (FP) equation,

$$\frac{\partial \mathbf{P}}{\partial t} = \frac{\Gamma}{2}\left(\frac{\partial}{\partial\alpha}\alpha + \frac{\partial}{\partial\alpha^*}\alpha^*\right)\mathbf{P} + D\frac{\partial^2\mathbf{P}}{\partial\alpha\partial\alpha^*}. \qquad (22.38)$$

Here $\mathbf{P}(\alpha, \alpha^*)$ is *any distribution*.

Under the conditions specified above, the WM interacts mainly with the near-resonant cold bath at temperature T_c. It therefore attains a steady state whose upper- and lower-level populations ρ_{ee} and ρ_{gg} thermally equilibrated at T_c:

$$\frac{\rho_{ee}}{\rho_{gg}} \simeq \frac{\bar{n}_c}{\bar{n}_c + 1}, \tag{22.39}$$

where the photon occupancy is

$$\bar{n}_j = \left[\exp\left(\frac{\hbar\omega_j}{k_B T_j}\right) - 1\right]^{-1} \quad (j = \text{c,h}). \tag{22.40}$$

The slow evolution of the piston (signal) field as a mediator of the interaction between the WM and the hot bath obeys the ME (22.35) in the form

$$\dot{\rho}_P = \gamma_h(\bar{n}_h + 1)\rho_{ee}([b\rho_P, b^\dagger] + [b, \rho_P b^\dagger]) \\ + \gamma_h \bar{n}_h \rho_{gg}([b^\dagger \rho_P, b] + [b^\dagger, \rho_P b]), \tag{22.41}$$

where $\gamma_h = \omega_h^2 g_R^2/\pi^3$ is the decay rate into the hot bath associated with the Raman coupling constant g_R^2 (under the Raman resonance condition for Δ).

22.2.3 Work Extraction Dependence on Piston-State Evolution

The piston mean-*energy (for either gain or loss)* satisfies

$$\langle H_P(t) \rangle = \Delta \frac{D}{\Gamma}(1 - e^{-\Gamma t}) + e^{-\Gamma t}\langle H_P(0) \rangle. \tag{22.42}$$

This mean energy increases (undergoes gain) for $\Gamma < 0$, *regardless of the passivity (or nonpassivity) of the initial state*. This *gain represents heat pumping* of P via absorption by S of a quantum from the hot bath at ω_h and its emission to the cold bath at $\omega_c \equiv \omega_a$ [Fig. 22.4(a)], endowing P with the energy $\omega_h - \omega_c = \Delta$. This process requires that the two bath-response spectra, $G_h(\omega)$ and $G_c(\omega)$, be *separated* [as in Fig. 22.4(b)], similarly to the semiclassical limit of this model in Chapter 20. This condition is, however, always realizable (Fig. 20.3) by bath-response spectral filtering (Ch. 20).

One can obtain a rate equation for the signal-mode intensity (mean energy) in the single-pass amplifier regime as

$$\dot{\bar{n}}_P = -2\gamma_h[\rho_{ee}\bar{n}_P(\bar{n}_h + 1) - \rho_{gg}(\bar{n}_P + 1)\bar{n}_h]. \tag{22.43}$$

In the semiclassical limit $\bar{n}_P \gg 1$, we then have for $I_P \equiv \dot{\bar{n}}_P$,

$$\dot{I}_P = -\Gamma I_P \\ -\Gamma = \gamma_h\left[\rho_{gg}\bar{n}_h - \rho_{ee}(\bar{n}_h + 1)\right], \tag{22.44}$$

where $-\Gamma$ is the laser/maser gain for an uninverted WM. The factor in the square brackets expresses the difference between the photon

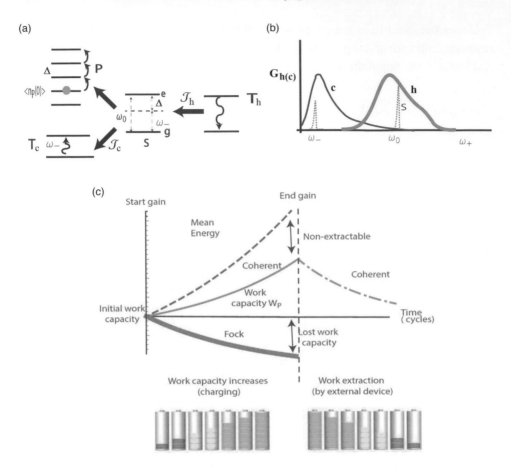

Figure 22.4 Autonomous quantized HE. (a) Energy and heat exchange between the piston (P), the WM (S), the hot and cold baths. Reversal of all arrows corresponds to refrigeration of the cold bath via energy investment by P. (b) Spectrally separated response of the cold (thin lines) and hot (thick lines) baths is engineered using the spectral filtering procedure described in the text. The dotted-line curves are the $q = -1, 0$ (ω_-, ω_a) harmonics of the bath response. The $q = 1$ (ω_h) harmonic is missing because it does not overlap with either of the (solid) bath spectra: namely, after the filtering $G_h(\omega_a) \gg G_c(\omega_a)$, and $G_c(\omega_-) \gg G_c(\omega_h)$, $G_h(\omega_\pm)$. Here, as before, $\omega_a \equiv \omega_0$. (c) Work capacity (ergotropy) as a function of time for initial coherent and Fock states in the HE gain regime. Ergotropy, whose increase is below that of the mean energy due to unavoidable heat production by the piston, "charges" the piston in a coherent state by work. Following piston-energy amplification (gain), this work can be extracted by an external device. (Adapted from Gelbwaser-Klimovsky and Kurizki, 2014. © 2014 American Physical Society)

stimulated-emission probability $\rho_{gg}\bar{n}_h$ and its absorption $\rho_{ee}(\bar{n}_h + 1)$ induced by the hot bath.

The gain and diffusion rates can be rewritten, using (22.39), as

$$-\Gamma = 2\gamma_h \frac{\bar{n}_h - \bar{n}_c}{2\bar{n}_c + 1}, \quad D = 2\gamma_h \frac{\bar{n}_h(\bar{n}_c + 1)}{2\bar{n}_c + 1}. \tag{22.45}$$

The amplification condition, $-\Gamma \geq 0$, requires $\bar{n}_{\mathrm{h}} \geq \bar{n}_c$ and corresponds to the condition

$$\frac{\omega_{\mathrm{h}}}{T_{\mathrm{h}}} \leq \frac{\omega_{\mathrm{c}}}{T_{\mathrm{c}}}. \tag{22.46}$$

By contrast to the main energy or intensity gain, *the ergotropy increase crucially depends upon on the nonpassivity of the initial phase-space distribution* that evolves according to the FP equation (22.38).

The evolution of any initial distribution is then given by

$$\mathbf{P}(r, \theta, t) = \frac{K(t)}{\pi} \int r_0 dr_0 d\theta_0 e^{-K(t)|re^{i\theta}-r_0 e^{i\theta_0}e^{-\Gamma t/2}|^2} \mathbf{P}_0(r_0, \theta_0), \tag{22.47}$$

where

$$\alpha = re^{i\theta}, \quad \alpha_0 = r_0 e^{i\theta_0}, \quad K(t) = \frac{\Gamma}{D(1 - e^{-\Gamma t})}. \tag{22.48}$$

We are interested in its radial derivative, $\frac{\partial \mathbf{P}(r,\theta,t)}{\partial r}$, which expresses its passivity or nonpassivity, as discussed below for generic cases:

(a) If $\Gamma > 0$ (*dissipative loss*), then at long times $e^{-\frac{\Gamma t}{2}} \to 0$, so that

$$\frac{\partial \mathbf{P}(r, \theta, t)}{\partial r} = 2r \frac{e^{-\frac{\Gamma r^2}{D(1-e^{-\Gamma t})}}}{\pi [\frac{D}{\Gamma}(1 - e^{-\Gamma t})]^2} < 0, \tag{22.49}$$

where we have normalized the distribution by $\int dr d\theta r \mathbf{P}(r, \theta) = 1$. The radial derivative (22.49) is *negative* for any $re^{i\theta}$ and *any distribution*. Hence, for $\Gamma > 0$, any evolving distribution $\mathbf{P}(r, \theta, t \to \infty)$ is passive and does not allow work extraction.

(b) If *the initial distribution is passive*, that is, *isotropic with monotonic decrease as a function of energy*: $\frac{\partial \mathbf{P}_0(r)}{\partial r} < 0$, then the distribution remains isotropic at $t > 0$ and, even in the gain ($\Gamma < 0$) regime we find that $\frac{\partial \mathbf{P}(r,t)}{\partial r}$ is *negative*. Hence, such $\mathbf{P}(r, t)$ remains passive, thereby prohibiting *work extraction: state-passivity is preserved* by the Fokker–Planck (FP) phase-plane evolution.

Specifically, for an initial thermal $\mathbf{P}$-distribution with the width σ, we have

$$\mathbf{P}(\alpha, \alpha^*, t) = \frac{1}{\pi} \frac{1}{\sigma e^{-\Gamma t} + D(1 - e^{-\Gamma t})/\Gamma} e^{-\frac{|\alpha|^2}{\sigma e^{-\Gamma t}+D(1-e^{-\Gamma t})/\Gamma}}. \tag{22.50}$$

This state remains always thermal, so that no work is extracted, but its mean energy *increases for negative* Γ. This example illustrates the difference between energy gain and work extraction [Figs. 22.4(c), 22.5(a)].

(c) If the *initial distribution is nonpassive* in the $\Gamma < 0$ regime, then work can be extracted. We here seek the conditions for maximal work extraction. A

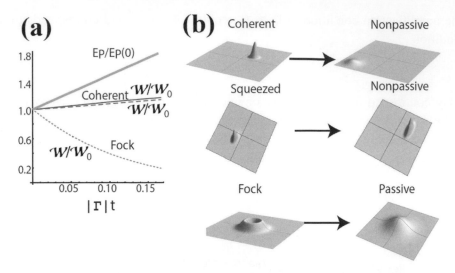

Figure 22.5 Ergotropy evolution. (a) Ergotropy change $\mathcal{W}/\mathcal{W}_0$ as a function of the gain factor $|\Gamma|t$ for coherent and Fock states of P with the same initial mean energy $E_P(0) = \Delta\langle n_P(0)\rangle$. Note their same mean-energy gain $E_P(t)/E_P(0)$ (solid thick curve). (b) The evolution of the phase-plane distribution shows that the high resilience of the initial coherent state against thermalization, as opposed to the fragility of a Fock state, is the key to their different work capacity evolution. The spectral separation and the response harmonics are as in Figure 22.4(b) to ensure gain in (22.36). (Adapted from Gelbwaser-Klimovsky and Kurizki, 2014. © 2014 American Physical Society)

clue is provided by the low-temperature approximation to the entropy production rate,

$$\dot{S}_P(T_P \approx 0) \approx (\Gamma + 2D)(\langle b^\dagger b\rangle - \langle b^\dagger\rangle\langle b\rangle) + D. \qquad (22.51)$$

While the diffusion rate D is constant and state-independent, the first term in (22.51) is strongly state-dependent.

Thus, an initially coherent state, $|\alpha_0\rangle$, evolves in the gain regime toward a distribution centered at an exponentially growing mean amplitude $\alpha(t)$ and is increasingly broadened by diffusion:

$$\mathbf{P}(\alpha, \alpha^*, t | \alpha_0, \alpha_0^*, 0) = \frac{1}{\pi\sigma^2(t)} \exp\left[-\frac{|\alpha - \alpha_0 e^{-\Gamma t/2}|^2}{\sigma^2(t)}\right]. \qquad (22.52)$$

That is, a coherent state at the input becomes a displaced thermal state [Fig. 22.6(a)] whose mean amplitude is shifted outward by $\alpha_0 e^{-\Gamma t/2}$ and its width grows (thermalizes) as

$$\sigma^2(t) = \frac{\bar{n}_h(\bar{n}_c + 1)}{\bar{n}_h - \bar{n}_c}(e^{-\Gamma t} - 1). \qquad (22.53)$$

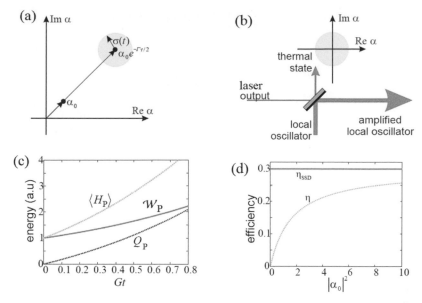

Figure 22.6 Amplification of the piston (signal) state in a two-level heat-powered laser/maser. (a) The Glauber–Sudarshan **P** distribution of an initial coherent state $|\alpha_0\rangle$ and the amplified state. The initial **P** distribution is singular while the final state is a displaced thermal state, with a finite width $\sigma(t)$. (b) Separation of the nonpassive (work-producing) and passive (heat-producing) components of the signal (piston) output by means of a beam splitter (BS) transformation that implements the displacement operator. (c) Small signal amplification: Q_P, the passive (thermal) energy, ergotropy $\mathcal{W}_P$, and total mean energy $\langle H_P \rangle$ for initial $|\alpha_0|^2 = 1$. (d) Efficiency η (work output divided by the heat input) as a function of the mean squared amplitude $|\alpha_0|^2$ tends to the semiclassical Scovil and Schultz–Dubois (SSD) efficiency $\eta_{\mathrm{SSD}} = \Delta/\omega_h$ which is limited by the Carnot bound for large initial-state amplitudes $|\alpha_0|$. The parameters are $\Delta/\omega_h = 0.3$, $\gamma_h/\Delta = 0.033$, $T_h = 2.33T_c$. (Adapted from Ghosh et al., 2018b.)

The mean energy of this state is

$$\langle H_P(t) \rangle = \Delta \left[|\alpha_0|^2 e^{-\Gamma t} + \sigma^2(t) \right], \tag{22.54}$$

the first term in the brackets corresponding to the coherent amplitude squared and the second to the thermal contribution.

The unitary displacement operator $\mathcal{D}(\alpha') = e^{\alpha' a^\dagger - \alpha'^* a}$, with $\alpha' = -\alpha_0 e^{-\Gamma t/2}$ applied to the amplified piston mode at t transforms it to a thermal (passive) state centered at the phase space origin, with energy $\sigma^2(t)$,

$$\mathbf{P}(\alpha, \alpha^*, t | \alpha_0, \alpha_0^*, 0) \mapsto \frac{\Gamma}{\pi D(1 - e^{-\Gamma t})} e^{-\frac{\Gamma |\alpha|^2}{D(1 - e^{-\Gamma t})}}. \tag{22.55}$$

Hence the ergotropy (extractable work) of the output state is

$$\mathcal{W}(t) = \Delta |\alpha_0|^2 e^{-\Gamma t}. \tag{22.56}$$

Thus, *the unitary operation transforms the nonpassive output distribution to a Gibbs state, thereby maximizing the work and power extraction,*

As long as our low-excitation assumption (22.22) holds, the ergotropy *exponentially increases under gain*, whereas the entropy-production (heat-up) term is minimized by this state, which constitutes the optimal case of work extraction. The long-time sustainable work extraction reflects the fact that an initial coherent state retains its nonpassivity and is never fully thermalized, since the distribution is increasingly peaked further away from the origin.

In contrast to the robust coherent state, the initial *Fock state* is highly fragile and quickly thermalizes, as shown in Fig. 22.5(b). While an initial Fock state of P has some ergotropy, $\mathcal{W}(0) = \Delta n_P(0)$, it is diminished by the HE action [Fig. 22.5(a)], so that work extraction by an input Fock state is always *negative*.

Thus, as opposed to energy gain, the extractable work strongly depends on the initial phase-plane distribution of the piston [Figs. 22.5(b), 22.7]: If the initial $P(\alpha, \alpha^*)$ does not fall off monotonically away from the center, but is rather centered at the origin, as in a Fock state, it rapidly becomes passive for any Γ and thereby terminates work extraction. By contrast, an initial coherent state, whose evolving distribution under gain $\Gamma < 0$ is centered at a growing distance from the origin, $|\alpha_P(t)| = |\alpha_P(0)|e^{-\frac{\Gamma t}{2}}$, *increases* its nonpassivity, owing to the low entropy

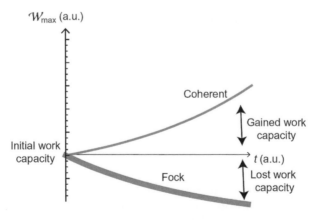

Figure 22.7 Comparison of work capacity in the QHM of Figure 22.1 for two different initial quantum states of the piston (with the same initial energy): While a coherent state may increase its work capacity, a Fock state will inevitably lose it. (Adapted from Gelbwaser-Klimovsky et al., 2015.)

production for $|\Gamma| \gg D$, and thus sustains highly efficient work extraction. This behavior persists as long as the linear amplification condition (22.22) holds, that is, until the onset of saturation for large $\alpha_P(t)$.

22.2.4 HE Efficiency Bound

The HE efficiency is evaluated from

$$\eta = \frac{\dot{\mathcal{W}}}{\mathcal{J}_h}, \tag{22.57a}$$

where the heat current from the hot bath to S+P is given by

$$\mathcal{J}_h = \hbar\omega_h(D - \Gamma\langle b^\dagger b\rangle). \tag{22.57b}$$

The energy ω_h of a hot-bath photon is shared between the signal-mode and cold-bath photons, with energies Δ and ω_c, respectively. This sets the limit on the HE efficiency, defined as the ratio of the extracted work in the signal mode to the heat input from the hot bath.

Since $\omega_c = \omega_h - \Delta$, the laser efficiency [conforming to (22.44)–(22.46)] satisfies

$$\eta = \eta_{SSD} = \frac{\Delta}{\omega_h} \leq 1 - \frac{T_c}{T_h}, \tag{22.58}$$

and we recover the Scovil and Schulz-DuBois (SSD) relation for three-level HE efficiency at the threshold point, where the signal (piston) to the hot-pump frequency ratio corresponds to the Carnot limit.

The three-level engine studied by Scovil and Schulz-DuBois was the first QHM model. Its principle of operation is to convert the WM population inversion into output power. The hot bath at T_h, mean frequency ω_h, induces transitions between the ground state $|0\rangle$ and the excited state $|2\rangle$ with energy E_2. The cold bath at T_c, near frequency ω_c, couples states $|0\rangle$ and $|1\rangle$ with E_1. The amplifier operates by coupling the energy levels E_2 and E_1 to the radiation field (piston) generating the output frequency resonant with $\Delta = E_2 - E_1$. The necessary condition for driven-WM amplification (positive gain) is population inversion:

$$p_2 - p_1 \geq 0. \tag{22.59}$$

This condition dictates the inequality

$$\frac{\omega_c}{\omega_h} \simeq \frac{E_1}{E_2} \geq \frac{T_c}{T_h}. \tag{22.60}$$

The maximal amplifier efficiency in the SSD scheme coincides with the Otto efficiency bound (Ch. 18),

$$\eta_{\text{SSD}} = \eta_{\text{Otto}} = \frac{\Delta}{E_2} = 1 - \frac{\omega_c}{\omega_h}. \tag{22.61}$$

The SSD efficiency bound (22.58) follows from (22.59) and (22.60) by the Carnot bound.

In contrast to the SSD model, in the present model the WM is uninverted and no other transition (or level) is involved. Thermal-occupancy imbalance $\bar{n}_h(\omega_h) > \bar{n}_c(\omega_a)$ at two different frequencies, ω_h and ω_c, rather than population inversion $\rho_{ee} > \rho_{gg}$ in a conventional (SSD) laser or maser, is the requirement for the output signal intensity to exceed its input counterpart in a single pass.

The present model, which uses Raman coupling to a virtual level, still shares common thermodynamic properties with the three-level SSD HE, in particular, (22.58) as the efficiency bound.

The efficiency of the HE with a coherent state $|\alpha_0\rangle$ at the input is found to be

$$\eta = \eta_{\text{SSD}} \frac{\mathcal{W}(t) - \mathcal{W}(0)}{\langle H_P(t)\rangle - \langle H_P(0)\rangle} = \frac{\Delta}{\omega_h} \frac{|\alpha_0|^2}{|\alpha_0|^2 + \frac{\bar{n}_h(\bar{n}_c+1)}{\bar{n}_h - \bar{n}_c}}. \tag{22.62}$$

Thus, efficiency, obtained by the fully quantum treatment, is always less than its classical (SSD) counterpart, and the two coincide for high input amplitudes α_0 [Fig. 22.6(c),(d)].

If the HE is operated by injecting coherent states with amplitudes that are distributed with weights $p(|\alpha_0|)$, the mean efficiency is bounded by

$$\langle \eta \rangle = \frac{\Delta}{\omega_h} \int \frac{|\alpha_0|^2}{|\alpha_0|^2 + \frac{\bar{n}_h(\bar{n}_c+1)}{\bar{n}_h - \bar{n}_c}} |\alpha_0| p(|\alpha_0|) \mathrm{d}|\alpha_0| \leqslant \eta_{\text{SSD}}. \tag{22.63}$$

Hence, the SSD efficiency limit is reached only if the mean initial amplitude $\langle|\alpha_0|\rangle$ dominates over the thermal contribution to the denominator of (22.63), $\bar{n}_h(\bar{n}_c + 1)/(\bar{n}_h - \bar{n}_c)$.

Work extraction requires a local oscillator (LO) in Figure 22.6(b) that is adapted to the amplitude and phase of the initial coherent state: As in Chapter 17, phase and amplitude estimation on the input is required to adjust the LO parameters.

The displacement operator that extracts the work can be realized by overlapping the output state of the piston on a highly reflective beam splitter with a coherent local oscillator field [Fig. 22.6(b)].

Work extraction from a noisy, but nonpassive, coherent-thermal output state of P is mathematically described as a displacement operation. It can be implemented by the transformation mixing the output with a coherent nearly classical local oscillator (LO) at a beam splitter that enacts

$$a_{\text{LO}} \to \mathcal{R} a_{\text{LO}} + \mathcal{T} a_{\text{P}},$$
$$a_{\text{P}} \to \mathcal{R} a_{\text{P}} - \mathcal{T} a_{\text{LO}}. \tag{22.64}$$

Here a_{LO} is the annihilation operator of the local oscillator field, which is treated as a c-number, assuming that the number of photons in the LO is high, $\langle a_{\text{LO}}^{\dagger} a_{\text{LO}} \rangle \gg 1$, much higher than in P, and $\mathcal{T}$ and $\mathcal{R}$ are the beam splitter transmissivity and reflectivity, respectively. By setting $\mathcal{T} a_{\text{LO}} = \mathcal{R} \alpha_{\text{P}}$, we obtain zero amplitude in the reflected channel of P. The local-oscillator intensity, on the other hand, will increase by the coherent-component energy according to

$$
\begin{aligned}
\Delta a_{\text{LO}}^2 &\rightarrow \Delta(\mathcal{R} a_{\text{LO}} + \mathcal{T} a_{\text{P}})^2 \\
&\approx \Delta[(1 - \mathcal{T}^2/2) a_{\text{LO}} + \mathcal{T}^2 a_{\text{LO}}]^2 \\
&\approx \Delta a_{\text{LO}}^2 (1 + \mathcal{T}^2) \\
&= \Delta(a_{\text{LO}}^2 + \alpha_0^2 e^{-\Gamma t}),
\end{aligned} \tag{22.65}
$$

where we have approximated $\mathcal{R} = \sqrt{1 - \mathcal{T}^2} \approx 1 - \mathcal{T}^2/2$, assuming that $\mathcal{T} \ll 1$, and have set $a_{\text{P}} = \alpha_0 e^{-\Gamma t/2}$.

22.2.5 Semiclassical Limit of Nonpassive Work Extraction

The fully quantized autonomous HE whose work extraction is determined by nonpassivity reproduces in the semiclassical limit the power extraction of an externally modulated heat engine that is governed, instead of $H_{\text{S+P}}$ in (22.12) and (22.14) by the Hamiltonian in Chapter 20. The two descriptions coincide when the piston is prepared in a large-amplitude coherent state $|\alpha_0| \gg 1$: The extracted power is then $-\hbar\Delta |\alpha_0|^2 \Gamma$. For an externally modulated HE, it is $-\hbar\Delta \frac{(\Delta\lambda)^2}{(2g)^2} \Gamma$. Thus, a parametric modulation amplitude $\lambda = \frac{2g}{\Delta} |\alpha_0|$ provides the same power extraction as the nonpassive coherent state $|\alpha_0\rangle$.

For an initial coherent state with large $|\alpha_0|^2$ and spectrally separated baths, such that only $G_{\text{c}}(\omega_{\text{c}})$ and $G_{\text{h}}(\omega_{\text{h}})$ are nonzero, we find for $|\Gamma| t \ll 1$,

$$
\begin{aligned}
\mathcal{J}_{\text{h}} &\approx \hbar \frac{g^2}{\Delta^2} |\alpha_0|^2 \omega_{\text{h}} G_{\text{h}}(\omega_{\text{h}}) G_{\text{c}}(\omega_{\text{c}}) \left(e^{-\frac{\hbar\omega_{\text{h}}}{k_{\text{B}} T_{\text{h}}}} - e^{-\frac{\hbar\omega_{\text{c}}}{k_{\text{B}} T_{\text{c}}}} \right), \\
\dot{W} &= -\hbar\Gamma e^{-\Gamma t} |\alpha_0|^2 \Delta \approx -\hbar\Gamma |\alpha_0|^2 \Delta,
\end{aligned} \tag{22.66}
$$

where

$$
\Gamma \simeq -\frac{g^2}{\Delta^2} G_{\text{h}}(\omega_{\text{h}}) G_{\text{c}}(\omega_{\text{a}}) \left(e^{-\frac{\hbar\omega_{\text{h}}}{k_{\text{B}} T_{\text{h}}}} - e^{-\frac{\hbar\omega_{\text{c}}}{k_{\text{B}} T_{\text{c}}}} \right). \tag{22.67}
$$

The corresponding efficiency bound becomes

$$
\eta = \frac{\dot{W}}{\mathcal{J}_{\text{h}}} \rightarrow \eta_{\text{Max}} = \frac{\Delta}{\omega_{\text{h}}}, \tag{22.68}
$$

which is the same as in the SSD model.

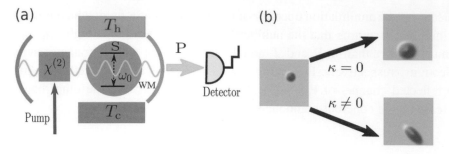

Figure 22.8 (a) A catalyzed quantum HE: The two-level WM (S) is continuously coupled to cold and hot thermal baths and to a nonlinearly pumped piston mode P. (b) Evolution of the phase-space distribution from an initial coherent state with ($\kappa \neq 0$) and without ($\kappa = 0$) a nonlinear pump. (Reprinted from Ghosh et al., 2018a.)

22.3 QHM Catalyzed by Piston Squeezing

Here we consider a quantum HE [Fig. 22.8(a)], where the S-P interaction is governed by the same Hamiltonian as in Section 22.2, except that a nonlinear pump acts on the piston P, so that

$$H_{\text{S+P}} \rightarrow H_{\text{S+P}} + H_{\text{pump}}(t). \tag{22.69}$$

As shown below, the pump can *catalyze* the HE power output by increasing the piston (P) ergotropy, though we subtract the power input by the pump from the output power. As in Section 22.2, the P mode is isolated from the baths. Yet, the baths indirectly modify its energy and entropy (Fig. 22.8) via S. This implies that the state of the piston cannot be fully cyclic; it must inevitably keep changing.

Specifically, we consider the coupling of P to a (degenerate) parametric amplifier via the following Hamiltonian that becomes time-dependent in the *interaction picture*,

$$H_{\text{pump}}(t) = \frac{i\hbar}{2} \kappa e^{-2i\Delta t} a^{\dagger 2} + \text{H.c.} \tag{22.70}$$

Here κ is the undepleted pumping rate. The quadratic form of $H_{\text{pump}}(t)$ generates squeezing of P. This is shown to enhance the ergotropy of P and its amplification by the S-B coupling.

The ME for P to second order in g/Δ is changed as follows compared to Section 22.2, where the nonlinear pump is absent ($\kappa = 0$),

$$\dot{\rho}_{\text{P}} \rightarrow \dot{\rho}_{\text{P}}|_{\kappa=0} + \frac{\kappa}{2}[b^{\dagger 2}, \rho_{\text{P}}] - \frac{\kappa^*}{2}[b^2, \rho_{\text{P}}]. \tag{22.71}$$

Here, we will use the well-known Wigner quasiprobability distribution function $\mathbf{W}(\alpha, \alpha^*)$, which is linearly related to $\mathbf{P}(\alpha, \alpha^*)$. The advantage of $\mathbf{W}(\alpha, \alpha^*)$ is that it is smoother than $\mathbf{P}(\alpha, \alpha^*)$, that is, lacks some of its singularities. The ME (22.71) can be recast as a Fokker–Planck (FP) equation for $\mathbf{W}(\alpha, \alpha^*)$,

$$\frac{\partial \mathbf{W}}{\partial t} = \left(\frac{\partial}{\partial \alpha} d_\alpha + \frac{\partial}{\partial \alpha^*} d_\alpha^* \right) \mathbf{W} + \left(D + \frac{\Gamma}{2} \right) \frac{\partial^2 \mathbf{W}}{\partial \alpha \partial \alpha^*}, \qquad (22.72)$$

where

$$d_\alpha = \frac{\Gamma}{2}\alpha - \kappa \alpha^*. \qquad (22.73)$$

This FP equation is similar to that for $\mathbf{P}(\alpha, \alpha^*)$, (22.38), the main difference being that in (22.73) there is a term proportional to κ due to the nonlinear pump.

For an initial coherent state $|\alpha_0\rangle$, corresponding to the Wigner distribution $\mathbf{W}(\alpha, \alpha^*, 0) = \frac{2}{\pi} e^{-2|\alpha - \alpha_0|^2}$, the solution is

$$\mathbf{W}(x_1, x_2, t) = \frac{1}{2\pi \sqrt{f_+ f_-}} \exp\left[-\frac{f_-(x_1 - x_{10} e^{\Gamma_+ t})^2 + f_+(x_2 - x_{20} e^{\Gamma_- t})^2}{2 f_+ f_-} \right]. \qquad (22.74a)$$

Here

$$\Gamma_\pm = -\Gamma/2 \pm |\kappa|, \qquad (22.74b)$$

the real variables x_1, x_2 are defined by

$$\alpha = x_1 + ix_2, \quad x_{10} = \operatorname{Re}\alpha_0, \quad x_{20} = \operatorname{Im}\alpha_0, \qquad (22.75)$$

and $f_\pm$ are the Gaussian widths,

$$f_\pm = \frac{e^{2\Gamma_\pm t}}{4} + \frac{\left(D + \frac{\Gamma}{2}\right)}{4\Gamma_\pm}(e^{2\Gamma_\pm t} - 1). \qquad (22.76)$$

Thus, under quadratic pumping in the gain regime $\Gamma < 0$, the initial coherent-state distribution becomes a Gaussian with maximal and minimal widths f_+ and f_- along the orthogonal axes x_1 and x_2 corresponding to the pump quadratures. The maximal width f_+ grows much faster than the minimal width f_- [Fig. 22.8(b)], signifying enhanced squeezing of the distribution.

22.3.1 Work Extraction under Nonlinear Pumping

We find from (22.2) and (22.5) the maximal power generated by P after subtraction of the pump power,

$$\mathcal{P}_{\mathrm{Max}} = \dot{\mathcal{W}} - \dot{W}_{\mathrm{pump}} = \langle \dot{H}_\mathrm{P} \rangle - k_\mathrm{B} T_\mathrm{P}(t) \dot{S}_\mathrm{P}(t) - \dot{W}_{\mathrm{pump}}, \qquad (22.77)$$

so that (22.77) is *the net rate* of extractable work converted from heat catalyzed by the pump. The first term in (22.77), $\langle \dot{H}_P \rangle$, is the power obtained for a perfectly non-passive state, the second term, $-T_P(t)\dot{S}_P$, expresses the passivity increase through the rise of the temperature $T_P(t)$ and the entropy $S_P(t)$, and the last term is the subtracted power supplied by the pump.

22.3.2 Efficiency Boost

To obtain an insight into the catalysis by pump squeezing, we consider its effect on the HE efficiency (22.57b), which can now be written as

$$\eta = \frac{\langle \dot{H}_P \rangle - k_B T_P \dot{S}_P - \dot{W}_{\text{pump}}}{\mathcal{J}_h}. \tag{22.78}$$

From (22.71), one finds

$$\langle \dot{H}_P \rangle = \hbar \Delta \text{Tr}(\dot{\rho}_P b^\dagger b) = \hbar \Delta (D + \kappa \langle b^{\dagger 2} \rangle - \Gamma \langle b^\dagger b \rangle) + \text{c.c.},$$

$$\dot{W}_{\text{pump}}(t) = \text{Tr} \left[\rho_P \frac{d}{dt} H_{\text{pump}}(t) \right] = \hbar \Delta \kappa \langle b^{\dagger 2} \rangle + \text{c.c.} \tag{22.79}$$

Combining (22.57b) and (22.79), we arrive at

$$\langle \dot{H}_P \rangle - \dot{W}_{\text{pump}} = \frac{\Delta}{\omega_h} \mathcal{J}_h. \tag{22.80}$$

From (22.78) and (22.80), the efficiency can be expressed as

$$\eta = \Delta \left(\frac{1}{\omega_h} - \frac{\hbar \dot{n}_{\text{pas}}}{\mathcal{J}_h} \right). \tag{22.81}$$

Here we have used the identity $k_B T_P \dot{S}_P = \hbar \Delta \dot{n}_{\text{pas}}$ for Gaussian states that relates the entropy increase to $n_{\text{pas}} = (e^{\Delta/T_P} - 1)^{-1}$, the thermal excitation of the passive state (22.5). From (22.81) it follows that the efficiency depends on the *ratio of* $\Delta \dot{n}_{\text{pas}}$ *to the incoming heat current* $\mathcal{J}_h$. These, in turn, depend on the evolving $T_P(t)$ and the squeezing parameter $r(t)$ of P, and on the expectation values x_{10}, x_{20}, of the quadrature operators $\hat{x}_1$ and $\hat{x}_2$ in the initial state of P.

The expression for η_{Max} can be simplified using the dependence of the squeezing parameter $r(t)$ on the quadrature widths $f_\pm$ for Gaussian states in (22.76),

$$(n_{\text{pas}} + 1/2) \cosh 2r(t) = f_+ + f_-. \tag{22.82}$$

Then one finds

$$\dot{n}_{\text{pas}} = -\Gamma(n_{\text{pas}} + 1/2) + (D + \Gamma/2) \cosh 2r(t),$$

$$\mathcal{J}_h = \hbar \omega_h \left\{ D + \Gamma/2 - \Gamma \left[(n_{\text{pas}} + 1/2) \cosh 2r(t) + x_{10}^2 e^{2\Gamma_+ t} + x_{20}^2 e^{2\Gamma_- t} \right] \right\}. \tag{22.83}$$

Both $\dot{n}_{\mathrm{pas}}$ and $\mathcal{J}_{\mathrm{h}}$ are enhanced by the squeezing. Remarkably, $\mathcal{J}_{\mathrm{h}}$ is more strongly enhanced, which yields both ergotropy and efficiency increase.

For Gaussian (squeezed) states, the efficiency can be evaluated, assuming that $n_{\mathrm{pas}} \gg D/|\Gamma|$, to be

$$\eta(t) \simeq \frac{\Delta}{\omega_{\mathrm{h}}} \left[1 - \frac{n_{\mathrm{pas}} + 1/2}{(n_{\mathrm{pas}} + \frac{1}{2}) \cosh 2r(t) + (x_{10}^2 e^{2\Gamma + t} + x_{20}^2 e^{2\Gamma - t})} \right]. \tag{22.84}$$

The maximal attainable efficiency η_{Max} is then found to be

$$\eta_{\mathrm{Max}} = \frac{\Delta}{\omega_{\mathrm{h}}}, \tag{22.85}$$

which is bounded from the above by the Carnot efficiency,

$$\eta_{\mathrm{Carnot}} = 1 - \frac{T_{\mathrm{c}}}{T_{\mathrm{h}}}. \tag{22.86}$$

It is seen from Figure 22.9 and the analysis of (22.84) that η tends to $\eta_{\mathrm{Max}} = \frac{\Delta}{\omega_{\mathrm{h}}}$ as the pumping rate κ increases.

Without the pump-induced squeezing [$\kappa = r(t) = 0$], the passivity term that limits the ergotropy (22.2) or the power (22.77) becomes small only in the semiclassical limit $x_{10}^2 + x_{20}^2 = |\alpha_0|^2 \gg 1$ and under the weak-coupling condition $(g/\Delta)|\alpha_0| \ll 1$, yielding the following efficiency in the (unpumped) gain regime $\Gamma < 0$

$$\eta_0 = \frac{\Delta}{\omega_{\mathrm{h}}} \frac{|\alpha_0|^2}{|\alpha_0|^2 - D/\Gamma} = \frac{\Delta}{\omega_{\mathrm{h}}} \frac{1}{1 + \frac{D}{|\Gamma||\alpha_0|^2}}. \tag{22.87}$$

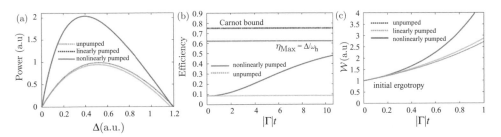

Figure 22.9 (a) Output power as a function of the piston frequency Δ for quadratic pumping, linear pumping, and no pumping (top to bottom) for $T_{\mathrm{c}} = 0.6 T_{\mathrm{h}}$. (b) Comparison of efficiency under weak quadratic pumping ($\kappa \neq 0$, and $|\kappa|/|\Gamma| \sim 0.1$) and without pumping ($\kappa = 0$). (c) The ergotropy drastically increases under quadratic pumping compared to its linear and unpumped (almost identical) counterparts (normalized by the initial work capacity) as a function of $|\Gamma|t$ for an initial coherent state with $|\alpha_0|^2 \sim 1$. (Adapted from Ghosh et al., 2018a.)

It can be verified that quadratic pumping may dramatically enhance the efficiency and the ergotropy (Fig. 22.9) for an initial piston nonpassive charging $|\alpha_0|^2 \sim 1$. The reason is that under the nonlinear pumping, any heat input in P is amplified by the squeezing as $\Delta(n_{\text{pas}} + 1/2) \cosh 2r(t)$, thereby enhancing the leading term in the denominator of (22.84). On the other hand, the terms depending on $\Gamma_\pm$ in (22.74a) and the heat-up (passive energy), Δn_{pas}, are unaffected by the squeezing. Consequently, the stronger the squeezing, the higher the efficiency.

The nonlinear (quadratic) pumping case should be contrasted with its *linear pumping* counterpart, generated by the Hamiltonian

$$H_{\text{pump}}(t) = i\hbar\kappa b^\dagger e^{-i\Delta t} + \text{H.c.} \qquad (22.88)$$

Its effect is to increase the ergotropy by

$$\mathcal{W}_{\text{L}} = \hbar\Delta|\alpha(t)|^2, \qquad (22.89)$$

via the displacement

$$\alpha(t) := \alpha_0 e^{-\frac{\Gamma t}{2}} + \frac{2\kappa}{|\Gamma|}(e^{-\frac{\Gamma t}{2}} - 1). \qquad (22.90)$$

This ergotropy contribution is *additive* to the passive energy, Δn_{pas}, unlike the quadratic pumping case. The efficiency for linear pumping is then limited by

$$\eta_{\text{L}} \xrightarrow[t \gg |\Gamma|^{-1}]{} \frac{\Delta}{\omega_{\text{h}}} \frac{|\alpha_0 + 2\kappa/|\Gamma||^2}{|\alpha_0 + 2\kappa/|\Gamma||^2 + n_{\text{pas}}(0) + D/|\Gamma|}. \qquad (22.91)$$

This efficiency never approaches the bound $\eta_{\text{Max}} = \frac{\Delta}{\omega_{\text{h}}}$ if $|\alpha_0|$ is much smaller than $n_{\text{pas}}(0) + D/|\Gamma|$. Consequently, the ergotropy increase generated for the same heat input linear pumping is much less significant than for nonlinear pumping, resulting in much lower power.

22.4 Discussion

We here put forward a fully quantized HE model, where the periodic driving is replaced by a quantum piston. This model differs from the vast majority of HE models that employ classical fields or forces to drive the WM and thus may be deemed *semiclassical*. In analogy to light–matter interaction where the quantization of light results in new effects, the same is true for quantum thermal machines.

The study of fully quantized heat machines must rely on a physically sound definition of work (Ch. 15). Here, the efficiency of fully quantized HE is shown to strongly depend on the nonpassivity (ergotropy – Ch. 15) of the initial quantum piston state, its subsequent thermalization, and entropy production. These properties of the quantum state are shown to determine the ability of the quantized piston to serve as a thermodynamic resource that can boost the HE power without breaking the standard thermodynamic bounds. The key to maximized power extraction is the resilience of the nonpassive state to thermalization, which ensures high ergotropy (work extraction).

If the piston is treated semiclassically, its entropy is negligible, hence it is referred to as work reservoir. However, a quantum treatment of a piston state must allow for its "heat-up." Part of the energy acquired by the piston leads to its entropy increase and cannot be extracted as useful work.

For a large-amplitude coherent state of the piston, the results presented in Chapter 20 are recovered, showing that the periodically driven model is a good approximation for a semiclassical piston (external field). As in the continuous-cycle semiclassical HE of Chapter 20, the fully quantized HE require spectral separation of the two baths.

It is instructive to compare the present analysis to that of a laser or a maser, which can also be described as a heat machine. A maser or a laser is typically based on a medium with three or more levels wherein one transition is pumped to establish population inversion on another (signal) transition. This results in the pump conversion into a coherent, amplified signal output. Scovil and Schulz-DuBois (SSD) related the maximal conversion efficiency of a three-level maser to the Carnot efficiency bound by assuming that one transition is pumped by a hot bath and another is coupled to a cold bath. The three-level SSD scheme has become a canonical microscopic model for HE and their quantum-mechanical characteristics. In our model, there is no population inversion in the system – the gain is provided by the hot bath, but the analogy is complete. Yet, the SSD semiclassical analysis does not discern between coherent- and Fock-state laser/maser gain underscored here.

The general thermodynamic analysis is followed by that of the simplest (minimal) autonomous model consisting of a TLS coupled to a quantized-oscillator piston.

It is remarkable *that as the initial coherent amplitude of the piston decreases, the resulting efficiency increases as well*, although the *entropy growth* of the piston might then be expected to reduce (rather than enhance) the efficiency. Work extraction obtained from an initial coherent state has been found to be superior to that of other states, because of its larger sustainable ergotropy or nonpassivity, conditioned on its low entropy production. This reflects the fact that the coherent state is the "pointer-state" of the evolution (see Ch. 9).

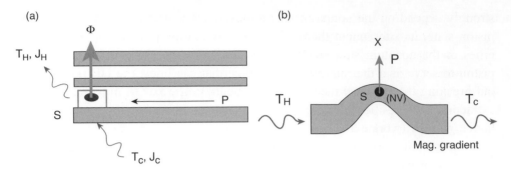

Figure 22.10 Realizations of the model in Figure 22.4(a) and (22.14). (a) A superconducting flux qubit in a coplanar resonator (cavity). The resonator field mode is the piston (P) that affects the magnetic flux threading the qubit acting as a two-level WM (S). The hot and cold baths are spectrally filtered by the cavity. (b) A NV-center defect in diamond mounted on a nanomechanical beam (cantilever). The strain field is the piston (P) that acts on S, which is subject to magnetic-field gradient. The nanomechanical beam spectrally filters the two baths. (Adapted from Gelbwaser-Klimovsky and Kurizki, 2014. © 2014 American Physical Society)

An alternative means of boosting HE performance is heat-to-work conversion catalysis by nonlinear (squeezed) pumping of the piston mode. The resulting power enhancement is due to the increased ability of the squeezed piston to store ergotropy. Since the catalyzed machine is still an HE (unlike those discussed in Ch. 16), it adheres to the Carnot bound.

The dispersive WM-piston (S-P) coupling (22.14) assumed in this chapter is realizable, for example, for a superconducting flux qubit that is off-resonantly coupled to P, a high-Q microwave cavity mode [Fig. 22.10(a)]. Another setup may be based on a NV defect that is subject to a magnetic-field gradient and coupled to the strain field (phonon mode) of a nanomechanical resonator [(Fig. 22.10(b)]. Both setups may attain large gain $|\Gamma|t_{\text{lea}} \gg 1$, since $t_{\text{lea}} \geq (\frac{\Delta}{Q})^{-1}$, the P-mode leakage time, may be long enough to evolve P from its initial state to the final state at a rate Γ, before it leaks out of the cavity. Equally promising is the Raman regime of a heat-powered atomic medium in a cavity that spectrally filters the hot and cold baths to be distinct.

To conclude, this chapter reveals quantum aspects of work production, outside the scope of studies pertaining to semiclassical HE, where energy or entropy of the driving-field changes in the course of work extraction are imperceptible. On the applied side, these quantum aspects are important for the design of efficient machines powered by work and heat that are stored in *autonomous* quantum devices at the nanoscale. On the foundational side, it provides better understanding of the rapport between thermodynamics and quantum mechanics, as it shows the existence of thermodynamic resources in quantum states: the role of quantum-state nonpassivity and its resilience to thermalization. Yet, we may conclude that it is not quantumness per se that improves the machine performance, but rather the properties of the WM and the piston that boost the ergotropy and minimize the wasted heat in both the input and the output.

23

Quantum Refrigerators and the Third Law

In this chapter, we study – in analogy to Chapters 20 and 22 – the semiclassical and fully quantized two-level refrigerator. The fully quantized refrigerator (QR) has a quantized piston and is fully autonomous, whereas the semiclassical refrigerator requires an external modulation (or an external heat bath) to power the refrigerator.

In Chapter 22, we found that work extraction by a fully quantized heat engine requires the quantum piston to be in a state with high resilience to thermalization (e.g., a coherent state). Here we show that efficient cooling (refrigeration) requires the opposite property of the piston state.

Refrigerators may be used to study the third law of thermodynamics in the quantum domain. The *dynamical* formulation of the third law of thermodynamics, the so called Nernst's *unattainability principle*, forbids cooling to attain absolute zero ($T = 0$) in a finite number of steps, or, more generally, *in finite time*. However, the universality of this principle has been postulated rather than proven. It is debatable whether this formulation is completely equivalent to Nernst's *heat theorem*, whereby the *entropy vanishes* at $T = 0$. Does the unattainability principle apply to quantum processes? Here we show for the minimal model of a refrigerator consisting of a periodically modulated TLS that the cooling rate of a quantized magnon bath does not vanish as $T \to 0$, thus challenging the unattainability principle.

23.1 Quantized Refrigerator (QR) Performance Bounds

23.1.1 Coefficient of Performance (COP) Bound of Refrigerators with Quantized Pistons

We wish to infer, in the most general form, the COP bound of a fully quantized refrigerator (QR), which requires the cold-bath current $\mathcal{J}_c > 0$. Upon substituting

(20.4) in (22.10) and dividing $\mathcal{J}_c$ by the *input* energy flow from the piston, $-\frac{d\langle H_P\rangle}{dt}$, we obtain the upper bound for the COP,

$$\text{COP} = \frac{\mathcal{J}_c}{-\frac{d\langle H_P\rangle}{dt}} \leq \frac{1}{\frac{T_h}{T_c}-1}\left(1 - \frac{k_B T_h \dot{S}_P}{\frac{d\langle H_P\rangle}{dt}}\right). \tag{23.1}$$

The first factor on the right-hand side of the inequality is the *standard* reciprocal Carnot bound, obtained under the reversibility condition:

$$\text{COP}(\dot{S}_P = 0) \leq \frac{1}{\frac{T_h}{T_c}-1}. \tag{23.2}$$

The factor in the brackets in (23.1) is remarkable, since for nonpassive states (which "store" work) we can have

$$\dot{S}_P \Big/ \frac{d\langle H_P\rangle}{dt} < 0. \tag{23.3}$$

This extra factor may increase the COP above the standard Carnot bound, provided that the sign of the extra factor is positive, which requires positive piston-entropy production (i.e., $\frac{d}{dt}S_P > 0$), while its energy is being reduced, $\frac{d}{dt}\langle H_P\rangle_{\rho P} < 0$, thereby powering the refrigerator. For this to occur, P must simultaneously receive heat and deliver power. Only certain quantum states of P that possess this ability, as shown below.

The lower bound of (23.1) is the maximum efficiency of an *absorption* refrigerator: It corresponds to a piston that is completely thermalized and delivers heat only, $\frac{d\langle H_P\rangle}{dt} = T_P \dot{S}_P$. Then the COP of a QR that is completely driven by heat becomes

$$\text{COP} = \frac{1}{\frac{T_h}{T_c}-1}\left(1 - \frac{T_h}{T_P}\right). \tag{23.4}$$

The refrigerator COP bounds (23.1) and (23.4), and the work-production HE efficiency bound (Ch. 22) require very different conditions. In what follows we inquire: How do these bounds depend on the initial quantum state of the piston and its evolution?

23.1.2 Refrigeration Dependence on the Piston State

Since a passive state has the minimal energy for a given entropy, any decrease in the energy of a passive state corresponds to its entropy reduction. Hence, only a quantum piston in a nonpassive state may simultaneously power the QR and absorb heat from the cold bath, thereby increasing its own entropy. The ability of the piston to absorb heat is not considered in the standard Carnot-bound derivation.

Therefore, the second law does not forbid the boost of the COP above the Carnot bound by such heat absorption.

The nonpassive states most suitable for cooling, namely, those that yield the highest COP according to (23.1)–(23.3), are the most unstable states that maximize entropy production, as opposed to states that maximize work extraction by minimizing entropy production. We may classify the COP according to the initial states of the piston as follows.

The cold-bath heat current (22.34) in the model of (22.14) with (again denoting the TLS resonant frequency $\omega_a \equiv \omega_0$)

$$\omega_c = \omega_0 - \Delta, \quad \omega_h = \omega_0 + \Delta \tag{23.5}$$

assumes the form

$$\mathcal{J}_c \propto e^{-\frac{\hbar\omega_c}{k_B T_c}} \bar{n}_P(t) - e^{-\frac{\hbar\omega_0}{k_B T_h}} [\bar{n}_P(t) + 1], \tag{23.6}$$

provided that the baths are spectrally separated, as in Figure 22.4(b). Here $\bar{n}_P(t)$ is the mean number of P quanta. The QR condition COP > 0 [see (23.1)] then assumes the form

$$0 < \frac{\omega_c}{\Delta} < \frac{1}{\frac{T_h}{T_c} - 1}, \tag{23.7}$$

provided

$$\bar{n}_P(t) > \bar{n}_{\min} = \frac{1}{e^{\frac{\hbar\omega_0}{k_B T_h} - \frac{\hbar\omega_c}{k_B T_c}} - 1}. \tag{23.8}$$

Hence, QR action is limited to $\bar{n}_P > \bar{n}_{\min}$.

The energy stored in the initial state of the piston can be used for the refrigeration of the cold bath, provided there is *a positive drift* in (22.38), $\Gamma > 0$ (dissipation). The resulting COP strongly depends on the initial states of P (Fig. 23.1). For an *initial pure state*, the work stored in P is equal to its energy so that P is the work source of the QR. The refrigerator stops cooling when Eq. (23.7) no longer holds, having provided an amount $\Delta\bar{n}_P(0)$ of work. Concurrently, P, initially at $T_P = 0$, absorbs heat from the refrigerator and ends up in a thermal state with a critical temperature that strongly depends on $\frac{\Delta}{\omega_0}$,

$$(T_P)_{\text{crit}} = T_h \frac{\Delta}{\omega_0} \frac{1}{1 - \frac{\omega_c}{\omega_0} \frac{T_h}{T_c}}. \tag{23.9}$$

This $(T_P)_{\text{crit}} \to \infty$ if the upper bound of (23.7) holds.

The analysis of (22.38) shows that the COP in (23.1) depends not only on the *purity* but also on the *type* of the initial quantum state:

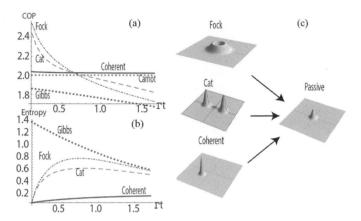

Figure 23.1 (a) COP of an autonomous QR as a function of time (in units of Γ^{-1}) for coherent (continuous), Fock (dot-dashed), and thermal (dashed) initial piston states. The Carnot COP bound (dotted, thin) is slightly below the coherent-state COP. Thermal-state and Fock-state COP asymptotically coincide at T_{crit}. At short times the Fock state has the largest COP bound, and the coherent state outperforms it at long times, with a COP above the Carnot. (b) Entropy (in units of k_B) of the same piston states as a function of time. The initially rapid entropy increase of the Fock state explains its high COP at short times. For the coherent state, the entropy increases more slowly but steadily, resulting in the highest COP at long times. (c) Phase-plane distribution evolution explains the COP evolution by the high resilience of a coherent state against thermalization, as opposed to the low resilience (fragility) of Fock and Schrödinger-cat states. The spectral separation conditions are as in Figure 22.4(b). The piston frequency is chosen to ensure cooling ($\Gamma > 0$). The initial mean energy satisfies (23.7). (Reprinted from Gelbwaser-Klimovsky and Kurizki, 2014. © 2014 American Physical Society)

(i) For an initial *coherent state*,

$$\frac{\dot{\mathcal{S}}_\text{P}}{\frac{d\langle H_\text{P}\rangle}{dt}} = \frac{D}{[D - \Gamma \bar{n}_\text{P}(0)]k_\text{B}T_\text{P}}. \tag{23.10}$$

Namely, its COP is determined by *the diffusion rate*, D, *the drift rate*, Γ, and the initial population $\bar{n}_\text{P}(0)$. Denoting $d(t) \equiv \frac{D}{\Gamma}(1 - e^{-\Gamma t})$, we find for a coherent state

$$\frac{1}{k_\text{B}T_\text{P}} = \frac{1}{\hbar\Delta}\log\frac{1 + d(t)}{d(t)},$$

$$\dot{\mathcal{S}}_\text{P} = \hbar\Delta\frac{De^{-\Gamma t}}{k_\text{B}T_\text{P}}. \tag{23.11}$$

For an initial coherent state, owing to its resilience to thermalization (i.e., the slow increase of its entropy), the COP is just above the Carnot bound.

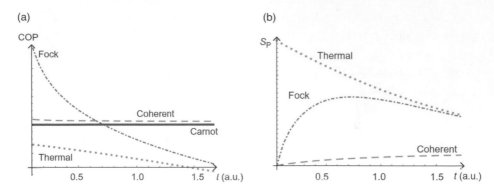

Figure 23.2 (a) COP and (b) entropy of the piston in a QR as a function of time for different initial states with the same initial mean energy. An initial Fock state yields higher COP than a coherent state at short times because its entropy increases faster. (Reprinted from Gelbwaser-Klimovsky et al., 2015.)

(ii) For an initial Fock state $|\bar{n}_P(0)\rangle$, the COP may *surpass at short times that of a coherent state with the same* $\bar{n}_P(0)$ [Fig. 23.1(a)], since a Fock state *is much more prone to thermalization* [Fig. 23.1(b)]. This allows it to both deliver power and absorb heat [Fig. 23.1(b)] to a higher degree than a coherent state. Contrary to work extraction, cooling requires the maximization of entropy production ($\dot{S}_P$). Among all the pure states, Fock state has the largest heat absorption capacity, as can be seen from (22.51). The COP then surpasses that of a coherent state with the same initial energy at short times, since the entropy of the initial Fock state first increases rapidly (Fig. 23.1,23.2). At longer times, its entropy saturates and starts decreasing, which reduces the COP below the Carnot bound.

(iii) Other types of nonclassical initial states, such as a Schröedinger-cat state, yield a COP [Fig. 23.1(a)] that is *below that of a Fock state* with the same $\bar{n}_P(0)$, but above that of the corresponding coherent state.

(b) For an initial thermal state, since the ergotropy is zero, (22.10) implies that the relation $T_P > T_h$ is required for QR action. In the course of its evolution, P remains in a thermal state, but T_P decreases until it attains $(T_P)_{crit}$ and stops cooling. Explicitly, an initial thermal state yields

$$\frac{1}{k_B T_P} = \frac{1}{\hbar\Delta} \log\left(\frac{1 + d(t)}{d(t)}\right),$$
$$\dot{S}_P = \hbar\Delta \frac{[D - \Gamma n(0)_P]e^{-\Gamma t}}{k_B T_P}. \tag{23.12}$$

For an initial thermal state at a high temperature ($T_P > T_h$), the COP is the same as for an absorption refrigerator: The piston then acts as an "ultrahot" bath that powers the refrigerator. The piston remains in a thermal state, with temperature T_P that decreases until it reaches a critical temperature at which the cooling stops.

Thus, instead of $G_j(\omega)$ we obtain the effective response spectrum (20.13), a "skewed Lorentzian" with *controllable* width and center that may be chosen to avoid an overlap between the coupling spectra of the two baths.

23.2 Performance of Semiclassical Minimal (Two-Level) Refrigerators

Here we extend the analysis in Chapter 19 of semiclassical heat machines based on frequently modulated TLS WM by considering the effect of modulation rate increase, which is the only control "knob" in such machines. The quantum speed limit Δ_{SL} is the highest modulation rate that allows this machine to operate as a heat engine (HE). Above this modulation rate, the machine starts operating as a refrigerator of the cold bath by dumping heat into the hot bath: The heat current flows from the cold bath to the hot bath, which requires net power supply by the piston modulation (Fig. 23.3). Heat currents (20.41) and the power then satisfy

$$\Delta > \Delta_{SL} = \frac{2\pi}{\tau_{SL}} = \omega_0 \frac{T_h - T_c}{T_h + T_c} \quad \Rightarrow \quad \mathcal{J}_c > 0, \ \mathcal{J}_h < 0, \ \overline{W} > 0. \qquad (23.13)$$

The crossover from the HE to the refrigerator regime changes by tuning T_h and T_c only, as shown in Figure 19.4(b) ($\Delta_{SL} = \omega_0/2$).

In the present continuous cycle (Fig. 23.4),

$$\text{COP} = \frac{\omega_0 - \Delta}{\Delta}. \qquad (23.14)$$

Thus, the Carnot bound for the COP is reached at $\Delta = \Delta_{SL}$, at which the machine switches over from HE to a refrigerator (Figs. 23.5 and 23.6).

The present model of refrigeration is reminiscent of sideband cooling: an optical Raman process implemented in solids and molecules, where the red- and blue-shifted TLS frequencies correspond to the Stokes and anti-Stokes lines of sideband cooling respectively. Heat is then pumped into an upshifted line in the h-bath spectrum, at the expense of a downshifted c-bath spectral line, the energy difference being supplied by the modulation control.

Under sufficiently fast modulation, the refrigerator may operate in a non-Markovian cycle of duration τ_S, provided (Ch. 20)

$$\tau_S < \tau_{SL} \ll t_c. \qquad (23.15)$$

The COP in the non-Markovian QR regime takes the form

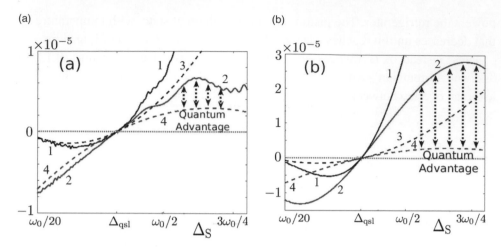

Figure 23.3 Power $\overline{W}$ (curves 1) and cold-bath heat current $\overline{\mathcal{J}_c}$ (curves 2) averaged over $n = 10$ modulation periods as compared to the counterparts under Markovian approximation for long cycles [i.e., $n \to \infty$ (curves 3 and 4, respectively)], versus the modulation rate Δ for (a) the spectral functions shown in Figure 20.4 and (b) the spectral functions shown in Figure 20.5. The enhanced overlap resulting from fast modulation (large Δ) enhances the heat current $\overline{\mathcal{J}_c}$ in the refrigerator regime by a factor larger than 2 for (a) and larger than 9 for (b) (dotted double-arrowed lines), signifying quantum advantage. The horizontal dotted line signifies zero power and currents. Here $\lambda = 0.2$, $\omega_0 = 20$, $\gamma_0 = 1$, $\Gamma_B = 0.2$, $\beta_h = 0.001$, $\beta_c = 0.002$. (Adapted from Mukherjee et al., 2020.)

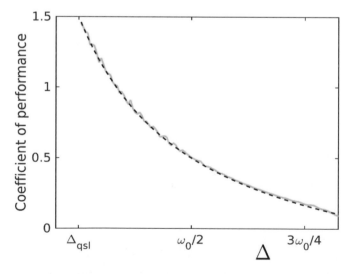

Figure 23.4 Coefficient of performance for the refrigerator in the long time $t \to \infty$ limit (dashed line). (Adapted from Mukherjee et al., 2020.)

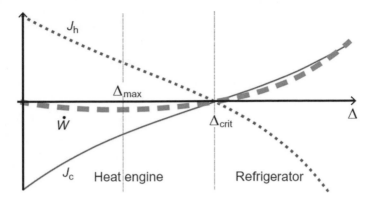

Figure 23.5 Heat currents for the QHM described by Figure 20.2 as a function of the modulation frequency. For low modulation rates (below Δ_{SL}), the QHM operates as an HE (negative power $\dot{W} < 0$) and for high rates (above Δ_{SL}) as a refrigerator ($\dot{W} > 0$). The heat currents change their signs at Δ_{SL}. (Adapted from Gelbwaser-Klimovsky et al., 2015.)

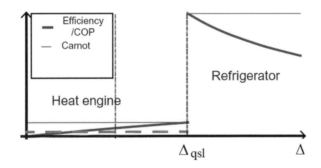

Figure 23.6 HE efficiency and/or COP as a function of the modulation rate Δ. At Δ_{SL}, the zero-power Carnot bound is attained, and the machine switches over from HE to refrigerator action. (Adapted from Gelbwaser-Klimovsky et al., 2015.)

$$
\begin{aligned}
\text{COP} &= \frac{\oint_{\tau_C} \mathcal{J}_c(t)dt}{\oint_{\tau_C} [\mathcal{J}_h(t) + \mathcal{J}_c(t)]dt} \\[2mm]
&= \frac{(\omega_0 - \Delta)\zeta_c \oint_{\tau_C} \mathcal{I}_c(\omega_0 - \Delta, t)dt}{(\omega_0 + \Delta)\zeta_h \oint_{\tau_C} \mathcal{I}_h(\omega_0 + \Delta, t)dt + (\omega_0 - \Delta)\zeta_c \oint_{\tau_C} \mathcal{I}_c(\omega_0 - \Delta, t)dt}.
\end{aligned}
\tag{23.16}
$$

AZE can then lead to quantum advantage by enhancing the cold-bath heat current $\bar{\mathcal{J}}_c$ and thus resulting in higher cooling power of the cold bath. As for HE, quasi-Lorentzian as well as super-Ohmic bath spectral functions can yield significant quantum advantage in cooling power in AZE regime (Fig. 23.3). On the other hand, the coefficient of performance,

$$\text{COP} = -\frac{\bar{\mathcal{J}}_c}{\bar{\bar{W}}}, \tag{23.17}$$

is not affected by the broadening of the sinc function and on average remains identical to that obtained under slow modulation in the Markovian regime [Fig. 20.7(b)].

The results (21.17)–(21.19), (21.22)–(21.24) apply not only to HE (that convert heat from the hot bath into power) but also to refrigerators that consume power to cool the cold bath and to heat distributors where both baths are heated up at the expense of the invested power.

23.2.1 Collective Refrigeration Regime

The ratio (21.25) of the collective energy currents to their individual counterparts is independent of whether the QHM is operated as an HE or a refrigerator. Hence, cooperative effects can enhance the performance of a refrigerator (i.e., the cold current $\mathcal{J}_c$). This enhancement, however, comes at the price of an equally enhanced external power supply.

As seen from Figure 21.5, the more the cold-bath heat current $\mathcal{J}_c$ increases, the colder is the hot bath. The largest boost, however, occurs for the smallest possible $\beta_{\text{eff}}\omega_0$ (see Figure 21.3), where the energy currents vanish. Just like in a HE for given bath temperatures, there is always a benefit of collective operation. However, whereas for the HE the boosted power is obtained for "free" from enhanced heat currents, enhanced cooling requires larger power investment compared to the individual-TLS case.

23.3 Cooling-Speed Scaling with Temperature

While the temperature of the hot bath remains constant throughout the time evolution, the temperature of the cold bath, whose heat capacity is finite, decreases due to heat being extracted to it via the work invested by the external (classical) driving field.

The refrigeration of a finite-capacity c-bath consists of small successive temperature changes over many modulation cycles, in accordance with the Born approximation underlying the ME whereby the bath is nearly invariant. The finite-capacity c-bath is supposed to have a continuous spectrum, otherwise bath-mode discreteness and the associated recurrences may render the Born approximation invalid and preclude bath thermalization.

From the definition of the heat capacity at constant volume for the cold bath, C_V, it follows that

$$C_V \frac{dT_c}{dt} = -\mathcal{J}_c, \tag{23.18}$$

where a positive sign of $\mathcal{J}_c$ corresponds to heat flowing from the cold bath to the working fluid.

Under spectral separation of the two baths, the TLS can be coupled to the hot bath at $\omega \simeq \omega_0 + \Delta$, and at $\omega - \Delta$ to both baths. The cold-bath heat current $\mathcal{J}_c$ is then maximized if Δ satisfies the condition

$$\hbar(\omega_0 - \Delta) \simeq k_B T_c. \qquad (23.19)$$

This requires Δ to be increased as T_c decreases.

In what follows, we assume that the baths consist of oscillators and examine the scaling of C_V and $\mathcal{J}_c$ with T_c: The constant-volume heat capacity of the cold bath, C_V, is given by

$$C_V = \frac{d}{dT_c}\langle H_{Bc} \rangle \simeq \hbar \frac{d}{dT_c} \int_0^\infty d\omega\, \omega \rho(\omega) \bar{n}_c(\omega). \qquad (23.20)$$

The heat capacity depends on the dimensionality of the cold (bosonic) bath. If $\rho(\omega) \simeq \omega^{d-1}$ is the d-dimensional density of the bath modes and $k_B T_c \ll \hbar\omega_{cut}$, that is, the temperature is low compared to the cutoff frequency, then the integral has an upper cutoff at $\hbar\omega \sim k_B T_c$, due to the factor $\bar{n}_c(\omega) = (e^{\omega/T_c} - 1)^{-1}$. As a result,

$$C_V \propto T_c^d \quad \text{for} \quad T_c \to 0. \qquad (23.21)$$

The closer is T_c to 0, the lower is $\omega_0 - \Delta$, hence the steady-state cooling condition must be continuously updated,

$$\bar{n}_c(\omega_0 - \Delta) > \bar{n}_h(\omega_0 + \Delta) \quad \Leftrightarrow \quad \frac{\hbar(\omega_0 + \Delta)}{k_B T_h} > \frac{\hbar(\omega_0 - \Delta)}{k_B T_c}. \qquad (23.22)$$

An analogous relation holds if $\bar{n}_{c(h)}(\omega)$ are Boltzmann rather than Bose factors (occupancies). Correspondingly, we parametrize $\mathcal{J}_c$ as

$$\mathcal{J}_c \simeq \hbar(\omega_0 - \Delta)G_{0c}(\omega_0 - \Delta)\frac{\bar{n}_c(\omega_0 - \Delta) - \bar{n}_h(\omega_0 + \Delta)}{2\bar{n}_h(\omega_0 + \Delta) + 1} \qquad (23.23)$$

or for the low-frequency range of the cold bath,

$$|\eta(\omega)|^2 \propto \omega^b, \quad \rho(\omega) \approx \omega^{d-1} \quad (0 \le \omega = \omega_0 - \Delta \ll \omega_{cut}). \qquad (23.24)$$

Here, $|\eta(\omega)|^2$ is the b-dependent system-coupling to the bosonic bath (discussed below). The maximized heat current then obeys the scaling

$$\mathcal{J}_c(T_c) \propto -T_c^{b+d}. \qquad (23.25)$$

Upon substituting (23.21) and (23.25) in (23.18), we observe that the T_c^d scaling of C_V is canceled by a similar scaling of the density of modes in (23.25). The resulting scaling is

$$dT_c/dt \propto -T_c^b. \qquad (23.26)$$

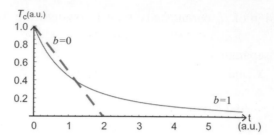

Figure 23.7 Cooling of a phonon bath ($b = 1$, solid line) and a magnon bath ($b = 0$, dashed line) in the scheme of Figure 23.5. The absolute zero is attained in this picture by a magnon bath at a finite time. (Adapted from Kolář et al., 2012. © 2012 American Physical Society)

Remarkably, the cooling speed (23.26) is determined by a single parameter b. For $b = 1$, the temperature T_c decreases exponentially with time, in accordance with Nernst's theorem. For smaller exponents, $0 \leq b < 1$, however, the temperature appears to reach absolute zero in finite time, in an apparent contradiction to Nernst's third law (within the Born–Markov approximation involved).

The value of b is determined by the bath dispersion relation. Two contrasting cases can be discerned for T_c cooling in this model (Fig. 23.7):

- The dispersion relation of acoustic phonons reads $\omega(\mathbf{k}) \simeq vk$, where v is the sound velocity and $k = |\mathbf{k}|$. Their coupling to the system qubit scales linearly with the frequency,

$$|\eta(\omega)|^2 \sim k^2(\omega)/\omega \sim \omega, \qquad (23.27)$$

  resulting in $b = 1$. Consequently, according to (23.26), a bath of acoustic phonons approaches $T_c = 0$ in an exponentially slow fashion, in accordance with Nernst's principle.
- Spin-wave excitations (or *magnons*) of a ferromagnetic spin lattice with nearest-neighbor interactions below the critical temperature. As discussed in Section 3.3, the Holstein–Primakoff transformation of the spin operators to boson operators represents the system as a set of interacting harmonic oscillators. At low T_c, the interaction between the oscillators that gives rise to the nonlinearity in the Holstein–Primakoff transformation is negligible. The system is then effectively a bosonic system governed by the Hamiltonian (3.47) or, equivalently,

$$H_0 = \sum_{\mathbf{k}} \omega(\mathbf{k}) a^\dagger(\mathbf{k}) a(\mathbf{k}), \qquad (23.28)$$

where $a(\boldsymbol{k})$ and $a^\dagger(\boldsymbol{k})$ are annihilation and creation operators of the collective spin-wave (magnon) modes. Their dispersion law is quadratic in the low-frequency region,

$$\omega(\boldsymbol{k}) \propto k^2 + \text{constant}. \qquad (23.29)$$

This system can be coupled to a TLS, as described in Section 4.3. The main difference between the dipolar coupling $\eta(\boldsymbol{k}(\omega))$ to acoustic phonons and magnons is the absence of the dispersive-coupling factor for the latter, as

$$|\eta(\omega)|^2 \sim \frac{k^2(\omega)}{\omega} \sim \text{constant} \qquad (23.30)$$

in the case of magnons. Here, the dipolar coupling of these bosonic magnos to the TLS is independent of the frequency (i.e., $b = 0$). Hence, absolute zero is approached *linearly in time*, thereby challenging Nernst's unattainability principle.

23.4 Discussion

We showed that a quantum piston can be used to power a QR in the same way that a battery is used: A piston "charged" by the fully quantized HE may be subsequently used to power the QR. This "quantum battery," once charged, does not require any external energy source to drive the QR.

Generally, nonpassivity allows a quantum state to power the QR while absorbing heat from the cold bath, thereby cooling it further. As opposed to work extraction, the more unstable a state becomes, the higher its COP at short times. Thus, a nonpassive quantum state is a thermodynamic resource that may endow the QR with extra efficiency. Such a resource may be viewed as a (small) *extra bath*, so that the transgression of the Carnot bound does not signify the breakdown of any thermodynamic principle.

Our general analysis of *quantized refrigerators*, both supplemented by the analysis of QR and in the same setup as the quantized heat engine (HE) analyzed in Chapter 22, has shown that it is *advantageous for the quantum piston to increase its temperature and entropy* during refrigeration, so that its energy exchange with the system is a combination of heat and work production. The instability (rapid entropy increase) of an initial Fock state compared to an initial coherent state makes Fock states temporarily superior in terms of cooling.

The work stored in an initial pure state of the piston may be used to run the machine as a Carnot QR (driven by work), as long as the piston is below the critical temperature (23.9). Alternatively, an initial thermal state of the piston may drive the machine as an absorption QR, provided the piston temperature is above (23.9).

Here we are concerned with the third law of thermodynamics in its dynamical formulation, known as Nernst's unattainability theorem, which forbids physical systems to cool down to absolute zero ($T = 0$) in finite time. In order to investigate this question, we reconsider semiclassical refrigerators, consisting of a TLS whose transition frequency is periodically modulated, while being simultaneously coupled to cold and hot heat baths. We assume the cold bath to possess a *finite heat capacity* such that its temperature is progressively reduced as heat is extracted from it, in the refrigerator regime.

In the QR regime ($\Delta > \Delta_{SL}$), AZE has been shown to yield a quantum advantage over Markovian dynamics in the form of higher heat current $\bar{J}_c$, or, equivalently, higher cooling rate of the cold bath, for the same COP. The latter effect leads to the enticing possibility of quantum-enhanced speedup of the cooling rate of systems as we approach absolute zero, and challenges the validity of the third law in the quantum non-Markovian regime, if we expect the vanishing of the cooling rate at zero temperature as a manifestation of the Nernst unattainability theorem.

A minimal model of a QR [i.e., a periodically phase-flipped TLS permanently coupled to a finite-capacity cold bath and to an infinite-capacity heat dump (hot bath)] has been used to investigate the cooling of the cold bath toward absolute zero. The cold-bath cooling rate has been shown *not to vanish* as $T \to 0$ for quantized baths composed of magnons in the Heisenberg spin-chain model. This result challenges Nernst's third-law formulation known as the unattainability principle.

Yet, a key issue to be resolved is: Does a nonvanishing zero-temperature cooling rate suffice to attain $T = 0$, and if not, what is the lowest attainable temperature in a quantum (particularly, magnon) bath? This issue is subtle because, under nonadiabatic cooling, such a bath may not thermalize to a well-defined temperature (i.e., to a Gibbs state). Rather, the bath may then have an energy distribution.

24

Minimal Quantum Heat Manager: Heat Diode and Transistor

Heat management in quantum systems, particularly heat-flow rectification or amplification, has the potential for diverse technological applications. Here we consider a minimal quantum heat manager composed of two interacting qubits capable of near-perfect heat-flow rectification, known as heat-diode (HD) action or heat-flow amplification, alias heat transistor (HT) action. We reveal the potentially important role of spectral filtering of the coupling between the heat baths involved and the quantum system that controls the heat flow. Such filtering is enabled by interfacing the system and the baths by means of harmonic-oscillator modes whose resonant frequencies and coupling strengths are used to control the system–bath coupling spectra. We show that such bath spectral filtering (BSF) boosts the performance of the minimal quantum heat manager, allowing it to attain either perfect HD action or enhanced HT action. The BSF is a genuinely quantum electrodynamic effect, which stands in contrast to most mechanisms employed in quantum heat management schemes that have classical counterparts.

24.1 Heat Rectification with BSF

Consider a multilevel system S that interacts with heat baths on its left (L) and right (R) sides [Fig. 24.1(a)]. Here we ask: Can such a system rectify the heat flow between the baths (i.e. block the heat flow from L to R or vice versa), if these baths differ only in temperature? To this end, we need to break the L-R symmetry by resorting to a system composed of two anisotropically interacting TLS with L-R asymmetric interaction. It is governed by the Hamiltonian

$$\hat{H}_S = \frac{\hbar\omega_L}{2}\hat{\sigma}_{Lz} + \frac{\hbar\omega_R}{2}\hat{\sigma}_{Rz} + \hbar g\,\hat{\sigma}_{Lz}\hat{\sigma}_{Rx}, \qquad (24.1)$$

435

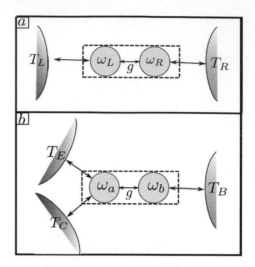

Figure 24.1 (a) Heat diode (HD), (b) heat transistor (HT) based on two coupled TLS with an anisotropic interaction strength g. For the HD, the resonant frequency of the TLS "L" ("R") is ω_L (ω_R), and it is coupled to a thermal bath at temperature T_L (T_R). These independent baths may have any distinct nonnegative temperatures. (Reprinted from Naseem et al., 2020.)

where $\hat{\sigma}_{ji}$ ($i = x, y, z$) are the Pauli matrices of the jth qubit. The subsystems L and R are coupled to corresponding baths,

$$H_{SB} = \hbar \hat{\sigma}_{Lx} \otimes \sum_k g_{Lk}(\hat{a}_{Lk} + \hat{a}_{Lk}^\dagger) + \hbar \hat{\sigma}_{Rx} \otimes \sum_k g_{Rk}(\hat{a}_{Rk} + \hat{a}_{Rk}^\dagger), \quad (24.2)$$

and the Hamiltonians for the baths are given by

$$\hat{H}_{B,j} = \hbar \sum_k \omega_k \hat{a}_{jk}^\dagger \hat{a}_{jk} \quad (j = L,R), \quad (24.3)$$

where $\hat{a}_{jk}^\dagger$ and $\hat{a}_{jk}$ are the creation and annihilation operators, respectively, of the kth mode of the jth bath, g_{jk} being the coupling amplitudes.

Upon changing the variables, $\tilde{\sigma}_{ji} = U \hat{\sigma}_{ji} U^\dagger$, by means of the unitary transformation

$$U = \exp\left(-i\frac{\theta}{2}\hat{\sigma}_{Lz}\hat{\sigma}_{Ry}\right) = \hat{I} \cos\frac{\theta}{2} - i\hat{\sigma}_{Lz}\hat{\sigma}_{Ry} \sin\frac{\theta}{2}, \quad (24.4)$$

where $\hat{I}$ is the identity operator and

$$\theta = \arctan(2g/\omega_R), \quad (24.5)$$

the Hamiltonian of the bipartite system acquires a seemingly uncoupled-particle form,

$$H_{\rm S} = \frac{\hbar\omega_{\rm L}}{2}\tilde{\sigma}_{{\rm L}z} + \frac{\hbar\Omega}{2}\tilde{\sigma}_{{\rm R}z}. \tag{24.6}$$

Here $\Omega = \sqrt{\omega_{\rm R}^2 + 4g^2}$, and the transformed operators read

$$\tilde{\sigma}_{{\rm L}z} = \hat{\sigma}_{{\rm L}z}\hat{I}_{\rm R}, \tag{24.7}$$

$$\tilde{\sigma}_{{\rm R}z} = c\hat{I}_{\rm L}\hat{\sigma}_{{\rm R}z} + s\hat{\sigma}_{{\rm L}z}\hat{\sigma}_{{\rm R}x}, \tag{24.8}$$

where

$$c = \cos\theta, \quad s = \sin\theta, \tag{24.9}$$

and $\hat{I}_j$ is the identity operator for the jth TLS.

The Hamiltonian (24.6) is diagonal in the basis

$$|n\rangle = U |i_{\rm L}, i_{\rm R}\rangle, \tag{24.10}$$

where

$$|i_{\rm L}, i_{\rm R}\rangle = |i_{\rm L}\rangle_{\rm L} |i_{\rm R}\rangle_{\rm R} \quad (i_j = 0, 1), \tag{24.11}$$

$|i_j\rangle_j$ is an eigenstate of $\hat{\sigma}_{jz}$ such that $\hat{\sigma}_{jz}|0\rangle_j = |0\rangle_j$, and

$$n = 2i_{\rm L} + i_{\rm R} + 1 \quad (n = 1, 2, 3, 4). \tag{24.12}$$

Denoting

$$c' = \cos(\theta/2), \quad s' = \sin(\theta/2), \tag{24.13}$$

we have

$$\begin{aligned}
|1\rangle &= c' |0, 0\rangle + s' |0, 1\rangle, & |2\rangle &= -s' |0, 0\rangle + c' |0, 1\rangle, \\
|3\rangle &= c' |1, 0\rangle - s' |1, 1\rangle, & |4\rangle &= s' |1, 0\rangle + c' |1, 1\rangle.
\end{aligned} \tag{24.14}$$

The corresponding eigenvalues $\hbar\omega_n$ are given by

$$\omega_{1,4} = \pm\frac{1}{2}(\omega_{\rm L} + \Omega), \quad \omega_{2,3} = \pm\frac{1}{2}(\omega_{\rm L} - \Omega). \tag{24.15}$$

In order to derive the master equation for this system, we move to the interaction picture, in which the system operators in (24.2) have the following time-dependent form in the $|n\rangle$ basis (24.14),

$$\begin{aligned}
\hat{\sigma}_{{\rm L}x}(t) &= c\sigma_{{\rm L}-}I_{\rm R}e^{-i\omega_{\rm L}t} - s\sigma_{{\rm L}-}\sigma_{{\rm R}+}e^{-i(\omega_{\rm L}-\Omega)t} \\
&\quad + s\sigma_{{\rm L}-}\sigma_{{\rm R}-}e^{-i(\Omega+\omega_{\rm L})t} + {\rm H.c.}, \\
\hat{\sigma}_{{\rm R}x}(t) &= s\sigma_{{\rm L}z}\sigma_{{\rm R}z} + cI_{\rm L}\sigma_{{\rm R}-}e^{-i\Omega t} + cI_{\rm L}\sigma_{{\rm R}+}e^{i\Omega t}.
\end{aligned} \tag{24.16}$$

Here and below the products of the identity and Pauli matrices, I_j and $\sigma_{j\pm} = (\sigma_{jx} \pm i\sigma_{jy})/2$, are the Kroenecker products, but they are not the tensor products of operators, since the basis $|n\rangle$ is not the tensor product of bases corresponding to the two TLS.

The system state ρ evolves according to the Liouville equation

$$\dot{\rho} = \mathcal{L}_L\rho + \mathcal{L}_R\rho, \tag{24.17}$$

where $\mathcal{L}_j$ ($j = $ L,R) is the superoperator corresponding to the jth bath. In the Markovian (Lindblad) approximation, the superoperators are given by (Ch. 11)

$$\mathcal{L}_j = 2\pi \sum_\alpha \left\{ G_j(\omega_\alpha)\mathcal{D}[A_{\omega_\alpha}] + G_j(-\omega_\alpha)\mathcal{D}[A_{\omega_\alpha}^\dagger] \right\}. \tag{24.18}$$

Here the dissipater, defined by

$$\mathcal{D}[O]\rho = \frac{1}{2}\left(2O\rho O^\dagger - O^\dagger O\rho - \rho O^\dagger O\right), \tag{24.19}$$

is a function of A_{ω_α} and $A_{\omega_\alpha}^\dagger$, the lowering and raising operators corresponding to the eigenstates of the system with energy difference ω_α. The temperature-dependent system–bath coupling spectra $G_j(\omega)$ are as in Chapter 11.

Here we are only interested in the equations for the populations (the diagonal elements of the density matrix). Accordingly, for example, the Lindblad superoperator for the R-bath reads

$$\mathcal{L}_R = 2\pi c^2 \left\{ G_R(\Omega)\mathcal{D}[I_L\sigma_{R-}] + G_R(-\Omega)\mathcal{D}[I_L\sigma_{R+}] \right\}. \tag{24.20}$$

Here we neglected the term responsible for proper dephasing, since it does not affect the populations. Thanks to θ in (24.5) being nonzero, the Lindblad superoperators are not L-R interchangeable, even if the temperatures are interchanged!

The quantities of interest, as in Chapters 20-23, are the heat currents

$$\mathcal{J}_j = \text{Tr}[(\mathcal{L}_j\rho)H_S] \quad (j = \text{L,R}). \tag{24.21}$$

The first law of thermodynamics requires that at steady state the heat currents satisfy $\mathcal{J}_R = -\mathcal{J}_L$. Therefore, it is sufficient to consider only $\mathcal{J}_R$ in what follows.

The ability to rectify heat flow between two baths is quantified by the rectification factor

$$\mathcal{R} = \frac{|\mathcal{J}_R(T_R, T_L) + \mathcal{J}_R(T_L, T_R)|}{\max\{|\mathcal{J}_R(T_R, T_L)|, |\mathcal{J}_R(T_L, T_R)|\}}, \tag{24.22}$$

with $\mathcal{J}_R(T_R, T_L)$ being the heat current for $T_L > T_R$, and $\mathcal{J}_R(T_L, T_R)$ its counterpart for $T_L < T_R$. The rectification factor varies between 1 for perfect rectification and 0 for complete reciprocity of the heat flows. It is seen from (24.18) that if $\mathcal{L}_L$ and $\mathcal{L}_R$ differ only in temperature [the mean quanta number $\bar{n}(\omega)$], then the L and R heat

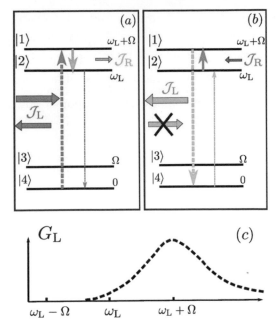

Figure 24.2 Rectification of the heat current in an HD based on two coupled TLS. In panel (a) the left bath is hotter than the right bath, and in panel (b) it is the opposite. The solid and dashed lines correspond to transitions induced by the right and left baths, respectively. Arrow thickness represents the magnitude of the transition rate. The transition at ω_L is weak due to the choice of a filtered left bath spectrum, whereas the right bath has flat spectral response (FSR). Heat can flow from left to right via the only possible "Raman-like" cycle (4124) in (a), whereas the opposite cycle (4214) in (b) is inhibited because the cold bath cannot excite the $|4\rangle \to |2\rangle$ transition. Accordingly, there is no heat flow in panel (b), and our HD gives perfect rectification. (c) Filtered spectral response function of the L bath with $G_L(\omega_L - \Omega) = 0$ is a skewed Lorentzian. (Reprinted from Naseem et al., 2020.)

currents are interchangeable when the respective temperatures are interchanged, leading to complete heat reciprocity. Nonreciprocity arises only when asymmetry exists between the L and R Lindblad superoperators, regardless of the respective temperatures.

For the choice of bath spectra in Figure 24.2(c), the anticipation of perfect rectification is confirmed by the analysis. The master equation for the bipartite system in the diagonalizing basis $|n\rangle$ (24.14) is then given by

$$
\begin{aligned}
\mathcal{L}_L \rho &= s^2 \, \Gamma_{14}^L (|4\rangle \langle 4| - |1\rangle \langle 1|) + c^2 \, \Gamma_{24}^L (|4\rangle \langle 4| - |2\rangle \langle 2|) \\
&\quad + c^2 \, \Gamma_{13}^L (|3\rangle \langle 3| - |1\rangle \langle 1|), \\
\mathcal{L}_R \rho &= c^2 \, \Gamma_{12}^R (|2\rangle \langle 2| - |1\rangle \langle 1|) + c^2 \, \Gamma_{34}^R (|4\rangle \langle 4| - |3\rangle \langle 3|).
\end{aligned} \tag{24.23}
$$

Here

$$\Gamma_{ik}^{j} = 2\pi [G_j(\omega_{ik})\rho_{ii} - G_j(-\omega_{ik})\rho_{kk}] \quad (j = \text{L,R}), \tag{24.24}$$

where $\bar{n}_j(\omega) = \left(e^{\frac{\hbar\omega}{k_B T_j}} - 1\right)^{-1}$ is the mean number of quanta at ω in the jth bath.

The steady-state heat currents can be evaluated upon noting that

$$\mathcal{L}_L + \mathcal{L}_R = 0. \tag{24.25}$$

The cumbersome expressions for them will not be presented here.

The standard assumption is that the bath response $G_j(\omega)$ ($j = \text{L,R}$) is spectrally flat. This assumption can put severe constraints on rectification by coupled subsystems, as can be seen upon considering the heat transfer channels between the baths (Fig. 24.2). The rectification factor may fall short of 1, depending on the coupling strength between the subsystems, their energy mismatch, and bath temperatures. In particular, for *identical TLS*, L and R, the rectification factor is typically much less than 1, as shown below.

To overcome these restrictions and attain perfect rectification for weakly asymmetric L-R subsystems, bath-spectral filtering (BSF) may be invoked. The ability to controllably shape $G_j(\omega)$ provides a key resource for rectification. This ability can be realized if each bath is supplemented with a harmonic-oscillator (HO) mode that serves as an interface between the bath and the respective subsystem (e.g. qubit). As shown in Chapters 20 and 22, a filter mode of an HO with resonant frequency $\tilde{\omega}_j$ that is coupled to the TLS with strength η_j and to the bath via coupling spectrum $G_j(\omega)$, yields the following modified (filtered) qubit-bath coupling spectrum

$$\tilde{G}_j = \frac{\eta_j}{\pi} \frac{\pi G_j(\omega)}{\left\{\omega - [\tilde{\omega}_j + \Delta_j(\omega)]\right\}^2 + [\pi G_j(\omega)]^2}, \tag{24.26}$$

where G_j is the unfiltered coupling spectrum, and

$$\Delta_j(\omega) = P \int_0^\infty d\omega' \frac{G_j(\omega')}{\omega - \omega'}, \tag{24.27}$$

P and $\Delta_j(\omega)$ being, respectively, the principal value and the bath-induced Lamb shift. The filtered spectrum (24.26) can drastically differ from its original (unfiltered) counterpart. The general spectral shape of the filtered spectrum is a *skewed Lorentzian*. For unfiltered spectrally flat $G_j(\omega)$ (flat spectral response – FSR), the filtered spectrum is a regular Lorentzian

$$\tilde{G}_j = \frac{\eta_j}{\pi} \frac{\kappa_j^2}{(\omega - \omega_j)^2 + (\pi\kappa_j)^2}, \tag{24.28}$$

whose width and center are, respectively, the controllable $\kappa_j = \pi G_j(\omega_j)$ and ω_j. However, (24.28) in general does not suffice for perfect rectification, since the tails

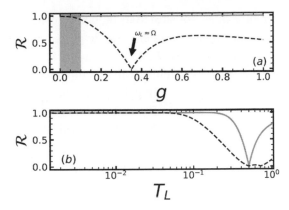

Figure 24.3 *Rectification $\mathcal{R}$ as a function of the (a) coupling strength g, (b) temperature T_L for the coupled TLS. The solid and dashed lines are for the filtered and unfiltered L bath spectrum, respectively. The L bath has filtered spectral density as shown in Figure 24.2(c), and the right bath has FSR. The shaded region is for $\omega_L \gg \Omega$. Parameters: $\omega_L = 1$, $\omega_R = 0.1$, $\kappa_L = \kappa_R = 0.001$, (a) $T_L = 2$, $T_R = 0.2$, and (b) $g = 0.35$, $T_R = 0.5$. All the system parameters are scaled to $\omega_L/2\pi = 10$ GHz. (Reprinted from Naseem et al., 2020.)*

of a regular Lorentzian fall off too slowly with frequency. Hence, one should strive for BSF that yields a strongly asymmetric skewed-Lorentzian with fast spectral drop-off on one wing.

In an HD composed of coupled TLS, one spectrally filtered bath couples to the transition at frequency Ω, while the other couples to the transition at frequencies ω_L and $\omega_L + \Omega$, provided the filtered coupling spectra satisfy

$$G_L(\omega_L + \Omega) \gg G_R(\omega_L + \Omega),$$
$$G_L(\omega_L) \gg G_R(\omega_L), \qquad (24.29)$$
$$G_L(\Omega) \ll G_R(\Omega).$$

Such BSF yields disjoint coupling spectra of the L- and R-subsystems to their respective baths. Under conditions (24.29), only one bath contributes to the heat flow in either direction, leaving only one unidirectional heat-transfer channel (Raman-like cycle) intact in Figure 24.2. This ensures perfect rectification, regardless of whether the two TLS are identical (resonant) or what the magnitude of g is (Fig. 24.3). By contrast, FSR allows for other, bidirectional channels due to spectral overlap of the two baths and therefore can yield weak rectification.

In certain situations BSF is not mandatory for rectification. An example is the case, where the R bath is only coupled to the transition at the frequency Ω such that $\Omega \ll \omega_L - \Omega$, whereas the L bath is coupled to the transitions with frequencies ω_L and $\omega_L + \Omega$. Therefore, as we change the bath temperatures, the cold bath (here R bath) temperature does not suffice to induce transitions except at the frequency Ω. In the absence of BSF, the other possible global cycle is associated with the transitions at ω_L and $\omega_L - \Omega$ by the L bath and at Ω by the R bath. For the parameter choice in Figure 24.3, since the R bath temperature is too low to induce the transition even at $\omega_L - \Omega$ (as it is much higher than Ω), there is no significant effect of BSF on the rectification. By contrast, if $\omega_L \sim \Omega$, the cold bath may induce transitions at ω_L, and $\omega_L - \Omega$. In this case, BSF drastically improves the rectification if we select our L bath spectrum such that the transition $\omega_L - \Omega$ is either completely filtered out or at least drastically suppressed, as in Figure 24.2.

24.2 Heat-Transistor Amplification with BSF

A three-terminal heat-transistor (HT) is comprised of three baths dubbed base (B), emitter (E), and collector (C) that are coupled via a controller system S comprised of coupled TLS a and b [Fig. 24.1(b)].

The master equation in the interaction picture now evaluates to

$$\dot{\hat{\rho}} = (\mathcal{L}_E + \mathcal{L}_C + \mathcal{L}_B)\,\hat{\rho}, \tag{24.30}$$

where, similarly to (24.20),

$$\mathcal{L}_B = 2\pi c^2 \left\{ G_B(\Omega)\mathcal{D}[I_a\sigma_{b-}] + G_B(-\Omega)\mathcal{D}[I_a\sigma_{b+}] \right\} \tag{24.31}$$

and

$$\mathcal{L}_j = \sum_{l=0,\pm1} \mathcal{L}_{l,j} \quad (j = \text{E,C}). \tag{24.32}$$

Here

$$\begin{aligned} \mathcal{L}_{0,j} &= 2\pi c^2 \left\{ G_j(\omega_L)\mathcal{D}[\sigma_{a-}I_b] + G_j(-\omega_L)\mathcal{D}[\sigma_{a+}I_b] \right\}, \\ \mathcal{L}_{\pm1,j} &= 2\pi s^2 \left\{ G_j(\omega_\pm)\mathcal{D}[w_{\pm1}] + G_j(-\omega_\pm)\mathcal{D}[w_{\pm1}^\dagger] \right\}, \\ w_1 &= \sigma_{a-}\sigma_{b-}, \quad w_{-1} = \sigma_{a-}\sigma_{b+}, \quad \omega_\pm = \omega_a \pm \Omega. \end{aligned} \tag{24.33}$$

In (24.31)–(24.33), we dropped the terms that affect only equations for the coherences and do not enter the equations for the populations.

For the choice of bath spectra shown in Figure 24.4(c), the superoperators in the master equation (24.30) in the diagonalizing basis $|n\rangle$ [cf. (24.6)–(24.15)] are given by

$$\begin{aligned} \mathcal{L}_E(\rho) &= s^2 \Gamma_{23}^E(|3\rangle\langle3| - |2\rangle\langle2|), \\ \mathcal{L}_C(\rho) &= c^2 \Gamma_{13}^C(|3\rangle\langle3| - |1\rangle\langle1|) + c^2 \Gamma_{24}^C(|4\rangle\langle4| - |2\rangle\langle2|), \\ \mathcal{L}_B(\rho) &= s^2 \Gamma_{12}^B(|2\rangle\langle2| - |1\rangle\langle1|) + s^2 \Gamma_{34}^B(|4\rangle\langle4| - |3\rangle\langle3|). \end{aligned} \tag{24.34}$$

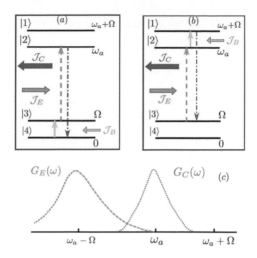

Figure 24.4 HT heat transfer via coupled TLS under BSF. The choice of filtered bath spectra for E and C baths is shown in panel (c), whereas the B bath has FSR. In panels (a) and (b) the Raman-like cycles (4324) and (3213), respectively, and their inverse are the only allowed cycles for heat transfer. The dashed, dashed-dotted, and solid arrows represent the transitions induced by the E, C, and B baths, respectively. All other cycles are prohibited. The B current is associated with the lowest transition frequency (Ω) in the system. The horizontal arrows denote the heat currents $\mathcal{J}_E$, $\mathcal{J}_C$, and $\mathcal{J}_B$. (Reprinted from Naseem et al., 2020.)

The steady-state heat currents then evaluate to

$$\mathcal{J}_E = s^2 \hbar \omega_- \Gamma_T, \quad \mathcal{J}_B = s^2 \hbar \Omega \Gamma_T, \quad \mathcal{J}_C = -s^2 \hbar \omega_a \Gamma_T. \qquad (24.35)$$

Significantly, the emitter and base heat currents are directed opposite to the collector current. The cumbersome exact expression for Γ_T is simplified as $T_E \to \infty$ to the form

$$\Gamma_T = \frac{c^2 \gamma_C \gamma_B (\epsilon_B - \epsilon_C)}{c^2 \gamma_B (1 + \epsilon_B)(1 + \epsilon_C) + s^2 \{\epsilon_C \gamma_B + \epsilon_B [\gamma_E + \epsilon_C (\gamma_E + \gamma_B)]\}}, \qquad (24.36)$$

where

$$\gamma_E = \omega_- G_E(-\omega_-), \quad \gamma_C = \omega_a G_C(-\omega_a), \quad \gamma_B = \Omega G_B(-\Omega),$$

$$\epsilon_B = e^{\frac{\hbar \Omega}{k_B T_B}}, \quad \epsilon_C = e^{\frac{\hbar \omega_a}{k_B T_C}}, \quad c = \cos\theta, \quad s = \sin\theta. \qquad (24.37)$$

The HT functions properly if a small change of the B temperature results in a highly amplified heat flow through E or C. The amplification factor is defined as

$$\alpha_{E,C} = \frac{\partial \mathcal{J}_{E,C}}{\partial \mathcal{J}_B}, \qquad (24.38)$$

or, equivalently, upon using energy conservation, as

$$\alpha_E = \left| \frac{\partial(\mathcal{J}_E)}{\partial(\mathcal{J}_E + \mathcal{J}_C)} \right| = \left| \frac{R_E}{R_E + R_C} \right|. \tag{24.39}$$

Here $R_E = (\partial \mathcal{J}_E / \partial \mathcal{J}_B)^{-1}_{T_E=\text{const}}$ and $R_C = -(\partial \mathcal{J}_C / \partial \mathcal{J}_B)^{-1}_{T_C=\text{const}}$ are dubbed *differential thermal resistances*. Similar relations can be written for α_C. Amplification factors larger than 1 require, according to (24.39), negative differential thermal resistance (NDTR), namely, $R_E \times R_C < 0$.

It is our purpose to find out what mechanisms may ensure that the amplification factors exceed 1 and that NDTR holds in the minimal models considered above. For the two coupled TLS, only 2 (Raman-like) cycles and their inverse (Fig. 24.4) are open channels for heat flow between all three baths, whereby energy absorption from B results in energy transfer from E to C [Fig. 24.4(a)] or vice versa [Fig. 24.4(b)].

The amplification factors are then found to be

$$\alpha_E = \frac{\partial \mathcal{J}_E}{\partial \mathcal{J}_B} = \frac{\omega_-}{\Omega}, \quad \alpha_C = \frac{\partial \mathcal{J}_C}{\partial \mathcal{J}_B} = -\frac{\omega_a}{\Omega}. \tag{24.40}$$

For the choice of bath spectra in Figure 24.5(c), the possible transitions induced by the baths are shown in Figure 24.5(a). The steady-state heat currents in this case are given by

$$\begin{aligned}
\mathcal{J}_E &= \Omega s^2 \Gamma_1, \\
\mathcal{J}_B &= \omega_- s^2 \Gamma_1 - \omega_a c^2 \Gamma_2, \\
\mathcal{J}_C &= -\omega_a s^2 \Gamma_1 + \omega_a c^2 \Gamma_2,
\end{aligned} \tag{24.41}$$

where

$$\Gamma_1 = \Gamma^E_{32}, \quad \Gamma_2 = \Gamma^E_{13} + \Gamma^E_{24}. \tag{24.42}$$

For the same parameters, this choice of bath spectra allows larger heat flow than that presented in Figure 24.4(c), because it allows more energy cycles to transfer heat between the baths, including direct transfer of heat between the emitter and the collector, which is indicated by slashed lines in Figure 24.5(a). This can also be seen upon comparing (24.35) and (24.41) and noting that Γ_1 and Γ_2 have always opposite signs.

In more detail, let us consider an energy cycle as in Figure 24.5(b), which shows that $\Gamma^C_{24} > 0$, $\Gamma^E_{24} < 0$. This cycle transfers the heat directly between the collector and the emitter. A similar cycle can be considered to show that $\Gamma^C_{13} > 0$ implies $\Gamma^E_{13} < 0$. Consequently, in all possible cycles, Γ_1 and Γ_2 must have opposite signs, which results in the increase of heat currents in the system compared to (24.35).

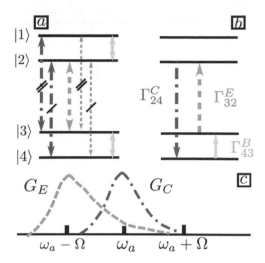

Figure 24.5 (a) The possible transitions induced by the three baths in an HT under suitable BSF. The filtered bath spectra are presented in (c). The solid, dashed, dot-dashed transitions are induced by the base, emitter, and collector baths, respectively. The slashed lines represent direct transfer of heat between the emitter and the collector without passing through the base. (b) An example of an energy cycle that transfers the heat between the three baths. It shows that Γ_{32}^E and Γ_{24}^E have opposite signs. (Reprinted from Naseem et al., 2020.)

The amplification factors in this case are given by

$$\alpha_E = \frac{\omega_-}{\Omega} - \frac{\omega_a c^2}{\Omega s^2} \frac{\partial \Gamma_2}{\partial \Gamma_1}, \quad \alpha_C = -\frac{\omega_a}{\Omega} + \frac{\omega_a c^2}{\Omega s^2} \frac{\partial \Gamma_2}{\partial \Gamma_1}. \qquad (24.43)$$

Heat amplification is reduced for the overlapping bath spectra shown in Figure 24.5(c) as compared to the separated bath spectra presented in Figure 24.4(c), as seen upon comparing (24.40) and (24.43), and noting that $\frac{\partial \Gamma_2}{\partial \Gamma_1} > 0$. The amplification factors α_E for the filtered bath spectra in Figure 24.4(c) are strongly boosted (Fig. 24.5) compared to those obtained for unfiltered bath spectra (FSR).

The expressions for the steady-state heat currents under the chosen BSF show that $\mathcal{J}_C$, $\mathcal{J}_E$, and $\mathcal{J}_B$ are associated with the transition frequencies ω_a, ω_-, and Ω, respectively. For weakly coupled TLS, we have $\omega_a > \omega_- \gg \Omega$, so that, accordingly, $\mathcal{J}_C$, $\mathcal{J}_E \gg \mathcal{J}_B$, and the amplification factors become $\alpha_E = \omega_-/\Omega$, and $\alpha_C = -\omega_a/\Omega$. This choice of the filtered bath spectra (BSF) [Figs. 24.4(c), 24.5(c)] ensures the optimal HT regime, wherein small increase in T_B results in only slight increase of $\mathcal{J}_B$, but in large increase of $\mathcal{J}_E$ and $\mathcal{J}_C$ (Fig. 24.6).

The amplification factor, which depends on the two-qubit coupling strength g, can be strongly boosted by resorting to BSF, as in the case of rectification. The amplification boost can be understood by noting that in the optimal HT regime we have, from energy conservation, opposite heat flows from E and C,

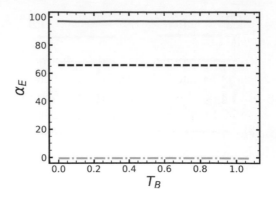

Figure 24.6 HT amplification α_E as a function of the base temperature T_B for coupled TLS. The solid, dashed, and dot-dashed lines are for BSF with the choice of bath spectra shown in Figure 24.4(c), Figure 24.5(c), and unfiltered baths, respectively. BSF gives rise to dramatic boost in amplification. Parameters: $\omega_a = 1$, $\omega_b = 0.01$, $g = 0.005$, $\kappa_C = \kappa_E = \kappa_B = 0.001$, $T_E = 1$, and $T_C = 0.01$. (Reprinted from Naseem et al., 2020.)

$$|\mathcal{J}_C| \approx |\mathcal{J}_E|. \tag{24.44}$$

To avoid oppositely directed heat flows from the emitter and the collector, which can strongly inhibit the amplification factors α_E and α_C, we must *spectrally isolate* $\mathcal{J}_C$ and $\mathcal{J}_E$. To this end, we should couple the transition frequencies associated with the E and C heat flows to separate baths: The E bath should couple only to the transition at frequency $\omega_E = \omega_a$, and the C bath only to that at $\omega_C = \omega_a - \Omega$. This condition amounts to the separation of the C and E coupling spectra by BSF, which again corresponds to skewed-Lorentzians with weakly overlapping tails (Fig. 24.4). By contrast, since the B current feeds on the much lower Ω and is much smaller than the other currents, it is unaffected by BSF: The B bath may conform to FSR.

The amplification boost caused by BSF can be dramatic, as shown in Figure 24.6. Similar boost is obtained for the amplification if the transitions at the frequencies Ω, ω_a and $\omega_a + \Omega$ are induced by the B, E, and C baths respectively.

24.3 Discussion

We have shown the ability of bath spectral filtering (BSF) to serve as a resource for boosting the performance of a multifunctional heat manager, particularly as heat diode (HD) or heat transistor (HT), when the manager is a quantum heat-control system with the minimal number of degrees of freedom – two coupled TLS. The key to achieving perfect HD action (the heat-flow rectification) is *asymmetry* in the coupling of the control system to left- and right-hand baths. Such asymmetry can arise for anisotropic two-TLS coupling. Yet, since the TLS coupling strength is restricted, the asymmetry required for HD or HT action is not always attainable, particularly for identical TLS.

By contrast, BSF can yield *strongly skewed-Lorentzian bath line shapes* that enable perfect rectification of an HD regardless of such restrictions. Similarly, skewed-Lorentzian BSF applied to the collector and emitter baths can boost HT amplification in the optimal regime where the base heat current is very small, so that the collector and emitter heat currents are nearly equal in magnitude and opposite in direction. The beneficial role of BSF in quantum heat management is akin to that discussed in Chapters 20, 22, and 23 for TLS-based minimal heat engines and refrigerators that must be coupled to spectrally separated hot and cold baths in order to attain high efficiency or power.

The guidelines for engineering the skewed-Lorentzian BSF to ensure HD and HT performance boost are as follows: Introduce a cutoff or at least a sharp drop-off on the required spectral wing of the Lorentzian by appropriately selecting the filter-mode frequency $\tilde{\omega}_j$ and the Lamb shift in (24.26), (24.27). Distributed Bragg reflectors (DBR) that possess bandgaps allow for such engineering.

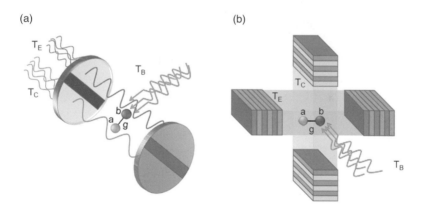

Figure 24.7 (a) Realization of the model in Figure 24.1 for coupled qubits in a bimodal cavity. The TLS (a and b) are coupled via dipole–dipole interaction with separation-dependent coupling strength g that can be engineered in a cavity. The qubits are coupled to the cavity modes that act as BSF: filters coupling the TLS a (b) to the thermal bath at temperature T_C (T_E). A heat-insulating strip (black) allows the coexistence of two regions with different temperatures. For HT configuration, the qubits are exposed to thermal radiations at a much lower frequency that plays the role of base bath B at temperature T_b. (b) Realization of the same model for distributed Bragg reflectors (DBR). The coupled qubits (a and b) are placed at the crossing region of the two cavity fields. (Reprinted from Naseem et al., 2020.)

The advantageous HD or HT schemes boosted by BSF described here can be experimentally realized in diverse setups:

A. Nitrogen-vacancy (NV) centers in diamond or similar defects with optical and microwave transitions can act as TLS embedded in a bimodal cavity. Each TLS can be near-resonantly coupled to another cavity mode that has an antinode at its location. That cavity mode acts as a filter coupling the TLS to one or another thermal bath with the temperature of the respective pair of mirrors [Fig. 24.7(a)]. Inter-TLS coupling can arise from their near-resonant dipole–dipole interaction. Heat-flow rectification may be observed upon interchanging the mirror temperatures. BSF takes place when the cavity finesse is high enough. The bath spectra associated with the two high-Q modes have suppressed overlap, and rectification is boosted. A HT configuration can be realized by exposing both TLS to the same thermal radiation at a much lower (say, microwave) frequency that plays the role of the base bath B [Fig. 24.7(a)]. Such a configuration can exhibit skewed-Lorentzian BSF when the TLS are placed within distributed Bragg reflectors (DBR) [Fig. 24.7(b)].

B. Electronic circuits composed of transmon qubits in superconducting cavities have been identified as potential HD. The modes in such circuits may be implemented by a transmission line.

Commonly, quantum HT and HD schemes have classical analogs. By contrast, the BSF is a uniquely quantum mechanical resource that is manifest in both the dissipative and dispersive (Lamb) rates of energy exchange between the system and the bath (Ch. 8).

Conclusions and Outlook

In this book, we have linked two burgeoning areas of quantum physics. The first one, expounded in Part I, is concerned with open quantum systems, namely, systems that interact with their environment (typically, a thermal bath or a meter), which can cause relaxation or decoherence of their quantum states (Chs. 1–7) or modify their energy spectra (Ch. 8). Part II of the book (Chs. 15–24) employs the concepts and tools developed in Part I to gain thorough understanding of thermodynamic processes in quantum systems and their control.

The problem that is most extensively discussed in Part I is: How does decoherence come about and how can it be controlled (Chs. 9–14)?. The basic description of decoherence can be traced back to von Neumann's projection postulate: The system is impulsively entangled with a bath or a meter and the latter is then ignored, traced out. The outcome is an irreversible change in the state of the system (Ch. 9). Essentially the same view underlies the notion of random, instantaneous "quantum jumps" between evolution paths in Hilbert space. Yet, the dichotomy between a pure state before the entanglement and a mixed state after the trace-out reveals the oversimplification of the "quantum jump" concept: it ignores the buildup of system–bath correlations and does not allow for interventions that can modify this process. This limitation holds true for the Markovian (memoryless) dynamics of the system caused by the traced-out bath (Ch. 11).

The tools and concepts provided in Chapters 9–14 cater to those interested in understanding and controlling the transition from unitary to classical-like or thermodynamic evolution of a system by venturing beyond the Markovian approximation:

- The *quantum Zeno effect* (QZE) of frequent measurements on an open system shows that one need not isolate a system from its environment to suppress its relaxation or decoherence, and conversely, the *anti-Zeno effect* (AZE) of less

449

frequent measurements shows that such processes may be augmented without increasing the system–bath coupling (Ch. 10).

- The *non-Markovian master equation* for a field- (dynamically) controlled system that interacts with thermal baths shows that the control must be faster than the bath correlation (memory) time in order to counter or change bath-induced decoherence.

- A unified *multitask framework for the dynamical control* of an open system is provided by a method dubbed bath-optimized task-oriented control (BOTOC), which is expounded in Chapter 12 and compared to the more standard method of dynamical decoupling (DD). BOTOC allows for controlling the interaction of a quantum system with multiple baths and is aimed at optimally executing a chosen operation, whether unitary or non-unitary, for example: entanglement and quantum information (QI) protection during quantum computation or QI processing (Ch. 13), state-transfer in hybrid systems with the highest fidelity possible (Ch. 14), or state purification (Ch. 16). To this end, one must design the temporal control of the Hamiltonian that governs the system, subject to variational minimization or maximization (as the case may be) of a functional that quantifies the success probability of the operation. For certain tasks, BOTOC may *benefit from the system–bath interactions* when the operation is non-unitary, for example, state purification (Ch. 16). BOTOC is not limited to pulsed control, as opposed to DD and thus may drastically reduce the energy and control errors if a smoothly varying field is employed instead of pulses.

Further development of the tools described above, both theoretically and experimentally, should aim at their challenging extension to the regime of strong system–bath coupling (beyond the Born approximation). On the conceptual side, such development should enable us to find out: What are the complexity limits of multipartite open quantum systems and their baths that may be effectively controlled by external interventions and thereby preserve their unitarity? On the applied side, an effort should be made to identify the most appropriate methods of coherence or entanglement protection and manipulations in QI processing and quantum sensing. Such methods are needed in view of the inherent restrictions of quantum error correction methods, which presume that only a few elements in a quantum ensemble are simultaneously affected by decoherence.

The major emphasis placed in Part II on the description of thermodynamic control in quantum mechanical terms must first fully resolve the quantum thermalization problem (Chs. 1, 2): the universal tendency of nonequilibrium states of a closed system to evolve toward thermal equilibrium despite the fundamental reversibility of quantum mechanics. These apparently contradictory time-evolutions are reconciled by means of von Neumann's quantum ergodic hypothesis and Srednicki's eigenstate thermalization hypothesis (ETH), which predicts that almost any super-position of realistic Hamiltonian-eigenstates eventually relaxes to a state that practically coincides with a thermal-equilibrium state. It is, however, as important to fully understand *noncompliance with thermalization*, which may lead to persist-ent (or long-time) nonclassical and (or) nonthermal behavior of quantum systems. Such persistent nonthermalization may be an alternative to irreversibility in closed, composite, quantum systems. In particular, it may violate the generalized Gibbs ensemble (GGE) behavior that represents modified thermalization under symme-try constraints. Nonthermalizing states dubbed "scars" were already identified by Srednicki. There are yet other scenarios that may hinder thermalization, which need be further explored:

- *Monitoring the system by quantum nondemolition (QND) measurements* (Ch. 16): QND monitoring is supposed to be nonintrusive for closed systems, yet Chapter 16 shows that it may strongly affect system–bath correlations and thereby prevent their mutual equilibration, provided the monitoring intervals are shorter than the bath memory time (i.e., they are in the non-Markovian domain). Although monitoring effectively decoheres the evolution and is com-monly assumed to be a means of attaining thermalization, this is not the case in the frequent-monitoring regimes: *the QZE and AZE regimes prevent thermaliza-tion* of a TLS immersed in a thermal bath (Ch. 16). Since these regimes represent *generic control paradigms,* we may ask: What are the general requirements for persistent nonthermalization of composite quantum systems under dynamical control?
- *Interactions of quantum systems with spectrally structured baths: If the bath spectra possess bandgaps or cutoffs, we may encounter nonthermalization* even if the bath modes are all thermalized. An example of such anomalous behavior for a TLS near a band-edge at zero temperature has previously been shown by us. What needs to be elucidated is: What are the general conditions on bath spectra to ensure or, conversely, avoid thermalization?
- *Strong system–bath coupling:* Traditional thermodynamics presupposes negligi-ble coupling (separability) of the system (working medium) and an inïñÁnite bath. Yet, a coupling that is too small would render quantum (nanoscale) heat machines almost useless for practical applications. Hence, their performance bound should be explored as a function of coupling, all the way to the strong-coupling limit. *But does thermalization occur in this limit, and will the machines function there?* Spohn's inequality (Ch. 15) is invalid in this non-Markovian limit, that is, the entropy production may oscillate between positive and neg-ative values (Ch. 16). Hence the question: How to quantify heat and entropy exchange under such conditions?

There are compelling practical reasons to resolve the foregoing issues: They arise in single- or few-atom machines that are now being realized. It is imperative to examine the conceptual compatibility of such machines with quantum mechanics on the one hand, and thermodynamics on the other. Key issues in their description still require elucidation:

- *Arrow of time:* According to the second law, the tendency of total entropy to increase defines the "arrow of time" of systems interacting with heat baths. Since there is usually no control of the bath, reversibility or irreversibility is determined by Spohn's inequality, which asserts that any quantum system coupled to any bath must irreversibly relax to its steady state. The universality of this inequality as a faithful rendition of the second law has been restricted here and replaced by a more general inequality (Ch. 15), motivated by the need to properly estimate the performance bounds of quantum machines.

- *Quantum-machine performance bounds:* There has been much surprise over the ability of engines to surpass the Carnot bound by exploiting nonthermal quantum baths. Efficiency bounds above Carnot have been conjectured to follow from "generalized quantum second laws." Yet, machines fueled by nonthermal baths adhere to rules that differ from those of heat engines (Ch. 18) and are not restricted by the second law. Work extraction depends on the nonpassivity (ergotropy – Chs. 15, 18–22) of the system state and the bath characteristics as well as on the *work-information* and the *heat-information* trade-offs achievable in the quantum domain (Ch. 17). The open questions are: What are the *performance bounds* of quantum machines that deviate from heat-machine operation? What are the trade-offs between *heat-to-work conversion* and the resources of *information and ergotropy* in quantum machines?

- *Heat machines comprising quantum ingredients:* machines whose working medium, the baths, or the work-storing piston/battery are quantized have been a subject of great interest as part of the quest for devices that may possess quantum supremacy. Such supremacy has been commonly linked to *quantum coherence or entanglement.* Yet, in Chapters 19–23 counterexamples are given whereby machines with classical phase coherent (of the piston) outperform their quantum-coherent counterparts. It is also noteworthy that, unlike heat machines wherein quantum coherence helps overcome detrimental frictional effects, the *frictionless, incoherent* minimal (TLS-based) heat engine in Chapter 20 actually increases its efficiency with the cycle speed, up to the Carnot bound. Hence the need to clearly define the criteria and conditions for quantum machine supremacy.

- *Quantum machine speed limit:* The dependence of machine efficiency and power on the speed (cycle rate) at which they operate have been outstanding issues since the inception of thermodynamics. Speed pertains to the nearly unfathomed aspect of *non-Markovianity:* If we keep increasing the machine speed, non-Markovianity becomes unavoidable and may entail advantages (Chs. 20 and 23), but the ultimate quantum speed limit of heat machines is still unknown. An intriguing illustration of the need to resolve this issue is given in Chapter 23, where the minimal quantum refrigerator model appears to challenge Nernst's unattainability principle for the cooling speed of a quantum magnon bath: this speed remains finite as the absolute zero is approached. However, the underlying assumption that must still be verified is that the bath equilibrates at every cooling step. This issue has strong practical implications, since the cooling speed of quantum baths determines the ability to maximize the information-gathering accuracy of quantum sensors by dynamical control.

It is seen from this summary that, despite its great progress, the endeavor to gain in-depth understanding of thermodynamics in quantized systems under control is still far from completion: Some of the principal and practical issues remain unresolved.

The resolution of outstanding issues in the control of quantum systems, with a focus on thermodynamic processes, is essential for the realization of quantum information processing and sensing, and, more generally, for sustaining the trend of device miniaturization toward the nanoscale. As the device size approaches the nanoscale, we are necessarily confronted with the coexistence of quantum and thermodynamic effects. Since cooling rate and power constraints become very severe in this domain, a major technological goal we may identify for control is the optimization of refrigerator performance. Such optimization becomes crucial in nanoscale electronics as the density of transistors on microchips grows, thereby increasing heat production that may impede further miniaturization of electronic devices unless the cooling problem is resolved by control methods.

The long-term goal should be to seamlessly unify the thermodynamic and quantum coherent descriptions of composite quantum systems, at least on mesoscopic scales where the Hilbert spaces of both systems and baths span a limited number of degrees of freedom. Such unification may allow us to test and possibly revise the ruling paradigms described above. A complementary goal should be to establish the limits of controllability of quantum systems, which are expected to be determined by their complexity.

Bibliography

Chapter 1

Batalhão, T. B., Souza, A. M., Sarthour, R. S., et al. 2015. Irreversibility and the arrow of time in a quenched quantum system. *Phys. Rev. Lett.*, **115**(19), 190601.

Geiger, R., Langen, T., Mazets, I. E., and Schmiedmayer, J. 2014. Local relaxation and light-cone-like propagation of correlations in a trapped one-dimensional Bose gas. *New J. Phys.*, **16**, 053034.

Gogolin, C., and Eisert, J. 2016. Equilibration, thermalisation, and the emergence of statistical mechanics in closed quantum systems. *Rep. Prog. Phys.*, **79**, 056001.

Langen, T., Erne, S., Geiger, R., et al. 2015. Experimental observation of a generalized Gibbs ensemble. *Science*, **348**(6231), 207–211.

Linden, N., Popescu, S., Short, A. J., and Winter, A. 2009. Quantum mechanical evolution towards thermal equilibrium. *Phys. Rev. E*, **79**(6), 061103.

Perarnau-Llobet, M., Riera, A., Gallego, R., Wilming, H., and Eisert, J. 2016. Work and entropy production in generalised Gibbs ensembles. *New J. Phys.*, **18**, 123035.

Polkovnikov, A., Sengupta, K., Silva, A., and Vengalattore, M. 2011. Colloquium: Nonequilibrium dynamics of closed interacting quantum systems. *Rev. Mod. Phys.*, **83**(3), 863–883.

Reimann, P. 2010. Canonical thermalization. *New J. Phys.*, **12**, 055027.

Rigol, M., and Srednicki, M. 2012. Alternatives to eigenstate thermalization. *Phys. Rev. Lett.*, **108**(11), 110601.

Rigol, M., Dunjko, V., and Olshanii, M. 2008. Thermalization and its mechanism for generic isolated quantum systems. *Nature*, **452**(7189), 854–858.

Srednicki, M. 1994. Chaos and quantum thermalization. *Phys. Rev. E*, **50**(2), 888–901.

Srednicki, M. 1999. The approach to thermal equilibrium in quantized chaotic systems. *J. Phys. A*, **32**(7), 1163–1175.

Trotzky, S., Chen, Y.-A., Flesch, A., et al. 2012. Probing the relaxation towards equilibrium in an isolated strongly correlated one-dimensional Bose gas. *Nature Phys.*, **8**, 325–330.

von Neumann, J. 1929. Proof of the ergodic theorem and the *H*-theorem in quantum mechanics. *Z. Phys.*, **57**, 30–70 (English translation: 2010, *Eur. Phys. J. H*, **35**, 201–237).

Chapter 2

Akulin, V. M., and Karlov, N. V. 1992. *Intense Resonant Interactions in Quantum Electronics*. Berlin: Springer.

Breuer, H.-P., and Petruccione, F. 2002. *The Theory of Open Quantum Systems*. Oxford: Oxford University Press.

Deffner, S., and Lutz, E. 2011. Nonequilibrium entropy production for open quantum systems. *Phys. Rev. Lett.*, **107**, 140404.

Gemmer, J., Michel, M., and Mahler, G. 2004. *Quantum Thermodynamics*. Berlin: Springer.

Kofman, A. G., Kurizki, G., and Sherman, B. 1994. Spontaneous and induced atomic decay in photonic band structures. *J. Mod. Opt.*, **41**(2), 353–384.

Lindblad, C. 1983. *Non-Equilibrium Entropy and Irreversibility*. Dordrecht: Springer.

Linden, N., Popescu, S., Short, A. J., and Winter, A. 2009. Quantum mechanical evolution towards thermal equilibrium. *Phys. Rev. E*, **79**(6), 061103.

Reimann, P. 2010. Canonical thermalization. *New J. Phys.*, **12**, 055027.

Schlosshauer, M. A. 2007. *Decoherence and the Quantum-to-Classical Transition*. Berlin: Springer.

Weiss, U. 2012. *Quantum Dissipative Systems*. 4th ed. Singapore: World Scientific.

Zurek, W. H. 2003. Decoherence, einselection, and the quantum origins of the classical. *Rev. Mod. Phys.*, **75**(3), 715–775.

Zurek, W. H. 2009. Quantum Darwinism. *Nat. Phys.*, **5**, 181–188.

Chapter 3

Andre, A., and Lukin, M. D. 2002. Manipulating light pulses via dynamically controlled photonic band gap. *Phys. Rev. Lett.*, **89**(14), 143602.

Berestetskii, V. B., Lifshitz, E. M., and Pitaevskii, L. P. 1982. *Quantum Electrodynamics*. 2nd ed. Amsterdam: Elsevier.

Friedler, I., Kurizki, G., and Petrosyan, D. 2005. Deterministic quantum logic with photons via optically induced photonic band gaps. *Phys. Rev. A*, **71**(2), 023803.

Giamarchi, T. 2003. *Quantum Physics in One Dimension*. Oxford: Oxford University Press.

Jones, W., and March, N. H. 1985. *Theoretical Solid State Physics*. 2 vols. Mineola, NY: Dover Publications.

Kerker, M. 1969. *The Scattering of Light and Other Electromagnetic Radiation*. New York: Academic.

Kittel, C. 1987. *Quantum Theory of Solids*. 2nd ed. Hoboken, NJ: Wiley.

Kurizki, G. 1990. Two-atom resonant radiative coupling in photonic band structures. *Phys. Rev. A*, **42**(5), 2915–2924.

Landau, L. D., and Lifshitz, E. M. 2011. *Statistical Physics, Part 1*. 3rd ed. Amsterdam: Elsevier.

Landau, L. D., Lifshitz, E. M., and Pitaevskii, L. P. 1984. *Electrodynamics of Continuous Media*. 2nd ed. Amsterdam: Elsevier.

Lifshitz, E. M., and Pitaevskii, L. P. 1980. *Statistical Physics, Part 2: Theory of the Condensed State*. 2nd ed. Oxford: Pergamon Press.

Chapter 4

Cohen-Tannoudji, C., Dupont-Roc, J., and Grynberg, G. 1992. *Atom-Photon Interactions: Basic Process and Applications*. New York: John Wiley & Sons.

Gelbwaser-Klimovsky, D., and Aspuru-Guzik, A. 2015. Strongly coupled quantum heat machines. *J. Phys. Chem. Lett.*, **6**, 3477–3482.

Gelbwaser-Klimovsky, D., and Kurizki, G. 2015. Work extraction from heat-powered quantized optomechanical setups. *Sci. Rep.*, **5**, 7809.

Kittel, C. 1987. *Quantum Theory of Solids*. 2nd ed. New York: John Wiley & Sons.

Liu, Y., and Houck, A. A. 2017. Quantum electrodynamics near a photonic bandgap. *Nature Phys.*, **13**, 48–52.

Mandl, F., and Shaw, G. 2010. *Quantum Field Theory*. 2nd ed. New York: John Wiley & Sons.

Mazets, I. E., Kurizki, G., Katz, N., and Davidson, N. 2005. Optically induced polarons in Bose-Einstein condensates: Monitoring composite quasiparticle decay. *Phys. Rev. Lett.*, **94**(19), 190403.

Rao, D. D. B., and Kurizki, G. 2011. From Zeno to anti-Zeno regime: Decoherence-control dependence on the quantum statistics of the bath. *Phys. Rev. A*, **83**, 032105.

Scully, M. O., and Zubairy, M. S. 1997. *Quantum Optics*. Cambridge: Cambridge University Press.

Chapter 5

Agarwal, G. S. 1974. *Quantum Statistical Theories of Spontaneous Emission*. Berlin: Springer-Verlag.

Fleischhauer, M., and Lukin, M. D. 2000. Dark-state polaritons in electromagnetically induced transparency. *Phys. Rev. Lett.*, **84**(22), 5094–5097.

Harris, S. E. 1997. Electromagnetically induced transparency. *Physics Today*, **50**(7), 36–42.

Kofman, A. G., and Kurizki, G. 1996. Quantum Zeno effect on atomic excitation decay in resonators. *Phys. Rev. A*, **54**(5), R3750–R3753.

Kofman, A. G., Kurizki, G., and Sherman, B. 1994. Spontaneous and induced atomic decay in photonic band structures. *J. Mod. Opt.*, **41**(2), 353–384.

Kofman, A. G., and Kurizki, G. 1998. Lasing without inversion in cavities via vacuum-field control of coherence and decay. *Opt. Commun.*, **153**, 251–256.

Lambropoulos, P., Nikolopoulos, G. M., Nielsen, T. R., and Bay, S. 2000. Fundamental quantum optics in structured reservoirs. *Rep. Prog. Phys.*, **63**(4), 455–503.

Scully, M. O. 1992. From lasers and masers to phaseonium and phasers. *Phys. Rep.*, **219**, 191–201.

Weisskopf, V., and Wigner, E. 1930. Calculation of the natural line width based on Dirac's light theory. *Z. Phys.*, **63**, 54–73.

Chapter 6

Beck, S., Mazets, I. E., and Schweigler, T. 2018. Nonperturbative method to compute thermal correlations in one-dimensional systems. *Phys. Rev. A*, **98**(2), 023613.

Bensky, G., Rao, D. D. B., Gordon, G., Erez, N., and Kurizki, G. 2010. Non-Markovian control of qubit thermodynamics by frequent quantum measurements. *Physica E*, **42**, 477–483.

Breuer, H.-P., and Petruccione, F. 2002. *The Theory of Open Quantum Systems*. Oxford: Oxford University Press.

Erez, N., Gordon, G., Nest, M., and Kurizki, G. 2008. Thermodynamic control by frequent quantum measurements. *Nature*, **452**(7188), 724–727.

Heims, S. P., and Jaynes, E. T. 1962. Theory of gyromagnetic effects and some related magnetic phenomena. *Rev. Mod. Phys.*, **34**(2), 143–165.

Horodecki, R., Horodecki, P., Horodecki, M., and Horodecki, K. 2009. Quantum entanglement. *Rev. Mod. Phys.*, **81**(2), 865–942.

Chapter 7

Agarwal, G. S. 1974. *Quantum Statistical Theories of Spontaneous Emission*. Berlin: Springer-Verlag.

Agarwal, G. S. 2013. *Quantum Optics*. Cambridge: Cambridge University Press.

Allen, L., and Eberly, J. H. 1987. *Optical Resonance and Two-Level Atoms*. Mineola, NY: Dover Publications.

Bensky, G., Petrosyan, D., Majer, J., Schmiedmayer, J., and Kurizki, G. 2012. Optimizing inhomogeneous spin ensembles for quantum memory. *Phys. Rev. A*, **86**(1), 012310.

Dicke, R. H. 1954. Coherence in spontaneous radiation processes. *Phys. Rev.*, **93**(1), 99–110.

Eberly, J. H. 2006. Emission of one photon in an electric dipole transition of one among N atoms. *J. Physics B*, **39**(15), S599–S604.

Krimer, D. O., Hartl, B., and Rotter, Stefan. 2015. Hybrid quantum systems with collectively coupled spin states: Suppression of decoherence through spectral hole burning. *Phys. Rev. Lett.*, **115**(3), 033601.

Landau, L. D., and Lifshitz, L. M. 1977. *Quantum Mechanics: Non-Relativistic Theory*. 3rd ed. Amsterdam: Elsevier.

Lehmberg, R. H. 1970a. Radiation from an N-atom system. I. General formalism. *Phys. Rev. A*, **2**(3), 883–888.

Lehmberg, R. H. 1970b. Radiation from an N-atom system. II. Spontaneous emission from a pair of atoms. *Phys. Rev. A*, **2**(3), 889–896.

Mazets, I. E., and Kurizki, G. 2007. Multiatom cooperative emission following single-photon absorption: Dicke-state dynamics. *J. Phys. B*, **40**(6), F105–F112.

Putz, S., Angerer, A., Krimer, D. O. et al. 2017. Spectral hole burning and its application in microwave photonics. *Nat. Photon.*, **11**, 36–39.

Svidzinsky, A. A., Chang, J.-T., and Scully, M. O. 2008. Dynamical evolution of correlated spontaneous emission of a single photon from a uniformly excited cloud of N atoms. *Phys. Rev. Lett.*, **100**(16), 160504.

Tavis, M., and Cummings, F. W. 1968. Exact solution for an N-molecule-radiation-field Hamiltonian. *Phys. Rev.*, **170**(2), 379–384.

Chapter 8

Agarwal, G. S. 1974. *Quantum Statistical Theories of Spontaneous Emission*. Berlin: Springer-Verlag.

Agarwal, G. S., Puri, R. R., and Singh, R. P. 1997. Atomic Schrödinger cat states. *Phys. Rev. A*, **56**(3), 2249–2254.

Davydov, A. S. 1965. *Quantum Mechanics*. 2nd ed. Oxford: Pergamon Press.

Diehl, S., Micheli, A. et al. 2008. Quantum states and phases in driven open quantum systems with cold atoms. *Nat. Phys.*, **4**(11), 878–883.

Eberly, J. H. 2006. Emission of one photon in an electric dipole transition of one among N atoms. *J. Physics B*, **39**(15), S599–S604.

Garraway, B. M., Knight, P. L., and Plenio, M. B. 1998. Generation and preservation of coherence in dissipative quantum optical environments. *Phys. Scr.*, **T76**, 152–158.

Kurizki, G. 1990. Two-atom resonant radiative coupling in photonic band structures. *Phys. Rev. A*, **42**(5), 2915–2924.

Kurizki, G., Kofman, A. G., and Yudson, V. 1996. Resonant photon exchange by atom pairs in high-Q cavities. *Phys. Rev. A*, **53**(1), R35–R38.

Lehmberg, R. H. 1970a. Radiation from an N-atom system. I. General formalism. *Phys. Rev. A*, **2**, 883–888.

Lehmberg, R. H. 1970b. Radiation from an N-atom system. II. Spontaneous emission from a pair of atoms. *Phys. Rev. A*, **2**, 889–896.

Mazets, I. E., and Kurizki, G. 2007. Multiatom cooperative emission following single-photon absorption: Dicke-state dynamics. *J. Phys. B*, **40**(6), F105–F112.

Rao, D. D. B., Bar-Gill, N., and Kurizki, G. 2011. Generation of macroscopic superpositions of quantum states by linear coupling to a bath. *Phys. Rev. Lett.*, **106**, 010404.

Sete, E. A., Svidzinsky, A. A., Eleuch, H. et al. 2010. Correlated spontaneous emission on the Danube. *J. Mod. Opt.*, **57**(14–15), 1311–1330.

Shahmoon, E., and Kurizki, G. 2013. Nonradiative interaction and entanglement between distant atoms. *Phys. Rev. A*, **87**(3), 033831.

Shahmoon, E., G. Kurizki, M. Fleischhauer, and Petrosyan, D. 2011. Strongly interacting photons in hollow-core waveguides. *Phys. Rev. A*, **83**(3), 033806.

Shahmoon, E., Grišins, P., Stimming, H. P., Mazets, I., and Kurizki, G. 2016. Highly nonlocal optical nonlinearities in atoms trapped near a waveguide. *Optica*, **3**(7), 725–733.

Sørensen, A., and Mølmer, K. 2000. Entanglement and quantum computation with ions in thermal motion. *Phys. Rev. A*, **62**(2), 022311.

Svidzinsky, A. A., Chang, J.-T., and Scully, M. O. 2008. Dynamical evolution of correlated spontaneous emission of a single photon from a uniformly excited cloud of N atoms. *Phys. Rev. Lett.*, **100**(16), 160504.

Verstraete, F., Wolf, M. M., and Cirac, J. I. 2009. Quantum computation and quantum-state engineering driven by dissipation. *Nat. Phys.*, **5**(9), 633–636.

Yurke, B., and Stoler, D. 1986. Generating quantum mechanical superpositions of macroscopically distinguishable states via amplitude dispersion. *Phys. Rev. Lett.*, **57**(1), 13–16.

Chapter 9

Breuer, H.-P., and Petruccione, F. 2002. *The Theory of Open Quantum Systems*. Oxford: Oxford University Press.

Erez, N., Gordon, G., Nest, M., and Kurizki, G. 2008. Thermodynamic control by frequent quantum measurements. *Nature*, **452**, 724–727.

Marshall, A. W., and Olkin, I. 1979. *Inequalities: Theory of Majorization and Its Applications*. New York: Academic Press.

Nielsen, M. A., and Chuang, I. L. 2000. *Quantum Computation and Quantum Information*. Cambridge: Cambridge University Press.

von Neumann, J. 1955. *Mathematical Foundations of Quantum Mechanics*. Princeton, New Jersey: Princeton University Press.

Wootters, W. K., and Fields, B. D. 1989. Optimal state-determination by mutually unbiased measurements. *Ann. Phys.*, **191**(2), 363–381.

Zurek, W. H. 1981. Pointer basis of quantum apparatus: Into what mixture does the wave packet collapse? *Phys. Rev. D*, **24**(6), 1516–1525.

Zurek, W. H. 2009. Quantum Darwinism. *Nat. Phys.*, **5**(3), 181–188.

Zurek, W. H. 2014. Quantum Darwinism, classical reality, and the randomness of quantum jumps. *Phys. Today*, **67**(10), 44–50.

Chapter 10

Ai, Q., Xu, D., Yi, S. et al. 2013. Quantum anti-Zeno effect without wave function reduction. *Sci. Rep.*, **3**(1), 1752.

Biagioni, P., Valle, G. Della, Ornigotti, M. et al. 2008. Experimental demonstration of the optical Zeno effect by scanning tunneling optical microscopy. *Opt. Express*, **16**(6), 3762–3767.

Bretschneider, C. O., Álvarez, G. A., Kurizki, G., and Frydman, L. 2012. Controlling spin-spin network dynamics by repeated projective measurements. *Phys. Rev. Lett.*, **108**, 140403.

Breuer, H.-P., and Petruccione, F. 2002. *The Theory of Open Quantum Systems*. Oxford: Oxford University Press.

Cook, R. J. 1988. What are quantum jumps. *Phys. Scr.*, **T21**, 49–51.

Do, H.-V., Lovecchio, C., Mastroserio, I. et al. 2019. Experimental proof of quantum Zeno-assisted noise sensing. *New J. Phys.*, **21**, 113056.

Facchi, P., Nakazato, H., and Pascazio, S. 2001. From the quantum Zeno to the inverse quantum Zeno effect. *Phys. Rev. Lett.*, **86**(13), 2699–2703.

Fischer, M. C., Gutierrez-Medina, B., and Raizen, M. G. 2001. Observation of the quantum Zeno and anti-Zeno effects in an unstable system. *Phys. Rev. Lett.*, **87**(4), 040402.

Fonda, L., Ghirardi, G. C., Rimini, A., and Weber, T. 1973. Quantum foundations of exponential decay law. *Nuovo Cim.*, **15A**, 689–704.

Fujii, K., and Yamamoto, K. 2010. Anti-Zeno effect for quantum transport in disordered systems. *Phys. Rev. A*, **82**(4), 042109.

Harel, G., Kofman, A. G., Kozhekin, A., and Kurizki, G. 1998. Control of non-Markovian decay and decoherence by measurements and interference. *Opt. Express*, **2**(9), 355–367.

Itano, W. M., Heinzen, D. J., Bollinger, J. J., and Wineland, D. J. 1990. Quantum Zeno effect. *Phys. Rev. A*, **41**(5), 2295–2300.

Khalfin, L. A. 1968. Phenomenological theory of K0 mesons and non-exponential character of decay. *JETP Lett.*, **8**, 65.

Kofman, A. G., and Kurizki, G. 1996. Quantum Zeno effect on atomic excitation decay in resonators. *Phys. Rev. A*, **54**(5), R3750–R3753.

Kofman, A. G., and Kurizki, G. 1999. Decay control in dissipative quantum systems. *Acta Phys. Slov.*, **49**(4), 541–548.

Kofman, A. G., and Kurizki, G. 2000. Acceleration of quantum decay processes by frequent observations. *Nature*, **405**(6786), 546–550.

Kofman, A. G., and Kurizki, G. 2001. Frequent observations accelerate decay: The anti-Zeno effect. *Z. Naturforsch.*, **56a**, 83–90.

Kurizki, A. G., Kofman G., and Opatrný, T. 2001. Zeno and anti-Zeno effects for photon polarization dephasing. *Phys. Rev. A*, **63**(4), 042108.

Misra, B., and Sudarshan, E. C. G. 1977. The Zeno's paradox in quantum theory. *J. Math. Phys.*, **18**(4), 756–763.

Peres, A. 1980. Zeno paradox in quantum theory. *Am. J. Phys.*, **48**(11), 931–932.

Schäfer, F., Herrera, I., Cherukattil, S. et al. 2014. Experimental realization of quantum zeno dynamics. *Nat. Commun.*, **5**(1), 3194.

Schulman, L. S. 1998. Continuous and pulsed observations in the quantum Zeno effect. *Phys. Rev. A*, **57**(3), 1509–1515.

Zhang, W., Kofman, A. G., Zhuang, J., You, J. Q., and Nori, F. 2013. Quantum Zeno and anti-Zeno effects measured by transition probabilities. *Phys. Lett. A*, **377**, 1837–1843.

Chapter 11

Breuer, H.-P., and Petruccione, F. 2002. *The Theory of Open Quantum Systems*. Oxford: Oxford University Press.

Gordon, G., Erez, N., and Kurizki, G. 2007. Universal dynamical decoherence control of noisy single- and multi-qubit systems. *J. Phys. B*, **40**(9), S75–S94.

Hashitsume, N., Shibata, F., and Shingu, M. 1977. Quantal master equation valid for any time scale. *J. Stat. Phys.*, **17**(4), 155–169.

Kofman, A. G., and Kurizki, G. 2004. Unified theory of dynamically suppressed qubit decoherence in thermal baths. *Phys. Rev. Lett.*, **93**(13), 130406.

Kurizki, G. 2013. Universal dynamical control of open quantum systems. *ISRN Optics*, **2013**, 783865.

Kurizki, G., and Zwick, A. 2016. From coherent to incoherent dynamical control of open quantum systems. *Adv. Chem. Phys.*, **159**, 139–217.

Nakajima, S. 1958. On quantum theory of transport phenomena. *Prog. Theor. Phys*, **20**(6), 948.

Shahmoon, E., and Kurizki, G. 2013. Engineering a thermal squeezed reservoir by energy-level modulation. *Phys. Rev. A*, **87**(1), 013841.

Shibata, F., Takahashi, Y., and Hashitsume, N. 1977. Generalized stochastic Liouville equation. Non-Markovian versus memoryless master equations. *J. Stat. Phys.*, **17**(4), 171–187.

Uhrig, G. S. 2007. Keeping a quantum bit alive by optimized pi-pulse sequences. *Phys. Rev. Lett.*, **98**(10), 100504.

Viola, L., and Knill, E. 2003. Robust dynamical decoupling of quantum systems with bounded controls. *Phys. Rev. Lett.*, **90**(3), 037901.

Viola, L., and Lloyd, S. 1998. Dynamical suppression of decoherence in two-state quantum systems. *Phys. Rev. A*, **58**(4), 2733–2744.

Vitali, D., and Tombesi, P. 2001. Heating and decoherence suppression using decoupling techniques. *Phys. Rev. A*, **65**(1), 012305.

Zwanzig, R. 1964. Ensemble method in the theory of irreversibility. *J. Chem. Phys.*, **33**(5), 1338–1341.

Chapter 12

Clausen, J., Bensky, G., and Kurizki, G. 2010. Bath-optimized minimal-energy protection of quantum operations from decoherence. *Phys. Rev. Lett.*, **104**(4), 040401.

Clausen, J., Bensky, G., and Kurizki, G. 2012. Task-optimized control of open quantum systems. *Phys. Rev. A*, **85**(5), 052105.

Gordon, G., Kurizki, G., and Lidar, D. A. 2008. Optimal dynamical decoherence control of a qubit. *Phys. Rev. Lett.*, **101**(1), 010403.

Kurizki, G. 2013. Universal dynamical control of open quantum systems. *ISRN Optics*, **2013**, 783865.

Kurizki, G., and Zwick, A. 2016. From coherent to incoherent dynamical control of open quantum systems. *Adv. Chem. Phys.*, **159**, 139–217.

Uhrig, G. S. 2007. Keeping a quantum bit alive by optimized pi-pulse sequences. *Phys. Rev. Lett.*, **98**(10), 100504.

Viola, L., and Lloyd, S. 1998. Dynamical suppression of decoherence in two-state quantum systems. *Phys. Rev. A*, **58**(4), 2733–2744.

Chapter 13

Cirac, J. I., and Zoller, P. 1995. Quantum computations with cold trapped ions. *Phys. Rev. Lett.*, **74**(20), 4091–4094.

Gordon, G. 2009. Dynamical decoherence control of multi-partite systems. *J. Phys. B*, **42**, 223001.

Gordon, G., Erez, N., and Kurizki, G. 2007. Universal dynamical decoherence control of noisy single- and multi-qubit systems. *J. Phys. B*, **40**(9), S75–S94.

Gordon, G., and Kurizki, G. 2006. Preventing multipartite disentanglement by local modulations. *Phys. Rev. Lett.*, **97**(11), 110503.

Gordon, G., and Kurizki, G. 2007. Universal dephasing control during quantum computation. *Phys. Rev. A*, **76**, 042310.

Gordon, G., and Kurizki, G. 2008. Dynamical protection of quantum computation from decoherence in laser-driven cold-ion and cold-atom systems. *New J. Phys.*, **10**, 045005.

Kofman, A. G., and Kurizki, G. 2001. Universal dynamical control of quantum mechanical decay: Modulation of the coupling to the continuum. *Phys. Rev. Lett.*, **87**(27), 270405.

Kurizki, G. 2013. Universal dynamical control of open quantum systems. *ISRN Optics*, **2013**, 783865.

Schmidt-Kaler, F., Häffner, H., Riebe, M. et al. 2003. Realization of the Cirac-Zoller controlled-NOT quantum gate. *Nature*, **422**(6930), 408–411.

Chapter 14

Bretschneider, C. O., Álvarez, G. A., Kurizki, G., and Frydman, L. 2012. Controlling spin-spin network dynamics by repeated projective measurements. *Phys. Rev. Lett.*, **108**(14), 140403.

Escher, B. M., Bensky, G., Clausen, J., and Kurizki, G. 2011. Optimized control of quantum state transfer from noisy to quiet qubits. *J. Phys. B*, **44**(15), 154015.

Kurizki, G., and Zwick, A. 2016. From coherent to incoherent dynamical control of open quantum systems. *Adv. Chem. Phys.*, **159**, 139–217.

Kurizki, G., Shahmoon, E., and Zwick, A. 2015. Thermal baths as quantum resources: more friends than foes? *Phys. Scr.*, **90**(12), 128002.

Zwick, A., Álvarez, G. A., Bensky, G., and Kurizki, G. 2014. Optimized dynamical control of state transfer through noisy spin chains. *New J. Phys.*, **16**(6), 065021.

Chapter 15

Alicki, R. 1979. The quantum open system as a model of the heat engine. *J. Phys. A*, **12**(5), L103–L107.

Alicki, R., and Kosloff, R. 2018. Introduction to quantum thermodynamics: History and prospects. Pages 1–33 of: Binder, F., et al. (eds.), *Thermodynamics in the Quantum Regime*. Cham, Switzerland: Springer.

Allahverdyan, A. E., Balian, R., and Nieuwenhuizen, T. M. 2004. Maximal work extraction from finite quantum systems. *EPL*, **67**(4), 565–571.

Brown, E. G., Friis, N., and Huber, M. 2016. Passivity and practical work extraction using Gaussian operations. *New J. Phys.*, **18**, 113028.

Erez, N., Gordon, G., Nest, M., and Kurizki, G. 2008. Thermodynamic control by frequent quantum measurements. *Nature*, **452**, 724–727.

Ghosh, A., Niedenzu, W., Mukherjee, V., and Kurizki, G. 2018. Thermodynamic principles and implementations of quantum machines. Pages 37–66 of: F. Binder, L. A. Correa, C. Gogolin, J. Anders, and G. Adesso (ed.), *Thermodynamics in the Quantum Regime*. Cham, Switzerland: Springer.

Jarzynski, C. 2011. Equalities and inequalities: Irreversibility and the second law of thermodynamics at the nanoscale. *Annu. Rev. Condens. Matter Phys.*, **2**(1), 329–351.

Lenard, A. 1978. Thermodynamical proof of the Gibbs formula for elementary quantum systems. *J. Stat. Phys.*, **19**, 575–586.

Lindblad, C. 1983. *Non-Equilibrium Entropy and Irreversibility*. Dordrecht: Springer.

Lindblad, G. 1975. Completely positive maps and entropy inequalities. *Commun. Math. Phys.*, **40**, 147–151.

Manzano, G., Calve, F., Zambrini, R., and Parrondo, J. M. R. 2016. Entropy production and thermodynamic power of the squeezed thermal reservoir. *Phys. Rev. E*, **93**(5), 052120.

Niedenzu, W., Mukherjee, V., Ghosh, A., Kofman, A. G., and Kurizki, G. 2018. Quantum engine efficiency bound beyond the second law of thermodynamics. *Nat. Commun.*, **9**(1), 165.

Pusz, W., and Woronowicz, S. L. 1978. Passive states and KMS states for general quantum systems. *Commun. Math. Phys.*, **58**(3), 273–290.

Sparaciari, C., Jennings, D., and Oppenheim, J. 2017. Energetic instability of passive states in thermodynamics. *Nat. Commun.*, **8**(1), 1895.

Spohn, H. 1978. Entropy production for quantum dynamical semigroups. *J. Math. Phys.*, **19**(5), 1227–1230.

Spohn, H., and Lebowitz, J. L. 1978. Irreversible thermodynamics for quantum systems weakly coupled to thermal reservoirs. Pages 109–142 of: Rice, S. A. (ed.), *Adv. Chem. Phys.*, vol. 38. Hoboken, NJ: John Wiley.

Uzdin, R., and Rahav, S. 2018. Global passivity in microscopic thermodynamics. *Phys. Rev. X*, **8**(2), 021064.

Chapter 16

Álvarez, G. A., Rao, D. D. B., Frydman, L., and Kurizki, G. 2010. Zeno and anti-Zeno polarization control of spin ensembles by induced dephasing. *Phys. Rev. Lett.*, **105**, 160401.

Bensky, G., Rao, D. D. B., Gordon, G., Erez, N., and Kurizki, G. 2010. Non-Markovian control of qubit thermodynamics by frequent quantum measurements. *Physica E*, **42**, 477–483.

Clausen, J., Bensky, G., and Kurizki, G. 2012. Task-optimized control of open quantum systems. *Phys. Rev. A*, **85**, 052105.

Erez, N., Gordon, G., Nest, M., and Kurizki, G. 2008. Thermodynamic control by frequent quantum measurements. *Nature*, **452**, 724–727.

Gordon, G., Bensky, G., Gelbwaser-Klimovsky, D. et al. 2009. Cooling down quantum bits on ultrashort time scales. *New J. Phys.*, **11**, 123025.

Gordon, G., Rao, D. D. B., and Kurizki, G. 2010. Equilibration by quantum observation. *New J. Phys.*, **12**, 053033.

Kurizki, G. 2013. Universal dynamical control of open quantum systems. *ISRN Optics*, **2013**, 783865.

Kurizki, G., and Zwick, A. 2016. From coherent to incoherent dynamical control of open quantum systems. *Adv. Chem. Phys.*, **159**, 139–217.

Rao, D. D. B., Momenzadeh, S. A., and Wrachtrup, J. 2016. Heralded control of mechanical motion by single spins. *Phys. Rev. Lett.*, **117**(7), 077203.

Chapter 17

Bennett, C. H. 1982. The Thermodynamics of computation – a review. *Int. J. Theor. Phys.*, **21**(12), 905–940.

Bérut, A., Arakelyan, A., Petrosyan, A. et al. 2012. Experimental verification of Landauer's principle linking information and thermodynamics. *Nature*, **483**(7388), 187–190.

Elouard, C.l, and Jordan, A. N. 2018. Efficient quantum measurement engines. *Phys. Rev. Lett.*, **120**(26), 260601.

Faist, P., Dupuis, F., Oppenheim, J., and Renner, R. 2015. The minimal work cost of information processing. *Nat. Commun.*, **6**(1), 7669.

Gardiner, C. W., and Zoller, P. 2000. *Quantum Noise*. Berlin: Springer.

Gelbwaser-Klimovsky, D., Erez, N., Alicki, R., and Kurizki, G. 2013. Work extraction via quantum nondemolition measurements of qubits in cavities: Non-Markovian effects. *Phys. Rev. A*, **88**(2), 022112.

Gelbwaser-Klimovsky, D., Niedenzu, W., and Kurizki, G. 2015. Thermodynamics of quantum systems under dynamical control. *Adv. At. Mol. Opt. Phys.*, **64**, 329–407.

Goold, J., Paternostro, M., and Modi, K. 2015. Nonequilibrium quantum Landauer principle. *Phys. Rev. Lett.*, **114**(6), 060602.

Landauer, R. 1961. Irreversibility and heat generation in the computing process. *IBM J. Res. Dev.*, **5**(3), 183–191.

Leff, H. S., and Rex, A. F. (eds). 2002. *Maxwell's Demon 2: Entropy, Classical and Quantum Information, Computing*. Boca Raton: CRC.

Leonhardt, U. 1997. *Measuring the Quantum State of Light*. Cambridge: Cambridge University Press.

Lutz, E., and Ciliberto, S. 2015. Information: From Maxwell's demon to Landauer's eraser. *Phys. Today*, **68**(9), 30–35.

Opatrny, T., Misra, A., and Kurizki, G. 2020. Work extraction from thermal phase-sensitive quantum observation. *(In preparation)*.

Parrondo, J. M. R., Horowitz, J. M., and Sagawa, T. 2015. Thermodynamics of information. *Nat. Phys.*, **11**(2), 131–139.

Sagawa, T., and Ueda, M. 2008. Second law of thermodynamics with discrete quantum feedback control. *Phys. Rev. Lett.*, **100**(8), 080403.

Schleich, W. P. 2001. *Quantum Optics in Phase Space*. Berlin: Wiley.

Szilárd, L. 2002. On the decrease of entropy in a thermodynamic system by the intervention of intelligent beings. Pages 110–119 of: Leff, H. S., and Rex, A. F. (eds.), *Maxwell's Demon 2: Entropy, Classical and Quantum Information, Computing*. Boca Raton: CRC.

Vidrighin, M. D., Dahlsten, O., Barbieri, M. et al. 2016. Photonic Maxwell's demon. *Phys. Rev. Lett.*, **116**(5), 050401.

Chapter 18

Abah, O., and Lutz, E. 2014. Efficiency of heat engines coupled to nonequilibrium reservoirs. *EPL*, **106**(2), 20001.

Dağ, C. B., Niedenzu, W., Müstecaplioğlu, Ö. E., and Kurizki, G. 2016. Multiatom quantum coherences in micromasers as fuel for thermal and nonthermal machines. *Entropy*, **18**(17), 244.

Dağ, C. B., Niedenzu, W., Ozaydin, F., Müstecaplioğlu, Ö. E., and Kurizki, G. 2019. Temperature control in dissipative cavities by entangled dimers. *J. Phys. Chem. C*, **123**, 4035–4043.

Dillenschneider, R., and Lutz, E. 2009. Energetics of quantum correlations. *EPL*, **88**(5), 50003.

Ghosh, A., Niedenzu, W., Mukherjee, V., and Kurizki, G. 2018. Thermodynamic principles and implementations of quantum machines. Pages 37–66 of: Binder, F., et al. (eds), *Thermodynamics in the Quantum Regime*. Cham, Switzerland: Springer.

Huang, X. L., Wang, T., and Yi, X. X. 2012. Effects of reservoir squeezing on quantum systems and work extraction. *Phys. Rev. E*, **86**(5), 051105.

Li, H., Zou, J., Yu, W.-L. et al. 2014. Quantum coherence rather than quantum correlations reflect the effects of a reservoir on a system's work capability. *Phys. Rev. E*, **89**(5), 052132.

Liberato, S. De, and Ueda, M. 2011. Carnot's theorem for nonthermal stationary reservoirs. *Phys. Rev. E*, **84**(5), 051122.

Niedenzu, W., Gelbwaser-Klimovsky, D., Kofman, A. G., and Kurizki, G. 2016. On the operation of machines powered by quantum non-thermal baths. *New J. Phys.*, **18**, 083012.

Niedenzu, W., Mukherjee, V., Ghosh, A., Kofman, A. G., and Kurizki, G. 2018. Quantum engine efficiency bound beyond the second law of thermodynamics. *Nat. Commun.*, **9**(1), 165.

Rossnagel, J., Abah, O., Schmidt-Kaler, F., Singer, K., and Lutz, E. 2014. Nanoscale heat engine beyond the Carnot limit. *Phys. Rev. Lett.*, **112**(3), 030602.

Scully, M. O., Chapin, K. R., Dorfman, K. E., Kim, M. B., and Svidzinsky, A. 2011. Quantum heat engine power can be increased by noise-induced coherence. *PNAS*, **108**(37), 15097–15100.

Scully, M. O., Zubairy, M. S., Agarwal, G. S., and Walther, H. 2003. Extracting work from a single heat bath via vanishing quantum coherence. *Science*, **299**(5608), 862–864.

Türkpençe, D., and Müstecaplioğlu, Ö. E. 2016. Quantum fuel with multilevel atomic coherence for ultrahigh specific work in a photonic Carnot engine. *New J. Phys.*, **93**(1), 012145.

Chapter 19

Alicki, R., and Kosloff, R. 2018. Introduction to quantum thermodynamics: History and prospects. Pages 1–33 of: Binder, F., et al. (eds.), *Thermodynamics in the Quantum Regime*. Cham, Switzerland: Springer.

Campisi, M., and Fazio, R. 2016. The power of a critical heat engine. *Nat. Commun.*, **7**(1), 11895.

Gardas, B., and Deffner, S. 2015. Thermodynamic universality of quantum Carnot engines. *Phys. Rev. E*, **92**(4), 042126.

Gelbwaser-Klimovsky, D., Alicki, R., and Kurizki, G. 2013. Minimal universal quantum heat machine. *Phys. Rev. E*, **87**(1), 012140.

Gelbwaser-Klimovsky, D., and Kurizki, G. 2014. Heat-machine control by quantum-state preparation: From quantum engines to refrigerators. *Phys. Rev. E*, **90**, 022102.

Gelbwaser-Klimovsky, D., Niedenzu, W., and Kurizki, G. 2015. Thermodynamics of quantum systems under dynamical control. *Adv. At. Mol. Opt. Phys.*, **64**, 329–407.

Gemmer, J., Michel, M., and Mahler, G. 2009. *Quantum Thermodynamics*. Berlin: Springer.

Kosloff, R. 2013. Quantum thermodynamics: A dynamical viewpoint. *Entropy*, **15**(6), 2100–2128.

Kosloff, R., and Levy, A. 2014. Quantum heat engines and refrigerators: Continuous devices. *Annu. Rev. Phys. Chem.*, **65**, 365–393.

Mukherjee, V., Kofman, A. G., and Kurizki, G. 2020. Anti-Zeno quantum advantage in fast-driven heat machines. *Comms. Phys.*, **3**(1), 8.

Mukherjee, V., Niedenzu, W., Kofman, A. G., and Kurizki, G. 2016. Speed and efficiency limits of multilevel incoherent heat engines. *Phys. Rev. E*, **94**(6), 062109.

Chapter 20

Alicki, R. 2014. Quantum thermodynamics. An example of two-level quantum machine. *Open Syst. Inf. Dyn.*, **21**(1–2), 1440002.

Gelbwaser-Klimovsky, D., Alicki, R., and Kurizki, G. 2013. Minimal universal quantum heat machine. *Phys. Rev. E*, **87**(1), 012140.

Gelbwaser-Klimovsky, D., Niedenzu, W., and Kurizki, G. 2015. Thermodynamics of quantum systems under dynamical control. *Adv. At. Mol. Opt. Phys.*, **64**, 329–407.

Ghosh, A., Gelbwaser-Klimovsky, D., Niedenzu, W. et al. 2018. Two-level masers as heat-to-work converters. *PNAS*, **115**(40), 9941–9944.

Mukherjee, V., Kofman, A. G., and Kurizki, G. 2020. Anti-Zeno quantum advantage in fast-driven heat machines. *Comms. Phys.*, **3**, 8.

Chapter 21

Agarwal, G. S. 1974. *Quantum Statistical Theories of Spontaneous Emission*. Berlin: Springer-Verlag.

Alicki, R. 2014. Quantum thermodynamics. An example of two-level quantum machine. *Open Syst. Inf. Dyn.*, **21**(1–2), 1440002.

Andreev, A. V., Emel'yanov, V. I., and Il'inskii, Yu. A. 1993. *Cooperative Effects in Optics: Superradiance and Phase Transitions*. Bristol: IOP Publishing.

Breuer, H.-P., and Petruccione, F. 2002. *The Theory of Open Quantum Systems*. Oxford: Oxford University Press.

Dicke, R. H. 1954. Coherence in spontaneous radiation processes. *Phys. Rev.*, **93**(1), 99–110.

Hardal, A. Ü. C., and Müstecaplioğlu, Ö. E. 2015. Superradiant quantum heat engine. *Sci. Rep.*, **5**, 12953.

Lalumiere, K., Sanders, B. C., van Loo, A. F. et al. 2013. Input-output theory for waveguide QED with an ensemble of inhomogeneous atoms. *Phys. Rev. A*, **88**(4), 043806.

Lehmberg, R. H. 1970. Radiation from an N-atom system. II. Spontaneous emission from a pair of atoms. *Phys. Rev. A*, **2**(3), 889–896.

Manatuly, A., Niedenzu, W., Román-Ancheyta, R. et al. 2019. Collectively enhanced thermalization via multiqubit collisions. *Phys. Rev. E*, **99**(4), 042145.

Mazets, I. E., and Kurizki, G. 2007. Multiatom cooperative emission following single-photon absorption: Dicke-state dynamics. *J. Phys. B*, **40**(6), F105–F112.

Meiser, D., and Holland, M. J. 2010. Steady-state superradiance with alkaline-earth-metal atoms. *Phys. Rev. A*, **81**(3), 033847.

Niedenzu, W., and Kurizki, G. 2018. Cooperative many-body enhancement of quantum thermal machine power. *New J. Phys.*, **20**, 113038.

Schmidt, R., Negretti, A., Ankerhold, J., Calarco, T., and Stockburger, J. T. 2011. Optimal control of open quantum systems: Cooperative effects of driving and dissipation. *Phys. Rev. Lett.*, **107**, 130404.

Svidzinsky, A. A., Chang, J.-T., and Scully, M. O. 2008. Dynamical evolution of correlated spontaneous emission of a single photon from a uniformly excited cloud of *N* atoms. *Phys. Rev. Lett.*, **100**(16), 160504.

Tavis, M., and Cummings, F. W. 1968. Exact solution for an N-molecule-radiation-field Hamiltonian. *Phys. Rev.*, **170**(2), 379–384.

van Loo, A. F., Fedorov, A., Lalumiere, K. et al. 2013. Photon-mediated interactions between distant artificial atoms. *Science*, **342**(6165), 1494–1496.

Chapter 22

Alicki, R., and Kosloff, R. 2018. Introduction to quantum thermodynamics: History and prospects. Pages 1–33 of: Binder, F., et al. (eds), *Thermodynamics in the Quantum Regime*. Cham, Switzerland: Springer.

Brandner, K., Bauer, M., and Seifert, U. 2017. Universal coherence-induced power losses of quantum heat engines in linear response. *Phys. Rev. Lett.*, **119**(17), 170602.

Dağ, C. B., Niedenzu, W., Ozaydin, F., Müstecaplıoğlu, Ö. E., and Kurizki, G. 2019. Temperature control in dissipative cavities by entangled dimers. *J. Phys. Chem. C*, **123**, 4035–4043.

Gelbwaser-Klimovsky, D., Alicki, R., and Kurizki, G. 2013. Work and energy gain of heat-pumped quantized amplifiers. *EPL*, **103**(6), 60005.

Gelbwaser-Klimovsky, D., and Kurizki, G. 2014. Heat-machine control by quantum-state preparation: From quantum engines to refrigerators. *Phys. Rev. E*, **90**(2), 022102.

Gelbwaser-Klimovsky, D., and Kurizki, G. 2015. Work extraction from heat-powered quantized optomechanical setups. *Sci. Rep.*, **5**, 7809.

Gelbwaser-Klimovsky, D., Niedenzu, W., and Kurizki, G. 2015. Thermodynamics of quantum systems under dynamical control. *Adv. At. Mol. Opt. Phys.*, **64**, 329–407.

Ghosh, A., Gelbwaser-Klimovsky, D., Niedenzu, W. et al. 2018b. Two-level masers as heat-to-work converters. *PNAS*, **115**(40), 9941–9944.

Ghosh, A., Latune, C. L., Davidovich, L., and Kurizki, G. 2017. Catalysis of heat-to-work conversion in quantum machines. *PNAS*, **114**(46), 12156–12161.

Ghosh, A., Mukherjee, V., Niedenzu, W., and Kurizki, G. 2019. Are quantum thermodynamic machines better than their classical counterparts? *Eur. Phys. J. Spec. Top.*, **227**(15), 2043–2051.

Ghosh, A., Niedenzu, W., Mukherjee, V., and Kurizki, G. 2018a. Thermodynamic principles and implementations of quantum machines. Pages 37–66 of: et al., F. Binder (ed.), *Thermodynamics in the Quantum Regime*. Cham, Switzerland: Springer.

Hovhannisyan, K. V., Perarnau-Llobet, M., Huber, M., and Acín, A. 2013. Entanglement generation is not necessary for optimal work extraction. *Phys. Rev. Lett.*, **111**(24), 240401.

Klaers, J., Faelt, S., Imamoglu, A., and Togan, E. 2017. Squeezed thermal reservoirs as a resource for a nanomechanical engine beyond the Carnot limit. *Phys. Rev. X*, **7**(3), 031044.

Klatzow, J., Becker, J. N., Ledingham, P. M. et al. 2019. Experimental demonstration of quantum effects in the operation of microscopic heat engines. *Phys. Rev. Lett.*, **122**(11), 110601.

Perarnau-Llobet, M., Hovhannisyan, K. V., Huber, M. et al. 2015. Extractable work from correlations. *Phys. Rev. X*, **5**(4), 041011.

Scovil, H. E. D., and Schulz-DuBois, E. O. 1959. Three-level masers as heat engines. *Phys. Rev. Lett.*, **2**(6), 262–263.

Scully, M. O., Chapin, K. R., Dorfman, K. E., Kim, M. B., and Svidzinsky, A. 2011. Quantum heat engine power can be increased by noise-induced coherence. *PNAS*, **108**(37), 15097–15100.

von Lindenfels, D., Gräb, O., Schmiegelow, C. T. et al. 2019. Spin heat engine coupled to a harmonic-oscillator flywheel. *Phys. Rev. Lett.*, **123**(8), 080602.

Chapter 23

Correa, L. A., Palao, J. P., Alonso, D., and Adesso, G. 2014. Quantum-enhanced absorption refrigerators. *Sci. Rep.*, **4**, 3949.

Gelbwaser-Klimovsky, D., and Kurizki, G. 2014. Heat-machine control by quantum-state preparation: From quantum engines to refrigerators. *Phys. Rev. E*, **90**(2), 022102.

Gelbwaser-Klimovsky, D., Niedenzu, W., and Kurizki, G. 2015. Thermodynamics of quantum systems under dynamical control. *Adv. At. Mol. Opt. Phys.*, **64**, 329–407.

Kolář, M., Gelbwaser-Klimovsky, D., Alicki, R., and Kurizki, G. 2012. Quantum bath refrigeration towards absolute zero: Challenging the unattainability principle. *Phys. Rev. Lett.*, **109**(9), 090601.

Kosloff, R., and Levy, A. 2014. Quantum heat engines and refrigerators: Continuous devices. *Annu. Rev. Phys. Chem.*, **65**, 365–393.

Mukherjee, V., Kofman, A. G., and Kurizki, G. 2020. Anti-Zeno quantum advantage in fast-driven heat machines. *Comms. Phys.*, **3**, 8.

Mukherjee, V., Niedenzu, W., Kofman, A. G., and Kurizki, G. 2016. Speed and efficiency limits of multilevel incoherent heat engines. *Phys. Rev. E*, **94**(6), 062109.

Naseem, M. T., Misra, A., and Müstecaplioğlu, Ö. E. 2020. Two-body quantum absorption refrigerators with optomechanical-like interactions. *Quantum Sci. Technol.*, **5**(3), 035006.

Niedenzu, W., and Kurizki, G. 2018. Cooperative many-body enhancement of quantum thermal machine power. *New J. Phys.*, **20**, 113038.

Vogl, U., and Weitz, M. 2009. Laser cooling by collisional redistribution of radiation. *Nature*, **461**(7260), 70–73.

Chapter 24

Du, J., Shen, W., Su, S., and Chen, J. 2019. Quantum thermal management devices based on strong coupling qubits. *Phys. Rev. E*, **99**, 062123.

Johansson, J. R., Johansson, G., and Nori, F. 2014. Optomechanical-like coupling between superconducting resonators. *Phys. Rev. A*, **90**(5), 053833.

Joulain, K., Drevillon, J., Ezzahri, Y., and Ordonez-Miranda, Jose. 2016. Quantum thermal transistor. *Phys. Rev. Lett.*, **116**(20), 200601.

Kargi, C., Naseem, M. T., Opatrný, T., Müstecaplioğlu, Ö. E., and Kurizki, G. 2019. Quantum optical two-atom thermal diode. *Phys. Rev. E*, **99**(4), 042121.

Karimi, B., Pekola, J. P., Campisi, M., and Fazio, R. 2017. Coupled qubits as a quantum heat switch. *Quantum Sci. Technol.*, **2**(4), 044007.

Kurizki, G., Kofman, A. G., and Yudson, V. 1996. Resonant photon exchange by atom pairs in high-Q cavities. *Phys. Rev. A*, **53**(1), R35–R38.

Liu, H., Wang, C., Wang, L.-Q., and Ren, J. 2019. Strong system-bath coupling induces negative differential thermal conductance and heat amplification in nonequilibrium two-qubit systems. *Phys. Rev. E*, **99**(3), 032114.

Man, Z.-X., An, N. B., and Xia, Y.-J. 2016. Controlling heat flows among three reservoirs asymmetrically coupled to two two-level systems. *Phys. Rev. E*, **94**(4), 042135.

Martínez-Pérez, M. J., Fornieri, A., and Giazotto, F. 2015. Rectification of electronic heat current by a hybrid thermal diode. *Nature Nanotech.*, **10**, 303–307.

Motz, T., Wiedmann, M., Stockburger, J. T., and Ankerhold, J. 2018. Rectification of heat currents across nonlinear quantum chains: a versatile approach beyond weak thermal contact. *New J. Phys.*, **20**(11), 113020.

Naseem, M. T., Misra, A., Müstecaplioğlu, Ö. E., and Kurizki, G. 2020. Minimal quantum heat manager boosted by bath spectral filtering. *Phys. Rev. Res.*, **2**(3), 033285.

Ordonez-Miranda, J., Ezzahri, Y., and Joulain, K. 2017. Quantum thermal diode based on two interacting spinlike systems under different excitations. *Phys. Rev. E*, **95**(2), 022128.

Petrosyan, D., and Kurizki, G. 2002. Scalable solid-state quantum processor using subradiant two-atom states. *Phys. Rev. Lett.*, **89**(20), 207902.

Ronzani, A., Karimi, B., Senior, J. et al. 2018. Tunable photonic heat transport in a quantum heat valve. *Nature Phys.*, **14**(10), 991–995.

Saira, O.-P., Meschke, M., Giazotto, F. et al. 2007. Heat transistor: Demonstration of gate-controlled electronic refrigeration. *Phys. Rev. Lett.*, **99**(2), 027203.

Segal, D. 2008. Single mode heat rectifier: Controlling energy flow between electronic conductors. *Phys. Rev. Lett.*, **100**(10), 105901.

Segal, D., and Nitzan, A. 2005. Spin-boson thermal rectifier. *Phys. Rev. Lett.*, **94**(3), 034301.

Senior, J., Gubaydullin, A., Karimi, B. et al. 2019. Heat rectification via a superconducting artificial atom. *arXiv:1908.05574*.

Shen, Y., Bradford, M., and Shen, J.-T. 2011. Single-photon diode by exploiting the photon polarization in a waveguide. *Phys. Rev. Lett.*, **107**(17), 173902.

Wang, C., Chen, X.-M., Sun, K.-W., and Ren, J. 2018. Heat amplification and negative differential thermal conductance in a strongly coupled nonequilibrium spin-boson system. *Phys. Rev. A*, **97**(5), 052112.

Wang, C., Xu, D., Liu, H., and Gao, X. 2019. Thermal rectification and heat amplification in a nonequilibrium V-type three-level system. *Phys. Rev. E*, **99**(4), 042102.

Werlang, T., Marchiori, M. A., Cornelio, M. F., and Valente, D. 2014. Optimal rectification in the ultrastrong coupling regime. *Phys. Rev. E*, **89**(6), 062109.

Zhang, Y., Yang, Z., Zhang, X. et al. 2018. Coulomb-coupled quantum-dot thermal transistors. *EPL*, **122**(1), 17002.

Index

anti-Zeno effect (AZE), xii, 134, 135, 137, 139, 141, 144, 146–152, 156–161, 205, 281, 284, 289, 303, 365, 371–378, 380, 429, 434, 449, 451

AZE, *see* anti-Zeno effect

bath spectral filtering (BSF), 435–448

bath-optimized task-oriented control (BOTOC), 200–210, 215, 216, 242, 450

bosonic baths, 22–24, 43, 54, 83, 110, 111, 118, 168, 182, 192, 193, 269, 431

BSF, *see* bath spectral filtering

canonical density operator, 3

canonical ensemble, 10, 11

collective coupling, 80, 109, 113, 226, 394

  to a single field-mode, 80

collective energy currents, 386–387

commensurate-frequency spectrum, 12

continuous cycles, 326, 343, 346–364, 427

control of steady states, 288–300

cooling speed, 453

  scaling with temperature, 430–432

cooperative decay, 80, 83, 85, 88, 92

cooperative emission, 85, 94

cooperative power enhancement, 387–392

correlation, 19, 39, 49, 69, 86, 88, 92, 126, 154, 156, 221, 223, 224, 226, 261, 272, 273, 306, 308, 311, 318, 320, 321, 324, 378, 380, 449, 451

correlation energy, 318, 320, 322, 324

correlation function, 19, 49, 55, 56, 139, 166–168, 178, 179, 181, 185, 188–191, 204, 220, 246, 251, 379

correlation matrix, 204

correlation spectrum, 237

correlation time, 115, 116, 126, 141, 142, 169, 176, 186, 188, 194, 208, 209, 223, 227, 231, 232, 237, 239, 240, 246, 256, 281, 285, 287, 299, 303, 321, 322, 324, 350, 365, 367, 384, 450

decay modified by measurements, 138, 140

decoherence, xii, 39, 110, 112–115, 119, 120, 125–133, 195, 196, 199, 217, 221–225, 227, 229–231, 234, 237, 239, 240, 242–245, 248, 249, 305, 346, 449, 450

  control, 161, 182, 186, 198, 205, 207, 211, 217, 218, 221–226, 228, 229, 231–236, 238–241, 450

    non-Markovian, 239

  factors, 226, 228, 230–232, 235

  matrix, 219–222, 224, 225, 228

  protection from, 215

  rate, 111, 116, 190, 196, 205, 207, 225, 229, 231, 233–237

  suppression, 161, 196, 198, 210, 219, 224, 240, 449

decoherence-free subspace (DFS), 221, 222, 226, 240

decoherence-free system, 221

defect mode, 65

detailed balance, 13

DFS, *see* decoherence-free subspace

dipole–dipole interactions, xii, 91, 93, 103, 106, 117, 447, 448

driven quantum system, 259–271

dynamical control, xi–xiii, 22, 161, 192, 195–198, 205, 210, 211, 217, 221, 224, 241, 242, 249, 250, 252, 254–256, 324, 450, 451, 453

dynamical decoupling (DD) control, 192–194, 196–198, 206–210, 240, 450

eigenstate thermalization hypothesis (ETH), 3, 11, 451

electron–phonon interactions, 45, 47, 348

engine performance, 340–342

entangled system–meter states, 121, 125, 126

entanglement, 91, 101–103, 112, 117, 118, 199, 226, 233–235, 238, 239, 241, 333, 338, 342, 393, 394, 450, 452

  between the system and the meter, 125

entanglement between the system and the bath, 69, 73, 132, 261, 276, 303, 321